T0176654

PHYSICAL AND CHEMICAL PROCESSES IN THE AQUATIC ENVIRONMENT

PHYSICAL AND CHEMICAL PROCESSES IN THE AQUATIC ENVIRONMENT

ERIK R. CHRISTENSEN

Department of Civil and Environmental Engineering
University of Wisconsin–Milwaukee

AN LI

School of Public Health
University of Illinois at Chicago

Published by John Wiley & Sons, Inc., Hoboken, New Jersey
Published simultaneously in Canada

For general information on our other products and services or for technical support, please contact our Customer Care Department within the United States at (800) 762-2974, outside the United States at (317) 572-3993 or fax (317) 572-4002.

Wiley also publishes its books in a variety of electronic formats. Some content that appears in print may not be available in electronic formats. For more information about Wiley products, visit our web site at www.wiley.com.

Library of Congress Cataloging-in-Publication Data:

Christensen, Erik R., 1943-
  Physical and chemical processes in the aquatic environment / Erik R. Christensen, PhD, distinguished professor emeritus, Department of Civil Engineering and Mechanics, University of Wisconsin–Milwaukee, An Li, PhD, professor, School of Public Health, University of Illinois at Chicago.
      pages cm
  Includes bibliographical references and index.
  ISBN 978-1-118-11176-5 (cloth)
1. Water–Pollution. 2. Water chemistry. I. Li, An, 1952- II. Title.
TD420.C525 2014
628.1–dc23
                        2014007294

Printed in the United States of America

ISBN: 9781118111765

10 9 8 7 6 5 4 3 2 1

CONTENTS

PREFACE

The idea of writing this book stems from the need in environmental research on fresh waters including rivers and lakes. The content of this work is based on our decades of research experience and teaching in this scientific and engineering area. Compared with other books on the topic, this book has a unique outline in that it follows pollution from sources to impact, considering source tracking, transport and mixing, physico-chemical properties of the pollutants, and ecotoxicological and human health effects. The book is primarily intended as a comprehensive reference text for professionals. However, it may be adapted as a textbook in graduate level courses on the fate and transport of pollutants in aquatic systems.

The emphasis on natural waters reflects a current trend to include the total environment rather than just water and wastewater treatment plants or industrial discharges. Although, the emphasis is on the physical and chemical phenomena underlying aquatic pollution, elements of biology and engineering approaches to minimize the impact are considered as well.

Another current trend is to use a multimedia approach instead of looking solely at the aqueous phase. This has been considered by including contaminated sediments and interactions with the atmosphere. Contaminated sediments can constitute a significant source of water pollution when other sources have been curtailed. Likewise, input to lakes from the atmosphere and loss by volatilization are relevant for many semivolatile organic pollutants in both "legacy" and "emerging" categories. In groundwater pollution, VOCs can be vented to the atmosphere or to above-ground treatment facilities.

A brief outline of the contents is as follows. Chapter 1 deals with pollutant transport modeled with the advection–diffusion reaction equation and, if necessary, hydrodynamic models. ^{137}Cs in sediment cores is modeled considering the difference in porewater distribution of small diameter clay particles carrying ^{137}Cs and larger diameter bulk sediment particles which are more influenced by compaction. Chapter 2 covers ^{210}Pb and ^{137}Cs sediment dating including deconvolution of the sedimentary record to remove mixing effects, using differential, integral, or frequency domain methods. Atmospheric interactions in Chapter 3 include theory and examples of dry and wet deposition as well as gas exchange of organic compounds such as PCBs, PAHs, and pesticides. Many PCB-contaminated areas such as Green Bay, Wisconsin exhibit PCB volatilization. The use of passive samplers for both the air and water phase is becoming more commonplace.

The important geochemistry processes occurring in natural waters are described in Chapter 4. Concerns over nuisance algae, *Cladophora*, in the Great Lakes area has prompted a renewed interest in phosphorus that appears to have become more available due to the invasion of zebra and quagga mussels during the last two decades. This and other matters relating to nutrients are described in Chapter 5. Chapter 6 on metals provides a discussion of metal dissolution from minerals, metal complexation including methylation of mercury, and the application of zero valent iron in remediation. The chapter on organic pollutants, Chapter 7, has a good deal of discussion on the "emerging" pollutants such as brominated and phosphate flame retardants, perfluorochemicals, and pharmaceutical and personal care products.

The chapter on pathogens, Chapter 8, looks at indicator organisms, cryptosporidiosis outbreaks, and inactivation kinetics among other topics. The importance of emerging molecular techniques such as quantitative polymerase chain reaction is highlighted. Chapter 9 on tracers considers radioisotopes, stable isotopes, and other techniques for source identification. This is a topic lacking in many other books on pollution control of natural waters. For example, stable isotopes of nitrogen and oxygen were used for nitrification assessment of the Illinois River. Mass balance modeling and factor analysis that traditionally have been used in air quality modeling is introduced with applications to the aquatic environment, for example to determine sources of PAHs. Brief accounts are given of positive matrix factorization, PMF, and Unmix, an eigenvalue and eigenvector based approach. PMF can be used to resolve patterns of PCB or PBDE congeners for both source analysis and degradation assessment.

Although the emphasis of this book is on the physical and chemical processes, a chapter on ecotoxicology, Chapter 10, is included due to the importance of this topic. It features molecular biological methods, nanoparticles, and comparison of the basis of the biotic ligand model with the Weibull dose–response model. The last chapter, Chapter 11, briefly summarizes the regulations on ambient water quality. In accordance with the focus of this book, the discussion does not include regulations and criteria for public water supplies, wastewater discharges and other effluents, and waters designated for special uses, although all these interact with the ambient water quality in a complex manner.

One of the challenges human society is facing today is to ensure the sustainability of water resources. To this end, the publication of this book is timely. Water pollution control is one of the most challenging issues in ensuring sustainable waters for our future. We hope this book adds to the current knowledge on the most relevant processes occurring in natural waters; thus helping readers who may range from policy makers, scientists, environmental engineers, and consultants to students.

We would like to thank our colleagues and students for input and for an opportunity to test our ideas and discuss the material for this book. Sections on sediment dating benefitted from discussions with David Edgington and John Robbins, while the insights of Eiji Fukumori contributed greatly to our understanding of sediment and tracer mixing. Modeling input on Unmix methodology for source apportionment was generously provided by Ronald Henry. Heartfelt thanks go to Neil Sturchio for discussions on stable isotope nitrogen tracers, and to Karl J. Rockne for working with us in several major projects on sediments of natural waters. The collaboration with Niels Nyholm and K. Ole Kusk was instrumental in refining our understanding of dose–response modeling in ecotoxicology. A special thank you is extended to our former and current graduate students Pichaya Rachdawong, Irwan A. Ab Razak, Puripus Soonthornnonda, Yonghong Zou, Hua Wei, Jiehong Guo, and others for help with specific tasks in connection with the writing of the manuscript.

ERIK R. CHRISTENSEN
AN LI

Milwaukee, Wisconsin
Chicago, Illinois

1

TRANSPORT OF POLLUTANTS

1.1 INTRODUCTION

Pollutants are dispersed in aquatic systems through advection and mixing. Although mixing occurs at the molecular level and by eddy diffusion, it is the eddy diffusion which is of primary interest in actual systems. Mixing is important for the effective dispersal of pollutants from a wastewater outfall into a lake, coastal zone, or estuary. Dispersal is aided by a high exit velocity, currents in the coastal zone, or river flow velocity. Horizontal mixing in lakes can be impeded by the development of thermal fronts, which are particularly prevalent during the spring warming period in nearshore areas lakes (Christensen et al., 1997). The thermal front is formed when lake water that is cooled below 4°C meets nearshore water that has been warmed to a higher temperature to form a vertical barrier of 4°C water. This barrier moves out into the lake and becomes less distinctive as spring warming progresses.

In most lakes and nearshore areas, thermal stratification will develop during the summer, and this effectively isolates the warm epilimnion, or upper layer, from the cold hypolimnion or bottom layer. The intensity of vertical mixing can be inferred from density gradients or by measurement of ^{222}Rn or the tritium–helium-3 pair.

Other phenomena such as up and down welling and inertial Kelvin waves (Wetzel, 1975) can influence water currents and temperatures, which is reflected in the temperature records of water intakes for the Great Lakes. For coastal areas, tides can be of great significance in creating nearshore currents. In the Bay of Fundy on the US east coast, the fjord configuration acts to amplify the difference in water levels between ebb and flood to a total of 18–22 m. While many ocean wastewater outfalls are beyond the direct influence of these currents, they have the potential to

Physical and Chemical Processes in the Aquatic Environment, First Edition. Erik R. Christensen and An Li.
© 2014 John Wiley & Sons, Inc. Published 2014 by John Wiley & Sons, Inc.

disperse pollutants during ebb, but also to exacerbate pollutant impact on coastal areas during flood.

Pollutant plumes of groundwater can occur as a result of BTEX (benzene, toluene, ethylbenzene, and xylene) compounds leaking from underground storage tanks (Frieseke and Christensen, 1995) or from low molecular weight organics used in dry-cleaning operations. Nitrates from fertilizers can also create significant groundwater pollution (Lee et al., 1995). For groundwater, the dispersion coefficient is proportional to the pore velocity, and plume retardation may occur as a result of partitioning to the solid organic phase.

Dissolved organic carbon (DOC), such as humic acid, can increase the rate of transfer of hydrophobic chemicals such as PCBs (polychlorinated biphenyls) and PBDEs (polybrominated diphenyl ethers) between the aqueous and sorbent phases (ter Laak et al., 2009). The DOC–water partition coefficient K_{DOC} (l/kg) increases with planarity of PCBs, apparently reflecting increased sorption for PCBs with one or no *ortho*-substituted carbon atoms. The facilitated transport can be important for uptake by organisms or passive samplers when equilibration is slow or occurs under time-varying conditions.

Mixing in aquatic sediments is of interest both for the distortion it can impart on historical records of particle-bound pollutants (Christensen and Karls, 1996) and because of the increased chance of pollutant release from the sediment to the water column of in-place pollutants (Christensen et al., 1993). The formulation of the advection–diffusion equation depends on whether or not the tracer (pollutant) is associated with settling particles or colloids (Fukumori et al., 1992).

1.2   ADVECTION–DIFFUSION EQUATION WITH REACTION

The general form of the 3-D advection–diffusion equation for aquatic systems is,

$$\frac{\partial c}{\partial t} + \frac{\partial uc}{\partial x} + \frac{\partial vc}{\partial y} + \frac{\partial wc}{\partial z} = \frac{\partial}{\partial x}\left(D_x\frac{\partial c}{\partial x}\right) + \frac{\partial}{\partial y}\left(D_y\frac{\partial c}{\partial y}\right) + \frac{\partial}{\partial z}\left(D_z\frac{\partial c}{\partial z}\right) + R \quad (1.1)$$

where

$$c = \text{concentration of a species,}$$
$$t = \text{time,}$$
$$u, v, w = \text{water velocities in the } x, y, \text{ and } z \text{ directions, respectively,}$$
$$D_x, D_y, D_z = \text{eddy diffusion coefficients in the } x, y, \text{ and } z \text{ directions,}$$
$$\text{respectively, and,}$$
$$R = \text{production rate of the species per unit volume.}$$

The molecular diffusivity in water is 2×10^{-5} cm^2/s (Weber and DiGiano, 1996). For oceans and lakes, the horizontal eddy diffusivities D_x and D_y are in the range of $10^4 - 10^8$ cm^2/s, and the vertical diffusivity D_z is between 10^1 and 10^3 cm^2/s.

The water velocities u, v, and w are either measured or may be obtained from the momentum conservation equation, i.e., the Navier–Stokes equation (Weber and

DiGiano, 1996). In the case of groundwater, this equation is reduced to the Darcy equation, which states that the average groundwater velocity is proportional to the gradient of the hydraulic head. For sediments, the velocity relative to the sediment–water interface is derived by dividing the mass sedimentation rate by the bulk sediment density ρ, where ρ can be modeled based on soil mechanics concepts (Fukumori et al., 1992).

As will be shown below, simple versions of Equation 1.1, especially one-dimensional cases, can be solved analytically. More realistic configurations can be addressed by finite differences or finite elements techniques.

A version of the one-dimensional advection–diffusion equation for solute transport, including advection, dispersion, and transient storage, was considered by Cox and Runkel (2008). The authors included both fixed grid (Eulerian) and traveling control volumes (Lagrangian). The Lagrangian approach has the advantage that advection is zero; therefore the mass losses that are often seen in Eulerian approaches are eliminated. However, the Lagrangian method has the numerical inconvenience of non-fixed and possibly deforming grids. The mass losses and oscillation and dispersion problems experienced in the Eulerian approach can be eliminated by reducing the grid Peclet numbers ($P = U\Delta x/D$) and Courant numbers (CN $= U\Delta t/\Delta x$) at the expense of increased computation time. Here U is flow velocity, Δt is time step, Δx is the grid size, and D is the dispersion coefficient. Cox and Runkel proposed a combined approach in which the grid is fixed through interpolation and back-tracking of moving cell flow paths.

Another example of the application of Equation 1.1 is shown by Li et al. (2008), where the one-dimensional advection equation, modified with a solid phase attachment term, was used to model transport and deposition of fullerene (C_{60}) nanoparticles in water-saturated porous media. These particles currently have several applications in, for example, biomedical technology and cosmetics, and, because of their toxicity to aquatic biota and human cells, an understanding of the fate and transport of these nanoparticles is important. They have negligible solubility in water but can form stable nanoscale aggregates (nC_{60}) by acquiring negative surface charge. Li et al. (2008) found that clean-bed filtration theory, modified to consider shadow zones and surface charge heterogeneity, could be used to predict (nC_{60}) transport in saturated porous media.

While the emphasis in this chapter is on the quantitative prediction of pollutant transport by means of equations such as Equation 1.1, one should realize that tracers can play an important role in the estimation of fate and transport of pollutants. An example of this is the use of stable ^{127}I and radioactive ^{129}I to evaluate the dispersion of ^{129}I from the French nuclear reprocessing facility in La Haque (Hou et al., 2007). From measurements of ^{127}I and ^{128}I in the English Channel and the North Sea it was found that the influence of the La Haque facility on ^{129}I distribution is clearly reflected in surface water of the North Sea. It was also concluded that reduction of iodate (IO_3^-) to iodide (I^-) in ^{129}I is a relatively fast reaction during transport to the European continental coast, while oxidation of I^- to IO_3^- is low between coastal areas and the open sea. The application of tracers to pollution plumes will be further explored in Chapter 9.

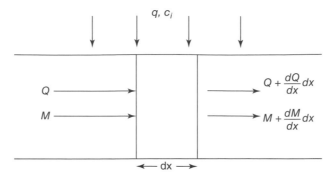

Figure 1.1. Definition sketch for a one-dimensional advection–diffusion equation with reaction.

1.3   STEADY-STATE MIXING IN ESTUARIES

Consider the simplified sketch of a stretch of an estuary or river in Figure 1.1.

Mass balance equations for water and pollutant of concentration c may be written as:

$$\text{Inflow} + \text{Discharge} = \text{Outflow},$$

or,

$$\left.\begin{array}{c} Q + q\,dx = Q + (dQ/dx)\,dx \\ M + qc_i\,dx = M + (dM/dx)dx \end{array}\right\} \tag{1.2}$$

where

Q = water flow m^3/s,
q = discharge per unit length m^3/(m s)
M = pollutant mass flow kg/s
c_i = concentration of pollutant in the discharge kg/m^3.

From Equation 1.2 we obtain,

$$\left.\begin{array}{c} \dfrac{dQ}{dx} = q \\[2mm] \dfrac{dM}{dx} = qc_i \end{array}\right\} \tag{1.3}$$

From Fick's law, the pollutant mass flow M may be expressed as,

$$M = Qc - AD\frac{dc}{dx} \tag{1.4}$$

where A (m^2) is the cross sectional area of the estuary and D (m^2/s) is the diffusion (dispersion) coefficient of diffusivity. If A and D are independent of x we obtain from

Equations 1.3 and 1.4,

$$\frac{Q}{A}\frac{dc}{dx} - D\frac{d^2c}{dx^2} = \frac{q}{A}(c_i - c)$$ (1.5)

In the case where $c_i \gg c$ the second term on the right hand side drops out. This result could also have been obtained directly from the 3-D advection–diffusion equation, Equation 1.1.

1.3.1 Determination of Diffusivity D from Salinity Measurements

Diffusivity D may be estimated from measurements of salinity in an estuary (Harremoës, 1978), or from conductivity in a freshwater estuary such as the Fox River, Wisconsin, United States, which empties into Green Bay. For an infinitely long estuary (Figure 1.2a), the net flow of salinity M at any value x must be zero under steady-state conditions. Thus, from Equation 1.4,

$$c = c_0 e^{\frac{Q}{AD}x}$$ (1.6)

where c_0 is the ocean salinity. If the measured value of the salinity is c_x at the position x, the diffusivity D may be calculated from Equation 1.6,

$$D = \frac{Qx}{A}\frac{1}{\ln\frac{c_x}{c_0}}$$ (1.7)

For the case of a uniform estuary with even discharge (Figure 1.2b), we obtain,

$$\frac{dc}{dx} = \frac{Q}{AD}c = \frac{qx}{AD}c$$ (1.8)

When this is inserted into Equation 1.4 with $M = 0$ we obtain,

$$c = c_L e^{\frac{q}{2AD}(x^2 - L^2)}$$ (1.9)

where c_L is the ocean salinity and L is the estuary. One example of a solution for c is shown in Figure 1.2b. Thus if the salinity is measured to be c_0 for $x = 0$, and diffusivity D can be expressed as,

$$D = \frac{qL^2}{2A\ln\frac{c_L}{c_0}}$$ (1.10)

1.3.2 Pollutant Prediction for an Estuary with Uniform Discharge

For a conservative, the net mass transfer across a boundary at x under steady-state conditions (Figure 1.3) is equal to the mass discharge into the estuary from the left of the boundary,

$$M = qxc - AD\frac{dc}{dx} = qxc_q$$ (1.11)

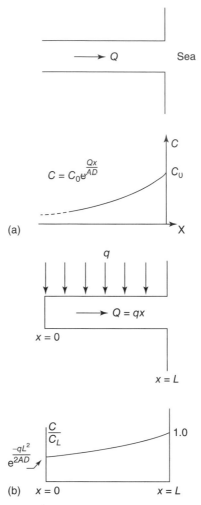

Figure 1.2. Determination of diffusivity D from salinity measurement in (a) an infinitely long estuary, and (b) a uniform estuary with even discharge. (Source: Adapted from Harremoës (1978))

This equation may be integrated to give,

$$\frac{c - c_q}{c_L - c_q} = e^{-\frac{qL^2}{2AD}\left(1 - \left(\frac{x}{L}\right)^2\right)}$$

(1.12)

where c_q is the pollutant concentration in runoff and c_L is the pollutant concentration at the mouth of the estuary. Pollutant concentrations according to this equation are plotted for both $c_L > c_q$ and $c_L < c_q$ in Figure 1.3.

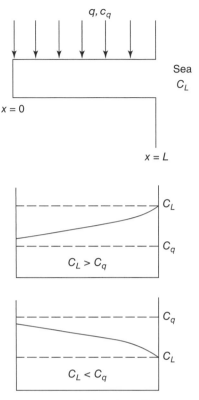

Figure 1.3. Conservative pollutants prediction in uniform estuary with even discharge. (Source: Adapted from Harremoës (1978))

In the case where the pollutant is non-conservative, e.g. nitrogen lost by de-nitrification processes to the sediments, Equation 1.11 may be modified as follows:

$$M = qxc - AD\frac{dc}{dx} = qxc_q - Wxr_a \tag{1.13}$$

where W is the width (m) of the estuary and r_a (kg/(m^2 s)) is a zero order removal rate. The solution to this equation is

$$\frac{c - \left(c_q - \dfrac{Wr_a}{q}\right)}{c_L - \left(c_q - \dfrac{Wr_a}{q}\right)} = e^{-\frac{qL^2}{2AD}\left(1 - \left(\frac{x}{L}\right)^2\right)} \tag{1.14}$$

In order to determine r_a one may measure the pollutant concentration c at any point in the estuary, and then solve the above equation for r_a.

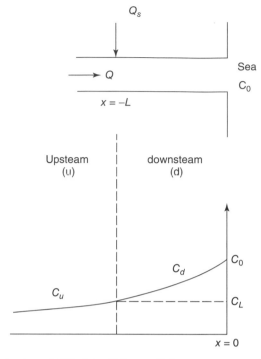

Figure 1.4. Determination of diffusivity D from salinity measurements in an infinite estuary with a large freshwater discharge. (Source: Adapted from Harremoës (1978))

1.3.3   Salinity in an Infinite Estuary with a Large Freshwater Discharge

The case of an infinitely long estuary with a large freshwater discharge, e.g. from a sewage treatment plant, is considered next (Figure 1.4). The net outflow of salinity into the ocean is still zero both upstream and downstream of the discharge point $(x = -L)$:

$$\text{Upstream,} \qquad M = Qc - AD\frac{dc}{dx} = 0 \qquad\qquad (1.15)$$

$$\text{Downstream,} \quad M = (Q + Q_s)c - AD\frac{dc}{dx} = 0 \qquad\qquad (1.16)$$

where Q_s (m^3/s) is the freshwater discharge from the point source. Equations 1.15 and 1.16 are solved using the following boundary conditions:

$$\left.\begin{array}{ll} x = 0 & c_d = c_0 \\ x = -L & c_u = c_d = c_L \\ x \to -\infty & c_u \to 0 \end{array}\right\} \qquad\qquad (1.17)$$

where c_d and c_u are the downstream and upstream solutions, respectively. The solution can be expressed as,

$$\frac{c_u}{c_L} = e^{\frac{Q}{AD}(x+L)} \qquad (1.18a)$$

$$\frac{c_d}{c_0} = e^{\frac{Q+Q_s}{AD}x} \qquad (1.18b)$$

A sketch of this solution is shown in Figure 1.4. From Equation 1.18b we obtain with $c_d = c_L$ for $x = -L$:

$$D = \frac{Q+Q_s}{A}L\frac{1}{\ln\frac{c_0}{c_L}} \qquad (1.19)$$

Therefore, diffusivity D can be calculated based on the measured salinity c_L at $x = -L$.

1.3.4 Conservative Pollutant Prediction for an Infinite Estuary with a Large Freshwater Discharge

With the diffusivity calculated from salinity as a natural tracer, e.g. according to Equation 1.19, it is now possible to predict the distribution of a conservative pollutant that is discharged from the point source with the concentration c_s (Figure 1.5). The net mass transfer of the pollutant of concentration c in the upstream area at any value of x is zero for steady-state conditions,

$$M = Qc - AD\frac{dc}{dx} = 0 \qquad (1.20)$$

However, for the downstream section, the mass transfer at any boundary must equal the pollutant mass discharged from the point source,

$$M = Qc - AD\frac{dc}{dx} = Q_s c_s. \qquad (1.21)$$

The upstream and downstream solutions to Equations 1.20 and 1.21 with the same boundary conditions as for the salinity distribution, Equation 1.17, can be expressed as,

$$c_d = \frac{Q_s}{Q+Q_s}c_s + \left[c_0 - \frac{Q_s}{Q+Q_s}c_s\right] * e^{\frac{Q+Q_s}{AD}x} \qquad (1.22)$$

$$c_u = \left[\frac{Q_s}{Q+Q_s}c_s + \left(c_0 - \frac{Q_s}{Q+Q_s}c_s\right)e^{-\frac{Q+Q_s}{AD}L}\right] * e^{\frac{Q}{AD}(x+L)} \qquad (1.23)$$

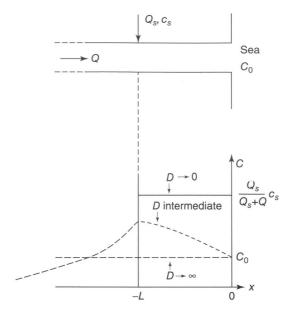

Figure 1.5. Conservative pollutants prediction in infinite estuary with a large freshwater discharge. (Source: Adapted from Harremoës (1978))

The above solutions are indicated in Figure 1.5 along with the limiting cases of $D \to \infty$ corresponding to vigorous mixing ($c = c_0$), and $D \to 0$ reflecting little mixing. In the latter case, the pollutant concentration is given by

$$c = \frac{Q_s}{Q + Q_s} c_s \qquad (1.24)$$

in the downstream area ($x > -L$) and $c = 0$ upstream ($x < -L$). In the downstream area, this is, of course, the same as would be obtained by simple dilution of the effluent with the estuary flow. The actual pollutant concentration falls between those of the limiting cases in the downstream area, and is also noticeable (> 0) upstream of the discharge point (Figure 1.5).

1.4   TIME-DEPENDENT MIXING IN RIVERS AND SOIL SYSTEMS

The one-dimensional (1-D) advection–diffusion equation may be written as,

$$-v\frac{\partial c}{\partial x} + D\frac{\partial^2 c}{\partial x^2} = \frac{\partial c}{\partial t} \qquad (1.25)$$

where the notation is the same as Equation 1.1 except that the velocity in the x-direction v has been named and the subscript for the diffusivity D has been dropped.

Consider the case where a slug of pollutant M (kg/m^2) is released into a river at $x = 0$, $t = 0$. Initially, $c = 0$ everywhere at $t = 0$, except at $x = 0$. The boundary conditions are that $c = 0$ for $x \rightarrow \pm\infty$ and that the following condition holds,

$$\frac{M}{v} = \int_0^\infty c(x, \tau)d\tau \tag{1.26}$$

This equation expresses the fact that the total mass per unit area passing through a boundary at x must equal M.

With these conditions, the solution to Equation 1.25 is (Haas and Vamos, 1995):

$$c = \frac{M}{\sqrt{4\pi Dt}} \exp\left(-\frac{(x - vt)^2}{4Dt}\right). \tag{1.27}$$

Typical solutions following the release of a conservative pollutant to a river are shown in Figure 1.6. When the diffusion coefficient is low (1.4×10^5 cm^2/s, Figure 1.6a), the peak is sharp and the max value high. As the slug moves downstream, the contaminated area widens and the peak drops. A similar picture is apparent when the diffusivity is higher, except that the bell-shaped curve is more spread out (Figure 1.6b).

From Equation 1.27 we can derive the following relationship,

$$D = \frac{1}{2}\frac{d\sigma^2}{dt}, \tag{1.28}$$

which can be used for an observational or experimental determination of D from the spread over time of a colored or otherwise identifiable pollution patch: This expression was, for example, used in a court case to estimate diffusivity based on the pollution plume in Lake Michigan from a wastewater treatment plant in Milwaukee, Wisconsin (Mortimer, 1981).

To illustrate the application of Equation 1.28, consider Figure 1.6a from which we estimate D as follows:

$$D = \frac{1}{2}\frac{(1.0)^2 - (0.4)^2 \text{ km}^2}{8 \text{ h}} = 0.053\frac{\text{km}^2}{\text{h}}, \tag{1.29}$$

which is close to the actual value of 0.05 km^2/h.

The horizontal diffusivity in the flow direction of a river may also be estimated from the equation (McQuivey and Keefer, 1974):

$$D = 0.058\frac{Q_0}{S_0 W_0} \tag{1.30}$$

where
      $Q_0 =$ volumetric flow rate,
      $W_0 =$ width of the river, and
      $S_0 =$ slope of the specific energy curve.

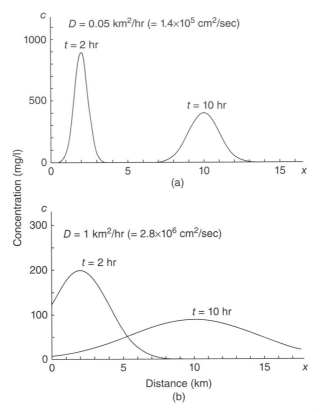

Figure 1.6. Transport and dispersion of a slug of pollutant of $M = 1000$ kg/m^2 in a river with $v = 1$ km/h and (a) $D = 0.05$ km^2/h and (b) $D = 1$ km^2/h.

Rather than an instantaneous release of a slug of a pollutant, a continuous discharge may occur. Consider, for example the continuous leakage of contaminated water into a semi-infinite system such as soil. For such a case, the solution to Equation 1.25 may take one of the following forms (Gershon and Nir, 1969):

$$\frac{c_x}{c_0} = \frac{1}{2}\text{erfc}\left[\frac{x-vt}{2\sqrt{Dt}}\right] + \frac{1}{2}\exp\left(\frac{vx}{D}\right)\text{erfc}\left(\frac{x+vt}{2\sqrt{Dt}}\right) \quad \text{(SINF1)}; \tag{1.31}$$

$$\frac{c_x}{c_0} = \frac{1}{2}\text{erfc}\left[\frac{x-vt}{2\sqrt{Dt}}\right] - \frac{1}{2}\exp\left(\frac{vx}{D}\right)\text{erfc}\left\{\left(\frac{x+vt}{2\sqrt{Dt}}\right) * \left[1 + \frac{v(x+vt)}{D}\right]\right\}$$

$$+ v\sqrt{\frac{t}{\pi D}}\exp\left(\frac{vx}{D} - \frac{1}{2}\left(\frac{x+vt}{\sqrt{Dt}}\right)^2\right) \quad \text{(SINF2)}; \tag{1.32}$$

where x is the depth below the surface. The initial condition for both equations is $c(x, t) = c(x, 0) = 0$, i.e. no contamination in the soil initially. The exit boundary condition is also the same for the two equations, i.e.

$$\frac{\partial c}{\partial x} \to 0 \text{ for } x \to \infty \text{ all } t$$

However, the two solutions differ with regards to the inlet boundary condition. For Equation 1.31, the inlet condition is,

$$c = c_0 \text{ for } x = 0 \text{ all } t$$

This would be the case in a horizontal column experiment with one end kept at constant concentration. Conversely, a vertical system, e.g. infiltration of leachate into the ground from a landfill is governed by the following flux condition:

$$\left(vc - D\frac{\partial c}{\partial x} \right)_{x=0^+} = J_0 = vc_0 \tag{1.33}$$

This is the inlet boundary condition that applies to Equation 1.32. Note that this is the same type of boundary condition that applies to the influx of a pollutant or tracer into a sediment from the water column (Guinasso and Schink, 1975; Christensen and Bhunia, 1986).

Example solutions, SINF1 and SINF2, according to Equations 1.31 and 1.32 are shown in Figure 1.7. The parameters are here $v = 10$ m/d, $x = 10$ m, and D equal to

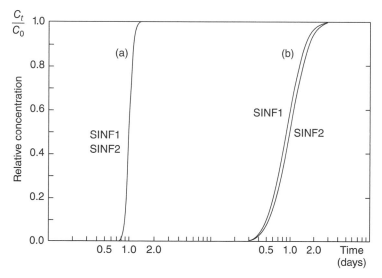

Figure 1.7. Relative concentration vs. time at $x = 10$ m for $v = 10$ m/d and (a) $D = 0.5$ m^2/d and (b) $D = 10$ m^2/d of pollution released at a constant rate for $t \geq 0$ at $x = 0$.

either 0.5 m^2/d (Figure 1.7a) or 10 m^2/d (Figure 1.7b). Note that the front is spread out as D increases. Also, SINF1 is higher than SINF2 when mixing is significant ($D = 10$ m^2/d), and the latter curve, for which the flux inlet condition applies, passes through $c_x/c_0 = 0.5$ for $t = 1$ d. Thus, the flux condition curve progresses into the soil with the velocity $v = 10$ m/d; whereas the mid-point of the concentration condition curve has reached $x = 10$ m in 0.92 days, i.e. travels 9% faster.

Even though velocities, distances, and diffusivities are much lower in the soil infiltration system (Figure 1.7) than in the river (Figure 1.6), the ratio between advective and diffusive transport, i.e. the Peclet number,

$$Pe = \frac{vz}{D} \tag{1.34}$$

is the same in the two cases. Therefore, the time taken to double the concentrations in the soil system may be inferred from the scaled distributions of the river system. The solutions shown in Figure 1.7 were obtained from a table of the error function. An abbreviated version is shown in Table 1.1. For high values of the argument x, the following series expansion was used (Lapidus and Amundson, 1952).

$$\text{erfc}(|x|) = \exp(-x^2)\frac{1}{\sqrt{\pi}}\left[\frac{1}{x} - \frac{1}{2x^3} + \frac{3}{4x^5} - \frac{5}{8x^7} + \cdots\right] \tag{1.35}$$

1.5   VERTICAL MIXING

The vertical diffusivity in a water body may be determined from the ^{222}Rn distribution in the water column, by using the tritium–helium-3 method, or from the vertical density gradient. For lakes, the vertical diffusivity reflects the degree of mixing, which has an impact on the distribution of dissolved oxygen and nutrients. Mixing can make nutrients in the hypolimnion available to bacteria and other organisms in the epilimnion, and the dissolved oxygen levels of the lake will generally increase so as to sustain the fish population and promote aerobic microbial decay.

In a stratified lake nutrients may not be readily available in the epilimnion, and low DO concentrations in the hypolimnion can cause fish deaths. In order to remedy this, artificial de-stratification can be used. Two types are commonly used: a diffuser system with injection of air near the bottom of the lake (Schladow and Fisher, 1995), and an axial flow system, with impellers placed below the water surface, moving warm, well-aerated lake water downward (Lawson and Anderson, 2007).

An axial flow system for de-stratification and mixing in Lake Elsinore, California, was investigated by Lawson and Anderson (2007). The lake is relatively shallow with typical depths from 3 to 7 m. It is polymictic, i.e. too shallow to develop stable thermal stratification. The axial pumps had little effect on stratification and DO levels in the lake. Measurement from an Acoustic Doppler current profiler showed inefficient lateral transmission of mixing energy into the water column. This was probably due to excessive turbulence near the axial pumps, the shallowness of the lake, and the

TABLE 1.1. Selected Values of erf(x)

x	0	1	2	3	4	5	6	7	8	9
0	0	0.11246	0.22270	0.32863	0.42839	0.52050	0.60386	0.67780	0.74210	0.79691
1	0.84270	0.88021	0.91031	0.93491	0.95229	0.96611	0.97635	0.98379	0.98909	0.99279
2	0.99532	0.99702	0.99814	0.99886	0.99931	0.99959	0.99976	0.99987	0.99992	0.99996

Note erfc(x) = 1 − erf(x)

erf(x) = −erf(x)

relatively flat lake bottom. Winter storms in 2005 had a larger effect on water column parameters than the axial pumps.

A shallow mixing depth can be of concern for drinking water supplies as a result of the potential for excessive algal growth, as recently documented for tributary bays of the Yangtze River near the Three Gorges Bay in China (Liu et al., 2012). Investigators studied the Xiangxi Bay of the Three Gorges Bay and suggested that blooms of blue-green algae, i.e. cyanobacteria, could be prevented from appearing during summer months by keeping Z_{eu}/Z_{mix}, where Z_{eu} and Z_{mix} are the depths of the euphotic and mixing zone, respectively, under a certain threshold value. Because Z_{eu} does not change much during the year, this is equivalent to making sure that the mixing depth is sufficiently large.

Short-term controlled fluctuations of the water level in the Three Gorges Reservoir may be an appropriate management strategy to prevent excessive algal growth in Xiangxi Bay. As shown by Liu et al. (2012), an increase in the water level followed by a comparatively rapid decrease will initiate vertical density currents in Xiangxi Bay that will increase the mixing depth, resulting in reduction of the chlorophyll content of the Bay.

Sufficient vertical mixing is important for wastewater outfalls, particularly ocean outfalls from large population centers such as Los Angeles or New York. Adequate vertical mixing will ensure a high degree of dilution of the effluent. This is necessary for minimizing coliform contamination of nearby beaches and any deleterious impact of the effluent plume on aquatic life.

1.5.1   The Radon Method

Application of the advection–diffusion equation, Equation 1.1 in the vertical direction for the water column of a lake gives (Imboden and Emerson, 1978),

$$D_z \frac{\partial^2 c}{\partial z^2} - \lambda_R c = 0 \tag{1.36}$$

where c is the activity of ^{222}Rn per volume unit, dpm/l, and λ_R is the decay constant. Since the half-life of ^{222}Rn is 3.83 d, the decay constant $\lambda_R = 0.181$ d^{-1}. The z-axis has its origin at the sediment–water interface and points up through the water column, and the origin is at the lake bottom.

The solution to Equation 1.36 with $c = c0$ at $z = 0$ and $c \to 0$ for $z \to \infty$ (deep lake) is,

$$c(z) = c_0 e^{-\alpha z}; \ \alpha = \sqrt{\frac{\lambda_R}{D_z}} \tag{1.37}$$

Thus, from a plot of log z vs. z the slope $-\alpha$, from which D_z may be determined, can be derived. Typical values of D_z for a Swiss lake were 0.05–0.57 cm^2/s (Imboden and Emerson, 1978).

1.5.2   The Tritium–helium-3 Method

This method is based on tritium ($T_{1/2} = 12.33$ yr) from bomb fallout (Torgerson et al., 1977). Tritium decays to helium-3 by emission of β particles,

$$\ce{^3_1H} \xrightarrow{\beta^-} \ce{^3_2He} \tag{1.38}$$

The helium-3 will then be present in excess levels relative to the amount that is in equilibrium with natural atmospheric helium-3, i.e. 0.00014% of 5.2 ppm. The amount of excess helium-3 relative to the tritium content is a measure of how long the water mass has been isolated from gas exchange with the atmosphere. Gas exchange is maximal during the spring and fall overturns. Build-up of helium-3 from tritium decay occurs during the winter under ice cover, or during summer when lake stratification is maximal and the thermocline presents an effective barrier to gas exchange.

The tritium–helium-3 method may be used to estimate water mass ages and vertical diffusivities. The following definitions apply,

$$\Delta\, 4\mathrm{He}\ (\%) = \left[\frac{(^4\mathrm{He})_{meas}}{(^4\mathrm{He})_{sat}} - 1 \right] * 100\% \tag{1.39a}$$

$$\delta^3\mathrm{He}(\%) = \left[\frac{(^3\mathrm{He} : {}^4\mathrm{He})_{meas}}{(^3\mathrm{He} : {}^4\mathrm{He})_{atm}} - 1 \right] * 100\% \tag{1.39b}$$

where $\Delta\,^4\mathrm{He}$ is the $^4\mathrm{He}_{excess}$, $\delta\,^3\mathrm{He}$ is the $^3\mathrm{He}$ excess and $^4\mathrm{He}_{meas}$, $^4\mathrm{He}_{sat}$, and $^4\mathrm{He}_{atm}$ are measured, saturation, and atmospheric values of $^4\mathrm{He}$ contents respectively. The ratios in Equations 1.39a and 1.39b are determined by mass spectrometry.

The water mass age t is the minimum time since water mass has exchanged gases with the atmosphere and it is determined from,

$$t = \frac{\ln\left[\dfrac{^3\mathrm{He}_{excess}}{\tau} + 1 \right]}{\lambda} \tag{1.40}$$

where
$\quad \mathrm{He}_{excess} =$ the atoms/g $^3\mathrm{He}$ over the saturation value,
$\quad\quad\quad\ \tau =$ atoms/g tritium content, and
$\quad\quad\quad\ \lambda =$ decay constant of tritium (0.05654 yr^{-1}).

For $t < 1$ yr, Equation 1.40 may be approximated with,

$$t = \frac{1}{\lambda}\frac{^3\mathrm{He}_{excess}}{\tau}, \tag{1.41}$$

or,

$$t = \frac{1}{0.002756}\frac{y}{T}(\delta + 1.4)\ \ (\text{days}) \tag{1.42}$$

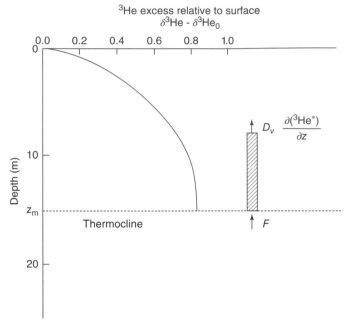

Figure 1.8. ^{3}He buildup and diffusion in the epilimnion of a lake. The graph shows $\delta\,^3$He$-\delta^3$He$_0$ % vs. depth. The curve is approximate for Lake Huron, August 1975 with $z_m \sim 17$ m, $D_v \sim 13$ m^2/d. (Source: Adapted from Torgerson et al. (1977))

where
$\qquad y = {}^4$He content (10^{-8} cc g^{-1} STP)
$\qquad T =$ tritium content in tritium units (1 tritium unit = a tritium atom per 10^{18}
$\qquad\qquad$ hydrogen atoms).

Torgerson et al. (1977) found water mass ages ranging from 75 days for Lake Huron up to 150 days for Lake Ontario. A deeper sample generally gives a higher water mass age.

The vertical diffusivity D_v may be derived from the profile of helium-3 vs. depth (Figure 1.8). This profile follows a parabolic curve, as may be seen from a box model for ^{3}He in a layer bounded at the bottom by the thermocline, and the top by a given z-value:

$$\frac{\partial[^3\text{He}^*]}{\partial t} = A_0 + D_v\frac{\partial^2[^3\text{He}^*]}{\partial z^2} \qquad (1.43)$$

where
$\qquad [^3\text{He}^*] = [^3\text{He}] - [^3\text{He}]_{\text{solubility}},$
$\qquad\quad A_0 = \lambda T$ is the production rate for ^{3}He, i.e. No. of ^{3}He atoms generated
$\qquad\qquad$ by tritium decay per cm^3 per second, and
$\qquad\quad \tau =$ tritium concentration, No. of tritium atoms per cm^3.

Assuming steady-state and the following boundary conditions,

$$D_v \frac{\partial [^3\text{He}^*]}{\partial z} = A_0(z_m - z) + F \text{ at depth } z,$$

where z_m is the depth of the box at the thermocline, F is the input flux to the box, and,

$$[^3\text{He}^*] = [^3\text{He}^*]_0 \text{ at } z = 0 \text{ (water surface)},$$

we obtain the solution,

$$[^3\text{He}^*] = [^3\text{He}^*]_0 + \left(\frac{A_0 z_m + F}{D_v} \right) z - \frac{A_0}{2D_v} z^2. \tag{1.44}$$

When the thermocline is present, it presents an effective barrier to ^3He flux such that $F = 0$. This solution is shown in Figure 1.8 in units of $\delta\,^3\text{He} - \delta\,^3\text{H}_0$ vs. depth where $\delta\,^3\text{He}_0$ is the $\delta\,^3\text{He}$ value at the surface. By fitting measured values of excess helium-3 vs. depth to the model Equation 1.44, it is possible to estimate the vertical diffusivity. Torgerson et al. (1977) found values of D_v ranging from 0.45 cm^2/s for Lake Erie to 1.56 cm^2/s for Lake Huron and 2.93 cm^2/s for Lake Ontario during the season of stratification, i.e. summer and early fall.

The tritium–helium-3 method can also be used to estimate groundwater age, which is important for recharge rates, groundwater residence times, and the calibration of flow models. However, because the aqueous phase molecular diffusion coefficient for ^3He is 3.6–5 times that of ^3H, the method can give misleading results if it is used for a dual-domain aquifer. For aquitards where diffusive mass transfer is the main transport mechanism, the tritium–helium-3 ages are likely to be artificially raised because of the slower dispersion of ^3H. Conversely, for a stream tube, i.e. preferential flow path in a higher conductivity aquifer, the estimated ages can be artificially lowered as a result of the preferential loss of ^3He by diffusion out of the tube (LaBolle et al., 2006; Neumann et al., 2008).

Neumann et al. (2008) conducted simulations of the effects of mobile–immobile domain mass transfer on the tritium–helium-3 dating method by considering the transport along stream tubes. They confirmed results of field experiments from the literature, and concluded that the tritium–helium-3 age is difficult to interpret when mass transfer is on the same time scale as the half-life of tritium and the time since the tritium peak was introduced; about 50 yrs.

1.5.3   Evaluation of Mixing Based on Density Gradients

The vertical diffusion coefficient D_z may also be estimated from the density gradient. Thibodeaux (1979) gave the following expression for D_z,

$$D_z = \frac{10^{-4}}{\left| \dfrac{1}{\rho} \dfrac{\partial \rho}{\partial z} \right|}, \tag{1.45}$$

which is valid within an order of magnitude. Here the depth z is in m and D_z in cm^2/s. For example, for $1/\rho(\partial\rho/\partial z) = 10^{-5}$ m^{-1}, which can be due to a temperature difference of about 2°C over 50 m in a coastal area (National Research Council, 1993), we obtain $D_z = 10$ cm^2/s.

For a submarine outfall, National Research Council (1993) gives the following equations for the maximum rise height of the plume h_{max} (m) and initial dilution S,

$$h_{max} = 2.84 \frac{q^{1/3}\left(\dfrac{g\Delta\rho}{\rho}\right)^{1/3}}{\left(\dfrac{-g}{\rho_a}\dfrac{\partial\rho_a}{\partial z}\right)^{1/2}} \tag{1.46}$$

and,

$$S = \frac{0.31\left(\dfrac{g\Delta\rho}{\rho}\right)^{1/3}h_{max}}{q^{2/3}} \tag{1.47}$$

where

$q =$ discharge rate per unit length of diffuser, m^3/(m s),
$g =$ acceleration as a result of gravity $= 9.81$ m/s^2,
$\dfrac{\Delta\rho}{\rho} =$ relative density difference between effluent and receiving water, and
$\dfrac{\partial\rho_a}{\partial z} =$ average ambient density gradient.

Consider a typical large discharge diffuser with a flow of 5 m^3/s (114 mg d), located in 60 m water depth, 10 km offshore. The z axis is vertical. With a 1 km long diffuser, a relative density difference $\Delta\rho/\rho = 0.027$ and an ambient density gradient $(-1/\rho_a)\partial\rho_a/\partial Z = 10^{-5}$ m^{-1}, we obtain $h_{max} = 31$ m and $S = 211$. For increased stratification, the ambient density gradient increases, reducing the plume rise and the dilution.

1.6   HYDRODYNAMIC MODELS

In order to implement an advection–diffusion equation such as Equation 1.1 for dispersion of pollutants in the coastal ocean or lakes, knowledge of the velocity field is required. The temperature and salinity profiles are also needed.

Blumberg and Mellor (1987) developed a three-dimensional hydrodynamic model to predict these variables on the basis of the water continuity equation and the Reynolds version of the Navier–Stokes equation for conservation of momentum, and conservation equations for temperature and salinity. The Earth's rotation is taken into account through the Coriolis effect. This model, with later improvements, is often referred to as the Princeton ocean model. Boundary conditions for the Reynolds equation include known surface wind stress and associated friction velocity. The friction stress and frictional velocity at the bottom must be specified.

The temperature and salinity at boundaries are derived from climatology. Known inflow and outflow determine open boundary conditions. The equations were solved by finite differences, and produced circulation predictions that were quite realistic when compared to available data.

The Princeton model was used to predict water circulation and mixing for the New York Harbor complex, Long Island Sound, and the New York Bight, with a variety of water-forcing as a result of wind, tides, and freshwater inflows in an area with varying topography (Blumberg et al., 1999). A major objective was to improve water quality management. The authors considered two 12-month periods; one from October 1988 to September 1989 for model calibration, and a second from October 1994 to September 1995 for model validation. The grid was based on an orthogonal curvilinear coordinate system. The model differs from most previous models in that it includes large-scale phenomena over annual cycles. On the basis of comparisons of results from an extensive monitoring program, it was found that the model was able to produce elevations and currents with fewer than 10% and 15% errors respectively, on time scales ranging from semidiurnal to annual.

The model was also used to estimate the effects of relocating sewage effluent from Boston Harbor to a new sewage discharge site 14 km offshore in Massachusetts Bay in a water depth of 30 m (Signell et al., 2000). The relocation occurred in year 2000. Predicted variables included effluent dilution, salinity, and circulation.

The simulations demonstrated an overall reduction in the anthropogenic impact on the marine environment following relocation of the outfall to Massachusetts Bay. Effluent levels in the Harbor dropped by a factor of 10 from 1–2% to 0.1–0.2%. Effluent concentrations only exceeded 0.5% within a few kilometer from the new outfall. During summer stratification, effluent released at the sea bed rises and is trapped beneath the pycnocline, i.e. the zone where the water density profile changes rapidly, limiting the local increase in effluent concentration to the lower layer; whereas surface concentrations decrease relative to the harbor outfall.

Liu et al. (2006) provided an example in which the Princeton model was applied to a freshwater environment. They considered a nearshore finite element transport model for fecal pollution in Lake Michigan. The model included a water continuity equation and a vertically integrated version of the momentum conservation equation. Coupled to this, the model included an advection–diffusion equation for the transport of *Escherichia coli*, *Enterococci*, and temperature. The latter equation included a source or sink term for temperature reflecting shortwave radiation, long wave back radiation, condensation, and heat flux from sensible heat transfer and heat inputs from tributaries.

The model produced good agreement between observed and simulated bacteria concentration at Mt. Baldy Beach, Indiana, except on Julian day 218, 2004, one of 3 days when the *esp* human pollution indicator was found. This suggested that there was a significant input of fecal contamination that did not come from Trail Creek or Kintzele Ditch. It was concluded that sunlight causes inactivation of bacteria in the surf-zone, and that sunlight, temperature, and sedimentation provide a better description of inactivation than first-order kinetics.

Hydrologic models can be adapted to simulate pathogen fate and transport as demonstrated by Dorner et al. (2006) for Canagagigue Creek, Ontario, Canada. The authors used an existing hydrologic model, WATFLOOD, that had been previously calibrated for the Grand River Watershed including Canagagiaue Creek. The model underwent revised calibration because of the inclusion of tile drainage and a smaller grid size. Several waterborne pathogens were considered: *Cryptosporidium* spp. *Giardia* spp., *Campylobacter* spp. and *E. coli* O157:H7. Calibration was carried out for *E. coli* by trial and error following calibration of the hydrologic model.

Application of the model to the pathogen *E. coli* O157:H7 produced results within about one order of magnitude of the observed data. The sparsity of pathogen data presents a challenge for model testing and prediction. The authors observed a rapid increase in measured *E. coli* concentrations during storm events, suggesting that re-suspended sediments can contribute significantly to measured bacteria concentrations.

1.7   GROUNDWATER PLUMES

Application of Equation 2.1 to a one-dimensional groundwater system with reaction gives,

$$\frac{\partial c}{\partial t} = D\frac{\partial^2 c}{\partial x^2} - v\frac{\partial c}{\partial x} - r \qquad (1.48)$$

where

c = concentration of a pollutant, $\mu g/cm^3$, and
r = rate of which the pollutant is removed, $\mu g/(cm^3\ s)$.

In the case where there is partitioning of a chemical between the aqueous and soil phase,

$$s = K_d c \qquad (1.49)$$

where

s = pollutant concentration per mass unit of solids, $\mu g/g$, and
c = pollutant concentration in the aqueous phase, $\mu g/cm^3$.

Thus, the unit of the partition coefficient K is pore volume per mass unit of solids, cm^3/g. The left hand side of Equation 1.48 becomes,

$$\varphi\frac{\partial c}{\partial t} + (1 - \varphi)\rho_s K_d\frac{\partial c}{\partial t} \qquad (1.50)$$

where ϕ is the porosity, ρ_s is the solids density, g/cm^3, and c is the pollutant concentration in the aqueous phase. Introducing the bulk sediment density, g/cm^3, $\rho = (1 - \phi)\rho_s$, Equation 1.48 is transformed into:

$$\left(1 + \frac{\rho K_d}{\varphi}\right)\frac{\partial c}{\partial t} = \frac{1}{\varphi}\left(D\frac{\partial^2 c}{\partial x^2} - v\frac{\partial c}{\partial x} - r\right) \qquad (1.51)$$

The right hand side refers to the system of water and soil. In this system, pollutant transport occurs only in the aqueous phase because the aquifer material is stationary. Assuming that degradation, r, only takes place in the aqueous phase, we obtain:

$$\frac{\partial c}{\partial t} = \frac{D}{1+R}\frac{\partial^2 c}{\partial x^2} - \frac{v}{1+R}\frac{\partial c}{\partial x} - \frac{k_d}{1+R} \tag{1.52}$$

where

> c = pollutant concentration in the aqueous phase, $\mu g/cm^3$,
> v = pore water velocity, cm/s,
> k_d = reaction rate in the aqueous phase = r/ϕ, $\mu g/(cm^3 s)$, and
> R = retardation = $(\rho\, K_d)/\phi$.

Equation 1.52 has the same form as Equation 1.1 except that diffusivity, velocity, and reaction rate have been multiplied with the retardation factor, $R.F. = 1/(1+R)$. Thus, solutions such as Equation 1.27 for a pulse or Equations 1.31 and 1.32 for a step function without reaction, apply to the above case when D and v are multiplied with the appropriate retardation factor. It may be shown that the retardation factor is equal to the mass of the pollutant in the mobile phase divided by the total mass of the pollutant. From Equation 1.52 it can be seen that $R.F.$ can also be expressed as,

$$R.F. = \frac{v_p}{v}, \tag{1.53}$$

where v_p is the velocity of the pollutant ($\leq v$).

High molecular weight compounds that are strongly particle-bound will be slowed down considerably; whereas low molecular weight compounds will experience minimal retardation, and will therefore follow the plume of groundwater with only a small delay. In fact, the groundwater system mimics a gas or liquid chromatograph, providing separation of various compounds according to their partition coefficients between the mobile and stationary phases.

1.8   SEDIMENT MIXING

For sediments, many compounds of interest, e.g. ^{210}Pb, Pb, ^{137}Cs, PAHs, and PCBs, are strongly particle-associated. The sediment particles themselves are mixed by bioturbation or physical action such that only the solid phase becomes of interest, except when the mixing intensity is very small (Robbins et al., 1979; Officer and Lynch, 1989; Christensen and Bhunia, 1986). Sediment mixing is of interest because it can distort historical records of pollutants in sediments, and can provide a means of returning pollutants from the sediments to the water column.

While most early papers recognized just one category of particles participating in the mixing process, Fukumori et al. (1992) found it to be necessary to distinguish between settling (≥ 50 μm) and colloidal (≥ 2 μm) particles. The system of equations for conservation of pollutant and sediment mass may be written as,

$$-\frac{\partial}{\partial z}(v\rho s) + \frac{\partial}{\partial z}\left(D\frac{\partial \rho s}{\partial z}\right) - \lambda \rho s = \frac{\partial \rho s}{\partial t} \qquad (1.54a)$$

$$-\frac{\partial}{\partial z}(\rho v) = \frac{\partial \rho}{\partial t} \qquad (1.54b)$$

with boundary conditions,

$$\left.\begin{array}{ll} \rho s v - D\dfrac{\partial \rho s}{\partial z} = J_0\left(t\right) & \text{at } z = 0 \\[4mm] \dfrac{\partial \rho s}{\partial z} = 0 & \text{at } z = l \end{array}\right\} \qquad (1.55)$$

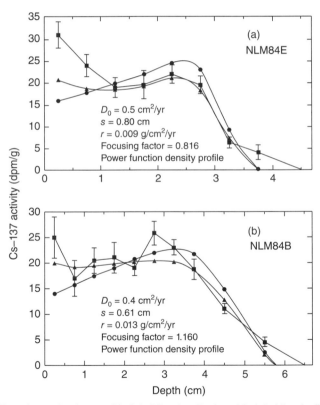

$\blacksquare$ Experimental points $\bullet$ Modeled line (settling) $\blacktriangle$ Modeled line (colloidal)

Figure 1.9. Comparison of experimental and calculated ^{137}Cs activities (a) in core NLM84E and (b) in core NLM84B. The calculated curves are from a finite difference method. The calculated curves apply to both the case where ^{137}Cs is associated with settling particles, and to the case where ^{137}Cs is associated with colloidal particles. (Source: Reproduced from Fukumori et al. (1992), with permission from Elsevier.)

where z (cm) is the depth below the sediment-water interface, v (cm/yr) is the burial velocity relative to the interface, t (y) is the time, D (cm^2/yr) is the diffusion coefficient, λ(yr^{-1}) is the radioactive decay constant, and $\rho = (1 - \phi)\rho_s$ is the bulk density of sediment solids. The porosity is designated ϕ and the solids density $\rho_s = 2.45$ (g/cm^3).

The tracer and bulk solids Equations 1.54a and 1.54b are valid for colloids, e.g. clay particles. For settling particles, the diffusive flux $D\, \partial(\rho s)/\partial z$ should be replaced by $D\, \rho\, \partial s/\partial z$.

As an example of the application of these equations, consider Figure 1.9, which shows measured and calculated values for ^{137}Cs activity in sediments from northern Lake Michigan (Fukumori et al., 1992). Distributions were calculated using a finite difference method under the assumptions that ^{137}Cs is associated with colloidal clay particles, which is known to be the case (Comans et al., 1991; Ab Razak et al., 1996), and the hypothetical case where ^{137}Cs would be bound to larger settling particles. The ^{137}Cs input flux $J_0(t)$ was taken from Health and Safety Laboratory (1977), and the bulk sediment density follows a power function (Fukumori et al., 1992). From Figure 1.9 it is clear that the colloidal interpretation for ^{137}Cs is the more accurate, thus supporting Equations 1.54a and 1.54b. The high ^{137}Cs activity values (dpm/g) near the top of the sediment in this case occur because of the low bulk density ρ in this region. For ^{210}Pb, which is associated with larger particles (Shimp et al., 1971; Ab Razak et al., 1996), the alternate form containing a diffusive flux of $D\, \rho\, \partial s/\partial z$ gives, as expected, better agreement between calculated and experimental activities (see Fukumori et al., 1992).

REFERENCES

Ab Razak, IA, Li, A, Christensen, ER. Association of PAHs, PCBs, ^{137}Cs, and ^{210}Pb with clay, silt, and organic carbon in sediments. *Wat. Sci. Technol.* 1996;34(7–8):29–35.

Blumberg, AF, Mellor, GL. A description of a three-dimensional coastal ocean circulation model. In: Heaps, NS, Editor. *Three dimensional coastal ocean models.* Washington, DC: American Geophysical Union; 1987. p 1–16.

Blumberg, AF, Khan, LA, St. John, JP. Three-dimensional hydrodynamic model of New York Harbor region. *J. Hydr. Eng. ASCE*, 1999;125(8):799–816.

Christensen, ER, Bhunia, PK. Modeling radiotracers in sediments: comparison with observations in Lakes Huron and Michigan. *J. Geophys. Res.* 1986;91(C7):8559–8571.

Christensen, ER, Edgington, DN, Giesy, JP. Contaminated Aquatic Sediments. *Wat. Sci. Technol.* 1993;28(8–9).

Christensen, ER, Karls, JF. Unmixing of lead, ^{137}Cs, and PAH records in lake sediments using curve fitting with first- and second order corrections. *Wat. Res.* 1996;30(11):2543–2558.

Christensen, ER, Phoomiphakdephan, W, Ab Razak, IA. Water quality in Milwaukee, Wisconsin versus intake crib location. *ASCE J. Environ. Eng.* 1997;123(5):492–498.

Comans, RNJ, Haller, M, De Preter, P. Sorption of cesium on illite: nonequilibrium behavior and reversibility. *Geochim. Cosmochim. Acta.* 1991;55:433–440.

Cox, TJ, Runkel, RL. Eulerian-Lagrangian numerical scheme for simulating advection, dispersion and transient storage in streams and a comparison of numerical methods. *J. Environ. Eng. ASCE*, 2008;134(12):996–1005.

Dorner, SM, Anderson, WB, Slawson, RM, Kouwen, N, Huck, PM. Hydrologic modeling of pathogen fate and transport. *Environ. Sci. Technol.* 2006;40(15):4746–4753.

Frieseke, RW, Christensen, ER. Groundwater remediation with granular collection system. *ASCE J. Environ. Eng.* 1995;122(6):546–549.

Fukumori, E, Christensen, ER, Klein, RJ. A model for ^{137}Cs and other tracers in lake sediments considering particle size and the inverse solution. *Earth Planet. Sci. Lett.* 1992;114:85–99.

Gershon, ND, Nir, A. Effects of boundary conditions of models on tracer distributions in flow through porous mediums. *Wat. Res.* 1969;5(4):830–839.

Guinasso, NL Jr., Schink, DR. Quantitative estimates of biological mixing rates in abyssal sediments. *J. Geophys. Res.* 1975;80:3032–3043.

Haas, CN, Vamos, RJ. *Hazardous and Industrial Waste Treatment.* Englewood Cliffs, New Jersey: Prentice Hall; 1995.

Harremoës, P. 1978. Analytical pollution models of well mixed estuaries. In: *Coastal Pollution Control. World Health Organization Training Course,* Vol. III, Geneva, Switzerland.

Health and Safety Laboratory. 1977. *Final Tabulation of Sr-90 Fallout Data: 1954-1976, Environ. Quarterly, HASL-329,* Energy Research and Development Admin., New York.

Hou, X, Aldahan, A, Nielsen, SP, Possnert, G, Nies, H, Hedfors, J. Speciation of [129]I and [127]I in seawater and implications for sources and transport pathways in the North Sea. *Environ. Sci. Technol.* 2007;41(17):5993–5999.

Imboden, DM, Emerson. Natural radon and phosphorus as limnologic tracers: Horizontal and vertical eddy diffusion in Greifensee. *Limnol. Oceanogr.* 1978;23:77–90.

LaBolle, EM, Fogg, GE, Eweis, JB. Diffusive fractionating of ^{3}H and ^{3}He in groundwater and its impact on groundwater age estimates. *Wat. Res.* 2006;42:W07202, doi: 10.1029/2005WR004WR004756.

Lapidus, L, Amundson, NR. Mathematics of adsorption in beds. VI The effect of longitudinal diffusion in ion exchange and chromatographic columns. *J. Phys. Chem.* 1952;56:984–988.

Lawson, R, Anderson, MA. Stratification and mixing in Lake Elsinore, California: An assessment of axial flows for improving water quality in a shallow eutrophic lake. *Wat. Res.* 2007;41:4457–4467.

Lee, YW, Dahab, MF, Bogardi, I. Nitrate-risk assessment using fuzzy-set approach. *ASCE J. Environ. Eng.* 1995;121(3):245–256.

Li, Y, Wang, Y, Pennell, KD, Abriola, LM. Investigation of the transport and deposition of fullerene (C60) nanoparticles in quartz sands under varying flow conditions. *Environ. Sci. Technol.* 2008;42(19):7174–7180.

Liu, L, Liu, D, Johnson, DM, Yi, Z, Huang, Y. Effects of vertical mixing on phytoplankton blooms in Xiangxi Bay of Three Georges Reservoir: Implications for management. *Wat. Res.* 2012;46:2121–2130.

Liu, L, Phanikumar, MS, Molloy, SL, Whitman, RL, Shively, DA, Nevers, MB, Schwab, DJ, Rose, JB. Modeling the transport and inactivation of *E.coli* and Enterococci in the nearshore region of Lake Michigan. *Environ. Sci. Technol.* 2006;40(16):5022–5028.

McQuivey, RS, Keefer, TN. Simple method for predicting dispersion in streams. *ASCE J. Environ. Eng. Div.* 1974;100:997–1011.

Mortimer, CH. *The Lake Michigan Pollution Case. A Review and Commentary on the Limnological and Other Issues.* Milwaukee, Wisconsin: Center for Great Lakes Studies, University of Wisconsin-Milwaukee; 1981.

National Research Council. Managing Wastewater in Coastal Urban Areas. Committee on Wastewater Management for Coastal Urban Areas. Water Science and Technology Board. *Commission on Engineering and Technical Systems,* pp 237–238. Washington, DC: National Academy Press; 1993.

Neumann, RB, Labolle, EM, Harvey, CF. The effects of dual-domain mass transfer on the Tritium-Helium-3 dating method. *Environ. Sci. Technol.* 2008;42(13):4837–4843.

Officer, CB, Lynch, DR. Bioturbation, sedimentation, and sediment-water exchanges. *Estuar. Coast. Shelf Sci.* 1989;28:1–12.

Robbins, JA, McCall, PL, Fisher, JB, Krezoski, JA. Effects of deposit feeders on migration of ^{137}Cs in lake sediments. *Earth Planet. Sci. Lett.* 1979;42:277–287.

Schladow, SG, Fisher, IH. The physical response of temperate lakes to artificial destratification. *Limnol. Oceanogr.* 1995;40(2):359–373.

Shimp, NF, Schleicher, JA, Ruch, RR, Heck, DB, Leland, HV. Trace Element and Organic Carbon Accumulation in the Most Recent Sediments of Southern Lake Michigan. Illinois State Geol. Surv. Environ. Geol. Notes No. 41; 1971.

Signell, RP, Jenter, HL, Blumberg, AF. Predicting the physical effects of relocating Boston's sewage outfall. *Estuar. Coast. Shelf Sci.* 2000;50:59–72.

ter Laak, TL, Van Eijkeren, JCH, Busser, FJM, Van Leeuwen, HP, Hermens, JLM. Facilitated cnd polybrominated diphenyl ethers by dissolved organic matter. *Environ. Sci. Technol.* 2009;43(5):1379–1385.

Thibodeaux, LJ. *Chemodynamics.* New York: John Wiley and Sons, Inc.; 1979.

Torgerson, T, Top, Z, Clarke, WB, Jenkins, WJ, Broecker, WS. A new method for physical limnology - tritium - helium-3 ages - results for Lakes Erie, Huron, and Ontario. *Limn. Oceanogr.* 1977;22(2):181–193.

Weber, WJ Jr, DiGiano, FA. *Process Dynamics in Environmental Systems.* New York: John Wiley and Sons, Inc.; 1996.

Wetzel, RG 1975. *Limnology.* Philadelphia: W.B. Saunders Company.

2

SEDIMENTATION PROCESSES

2.1 INTRODUCTION

Sediments often serve as a near-final repository of particle-associated pollutants in lakes or oceans. This implies that even when pollution from various point and non-point sources has been curtailed, the sediments may still be contaminated from past discharges. The sediments contain pollutants that under appropriate circumstances can be released back into the water column where they can present a threat to the biota and humans. When the rate of sedimentation is fast and sediment mixing is minor, clean sediment can accumulate on top of contaminated material, effectively isolating the pollution. If water depth for navigation is not a factor, and if some natural degradation of organic pollutants, such as polychlorinated biphenyls (PCBs), is possible, then the situation is often allowed to persist because natural cleansing is deemed to have occurred. If not, some type of remedial action, such as capping with clean sediment, injection of oxidizers, or hydraulic dredging and land disposal, may be required.

While sediment pollution can be quite difficult to cope with because of the recalcitrant nature of buried pollutants and the cost involved in cleaning up contaminated sites, the presence of these contaminants can also present an opportunity to determine sources through geochronology, chemical mass balance modeling, and related methodologies. Sediment dating based on ^{210}Pb and ^{137}Cs has proven particularly useful in evaluating the impact of pollutants introduced into the environment as a result of the industrial revolution. The ^{210}Pb dating technique will be described in some detail below.

Physical and Chemical Processes in the Aquatic Environment, First Edition. Erik R. Christensen and An Li.
© 2014 John Wiley & Sons, Inc. Published 2014 by John Wiley & Sons, Inc.

2.2 210Pb DATING OF SEDIMENTS

This dating method was first applied to snowfields in Greenland (Goldberg, 1963). Subsequently, it was demonstrated that it could be used to date lake (Krishnaswami et al., 1971) and ocean (Koide et al., 1972) sediments. The method is based on ^{210}Pb ($T_{1/2}$ = 22.26 yr), a member of the natural ^{238}U radioisotope series (Friedlander et al., 1964). Disequilibrium in this series is created by the emanation of the gas ^{222}Rn into the atmosphere from the earth's crust. Following this, ^{222}Rn decays, through a series of short-lived nuclides, into ^{210}Pb, which is carried by dry and wet fallout onto the surface of the earth, including lakes and oceans. After particle-associated ^{210}Pb settles on the sediment–water interface, it may be deemed a clock, which can be used to track the time since the activity was first deposited as part of the top sediment layer. For example, once the activity (in dpm/g) of this layer has declined to one half of its original value, after deposition of further sediment materials, the layer has an age of one half-life, i.e. 22.3 yrs. The ^{210}Pb activity at depth z (in cm) may be written,

$$A(z) = A_0 e^{-\frac{\lambda z}{v}} \tag{2.1}$$

where
$\lambda = 0.03114$ per year is the decay constant for ^{210}Pb
v = constant sedimentation rate (in cm/year)
$A_0 = {}^{210}$Pb activity initially, i.e. at the sediment–water interface (in dpm/g)

The age $t = z/v$ (years) of the layer at z can then be expressed as

$$t = \frac{1}{\lambda} \ln \left(\frac{A_0}{A} \right) \tag{2.2}$$

In the more common case where the linear sedimentation rate v is not constant, i.e. decreases vs. depth due to compaction, the mass sedimentation rate r (g/(cm^2 year)) may often still be deemed a constant, and Equation 2.1 can be written,

$$A(z) = A_0 e^{-\frac{\lambda m}{r}} \tag{2.3}$$

where m (in g/cm^2) is the cumulative sediment mass at depth z.

In this case the age of the layer at m can be calculated either from Equation 3.2 or from $t = m/r$. In the model presented, the activity at the sediment–water interface should be evaluated at $t = 0$. However, in practice, A_0 is measured at the same time $A(z)$ is measured, assuming that A_0 has not changed. Therefore, this model is often known as the constant initial concentration model (CIC model).

Some watersheds are characterized by a variable erosion rate, which causes a variable mass sedimentation rate with a near-constant rate of ^{210}Pb supply. For this case, the constant rate of supply (CRS) model may be appropriate. This model was investigated by Appleby and Oldfield (1978) and Hermanson and Christensen (1991). The layer age of depth z is here determined by

$$t = \frac{1}{\lambda} \ln \left(\frac{A_\infty}{A_z} \right) \tag{2.4}$$

where

A_∞ is the integrated activity (in dpm/cm^2) for the whole core and

A_z is the integrated activity (in dpm/cm^2) below layer of depth z.

The basic assumption of the CRS model is constant rate of supply of ^{210}Pb or constant ^{210}Pb inventory.

In order to derive Equation 2.4, constant influx J_0 of ^{210}Pb to a sediment from the overlying water column must be considered. The age t of a thin sediment layer at sediment depth z is determined. The integrated ^{210}Pb activity in the sediment above this boundary is $A = A_\infty - A_z$. The rate of change of A is given by

$$\frac{dA}{dt} = J_0 - \lambda A \tag{2.5}$$

with the solution

$$A = A_1 e^{-\lambda t} + \frac{J_0}{\lambda} \tag{2.6}$$

where $A = 0$ for $t = 0$ such that $A_1 = -J_0/\lambda$ and

$$A = \frac{J_0}{\lambda}(1 - e^{-\lambda t}) \tag{2.7}$$

with $A_\infty = J_0/\lambda$.

From Equation 2.7, we obtain Equation 2.4 for the age t of the thin layer.

2.2.1 Measurement of ^{210}Pb Activity

Determination of ^{210}Pb activity is carried out by gamma counting (46.5 keV), beta counting of the daughter ^{210}Bi, or alpha counting of the granddaughter ^{210}Po. Although all of these methods are used, ^{210}Po counting using energy-specific surface barrier detectors is often the method of choice because of its specificity and virtually zero background counts.

Some chemical separations are involved in the ^{210}Bi and ^{210}Po methods. In the ^{210}Bi method, Ra, Ba, Sr, and Pb are first separated from Th and U, and then Pb is separated from Ra, Ba, and Sr. Finally, ^{210}Pb along with a stable lead carrier is precipitated as PbSO$_4$, filtered and transferred to a copper planchet, which is placed in a beta counter for determination of ^{210}Bi beta activity (Koide and Bruland, 1975).

The method based on ^{210}Po consists of an initial step of digestion, volume reduction and pH adjustment followed by plating of ^{210}Po along with a yield tracer, e.g. ^{208}Po or ^{209}Po, onto copper plates. These plates are then counted for alpha activity with surface barrier detectors (Gin, 1992).

The basis for these methods of determining the ^{210}Pb activity is that the ^{210}Bi and ^{210}Pb activity are nearly equal to that of ^{210}Pb. In the ^{210}Bi method, this will be true after PbSO$_4$ precipitation and after several ^{210}Bi half-lives (5.01 d). In the ^{210}Po method, there is usually near-equilibrium between ^{210}Pb and ^{210}Po at the time of collection of the sediment core. Equilibrium is approached even more closely if a waiting period comparable to the half-life of ^{210}Po (138.1 d) is imposed between collection of the core and Po plating. The relationship between the activities of ^{210}Pb, ^{210}Bi, and ^{210}Po may be derived from the equations governing the sequential decay of these nuclides:

$$N_1 \xrightarrow{\lambda_1} N_2 \xrightarrow{\lambda_2} N_3 \xrightarrow{\lambda_3} N_4 \text{ (stable)} \tag{2.8}$$

or,

$$^{210}Pb \xrightarrow[\substack{\beta^- \\ 22.3 \text{ yr}}]{} {}^{210}Bi \xrightarrow[\substack{\beta^- \\ 5.01 \text{ d}}]{} {}^{210}Po \xrightarrow[\substack{\alpha \\ 138.1 \text{ d}}]{} {}^{206}Pb \tag{2.9}$$

where λ_1, λ_2, and λ_3 represent the decay constants, and the corresponding half-lives are indicated in Equation 2.9. The equations are

$$\frac{dN_1}{dt} = -\lambda_1 N_1 \tag{2.10}$$

$$\frac{dN_2}{dt} = \lambda_1 N_1 - \lambda_2 N_2 \tag{2.11}$$

$$\frac{dN_3}{dt} = \lambda_2 N_2 - \lambda_3 N_3 \tag{2.12}$$

$$\frac{dN_4}{dt} = \lambda_3 N_3 \tag{2.13}$$

With the initial conditions $N_1 = N_{10}$, $N_2 = 0$ and $N_3 = 0$ at $t = 0$ we obtain,

$$R_2 = R_0 \frac{\lambda_2}{\lambda_2 - \lambda_1} \left(e^{-\lambda_1 t} - e^{-\lambda_2 t} \right) \tag{2.14}$$

and,

$$R_3 = R_0 \frac{\lambda_1 \lambda_2 \lambda_3}{(\lambda_1 - \lambda_2)(\lambda_1 - \lambda_3)(\lambda_2 - \lambda_3)}$$

$$\times \left[(\lambda_2 - \lambda_3) e^{-\lambda_1 t} + (\lambda_3 - \lambda_1) e^{-\lambda_2 t} + (\lambda_1 - \lambda_2) e^{-\lambda_3 t} \right] \tag{2.15}$$

where $R_2 = \lambda_2 N_2$ and $R_3 = \lambda_3 N_3$ are the activities of ^{210}Bi and ^{210}Po respectively, and $R_0 = \lambda_1 N_{10}$ is the initial activity at $t = 0$ of ^{210}Pb. From $\lambda = \ln 2/T_{1/2}$, the numerical values of the decay constants are

$$
\left.
\begin{array}{ll}
\lambda_1 = 8.52 \cdot 10^{-5} \text{ d}^{-1} & ^{210}\text{Pb} \\[6pt]
\lambda_2 = 1.38 \cdot 10^{-1} \text{ d}^{-1} & ^{210}\text{Bi} \\[6pt]
\lambda_3 = 5.02 \cdot 10^{-3} \text{ d}^{-1} & ^{210}\text{Po}
\end{array}
\right\}
\qquad (2.16)
$$

A plot of R_2/R_0 is shown in Figure 2.1. This is known as an in-growth curve for ^{210}Bi activity. For $t > 30$ d, R_2 reduces to

$$
R_2 \cong R_0 e^{-\lambda_1 t} \qquad (2.17)
$$

meaning that the ^{210}Bi and ^{210}Pb activities are nearly equal. If the copper planchets are counted earlier, e.g. 10 d after PbSO$_4$ precipitation, the original ^{210}Pb activity can be calculated by multiplying the measured ^{210}Bi activity with the correction factor $R_0/R_2 = 1.337$.

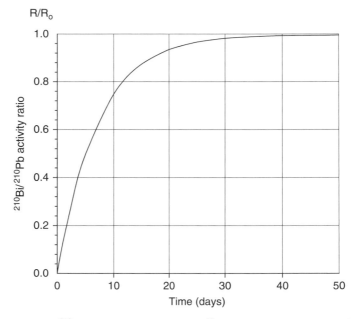

Figure 2.1. Ratio of ^{210}Bi activity (R) over initial ^{10}Pb activity (R_0) vs. time since PbSO$_4$ precipitation.

In a similar fashion, for $t > 1$ yr, we obtain

$$R_3 \cong \frac{\lambda_1 \lambda_2 \lambda_3}{(\lambda_1 - \lambda_2)(\lambda_1 - \lambda_3)} e^{-\lambda_1 t} \cong \lambda_1 N_{10} \, e^{-\lambda_1 t}$$

$$= R_0 \, e^{-\lambda_1 t} \tag{2.18}$$

indicating that the ^{210}Po activity is approximately equal to the ^{210}Pb activity. Because Equation 2.18 was derived under the assumption of extreme disequilibrium ($N_2 = N_3 = 0$ at $t = 0$), the actual time to achieve equilibrium, and thus $R_3 = R_0 e^{-\lambda_1 t}$, is considerably less than 1 yr and is more likely to be a few months or less. Thus the necessary waiting time between sediment core collection and chemical separations for ^{210}Pb is of the order of a few months.

2.2.2 ^{210}Pb Activity Profiles

In order to develop analytical solutions for ^{210}Pb profiles, certain simplifying assumptions are in order. First, steady-state must be assumed. The mass sedimentation rate r (in g/(cm^2 year)) is then constant,

$$r = \rho v \tag{2.19}$$

where ρ (g/cm^3) is the bulk sediment density and v (in cm/year) is the sedimentation rate. Let us further assume that there is no compaction; therefore ρ and v are constant vs. depth.

The advection–diffusion equation with the depth coordinate pointing down in the sediment from the sediment–water interface ($z = 0$) is

$$-v\frac{dc}{dz} + D\frac{d^2c}{dz^2} - \lambda c = 0 \tag{2.20}$$

where c (in dpm/cm^3) is the ^{210}Pb activity and λ the decay constant. The solution to this equation is

$$c = c_0 \exp\left(\frac{v}{2D}\left(1 - \sqrt{1 + \frac{4\lambda D}{v^2}}\right)z\right) \tag{2.21}$$

where c_0 is the ^{210}Pb activity at $z = 0$. Consider two cases,

1. $4\lambda D \ll v^2$

$$c \cong c_0 \exp\left(-\frac{\lambda}{v}z\right) \tag{2.22}$$

which is similar to Equation 2.1, and

2. $4\lambda D \gg v^2$

$$c \cong c_0 \exp\left(-\sqrt{\frac{\lambda}{D}} \, z\right) \tag{2.23}$$

which is similar to Equation (1.37) for diffusion of radon in the water column. Note that for known values of z, c_0, and c, the diffusivity D can be determined from

$$D = \lambda \frac{z^2}{\left(\ln \dfrac{c_0}{c}\right)^2} \tag{2.24}$$

To improve the accuracy of the determination, one would consider the straight-line log-transformed version of Equation 2.23 and then calculate D from the slope, $\alpha = -\sqrt{\lambda/D}$.

Both v and D may be determined from one core because mixing is usually confined to the upper layers where condition (2) may hold; whereas condition (1) applies to the region below the mixing layer. One difficulty with this approach is that it is only relevant for the uppermost part of the mixing layer because the method is only valid for a semi-infinite sediment. In areas of high sedimentation rate, such as Milwaukee Inner Harbor (Table 2.1), condition (1) holds throughout the core, and mixing can then only be determined by a radionuclide of shorter half-life, such as the cosmogenic ^{7}Be (Krishnaswami and Lal, 1978). With a half-life of 53 d, the decay constant for ^{7}Be is 4.77 per yr, compared to 0.03114 per yr for ^{210}Pb. Values of $4\lambda D/v^2$ for these two radionuclides are listed in Table 2.1 for both Inner and Outer Milwaukee Harbor. This table shows that the ^{210}Pb profiles are influenced by mixing in the Outer, but not the Inner, Harbor, and ^{7}Be is suitable for the determination of diffusivities in both areas.

Another short-lived radionuclide that is useful for estimating sediment mixing is ^{234}Th ($T_{1/2} = 24.1$ d). It is especially suitable for the marine environment, where a disequilibrium between ^{238}U and ^{234}Th exists during storm events. ^{238}U is dissolved in seawater under oxidizing conditions as the anion uranyl carbonate complex and

TABLE 2.1. Estimated Values of $4\lambda D/v^2$ (see text) for ^{210}Pb and ^{7}Be in Top Sediment Layers from the Inner and Outer Milwaukee Harbor

	Nuclide	
Area	^{210}Pb	^{7}Be
Inner Harbor $v = 2$ cm/yr (measured) $D \sim 10$ cm^2/yr (assumed)	0.31	47.7
Outer Harbor $v = 0.1$ cm/yr (measured) $D \sim 14$ cm^2/yr (measured)	174	$2.67 \cdot 10^4$

the daughter nuclide ^{234}Th takes the form of the particle-reactive hydrolysis product Th(OH)$_n$$^{(4-n)+}$. The much larger distribution coefficient for the thorium complex $\sim10^6$ L/kg compared to that of ^{238}U $\sim10^2$ L/kg (Kersten et al., 2005) induces rapid scavenging of thorium by suspended particular matter in excess of the background fraction supported by ^{238}U.

Kersten et al. (2005) used ^{234}Th to study the dispersal of zinc and lead from fly ash sludges dumped in Meclenburg Bay in northern Germany before 1971. On the basis of excess ^{234}Th activities in the upper 10 cm of sediment they concluded that rapid mixing was present during wave-driven storm re-suspension events. They estimated that half of the dumped Zn (455 t) and Pb (173 t) had spread from an initial 0.5 km^2 area into an affected area of 360 km^2. However, for each additional cm of natural sediment deposition corresponding to a 5-yr period, the amount of remobilization would be reduced by about 50%.

Two-Layer Model. Goldberg and Koide (1962) developed a two-layer steady-state model for the activity profile with an upper mixing layer (region I) and a historical layer below (region II). The equations are

$$\text{Region I}: \quad -v\frac{dc}{dz} + D\frac{d^2c}{dz^2} - \lambda c = 0 \tag{2.25}$$

with boundary conditions

a) $c_I = c_0, \quad z = 0$

b) $J_I = J_{II}; z = z_m \Rightarrow \left.\dfrac{dc}{dz}\right|_{z=z_m} = 0$

where the tracer flux $J = vc - D(dc/dz)$ and z_m represents the mixed layer thickness.

$$\text{Region II}: \quad -v\frac{dc}{dz} - \lambda c = 0 \tag{2.26}$$

With boundary conditions,

a) $c = c_I; z = z_m$

b) $c \to 0;$ for $z \to \infty$

The solution to these equations is

$$\left. \begin{aligned} c &= c_2\, [K\, e^{\alpha_1 z} + e^{\alpha_2 z}] \quad 0 \leq z \leq z_m \\[2mm] c &= c_3\, e^{-\frac{\lambda}{v} z} \qquad\qquad\quad z > z_m \end{aligned} \right\} \tag{2.27}$$

with

$$c_0 = c_2[K + 1]$$

$$c_3\, e^{-\frac{\lambda}{v}z_m} = c_2[K\, e^{\alpha_1\, z_m} + e^{\alpha_2\, z_m}]$$

$$K = -\frac{\alpha_2}{\alpha_1} \exp\left(\frac{v}{D}\left(\sqrt{1 + \frac{4\lambda D}{v^2}}\right) z_m\right)$$

$$\alpha_1 = \frac{v}{2D}\left[1 - \sqrt{1 + \frac{4\lambda D}{v^2}}\right]$$

$$\alpha_2 = \frac{v}{2D}\left[1 + \sqrt{1 + \frac{4\lambda D}{v^2}}\right]$$

A set of solutions for ^{210}Pb profiles of two Green Bay cores is shown in Figure 2.2 (Christensen, 1982). Note that the flux boundary condition requires the profile to be flat at $z = 4$ cm in region I. This feature is not seen in the measured profiles, indicating that D is not constant in the mixing layer, but rather gradually decreases with depth as, for example a half-Gaussian function (Christensen and Bhunia, 1986, Robbins, 1986, Fukumori et al., 1992).

Compaction. In the above equations, except Equation 2.3, it is assumed that there is no compaction vs. depth, i.e. that the sediment bulk density ρ and velocity v are constant. However, in most actual cores ρ increases with depth, meaning that the porosity ϕ and velocity v decrease with depth. The porosity may be determined from the expression for weight fraction of water w_i,

$$w_i = \frac{\phi \rho_w}{\phi \rho_w + (1 - \phi)\rho_s} \tag{2.28}$$

where

ϕ = porosity, i.e. volume fraction of sediment occupied with water,
ρ_s = solid sediment density (in g/cm^3),
$\rho = (1 - \phi)\rho_s$ = bulk sediment density (in g/cm^3), and
ρ_w = water density ≈ 1 g/cm^3.

For $\rho_w = 1$ g/cm^3 and $\rho_s = 2.45$ g/cm^3, we obtain

$$\varphi = \frac{2.45 w_i}{1 + 1.45 w_i} \tag{2.29}$$

The expression for the ^{210}Pb activity, Equation 2.3, in which compaction is considered may be derived from the advection–diffusion equation, Equation 2.25 with

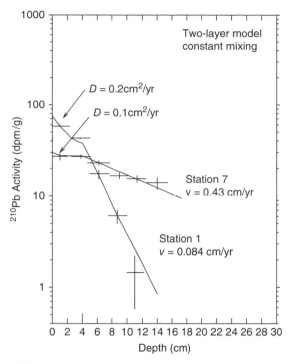

Figure 2.2. Excess ^{210}Pb activity vs. depth in Green Bay cores 1 and 7. Measured activities are compared to results from a two-layer model with constant mixing in the upper layer of 4 cm thickness (Christensen 1982).

$D = 0$ and $c = \rho s$ where d (in dpm/g) is the activity per mass unit of sediment. Therefore,

$$\frac{d}{dz}(\upsilon \rho s) + \lambda \rho s = 0 \qquad (2.30)$$

However, since the mass sedimentation rate $r = \rho \upsilon$ (in g/(cm^2 year)) is constant under steady-state conditions, Equation 2.30 becomes

$$r\frac{ds}{dz} + \lambda \rho s = 0 \qquad (2.31)$$

which may be integrated to yield

$$s = s_0 \exp\left(-\frac{\lambda}{r}m\right) \qquad (2.32)$$

where

$s = {}^{210}$Pb activity (in dpm/g) at depth z or cumulative mass m (in g/cm^2), and
$s_0 = {}^{210}$Pb activity (in dpm/g) at the sediment–water interface.

This equation is identical to Equation 2.3 with $s = A$ and $s_0 = A_0$.

An analytical expression for a ^{210}Pb profile influenced by compaction but not by mixing may be obtained based on the following empirical expression for the bulk sediment density

$$\rho = \rho_0 - \rho_1 \, e^{-az} \tag{2.33}$$

where

$\rho_0 - \rho_1$ is the bulk density at the sediment–water interface,

ρ_0 is the density at large depths, and

$a > 0$ is an adjustable parameter.

The cumulative mass m is then

$$m = \int_0^z \rho dz = \rho_0 z - \frac{\rho_1}{a}[1 - e^{-az}] \tag{2.34}$$

yielding the following expression for the activity s vs. depth,

$$s = s_0 \exp\left[-\frac{\lambda}{r}\left(\rho_0 z - \frac{\rho_1}{a}(1 - e^{-az})\right)\right] \tag{2.35}$$

An example of a plot of log activity vs. depth for a Lake Ontario core is shown in Figure 2.3. It is seen that compaction gives rise to flattening of the ^{210}Pb profile in the top layers in much the same way as mixing (Figure 2.2). To distinguish between these two influences, it is useful to plot log activity vs. cumulative mass m. According to Equation 2.32, this plot should be a straight line, regardless of any compaction effects. The mass sedimentation rate can be found from the slope of this line. Deviation from a straight line in the top layers (flattening) is indicative of mixing.

Measured ^{210}Pb profiles for eight cores in the Milwaukee Basin, plotted according to this procedure, are shown in Figure 2.4. It is clear that some mixing takes place in the upper layers (1.5–2 cm). The equivalent times of the mixed layers, $t = m/r$, vary from 3 ys (core A) to 11 yrs (core E), and the total time spans represented are between 57 yrs (core F) and 114 yrs (core E).

The modeling of ^{210}Pb profiles described here is fairly basic in that either mixing or compaction is considered, but not both of these phenomena. More advanced models based on numerical methods have been developed by Christensen (1982), Officer (1982), and Robbins (1986). Fukomori et al. (1992) also proposed a model that distinguishes between colloidal particles, e.g. clay with ^{137}Cs, and settling particles, e.g. silt and organic material, with ^{210}Pb.

2.3 ^{137}Cs AND $^{239+240}$Pu DATING OF SEDIMENTS

The first study of ^{137}Cs dating of lake sediments was undertaken by Pennington et al. (1973) who used the method to date sediments of five lakes in the English Lake

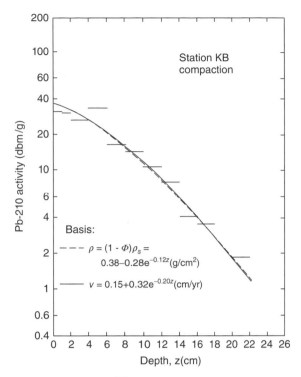

Figure 2.3. Comparison of measured ^{210}Pb activities vs. depth for a Lake Ontario core with calculated activities based on two different compaction models. (Source: Christensen (1982) with permission from American Geophysical Union.)

District. This method is based on the fallout nuclide ^{137}Cs, which was formed following atmospheric testing of nuclear weapons. The first major activity came in 1954 and maximum activity was recorded in 1963 (HASL-329, 1977). The method has since been used successfully to date lake sediments in the northern hemisphere but not ocean sediments, apparently because of the possibility of Cs mobility in the marine environment caused by cation exchange of Cs with alkali (Na, K) and alkaline earth (Ca, Mg, Sr) elements.

Fallout from the Chernobyl disaster in April 1986, which is noticeable in Europe but not the United States due to the lower altitude dispersion compared to nuclear weapons fallout, is a second source of ^{137}Cs. The Chernobyl material also contains ^{134}Cs, which distinguishes it from weapons fallout. Both ^{137}Cs and ^{134}Cs can be measured by a high purity germanium detector.

Richter et al. (1993) used ^{137}Cs from both sources to track the movement and settling of clay particles from the River Rhine into Lake Constance. A certain base level of Cs is deposited with calcite precipitation events in the water column (Robbins et al., 1992), but most of the ^{137}Cs near the mouth of the river (sites A, B, and C, Figure 2.5) is carried with fluvial inputs from the River Rhine. The minimum during

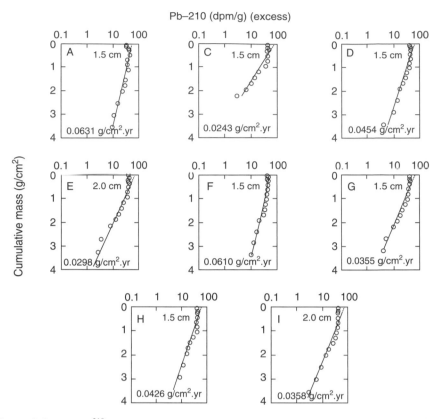

Figure 2.4. Excess ^{210}Pb activity vs. cumulative mass for eight cores collected in 1995 from the Milwaukee Basin of Lake Michigan with indication of mixing depth (cm) and mass sedimentation rate (g/cm^2·yr) (Rachdawong 1997).

the high water event of 1987 (hw 87, Figure 2.6) is thought to be caused by dilution of the clay particles in the river with sandy alluvial material with lower specific ^{137}Cs activity. Site D, situated in one of the central basins of the lake, does not appear to be impacted by this high water event because of its location far away from the streaming path of the River Rhine in the lake. Both the 1986 Chernobyl event and the 1963 fallout maximum are clearly visible at this site, but at shallower depths than for sites A–C.

^{137}Cs is often used in conjunction with ^{210}Pb dating to provide a check on the ^{210}Pb dates (Robbins and Edgington, 1975; Hermanson and Christensen, 1991). The maximum in 1963 is considered in addition to the onset of major ^{137}Cs activity in 1954. In a case with a high sedimentation rate (2.47 cm yr or 2.38 g/(cm^2 yr)), i.e. core VC94-6 from the Kinnickinnic River, Wisconsin, the ^{137}Cs profile (Figure 2.7) is very similar to the atmospheric input record (Christensen and Klein, 1991). Here, both the peak and the onset of ^{137}Cs activity are well dated. However, in an area such as northern Lake Michigan (Hermanson and Christensen, 1991) with a much lower

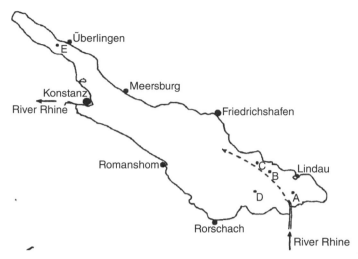

Figure 2.5. Locations of sediment sampling sites in Lake Constance, inflow and outflow of River Rhine, and flow path in the lake. (Source: Reproduced from Richter et al. (1993) with permission from IWA Publishing.)

typical mass sedimentation rate (0.0133 g/(cm^2 yr), NLM-E), mixing can become an important influence on the ^{137}Cs profile (Figure 2.8). The 1963 maximum becomes less distinct, the whole profile moves further down in the sediment, and there appears to be significant ^{137}Cs activity near the surface (NLM-B, NLM-E). As explained in Chapter 1, this maximum is consistent with the fact that ^{137}Cs is bound to colloidal ($<$ 2 μm diameter) clay mineral particles kept in suspension; whereas ^{210}Pb is primarily associated with organic carbon and silt particles that are more likely to settle.

In addition to ^{137}Cs, one can also use $^{239+240}$Pu to identify the global maximum fallout from nuclear weapons testing in 1963. The plutonium isotopes will remain tracers for a long time because of the long half-lives, $T_{1/2} = 6580$ years for ^{240}Pu and $T_{1/2} = 2.44\ 10^4$ for ^{239}Pu, compared to the short half-life of 30.17 yrs for ^{137}Cs. Wu et al. (2010) used both plutonium and ^{137}Cs activities in sediment cores from Lake Sugan, Shuangta Reservoir, and Lake Sihailongwan in northwestern China to identify this maximum and also to estimate the contribution from the Chinese nuclear test (CNT) site at Lop Nor since 1970. They achieved this by subtracting the global fallout maximum from the measured activities. As expected, the lake nearest the CNT, Lake Sugan, had the largest contributions from CNT, i.e. 40 and 27% of ^{137}Cs and $^{239+240}$Pu respectively.

The ^{240}Pu/^{239}Pu atom ratio has been used to characterize the source of plutonium. For weapons-grade plutonium the ratio ranges from 0.01 to 0.07, for reactor-grade plutonium it is 0.24–0.80, and for global fallout plutonium 0.180 ± 0.014 within the latitude zone 71–30°N (Wu et al., 2010). The authors measured the ^{240}Pu/^{239}Pu ratio in cores from the three lakes and found that values were within the above range for the global fallout, except for one point in a core from Lake Sugan where it measured 0.103 ± 0.007, possibly reflecting fallout from CNT. They concluded that use of the

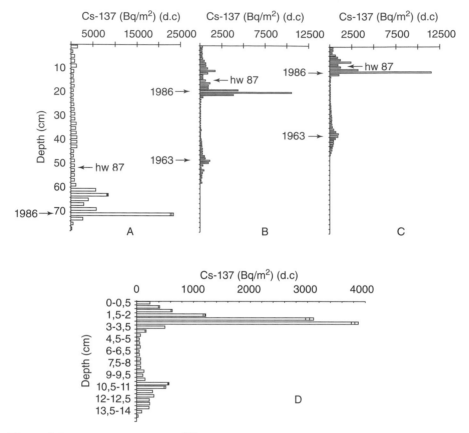

Figure 2.6. Depth distributions of ^{137}Cs in sediment cores of Lake Constance taken at four different locations relative to the mouth of River Rhine. The activities are corrected with respect to nuclear decay since May 1, 1986. The high water event of 1987 is marked hw 87. (Source: Reproduced from Richter et al. (1993) with permission from IWA Publishing.)

^{240}Pu/^{239}Pu ratio from the Lop Nor test site is insufficient to identify the regional fallout source.

2.4   DATED RECORDS OF METALS AND ORGANIC POLLUTANTS

The atmosphere can be a significant source of pollution for sediments in remote areas such as the upper Great Lakes or some Canadian lakes. Golden et al. (1993) found elevated levels of PCBs and DDT in upper sediment layers of cores from Lakes Superior, Michigan, and Ontario.

PCB levels have declined since 1970–1980, and DDT levels since 1965–1975. The PCB decline was also found in northern Lake Michigan sediments by Hermanson and Christensen (1991). Hermanson (1993) studied Pb, Cd, Zn, and Hg in Imitavik

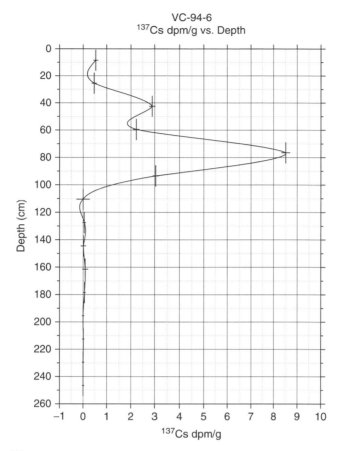

Figure 2.7. [137]Cs activity vs. depth in a core from the Kinnickinnic River, Wisconsin (Ab Razak 1995).

Lake on the Belcher Islands in Hudson Bay. The results (Figures 2.9 and 2.10) also demonstrate elevated fluxes of these substances in the top layers, indicating atmospheric input of these metals to this remote Arctic lake. It is not clear why the Pb flux does not appear to decline after 1970 in the same way as in other regions, including the Great Lakes (Christensen and Karls, 1996). Note, however, that as outlined in this chapter, the lead fallout record measured at Lake Espejo Lirios near Mexico City displays a higher 16 µg/(cm[2] yr) and later 1980 maximum than that of the record for the Laurentian Great Lakes, which peaks in 1970 with a lower maximum of 2.5 µg/(cm[2] yr). Thus, long range lead transport from both regions via the atmosphere may explain the lack of a distinct maximum for Pb (Figure 2.9).

Atmospheric input of pollutants to remote regions was also demonstrated by Lockhart et al. (1993). They determined ^{210}Pb and ^{137}Cs dated profiles of total polycyclic aromatic hydrocarbons (PAHs) and of Hg in cores from Lake 375 in the experimental lakes area, Ontario (49°45′N, 93°45W) and Hawk Lake, North West

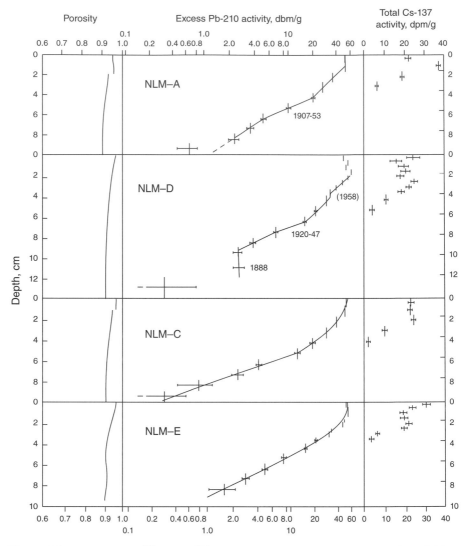

Figure 2.8. Porosities and ^{210}Pb and ^{137}Cs activities vs. depth for northern Lake Michigan cores NLM A, B, C, and E. (Source: Hermanson and Christensen (1991) with permission from ASCE.)

Territories (63°38′N, 90°40′W). The Hg flux increases toward the surface, except for the top sediment layer of Hawk Lake. This is similar to the trend for Imitavik Lake (Figure 2.10). The Hg fluxes to the sediment surface were 5 μg/(m^2 yr) for Hawk Lake and 21 μg/(m^2 yr) for Lake 375. This is consistent with the value of 16 μg/(m^2 yr) for Imitavik Lake (Hermanson, 1993), which lies at an intermediate latitude. The total PAH profiles show maxima in the mid (Hawk Lake) to late (Lake 375) 1940s, with surface fluxes (1988) of 6 μg/(m^2 yr) for Hawk Lake and 63 μg/(m^2 yr) for Lake 375.

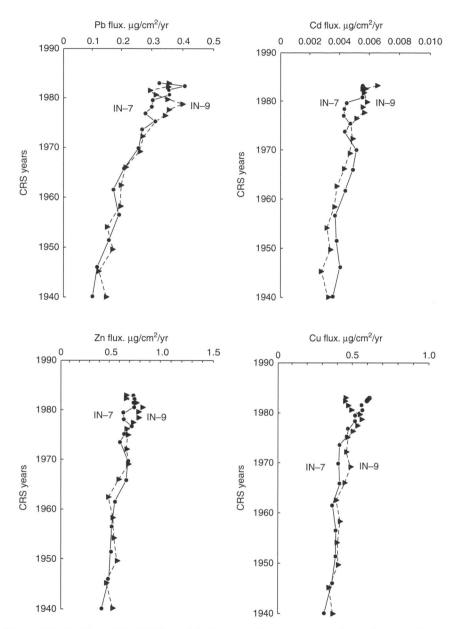

Figure 2.9. Profiles of Pb, Cd, Zn, and Cu flux to sediment cores from Imitavik Lake. (Source: Reproduced from Hermanson (1993) with permission from IWA Publishing.)

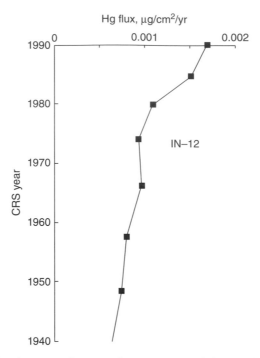

Figure 2.10. Profile of mercury flux to sediment core IN-12 from Imitavik Lake. (Source: Reproduced from Hermanson (1993) with permission from IWA Publishing.)

Mercury deposition rates of similar magnitude (16–20 $\mu g/(m^2$ year)) as Imitavik Lake and Lake 375 were found by Van Metre and Fuller (2009) for Hobbs Lake in west-central Wyoming. Hobbs Lake is at 3050 m elevation, 8 km west of the Continental Divide and oriented in a north-south direction, 150 m wide and 550 m long, with two streams entering from the south-east and outflow from the north. The ratio between areas of watershed and lake surface is 43. The lake is located in the Bridger Wilderness, distant from logging and mining activities. The authors developed a dual-core mass-balance approach in order to determine atmospheric fallout and concentration of ^{210}Pb or Hg in the deep parts of the core prior to enrichment by atmospheric fallout. They considered two cores, one mostly fallout-dominated down lake from the inflow, HOB.DN, and one, HOB.UP, which is more fluvial impacted. The model-calculated Hg fallout vs. deposition date was similar to the fallout record obtained from HOB.DN using a focusing factor correction based on ^{210}Pb inventories (in dpm/cm^2) of HOB.DN and a soil core. The fallout record shows small increases in the late 1800s and large increases beginning in 1940. The latter part is similar to the trend in Imitavik Lake (Figure 2.10), suggesting that the records reflect a regional or global input.

Focusing factors greater than 1 reflect sediment carried horizontally within the lake, from the watershed or on the lake bottom toward the core in addition to direct fallout. The focusing factor can be estimated from ^{210}Pb inventories in the core and from direct fallout. If ^{137}Cs activities are available, one can also calculate a ^{137}Cs-based focusing factor (Hermanson and Christensen, 1991). The measured ^{137}Cs inventory M_m is

$$M_m = \sum_{i=1}^{N} \rho_i s_i \Delta z_i \qquad (2.36)$$

where ρ_i, s_i, and Δz_i are sediment density (in g/cm^3), ^{137}Cs activity (in dpm/g), and N interval depth. The indices 1 and N correspond to the top and deepest part of the core showing ^{137}Cs activity. The calculated inventory ^{137}Cs M_a from the atmosphere can be expressed as

$$M_a = \sum_{j=1}^{N} F_j \exp(-\lambda t_j) \qquad (2.37)$$

where F_j, λ, and t_j represent atmospheric ^{137}Cs fallout in year j (in dpm/(cm^2 year)), decay constant for ^{137}Cs (per year), and t_j represents time difference between deposition and measurement (years). Fallout numbers for ^{137}Cs can be obtained from ^{90}Sr activities listed in Health and Safety Laboratory (1977) assuming a ^{137}Cs/^{90}Sr ratio of 1.5. The focusing factor (FF) is then calculated as

$$FF = \frac{M_m}{M_a} \qquad (2.38)$$

The Grand Calumet River, Illinois and Indiana is an example of an aquatic system that is impacted more directly by industrial discharges than by atmospheric inputs (Cahill and Unger, 1993). A portion of the river is designated by the International Joint Commission as one of the Areas of Concern within the Great Lakes. The sediments in the west branch of the Grand Calumet River are significantly polluted with Cr, Ni, Cu, Zn, Cd, Sn, and Pb. Cores up to 2 m in length were obtained with a vibra corer, and sedimentation rates were determined by ^{137}Cs dating to be in the range of less than 0.2 to 5.8 cm/year.

Metal pollution in the sediments of the Golden Horn, Turkey was investigated by Tuncer et al. (1993) using a 2.7 m long core obtained with a piston-type gravity corer. The sedimentation rate was found to be 3.5 cm/yr based on ^{210}Pb dating, producing a time span from 1913 to 1988. Significant enrichment with Cd, Ag, Cr, Ni, Pb, Cu, Mo, and Zn was found in this area, probably from industrial or municipal discharges. The results of factor analysis with varimax rotation indicated two crustal sources; an anthropogenic source and a source associated with sea salt.

The historical fallout record of lead in Mexico was investigated by Soto-Jimenez et al. (2006). They considered sediment cores from two coastal estuaries and two lakes. The cores were analyzed for stable Pb isotopes ^{204}Pb, ^{206}Pb, ^{207}Pb, and ^{208}Pb in order to characterize sources of lead. They had been previously ^{210}Pb dated and analyzed for several trace elements including lead. While many developed countries phased out lead in gasoline in the late 1960s and early 1970s, Mexican sales of leaded gasoline reached a peak in 1980 at 26,000 t/yr. This is reflected in the lead fallout to the Great Lakes region of North America, which peaked in 1970 at 2.5 $\mu g/(cm^2$ yr) (Christensen and Osuna, 1989) compared with the lead record from Lake Espejo Lirios near Mexico City showing a maximum of 16 $\mu g/(cm^2$ yr) in 1980.

The results of the stable Pb isotope analysis showed relatively narrow intervals for the $^{206}Pb/^{207}Pb$ (1.193–1.206) and $^{208}Pb/^{207}Pb$ (2.452–2.471) ratios of sediments between 1930 and 1997. The narrowness of these ranges suggests a single lead source, tetraethyl lead (TEL) in leaded gasoline. However, between 1810 and 1900 the $^{206}Pb/^{207}Pb$ ratio was 1.177–1.190 while the range for the $^{208}Pb/^{207}Pb$ ratio was 2.435–2.457. The latter ranges are similar to those characterizing Mexican bedrock. Other geographical areas have wider ranges of these ratios: For example, Lake Erie and Lake Ontario with isotopic compositions $^{206}Pb/^{207}Pb$ (1.195–1.247) and $^{208}Pb/^{207}Pb$ (2.442–2.507) (Ritson et al., 1994).

Mexican lead ores that were used for TEL consumed in Mexico have ratios $^{206}Pb/^{207}Pb$ (1.186–1.211) and $^{208}Pb/^{207}Pb$ (2.465–2.484). Thus the lead isotope ratios found in the four dated sediment cores appear to be the result of linear mixing with Mexican Bedrock and Mexican lead ores as endmembers, possibly influenced by a contribution from US sources, reflected in high $^{206}Pb/^{207}Pb$ (> 1.203) ratios at typical $^{208}Pb/^{207}Pb$ ratios 2.463 ± 0.004.

An example of a method for determination of historical input records of an organic pollutant was provided by Yang et al. (2007). They studied chlordane in ^{210}Pb and ^{137}Cs dated sediments from Long Island Sound (LIS), New York. Chlordane was used in the United States from 1948 as a pesticide and to control termites, but was banned in 1983 and all use was terminated in 1988. Yang et al. (2007) examined surficial sediment and four sediment cores from Long Island Sound and found chlordane concentrations exceeding probable effect levels in the four cores near the western end of the Sound near New York City.

The most contaminated core from Manhasset Bay (LIMB) had an estimated sedimentation rate of 0.27 $g/(cm^2$ year) (0.68 cm/year) and a bioturbation-induced dispersion coefficient of 3.4 cm^2/year. Sediment mixing was assumed to take place at a constant rate down to 19 cm depth. Both *trans*-chlordane and *cis*-chlordane was measured. The modeled surface concentration of chlordane in the LIMB core closely followed the US usage record from the onset of chlordane input in 1948 with a peak in 1971. However, the modeled concentration was below that estimated from the usage record in the most recent years, near the top of the core. While this was interpreted as indicating continued input of chlordane since 1995, the source or magnitude of the recent chlordane was not identified.

Possible sources include sediment focusing, runoff from house foundations and agricultural soils. The chlordane measured in sediment of LIS during the past 60 years was racemic, i.e. concentration ratios of two enantiomers equal to one. Because chlordane applied to house foundations as a termiticide was racemic but chlordane in agricultural soil was non-racemic, it was suggested that the chlordane mainly originated from house foundation soils.

Wei et al. (2012) examined dated records of polybromodiphenyl ethers (PBDEs) and decabromodipenyl ethane (DBDPE) in sediments from six water bodies in Arkansas, United States. Three of the sites were close to manufacturing facilities in El Dorado (Chemtura Corp.) and Magnolia (Albemarle Corp.) in the southern part of the state. Because of their toxicity, PBDEs were recently phased out in the United States; first the penta and octa, and, after 2012, the deca compound BDE209. There is a concern that PBDEs in the environment can debrominate from the deca compound to the more toxic penta compounds.

A total of 49 PBDEs were detected with concentrations as high as 57,000 and 2400 ng/g of BDE209 and DBDPE respectively (Wei et al., 2012). A straight-line regression equation of log surface concentration vs. log distance from the manufacturing facilities based on a Gaussian Plume Dispersion Model for both PBDEs and DBDPE provided a good fit for the experimental points, supporting the hypothesis that the manufacturing facilities were the major source of these pollutants. Measurements of PBDEs in a wastewater sludge retention pond near the Albemarle plant revealed significant concentrations of compounds that are not present in technical mixtures. This indicates that debromination of PBDEs may take place in the environment either by photolysis or biological debromination.

2.5 DECONVOLUTION OF SEDIMENTARY RECORDS

In order to improve the accuracy of a dated pollutant profile, records from several cores can be combined. In addition, the influence of mixing, which can be substantial when the sedimentation rate is low, can be removed by deconvolution or unmixing (Berger et al., 1977; Christensen and Goetz, 1987) of the individual records. This will further enhance the correctness of the dated input record. The following outlines the principles and application of an integral method of unmixing (Christensen and Goetz, 1987), a frequency domain method (Christensen and Osuna, 1989), and a differential approach (Berger et al., 1977; Christensen and Klein, 1991; Christensen et al., 1999).

In the integral method (Christensen and Goetz, 1987), the concentration of a pollutant $s(z, t)$ (in µg/g) at sediment depth z and time t may, for a linear time-invariant system, be written as a convolution integral of the product of the unit pulse input $f(\tau)$ with the impulse response $s_0(z, t - \tau)$ for all times τ from the start of the input to the current time t,

$$s(z, t) = \int_0^t s_0(z, t - \tau) f(\tau) d\tau \tag{2.39}$$

Thus an early input $f(\tau)$ with a small value $\tau = 10$ yr, e.g. lead input to the Great Lakes in 1886, assuming zero input before 1876, is multiplied by an impulse response at a long time, $t - \tau = 116$ yrs, such that the current time t is the sum of these times, $t = 126$ yrs past 1876, i.e. 2002. Conversely, a more recent input with higher τ must be combined with an impulse response of shorter time argument.

A discretized version of Equation 2.39 may be written as

$$S_{in} = \sum_{j=1}^m E_{ij} F_{n+1-j}, \quad i = 1, 2, 3, \dots, n \tag{2.40}$$

where S_{in} is the pollutant concentration at space index i and time index $i = n$. E_{ij} is a matrix of elementary contributions at space and time intervals i and j respectively. F_{n+1-j} is the input function to be determined. In order to obtain a non-trivial solution to Equation 2.40, the number of space intervals n must equal or exceed the number of time intervals m, $n \geq m$.

A number of solutions to this equation, assuming an equal number of space and time intervals, were obtained by Christensen and Goetz (1987). They considered inputs of Pb, Zn, and Cd to Lake Michigan sediments. The input F was determined by inverting E and multiplying with S. It was found that the method generated a maximum for lead from 1954–1969 in agreement with atmospheric loading data. This maximum was not visible in the measured data. Similar results were obtained for zinc and cadmium.

The frequency domain method (Christensen and Osuna, 1989) considers the convolution integral Equation 2.39 at the boundary z_m between a mixing layer and a historical layer below,

$$s(z_m, t) = \int_0^t s_0(z, t - \tau) f(\tau) d\tau \qquad (2.41)$$

In contrast to Equation 2.39, Equation 2.41 assumes that there is no further mixing below the lower boundary z_m of the mixing layer. The time function $s(z_m, t)$ is available as the historical record in the sediment below z_m for all times from the beginning of the record until the present time t. By taking the Fourier transform of Equation 2.41, we obtain

$$S(z_m, \omega) = S_0(z_m, \omega) F(\omega), \quad \text{or} \qquad (2.42)$$

$$F(\omega) = \frac{S(z_m, \omega)}{S_0(z_m, \omega)} \qquad (2.43)$$

where $S(z_m, \omega)$ is the Fourier transform of the mixing impacted historical record in the sediment and $S_0(z_m, \omega)$ and $F(\omega)$ are the Fourier transforms of the impulse response and input function respectively. Equation 2.39 determines the desired input function $f(\tau)$ by taking the inverse Fourier transform of $F(\omega)$.

Input functions of lead, zinc, and cadmium to Lake Michigan were calculated on the basis of this procedure, using the same experimental data as in Equation 2.40. The solution for Pb shows a distinct maximum in 1969, and this is also reflected in the atmospheric loading data in the Great Lakes region. This is a more accurate result than the step-wise constant input function with a maximum during 1954–1969 calculated from Equation 2.40. Comparison of the reconstructed input function for cadmium with the US consumption record, using frequency domain deconvolution, shows satisfactory agreement. There is good agreement for zinc prior to 1955, but the maximum zinc level occurs in 1961, 9 years earlier than the 1970 maximum from the zinc consumption data. This discrepancy could be caused by a shift in industrial processes, or the atmospheric release of zinc by means of non-point pollution during the early 1960s. In addition, although the input of zinc to the Great Lakes region is generally expected to follow the US consumption record, there can be regional differences, which could account for variances between the two records.

In order to determine the historical ^{18}O record in ocean sediment from the meltwater spike after the last ice age, Berger et al. (1977) developed an unmixing equation. This was based on a simple mass balance model for a particle-associated substance entering the sediment from the water column. Full mixing within a sediment layer and no mixing below, with uncompacted sediments, was assumed.

This unmixing model may be written as

$$\frac{d(cz_{m})}{dt} = J - vc \tag{2.44}$$

where c (in $\mu g/cm^3$ or dpm/cm^3) is the concentration of ^{18}O in the mixing layer, J (in $\mu g/(cm^2\ year)$) is the unknown input of ^{18}O to the sediment from the overlying water column, z_m is the thickness of the mixing layer, v is the sediment velocity just below the mixing layer, relative to the sediment–water interface, and t is time. The unknown ^{18}O flux J is then determined from Equation 2.44 by inserting values of vc and z_m, dc/dt determined from the observed ^{18}O record below the sediment–water interface.

The model was extended to include compaction and was subsequently applied to determine unmixed input records of ^{137}Cs, Pb, Zn, and Cd (Christensen and Klein, 1991) and PAHs (Christensen et al., 1999) to lake sediments. By writing $c = \rho s$ where ρ (in g/cm^3) and s (in $\mu g/g$) are sediment density and pollutant concentration per gram of sediment respectively, the unknown input flux J may be written as

$$J = rs + z_m \rho_m \frac{ds}{dt} \tag{2.45}$$

where ρ_m is the average sediment density in the mixing layer and $r = \rho v(g/(cm^2\ yr))$ is the mass sedimentation rate. If the same type of decay is present as for ^{137}Cs, there will be an additional positive term $\lambda \rho s$ on the right-hand side of Equation 2.41, where λ is a decay constant.

An application of this model to PAH input to seven sediment cores (A, C, D, E, F, G, H, Figure 2.4) from the Milwaukee Basin of Lake Michigan was carried out by Christensen et al. (1999). Assuming that the year of sampling is 1995, the year Y_i of a section of the historical layer below the sediment–water interface may be calculated from

$$Y_i = 1995 - (\overline{T}_i - t_m) \tag{2.46}$$

where $\overline{T}_i$ is the time it took to deposit the sediment above the midpoint of this section, and t_m is the time of deposition of the sediment above the lower boundary of the mixing layer. The curve fitted to experimental data is shown, along with the reconstructed input curve for core CLM-G, in Figure 2.11. The curve connecting observed values is generated by cubic splines connecting adjacent points such that slope and curvature (second derivative) are equal at the points where they join. The reconstructed or unmixed curve is generated by Equation 2.45.

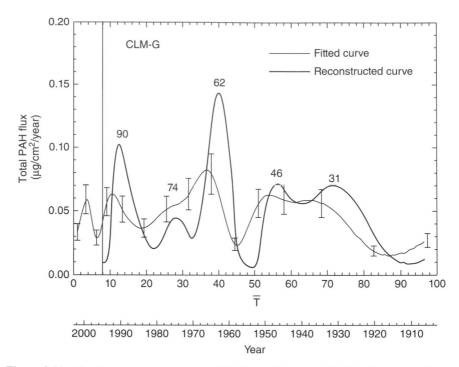

Figure 2.11. Fitted curve and reconstructed PAH record for core CLM-G. The vertical line at the sampling year (1995) indicates the bottom of the mixing layer or top of the historical layer. The fitted curve includes observed values. (Source: Christensen et al. (1999) with permission from ASCE.)

Note that the reconstructed curve intersects the fitted curve at points where the slope of the fitted curve is equal to zero. Maxima and minima in the original record are amplified in the reconstructed curve, and can even be generated from shoulders in the original record, as illustrated by the 1974 maximum. Characteristic features become more recent by unmixing as seen by the maximum in 1962, which before unmixing, using a deposition year as $1995 - \overline{T}$, is dated to 1959.

Figure 2.12(b) shows such a case, where the historical PAH flux has been determined by combining unmixed and ^{210}Pb focus-corrected PAH records from seven cores in the Milwaukee Basin (Christensen et al., 1999). The input record based on the original data points without unmixing is shown in Figure 2.12(a) for comparison. Here, each of seven original records has been constructed by fitting cubic splines through the actual data points. Unmixing sharpens the peaks and valleys and shifts their occurrence time forward by approximately 3–4 yrs. Note also that peaks are created (1977 and 1930) where only shoulders were present in the original record. Maxima occur in 1931 and 1950 because of coke burning, and in 1970 because of highway emissions, and coal and coke burning. The 1977 maximum coincides with

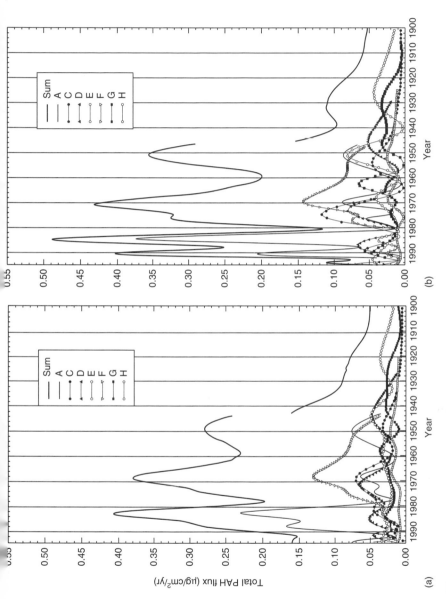

Figure 2.12. ^{210}Pb focus-corrected PAH fluxes to cores A, C, D, E, F, G, and H: (a) Curves fitted to the original data with year = 1995 − $\bar{T}$; (b) reconstructed or unmixed curves with year determined from Equation 2.46. Sum curve B is based on all seven cores from 1946–1995. Before 1946, all cores except A and F are considered, with estimated extrapolations for cores D and H before 1930 and 1925 respectively. Sum (a) is constructed similarly. (Source: Christensen et al. (1999) with permission from ASCE.)

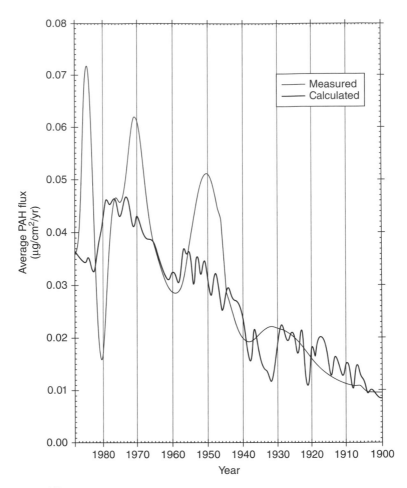

Figure 2.13. [210]Pb focus-corrected total average PAH flux to the Milwaukee Basin vs. year of deposition. Measured data are reconstructed from observed data and calculated fluxes are based on US consumption data (Btu). (Source: Christensen et al. (1999) with permission from ASCE.)

the height of US petroleum consumption. Minima occur at the time of the Great Depression (1935–1940), a low point in coke production during 1958–1959, and low points of both coke and petroleum consumption during 1982–1983.

The uncertainty level of the PAH flux is 11% from 1946–1981 with an estimated dating error of ±1.1 yr. The uncertainty level before this period is greater because the results of only five cores are included in the averaging. It is also greater after 1981 because one core (A) dominates the PAH input.

Average measured PAH fluxes, refined by unmixing, are compared to PAH fluxes calculated from US energy production figures in Figure 2.13. The calculated data are normalized to produce an optimal fit to the measured data. Due to uncertainty of recent fluxes (Figure 2.12b), the comparison is mainly valid for the period before 1980. The time records of peaks and valleys are modeled well, although peaks in 1950 (coke for steel production) and 1971 (highway dust) are larger than suggested from consumption data. This may be due to variances between Great Lakes data and US consumption figures, or more likely to poor implementation of pollution control technologies during peak periods.

2.6 CHEMICAL AND BIOLOGICAL DEGRADATION

PCBs can be degraded chemically by hydroxyl radicals generated with Fenton's reagent without (Sedlak and Andren, 1991) or with (Pignatello and Chapa, 1994) UV light. The reaction proceeds via addition of a hydroxyl group to one of the non-halogenated sites. Although the reactions also take place when particulate matter is present, the oxidation rate is very slow for highly substituted PCBs such as $2,2',4,5,5'$, which are strongly associated with the particulate phase (Sedlak and Andren, 1994).

Anaerobic dechlorination of PCBs (Aroclor 1242) in a laboratory environment takes place under methanogenic conditions, but is partly inhibited by sulfate and completely inhibited by nitrate (Rhee et al., 1993a). The rate is faster for m-Cl, followed by p-Cl, and no o-Cl is removed. Anaerobic dechlorination was found to be concentration-dependent in that it favored a range between 500 and 800 ppm. However, contaminated sediments that do not contain essential organic matter and/or nutrients may remain immune to dechlorination (Rhee et al., 1993b).

Conditions for anaerobic dechlorination of PCBs in lab experiments were further explored by Alder et al. (1993). They observed dechlorination at 100 ppm PCB concentration of Hudson River sediment. They also found a preference for meta- and para-dechlorination as compared to ortho-dechlorination.

Sokol et al. (1994) assessed in situ PCB reductive dechlorination in St. Lawrence River sediments. The extent of dechlorination varied widely from site to site, and there was no correlation between sediment PCB concentration (2–10,000 ppm) and the extent of dechlorination. However, some concentration dependence in the range 4–35 ppm was identified by Sokol et al. (1995). Cessation of dechlorination at a plateau appears to be due to the accumulation of daughter congeners with chlorine substitution patterns that are not amenable to further degradation by the present microbial consortium (Liu et al., 1996).

Biodegradation of aromatic hydrocarbons by bacteria is now well established (Smith 1990). Aerobic degradation of benzene begins with the formation of catechol, which is further catabolized by either 1,2-dioxygenase (ortho- or intradiol-cleavage) and subsequently via the β-ketoadiapate pathway or catechol 2,3-dioxygenase (meta- or extradiol-cleavage). Degradation of arenes, biphenyls, and PAHs by single bacterial isolates is also well understood. Most routes of degradation are aerobic, although anaerobic biodegradation has been documented (Smith, 1990).

Generally, the rate of degradation of PAHs is inversely proportional to the number of rings in the PAH molecule (Cerniglia, 1993). While co-metabolism has been found, less is known about microorganisms that are capable of utilizing high molecular weight PAHs as sole sources of carbon and energy. Also, the potential biodegradation rates are higher in PAH-contaminated sediments than in pristine sediments (Cerniglia, 1993). PAHs can be degraded by algae and fungi as well as by bacteria (Cerniglia, 1993).

Erickson et al. (1993) showed that while PAHs added to manufactured gas plant soils were rapidly degraded, the indigenous PAHs in these soils were not readily degraded even under optimum conditions. PAHs were degraded in soils amended by sewage sludge, with most half-lives registering between 50 and 500 days. High molecular weight compounds were most persistent and half-lives correlated positively with log K_{ow} and inversely with the logarithm of water solubility (Wild and Jones, 1993). The bacterial transformation of pyrene in an estuary environment of British Columbia, Canada was supported by the identification of a metabolite *cis*-4,5-dihydroxy-4,5-dihydropyrene in sediment and porewater (Li et al., 1996).

Experiments by Ye et al. (1996) demonstrated that *Sphingomonas paucimobilis* strain EPA 505 has the ability to degrade several 4- and 5-ring PAHs cometabolically. Also, BaP (5-ring) and chrysene (4-ring) were degraded to a similar extent, indicating that factors other than the number of rings may influence the microbial degradation of PAHs. Low molecular weight unsubstituted PAHs are amenable to microbial degradation in soil–water systems under denitrification conditions (Mihelcic and Luthy, 1988). A recent report indicates that PAHs can be oxidized in sediments that are heavily contaminated with PAHs, but not in more lightly contaminated sediments (Coates et al., 1996).

Other aspects of chemical and biological degradation include the following: Benzonitriles in sediments are abiotically degraded to benzoic acid at a rate constant that correlates with a hydrophobicity parameter (Masunaga et al., 1993), and biotic processes appear to be responsible for degradation of tributyl tin in aerobic sediment to the less toxic dibutyl and monobutyl tin (Dowson et al., 1993). Dehalogenation of monosubstituted anilines can occur in anaerobic sediments (Susarla et al., 1996). The para-substituted compounds is transformed faster than the meta-substituted compounds, and the rate constant was found to correlate with a lipophilic constant. Haloanilines can pose a risk to humans and the environment because of their toxicity and resistance to microbial degradation.

2.7   SEDIMENTS AS A SOURCE OF POLLUTANTS

The rate of release of tetra chlorinated dibenzo dioxin/furans from sediment to the water column is affected by the release of dissolved and colloidal-sized organic carbon materials from the sediment, and to a lesser degree by exposure time (Hinton et al., 1993). In a study of desorption of chlorobenzenes from particles in water, Koelmans et al. (1993) found that the desorption rate was faster for natural suspended solids than for sediments. The difference was attributed to rapid desorption from the algal cells and cell fragments that form a major part of the organic matter of suspended particles. At non-equilibrium conditions, faster desorption was found for the more hydrophobic chlorobenzenes.

For heavy metals, it was found that solubilization by the formation of nitrilo-triacetic acid (NTA)–metal complexes is less significant in hard waters where the strongly bound Ca–NTA complex is quickly biodegraded (Lo and Huang, 1993).

2.7.1 Phosphorus

The classical problem of phosphorus release from sediments was investigated by considering two tropical reservoirs in Singapore: The Kranji reservoir, which is highly polluted, and the Mac Ritchie Reservoir, which is located in a protected catchment area with very little pollution (Appan and Ting, 1996). The phosphorus was categorized as loosely bound P, iron and aluminum bound P, Ca bound P, and organic P. The Fe and Al bound fraction amounted to 53% for the Kranji reservoir and only 18–25% for the Mac Ritchie Reservoir. Laboratory experiments with water and sediments from the study area showed significant *ortho*-P and total P releases under anoxic conditions and high pH, particularly from sediments of the Kranji Reservoir. Similarly, organic bound P and Al and Fe bound P (NaOH–P) in the sediments of Masan Bay, Korea, appear to be linked to outbreaks of red-tides, a type of phytoplankton bloom. This occurs during the summer when the water becomes anoxic, causing an increased flux of P from the sediments (Kim and Lee, 1993).

2.7.2 Metals

Metal contamination of sediment occurs in remote areas as a result of atmospheric inputs (Hermanson, 1993, Lockhart et al., 1993), and in industrialized areas as documented, for example, by Cahill and Unger (1993) and Tuncer et al. (1993).

Oxidation of sulfur, iron, and nitrogen can lead to the production of hydrogen ions, i.e. acidic conditions (Table 2.2). This is relevant during dredging operations

TABLE 2.2. Acid Producing Oxidation Reactions in Aquatic Systems

Elements		Reactions
Inorganic	S	$H_2S + 2\,O_2 = SO_4^{2-} + 2\,H^+$
	S	$S^0 + 1.5\,O_2 + H_2O = SO_4^{2-} + 2H^+$
	S, Fe	$FeS + 2.25\,O_2 + 1.5\,H_2O = FeOOH + SO_4^{2-} + 2\,H^+$
	S, Fe	$FeS_2 + 3.75\,O_2 + 2.5\,H_2O = FeOOH + 2\,SO_4^{2-} + 4\,H^+$
	Fe	$Fe^{2+} + 0.25\,O_2 + 2.5\,H_2O = Fe(OH)_3 + 2\,H^+$
	N	$NH_4^+ + 2\,O_2 = NO_3^- + H_2O + 2\,H^+$
	N	$NO_x + 0.25(5 - 2x)\,O_2 + 0.5\,H_2O = NO_3^- + H^+$
Organic	N	$R - NH_2 + 2\,O_2 = R - OH + NO_3^- + H^+$
	S	$R - SH + H_2O + 2\,O_2 = R - OH + SO_4^{2-} + 2\,H^+$

(Source: Adapted from Calmano et al. (1993)).

TABLE 2.3. Sequential Selective Extraction Scheme for Determination of Metal Phases in Sediments

Extraction reagent	Extracted sediment phase
Extracted with 1 N pH 7 NH$_4$OAc	EP metals
Extract with 0.1 M HN$_2$OH · HCl in 0.01 M HNO$_3$	ERP (hydrous manganese oxides)
Acidified (pH 3) 30% H$_2$O$_2$ at 50 °C extract with pH 3 NH$_4$OAc	OSP
Extract with sodium citrate in sodium bicarbonate solid Na$_2$S$_2$O$_4$ (heat to 75 °C)	MRP (hydrous iron oxides)
Extract with conc. HNO$_3$ + 30% H$_2$O$_2$ (heat to 110 °C)	AEP

Abbreviations: AEP, acid-extractable phase; EP, exchangeable phase; ERP, easily reducible phase; MRP, moderately reducible phase; OSP, organics-sulfides phase.
(Source: Rule and Alden III (1996)).

to remove contaminated anaerobic sediment from harbors and ship canals because pH levels may drop, possibly resulting in the release of metals such as Cd and Zn (Calmano et al., 1993). Both the decrease in pH and the increase in the concentration of dissolved metals can be detrimental to the biota. In aquatic systems, variations in pH are moderated by the presence of buffers, especially CaCO$_3$, Al$_2$O$_3$, and Fe$_2$O$_3$. However, when dredged material is disposed of on land, pH drops can be significant (Calmano, personal communication, 1997). During oxidation, the easily reducible phase (ERP) and moderately reducible phase (MRP) fractions generally increase, while the organic-sulfide phase (OSP) decreases. A typical sequential selective extraction scheme for determination of metal phases in sediments is shown in Table 2.3.

2.7.3 Acid-volatile Sulfides

Acid-volatile sulfide (AVS) is defined as the fraction of sulfides extracted by cold hydrochloric acid. AVS consists primarily of iron monosulfides that are found in anoxic freshwater and marine sediments (Pesch et al., 1995). Many divalent cationic metals (Cd, Cu, Hg, Ni, Pb, and Zn) form sulfides that are less soluble than ferrous sulfide (Table 2.4). These metals may undergo substitution reactions with iron as follows:

$$Me^{2+} + FeS(s) \rightarrow MeS(s) + Fe^{2+} \tag{2.47}$$

thereby removing the potentially toxic metal ion Me^{2+} from the water column and replacing it with the generally less harmful ferrous ion Fe^{2+}. A less soluble ion, e.g. Hg^{2+}, is likely to replace a more soluble ion, e.g. Zn^{2+},

$$Hg^{2+} + ZnS = Zn^{2+} + HgS \tag{2.48}$$

TABLE 2.4. Solubility Products of Metal Sulfides

Metal sulfide	Log K
MnS (s)	−19.15
FeS (amorphic)	−21.80
FeS (s)	−22.39
NiS (s)	−27.98
ZnS (s)	−28.39
CdS (s)	−32.85
PbS (s)	−33.42
CuS (s)	−40.94
HgS (s)	−57.25

(Source: Di Toro et al. (1990)).

At molar metal/AVS ratios greater than 1.0, excess metal may be released to the interstitial water, where it may be toxic or bound by other compounds in the sediment. When evaluating the metal/AVS ratio, the metal fraction is defined as the simultaneously extracted metals (SEM) (along with AVS), and the ratio is referred to as the SEM/AVS ratio.

In a study of marine sediments, Casas and Crecelius (1994) found that Zn, Pb, and Cu were not detected in the porewater until the AVS was exceeded, and that mortalities of *Capitella capitata* occurred when SEM/AVS was greater than 1. They also suggested that (porewater concentration)/LC50 could be used to predict metal toxicity because mortalities occurred in all cases where this ratio was greater than 1. The ratio is referred to as the interstitial water toxic unit (IWTU). Similar results have been reported by Berry et al. (1996) for metal-spiked laboratory sediments and by Hansen et al. (1996) for metal-contaminated field sediments. Hansen et al. (1996) considered sediments from five saltwater and four freshwater locations in the United States, Canada, and China.

The SEM/AVS approach to predicting the biological availability and toxicity of Cd, Cu, Pb, Ni, and Zn is applicable only to anaerobic sediments that contain AVS, because other binding factors control bioavailability in aerobic sediments (Berry et al., 1996).

2.7.4 Organics

Compounds such as PCBs and PAHs are strongly particle bound. Other compounds, for example atrazine and degradation products, are found mainly in the aqueous phase. PCBs can cause cloracne and birth defects, and DDT is an endocrine disruptor. PAHs can cause acute toxicity and some PAH compounds are known carcinogens. DDT was previously used to eliminate malaria carrying mosquitoes, and PCBs were used in capacitors, transformers, and as hydraulic fluids until the US ban in the late 1970s. PAHs are mainly generated as products of incomplete combustion, although some, especially low molecular weight compounds, are present in petroleum (Zhang et al., 1993). PAHs can also originate in nature, e.g. perylene, which may be formed by biosynthesis (Youngblood and Blumer 1975), although these compounds are not very abundant (Helfrich and Armstrong, 1986). PCBs, DDT, and PAHs are ubiquitous in contaminated sediments. PCBs and DDT enter the aquatic food chain through benthos and are biomagnified through the aquatic food chain from fish to humans.

Determination of interstitial water concentration of nonionic organic compounds such as PAHs can be performed using the equilibrium partitioning approach (Swartz et al., 1995). The concentration PAH_{iw} (e.g. µg PAH/mL) can be expressed as

$$PAH_{iw} = \frac{PAH_b}{K_{oc} f_{oc}} \qquad (2.49)$$

where PAH_b is the bulk sediment PAH concentration, in $\mu g\,PAH/g_{sed}$, K_{oc} is the organic carbon partition coefficient, in mL/g_{carb}, and f_{oc} is the organic carbon fraction of the sediment, in g_{carb}/g_{sed}.

Once the interstitial water PAH concentration has been determined, the toxicity of the PAHs to benthic organisms can be estimated by calculating the LC50, i.e. lethal concentration for 50% of a population in a given time, e.g. 48 h. This can be calculated using a quantitative–structure–activity relationship (QSAR) such as that shown by Schwarz et al. (1995),

$$\log LC50 = 5.92 - 1.33 \log K_{ow} \qquad (2.50)$$

where K_{ow} (dimensionless) is the octanol–water partition coefficient. The toxic unit $PAH_{iw}/LC50$ results and, assuming concentration addition, the toxic response is obtained if the sum of toxic units for all PAH compounds is greater than 1. This model was successful in predicting the toxicity to marine and estuarine amphipods of field-collected sediments.

REFERENCES

Ab Razak, IA. Sedimentation and PAH sources of the Kinnickinnic River between the Becher Street bridge and the Wisconsin Wrecking Company. M.S. thesis, University of Wisconsin-Milwaukee; 1995.

Alder, AC, Haggblom, MM, Oppenheimer, SR, Young, LY. Reductive dechlorination of polychlorinated biphenyls in anaerobic sediments. *Environ. Sci. Technol.* 1993;27(3):530–538.

Appan, A, Ting, D-S. A laboratory study of sediment phosphorus flux in two tropical reservoirs. Water Quality International '96, IAWQ 18th Biennial International Conference, 23–28 June 1996. Conference preprint book 6, pp. 314–321.

Appleby, PG, Oldfield, F. The calculation of lead-210 dates assuming a constant rate of supply of unsupported Pb-210 to the sediments. *Catena* 1978;5:1–8.

Berger, WH, Johnson, RF, Killingley, JS. Unmixing of the deep-sea record and the declacial meltwater spike. *Nature* 1977;269:661–663.

Berry, WJ, Hansen, DJ, Mahony, JD, Robson, DL, Di Toro, DM, Shipley, BP, Rogers, B, Corbin, JM, Boothman, WS. Predicting the toxicity of metal-spiked laboratory sediments using acid-volatile sulfide and interstitial water concentrations. *Environ. Toxicol. Chem.* 1996;15(12):2067–2079.

Calmano, W, Hong, J, Förstner, U. Binding and mobilization of heavy metals in contaminated sediments affected by pH and redox potential. *Wat. Sci. Tech.* 1993;28(8–9):223–225.

Cahill, RA, Unger, MT. Evaluation of the extent of contaminated sediments in the west branch of the Grand Calumet River, Indiana-Illinois, USA. *Wat. Sci. Tech.* 1993;28(8–9):53–58.

Casas, AM, Crecelius, EA. Relationship between acid volatile sulfide and the toxicity of zinc, lead and copper in marine sediments. *Environ. Toxicol. Chem.* 1994;3:529–536.

Cerniglia, CE. Biodegradation of polycyclic aromatic hydrocarbons. *Biodegradation* 1993;3: 351–368.

Christensen, ER. A model for radionuclides in sediments influenced by mixing and compaction. *J. Geophys. Res.* 1982;87(C1):566–572.

Christensen, ER, Bhunia, PK. Modeling radiotracers in sediments: comparison with observations in Lakes Huron and Michigan. *J. Geophys. Res.* 1986;91C:8559–8571.

Christensen, ER, Goetz, RH. Historical Fluxes of Particle-Bound Pollutants from Deconvolved Sedimentary Records. *Environ. Sci. Technol.* 1987;21:1088–1096.

Christensen, ER. and Osuna, J. Atmospheric Fluxes of Pb, Zn, and Cd to Lake Michigan from Frequency Domain Deconvolution of Sedimentary Records. *J. Geophys. Res.* 1989; 94 (C10):14585–14597.

Christensen, ER, Klein, RJ. 'Unmixing' of ^{137}Cs, Pb, Zn, and Cd records in lake sediments. *Environ. Sci. Technol.* 1991;25:1627–1637.

Christensen, ER, Karls, JF. Unmixing of lead, ^{137}Cs, and PAH records in lake sediments using curve fitting with first- and second-order corrections. *Wat. Res.* 1996;30(11):2543–2558.

Christensen, ER, Rachdawong, P, Karls, JF, Van Camp, RP. PAHs in Sediments: Unmixing and CMB Modeling of Sources. *ASCE Journal of Environmental Engineering.* 1999;125(11): 1022–1032.

Coates, JD, Anderson, RT, Lovley, DR. Oxidation of polycyclic aromatic hydrocarbons under sulfate-reducing conditions. *Appl. Environ. Microbiol.* 1996;62(3):1099–1101.

Di Toro, DM, Mahony, JD, Hansen, DJ, Scott, KJ, Hicks, MB, Mayr, SM, and Redmond, MS. Toxicity of cadmium in sediments: The role of acid volatile sulfide. *Environ. Toxicol. Chem.* 1990;9(12):1487–1502.

Dowson, PH, Bubb, JM, Williams, TP, Lester, JN. Degradation of tributyltin in freshwater and estuarine marina sediments. *Wat. Sci. Tech.* 1993;28(8–9):133–137.

Erickson, DC, Loehr, RC, Neuhauser, EF. PAH loss during bioremediation of manufactured gas plant site soils. *Wat. Res.* 1993;27(5):911–919.

Friedlander, G, Kennedy, JW, Miller JM. *Nuclear and Radiochemistry*, 2nd ed. New York: John Wiley & Sons, Inc.; 1964.

Fukumori, E, Christensen, ER, and Klein, RJ. A model for ^{137}Cs and other tracers in lake sediments considering particle size and the inverse solution. *Earth Planet. Sci. Lett.* 1992;114:85–97.

Gin, MF. Sedimentation patterns of the Milwaukee Harbor Estuary determined from TOC, Pb-210, and Cs-137 measurements. M.S. Thesis. Department of Civil Engineering and Mechanics, University of Wisconsin-Milwaukee; 1992.

Goldberg, ED. Geochronology with [210]Pb. In: *Radioactive Dating*. Vienna: International Atomic Energy Agency; 1963, pp. 121–131.

Goldberg, ED, Koide, M. Geochronological studies of deep sea sediments by the ionium-thorium method. *Geochim. Cosmochim. Acta* 1962;26:417–445.

Golden, KA, Song, CS, Jeremiason, JD, Eisenreich, SJ, Sanders, G, Hallgren, J, Swackhamer, DL, Engstrom, DR, Long, DT. Accumulation and preliminary inventory of organochlroines in Great Lakes sediments. *Wat. Sci. Tech.* 1993;28(8–9):19–31.

Hansen, DJ, Berry, WJ, Mahony, JD, Boothman, WS, Di Toro, DM, Robson, DL, Ankley, GT, Ma, D, Yan, Q, Pesch, CE. Predicting the toxicity of metal-contaminated field sediments using interstitial concentration of metals and acid-volatile sulfide normalizations. *Environ. Toxicol. Chem.* 1996;15(12):2080–2094.

Health and Safety Laboratory, Environmental Quarterly. *Final tabulation of monthly ^{90}Sr fallout data: 1954–1976.* HASL-329, Energy Research and Development Administration, New York; 1977.

Helfrich, J, Armstrong, DE. Polycyclic aromatic hydrocarbons in sediments of the southern basin of Lake Michigan. *J. Great Lakes Res.* 1986;12(3):192–199.

Hermanson, MH. Historical accumulation of atmospherically derived pollutant trace metals in the Arctic as measured in dated sediment cores. *Wat. Sci. Tech.* 1993;28(8–9):33–41.

Hermanson, MH, Christensen, ER. Recent sedimentation in Lake Michigan. *J. Great Lakes Res.* 1991;17(1):33–50.

Hinton, SW, Brunck, R, Walbridge, L. Mass transfer of TCDD/F from suspended sediment particles. *Wat. Sci. Tech.* 1993;28(8–9):181–190.

Kersten, M, Leipe, T, Tauber, F. Storm disturbance of sediment contaminants at a hotspot in the Baltic Sea assessed by ^{234}Th radionuclide profiles. *Environ. Sci. Technol.* 2005;39(4):984–990.

Kim, B-J, Lee, C-W. Forms of phosphorus in the sediments of Masan Bay, Korea. *Wat. Sci. Tech.* 1993;28(8–9):195–198.

Koelmans, AA, de Lange, HJ, Lijklema, L. Desorption of chlorobenzenes from natural suspended solids and sediments. *Wat. Sci. Tech.* 1993;28(8–9):171–180.

Koide, M, Soutar, A, Goldberg, ED. Marine geochronology with ^{210}Pb. *Earth Planet. Sci. Lett.* 1972;14:442–446.

Koide, M, Bruland, KW. The electrodeposition and determination of radium by isotopic dilution in sea water and in sediments simultaneously with other natural radionuclides. *Anal. Chim. Acta* 1975;75:1–19.

Krishnaswami, S, Lal, D, Martin, JM, Meybeck, M. Geochronology of lake sediments. *Earth Planet. Sci. Lett.* 1971;11:407–414.

Krishnaswami, S, Lal, D. *Radionuclide Limnochronology.* In: Lerman, A. (Ed.) *Lakes; Chemistry, Geology, Physics.* New York: Springer-Verlag; 1978.

Li, X-F, Le, X-C, Simpson, CD, Cullen, WR, Reimer, KJ. Bacterial transformation of pyrene in a marine environment. *Environ. Sci. Technol.* 1996;30(4):1115–1119.

Liu, X, Sokol, RC, Kwon, O-S, Bethoney, CM, Rhee, G-Y. An investigation of factors limiting the reductive dechlorination of polychlorinated biphenyls. *Environ. Toxicol. Chem.* 1996;15(10):1738–1744.

Lo, SL, Huang, LJ. Effects of NTA on the fate of heavy metals in sediments. *Wat. Sci. Tech.* 1993;28(8–9):191–194.

Lockhart, WL, Wilkinson, P, Billeck, BN, Brunskill, GJ, Hunt, RV, Wagemann, R. Polycyclic aromatic hydrocarbons and mercury in sediments from two isolated lakes in Central and Northern Canada. *Wat. Sci. Tech.* 1993;28(8–9):43–52.

Masunaga, S, Wolfe, NL, Carriera, L. Transformation of para-substituted benzo-nitriles in sediment and sediment extract. *Wat. Sci. Tech.* 1993;28(8–9):123–132.

Mihelcic, JR, Luthy, RG. Degradation of polycyclic aromatic hydrocarbons compounds under various redox conditions in soil-water systems. *Appl. Environ. Microbiol.* 1988;54(5):1182–1187.

Officer, CB. Mixing, sedimentation rates and age dating for sediment cores. *Mar. Geol.* 1982;46:261–278.

Pennington, W, Cambray, RS, Fisher, EM. Observations on lake sediments using fall-out 137Cs as a tracer. *Nature* 1973;242:324–326.

Pesch, CE, Hansen, DJ, Boothman, WS, Berry, WJ, Mahony, JD. The role of acid-volatile sulfide and interstitial water metal concentrations in determining bioavailability of cadmium and nickel from contaminated sediments to the marine polychaete *Neanthes arenaceodentata. Environ. Toxicol. Chem.* 1995;14(1):129–141.

Pignatello, JJ, Chapa, G. Degradation of PCBs by ferric ion, hydrogen peroxide and UV light. *Environ. Toxicol. Chem.* 1994;13(3):423–427.

Rachdawong, P. Receptor models for source attribution of PAHs and PCBs in Lake Michigan sediments. Ph.D. thesis, University of Wisconsin-Milwaukee; 1997.

Rhee, G-Y, Bush, B, Bethoney, CM, DeNucci, A, Oh, H-M., Sokol, RC . Reductive dechlorination of Aroclor 1242 in anaerobic sediments: pattern, rate and concentration dependence. *Environ. Toxicol. Chem.* 1993a;12:1025–1032.

Rhee, G-Y, Bush, B, Bethoney, CM, DeNucci, A, Oh, H-M., Sokol, RC. Anaerobic dechlorination of Aroclor 1242 as affected by some environmental conditions. *Environ. Toxicol. Chem.* 1993b;12:1033–1039.

Richter, T, Schröder, G, Kaminski, S, Lindner, G. Transport of particles contaminated with cesium radonuclides into Lake Constance. *Wat. Sci. Tech.* 1993;28(8–9):117–121.

Ritson, PI, Esser, BK, Niemeyer, S, Flegal, AR. Lead isotopic determination of historical sources of lead to Lake Erie, North America. *Geochim. Cosmochim. Acta* 1994;58:3297–3305.

Robbins, JA. A model for particle-selective transport of tracers in sediments with conveyor-belt deposit feeders. *J. Geophys. Res.* 1986;91(C7):8542–8558.

Robbins, JA, Edgington, DN. Determination of recent sedimentation rates in Lake Michigan using ^{210}Pb and ^{137}Cs. *Geochim. Cosmochim. Acta* 1975;37:285–304.

Robbins, JA, Lindner, G, Pfeiffer, W, Kleiner, J, Stabel, HH, Frenzel, P. Epilimnetic scavenging of Chernobyl radionuclides in Lake Constance. *Geochim. Cosmochim. Acta* 1992;55:2339–2361.

Rule, JH, Alden III, RW. Interaction of cadmium and copper in anaerobic estuarine sediments: I. *Partitioning in geochemical fractions of sediments. Environ. Chem.* 1996;15(4):460–465.

Sedlak, DL, Andren, AW. Aqueous-phase oxidation of polychlorinated biphenyls by hdroxyradicals. *Environ. Sci. Technol.* 1991;25:1419–1427.

Sedlak, DL, Andren, AW. The effect of sorption on the oxidation of polychlorinated biphenyls (PCBs) by hyroxyradicals. *Wat. Res.* 1994;28(5):1207–1215.

Smith, MR. The biodegradation of aromatic hydrocarbons by bacteria. *Biodegradation* 1990;1:191–206.

Sokol, RC, Kwon, D-S, Bethoney, CM, Rhee, G-Y. Reductive dechlorination of polychlorinated biphenyls in St. Lawrence River sediments and variations in dechlorination characteristics. *Environ. Sci. Technol.* 1994;28(12):2054–2064.

Sokol, RC, Bethoney, CM, Rhee, G-Y. Effect of PCB concentration on reductive dechlorination and dechlorination potential in natural sediments. *Wat. Res.* 1995;29(1):45–48.

Soto-Jimenez, MF, Hibdon, SA, Rankin, CW, Aggarawl, J, Ruiz-Fernandez, AC, Paez-Osuna, F, Flegal, AR. Chronicling a century of lead pollution in Mexico: Stable lead isotopic composition analyses of dated sediment cores. *Environ. Sci. Technol.* 2006;40(3):764–770.

Susarla, S, Masunaga, S, Yoshitaka, Y. Kinetics of halogen substituted anilin transformation in anaerobic estuarine sediment. Water Quality International '96. IAWQ 18th Biennial International Conference, 23–28 June 1996. Conference preprint book 6, pp. 314–321.

Swartz, RC, Schults, DW, Ozretich, RJ, Lamberson, JO, Cole, FA, DeWitt, TH, Redmond, MS, and Ferraro, SP. ΣPAH: a model to predict the toxicity of polynuclear aromatic hydrocarbon mixtures in field-collected sediments. *Environ. Toxicol. Chem.* 1995;14(11):1977–1987.

Tuncer, GT, Tuncel, SG, Tuncel, G, Balkas, TI. Metal pollution in the Golden Horn, Turkey: contribution of natural and anthropogenic components since 1913. *Wat. Sci. Tech.* 1993;28(8–9):59–64.

Van Metre, PC, Fuller, CC. Dual-core mass-balance approach for evaluating mercury and ^{210}Pb atmospheric fallout and focusing to lakes. *Environ. Sci. Technol.* 2009;43(1):26–32.

Wei, H, Aziz-Schwanbeck, AC, Zou, Y, Corcoran, MB, Poghosyan, A, Li, A, Rockne, KJ, Christensen, ER, Sturchio, NC. Polybromodiphenyl ethers and decabromodiphenyl ethane in aquatic sediment from southern and eastern Arkansas, United States. *Environ. Sci. Technol.* 2012;46(15):8017–8024.

Wild, SR, Jones, KC. Biological and abiotic losses of polynuclear aromatic hydrocarbons (PAHs) from soils freshly amended with sewage sludge. *Environ. Toxicol. Chem.* 1993;12:5–12.

Wu, F, Zheng, J, Liao, H, Yamada, M. Vertical distributions of plutonium and ^{137}Cs in lacustrine sediments in Northwestern China: Quantifying sediment accumulation rates and source identifications. *Environ. Sci. Technol.* 2010;44(8):2911–2917.

Ye, D, Siddiqi, MA, Maccubbin, AE, Kumar, S, Sikha, HC. Degradation of polynuclear aromatic hydrocarbons by Sphingomonas paucimobilis. *Environ. Sci. Technol.* 1996;30(1):136–142.

Yang, L, Li, X, Crusius, J, Jans, U, Melcer, ME, Zhang, P. Persistent chlordane concentrations in Long Island sound sediment: Implications from chlordane, ^{210}Pb, and ^{137}Cs profiles. *Environ. Sci. Technol.* 2007;41(22):7723–7729.

Youngblood, WW, Blumer, M. Polycyclic aromatic hydrocarbons in the environment: homologous series in soils and recent marine sediments. *Geochim. Cosmochim. Acta* 1975;39:1303–1314.

Zhang, X, Christensen, ER, Yan, L-Y. Fluxes of polycyclic hydrocarbons to Green Bay and Lake Michigan Sediments. *J. Great Lakes Res.* 1993;19(2):429–444.

3

ATMOSPHERIC INTERACTIONS

3.1 INTRODUCTION

The atmosphere is a source of many chemicals that are present in water. All the components of air – nitrogen (N_2), oxygen (O_2), carbon dioxide (CO_2), and other minor components – dissolve in water to some extent. Dissolved oxygen (DO) is critical to water quality, aquatic life, and various chemical reactions that occur in water. Some gases, e.g. N_2 and Ar, are inert and do not react with water; while others (e.g. CO_2, SO_2, and NO_x) may alter the characteristics of the water into which they enter. In addition, the atmosphere is a major environmental compartment for volatile and semi-volatile chemical pollutants, and a major vehicle for their long range transport. Atmospheric pollutants are present either in the form of gaseous molecules, or as attachments to various particulate matter suspended in the air (see Wei and Li, 2010 for a review of the topic). They can enter aquatic ecosystems by means of wet or dry deposition of the particles, or as a result of partitioning from the gas phase. Conversely, volatilization to air is a major elimination process for certain hydrophobic organic contaminants (HOCs) from estuaries (Nelson et al., 1998) or lakes (Achman et al., 1993, Hornbuckle et al., 1993). Volatilization is more significant during the summer because the Henry's law constant is higher, favoring the gas phase.

This chapter begins with a discussion of dry and wet depositions as well as the gas exchange of chemicals from air to water. Following this discussion, the microlayers at the air–water interface are described because they play important roles in air–water exchange rates. Examples of metal deposition from the atmosphere, and deposition and gas exchange of organic contaminants, are also provided. A case study, relating to

Physical and Chemical Processes in the Aquatic Environment, First Edition. Erik R. Christensen and An Li.
© 2014 John Wiley & Sons, Inc. Published 2014 by John Wiley & Sons, Inc.

emission of volatile organic compounds (VOCs) from wastewater treatment plants, is presented to illustrate the applications of the theories. All these theoretical discussions will be needed in later chapters on specific chemical categories (nutrients, metals, and organic pollutants) in natural waters, because the atmosphere is a significant source of these chemicals; and for some of them, an important sink as well.

The driving force for a chemical to "move" across a boundary of environmental compartments is the difference in its fugacities between various phases. The last section of this chapter focuses on the applications of the fugacity concept, which provides insights into the partitioning and transport of chemicals between air and water and among other environmental compartments.

3.2   ATMOSPHERIC DEPOSITION PROCESSES

3.2.1   Gaseous vs. Particulate Chemicals in the Atmosphere

Various solid or liquid particles are suspended in air, forming aerosols. The individual particles are usually invisible to the naked eye, yet together they form a "phase" (or phases) in the atmosphere (Wei and Li, 2010). Chemical pollutants may exist in the atmosphere either as individual gas molecules or as attachments to suspended particulate matter. Needless to say the behavior and, as a consequence, health effects of the gaseous and particulate forms of the same chemical are very different. Their exchange methods between air and water are also different. Partitioning into water or volatilization from the aquatic phase are basic behaviors of volatile chemicals. In addition, we focus on the deposition of the particulate form of chemical pollutants from air to water.

Organic Pollutants.   Gaseous and particulate HOCs are often simultaneously collected using active samplers, which draw a large known volume of air through a glass fiber filter to catch airborne particles, followed by one or two polyurethane foam plugs (PUFs) to absorb HOCs in the gas phase. The PUF can be supplemented or substituted by a polymeric resin such as XAD-2 (U.S. EPA, 1999; Holsen et al., 1991, Hornbuckle et al., 1993).

Modified Anderson high volume active samplers were used to measure atmospheric concentrations of polychlorinated dibenzo-*p*-dioxins and dibenzofurans (PCDD/Fs) (Venier et al., 2009) and biphenyls (PCBs), PAHs and organochlorine pesticides (Venier and Hites, 2010) around the Great Lakes. The total volume of air sampled at each site ranged from $820 \, m^3$ and $6500 \, m^3$. Maximum concentrations of PCDD/Fs occurred during early December, pointing to residential heating as a source, but also during warm summer months, indicating high gas phase concentrations due to high Henry's law constant. PCDD/F concentrations were expressed as toxic equivalent concentrations TEQ_{DF} with 2,3,4,7,8-PeCDF and 1,2,3,7,8-PeCDD as major contributors. Atmospheric levels of lindane (γ-hexachlorocyclohexane, γ-HCH), DDTs, endosulfans, and chlordanes varied with the seasons with much higher levels in the summer. Concentrations of these organics were positively correlated with population density, with a less significant correlation for $\sum$ DDTs and chlordanes.

In the last decade, the use of passive air samplers (PAS) for sampling trace semivolatile chemical pollutants has gained increasing popularity due to their apparent advantages of low cost and no power supply requirement. Passive air sampling depends on molecular diffusion of the target chemicals from ambient air into the sampling medium, which can take the form of semipermeable membranes (SPMDs), polyurethane foam (PUF) disks, Amberlite XAD adsorbent resins, or PUF impregnated with sorbents (Zhang et al., 2012a; Westgate and Wania, 2011; Gouin et al., 2008; Wania et al., 2003; Shoeib and Harner, 2002; Moeckel et al., 2009). Various designs have been developed to collect the gas phase only (Wania et al., 2003) or both gas and particles (Tao et al., 2007, 2009).

Gouin et al. (2007) used passive air samplers to assess spatial and temporal trends of the chiral pesticides chlordanes and α-HCH around the Great Lakes. The origin of these compounds can often be inferred from their chiral signatures due to enantioselectivity of microbial degradation. Thus a racemic signature indicates fresh sources of chemicals that have not been subjected to degradation, whereas an aged source may be non-racemic. The authors found that α-HCH and especially *trans*-chlordane (TC), but possibly also *cis*-chlordane (CC), were non-racemic from remote areas such as Eagle Harbor but near racemic from the urban areas of Chicago and Toronto reflecting fresher or non-degraded sources. Low TC/CC ratios at the remote sites supported this interpretation.

PAS has enabled air monitoring in remote areas, and has been used in the Global Atmospheric Passive Sampling (GAPS) Network under the Stockholm Convention on Persistent Organic Pollutants (Koblizkova et al., 2012; Genualdi et al., 2010; Pozo et al., 2009; Jaward et al., 2005). Pozo et al. (2009) reported concentrations of persistent organic pollutants in air from the first full year of the GAPS network. They used a PUF passive sampler disk for the measurements. The highest geometric mean concentrations were $82 \, pg/m^3$ and $26 \, pg/m^3$ for the pesticide endosulfan and PCBs respectively. Other monitored chemicals included α- and γ-HCH, chlordanes, heptachlor, heptachlor epoxide, dieldrin, p,p'-DDE, and polybrominated diphenyl ethers (PBDEs). Seasonal variation was exhibited primarily through low first-quarter (December–March) concentrations. Most measurements were taken in the northern hemisphere, and the highest concentrations were observed in the mid-latitudes.

The concentration of particles in air samples is usually much higher in summer than in winter because the ground is not snow-covered or frozen and field plowing is in progress. As a result, atmospheric concentrations of chemicals in the particulate phase are much lower in summer because uncontaminated soil particles dilute the contaminants.

Summer concentrations in the particulate phase may also be lower because of evaporation from the particles into the gas phase, as gas–particle partitioning of volatile and semi-volatile chemicals depends strongly on temperature. In addition, the ratio between gaseous and particle-bound HOCs also varies diurnally. For instance, particulate PBDEs were more abundant at night than in daytime above the city of Guangzhou, China. Similarly, gaseous PCBs were higher at night in Zürich, Switzerland, due to an inversion layer effectively trapping the pollutants (Gasic et al., 2009).

They used high volume Tisch air samplers with polyurethane foam (PUF), but without glass fiber filters, to maximize the air volume sampled and because PCBs mainly occur in the gas phase at the sampling temperatures.

Inorganic Pollutants. In some cases, high NO_x concentrations in the atmosphere can contribute to aquatic pollution. In Chesapeake Bay, for example, it has been estimated that up to 35% of the nutrient nitrogen inputs come from the atmosphere (Hicks, 1998). Since Chesapeake Bay is nitrogen limited, this is a significant input. Naturally occurring nitrogen in the atmosphere takes the form of inert N_2, except for nitrogen-fixing algae, so it is the NO_x pollution from cars in the Washington, DC, and Baltimore area that is the problem. Other eastern estuaries have similar atmospheric nitrogen inputs. In particular, deposition of NO_x and sulfate contributes to the acidity of small, poorly buffered seepage lakes in the eastern United States. However, with the improved sulfur reduction technology for coal-fired power plants, this problem seems to have decreased in importance. Whereas nitrogen inputs to the coastal estuaries are of major concern, atmospheric HOCs and metals have a greater impact on inland lakes that are phosphorus limited. For the mid-Atlantic bight, less than 50% of Cd, Cu, and Mn is of atmospheric origin, whereas the majority of Ni, Zn, Fe, Al, and Pb comes from the atmosphere (Hicks, 1998). The atmosphere is known to be a significant source of Hg to Wisconsin inland lakes. The Hg is present in fly ash from coal-fired power plants.

3.2.2   Dry Deposition with Aerosols

To calculate a dry deposition flux, one can multiply the atmospheric particle concentration with an estimated deposition velocity. Alternatively, the dry deposition flux can be measured using a smooth plate with Mylar strips covered with Apezion L grease (Holsen et al., 1991). If the particle size distribution of dry fallout is to be measured, one would use a cascade impactor such as a Hering low pressure impactor with Teflon filter (Blando et al., 1998), which is capable of measuring several size fractions between $1-2\,\mu m$ to $0.05\,\mu m$.

Deposition velocity of dry particles is usually assumed to be constant, even recognizing the influence of a large number of variables including particle size, wind speed, air temperature, and humidity. This applies only to small particles ($< 5\,\mu m$), because, according to Stokes' Law, at larger diameters the particle velocity is proportional to the diameter squared. In reality, however, there is a minimum deposition velocity at a particle diameter of $\sim 0.5\,\mu m$ as a result of a balance between gravitational settling and other factors such as Brownian diffusion (Slinn and Slinn, 1981).

The particle deposition velocity v_d may be written as a sum of two terms (Slinn and Slinn, 1981),

$$v_d = v_g + \alpha \frac{D}{\nu} u^* \tag{3.1}$$

where the first term indicates gravity settling according to Stokes' law,

$$v_g = \frac{g\rho_p d^2}{18\mu} \tag{3.2}$$

where g is the acceleration due to gravity, ρ_p is the particle density, d it the particle diameter, and μ is the dynamic viscosity of air at the given temperature.

The second term in Equation 3.1 represents Brownian diffusion. Here, α is a constant, D is the particle diffusivity in air, v is the kinematic viscosity of air at the given temperature, and u^* is the air friction velocity that is related to the air velocity at 10 m elevation (u_{10}) over open water through the equation (Slinn and Slinn, 1981),

$$u^* = 0.0361 \cdot u_{10} \tag{3.3}$$

The particle diffusivity in air may be determined from the Stokes–Einstein equation,

$$D = \frac{kT}{3\pi\mu d} \tag{3.4}$$

where k is Boltzman's constant (1.38×10^{-23} J/s) and T is the temperature in K.

The deposition velocity v_d according to Equation 3.1 is plotted in Figure 3.1 along with experimental values (Zufall et al., 1998) and velocities obtained from a mass balance model (Caffrey et al., 1998). The plot is for 25°C air temperature and a velocity at 10 m elevation $u_{10} = 7.5$ m/s. The value of α is 19.5. There is a minimum at $d_p = 0.4$ μm. The corresponding velocity depends on temperature, particle diameter, and friction velocity.

Deposition velocities v_{dc} for the coarse fraction (> 6.5 μm aerodynamic diameter) of semivolatile compounds such as PCBs may be determined by a combination of flux measurements for both gas and particle phases and concentration measurements using known deposition velocities for the gas phase v_{dg} and fine particles v_{df} (< 6.5 μm aerodynamic diameter (Holsen et al., 1991). This may be expressed as follows:

$$F = v_{dg}c_g + v_{df}c_f + v_{dc}c_c \tag{3.5}$$

where,

F = total PCB flux,

c_g = PCB concentration in the gas phase

c_f = PCB concentration in the fine particulate fraction.

The total PCB flux F for both particles and gas was measured by means of Mylar strips (7.6 cm × 2.5 cm) coated with ~8 mg Apezion L grease (thickness ~8 μm).

The gas and fine particle phase concentrations c_g and c_f were determined based on PCBs collected in the polyurethane foam (PUF) plug and glass fiber filter of a standard sampling train. By contrast, the coarse fraction concentration c_c was evaluated based on material collected in stage A of a Noll rotary impactor. Stage A includes particles of aerodynamic diameter > 6.5 μm. Other stages collect particles greater than 11.5, 24.7, and 36.5 μm. The mass median diameter (MMD) is calculated based on mass collected in these four stages. For Chicago air, the MMD measured 26.8 μm.

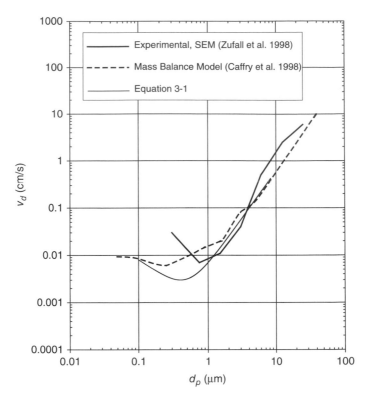

Figure 3.1. Deposition velocity v_d vs. physical particle diameter d_p. Model estimates are based on Equation 3.1 and the mass balance model of Caffrey et al. (1998). Experimental results (Zufall et al., 1998) are derived from concentration and flux measurements coupled with scanning electron microscopy (SEM) of fallout particles. (Source: Modified from Zufall et al., 1998)

As an example calculation of v_{dc}, consider the following parameters:

$$F = 7.0 \times 10^{-6} \text{ ng/(cm}^2 \text{ s)} \quad v_{dg} = 0.1 \text{ cm/s}$$

$$v_{df} = 0.5 \text{ cm / s} \quad c_g = 10.4 \text{ ng / m}^3$$

$$c_f = 3.2 \text{ ng / m}^3 \quad c_c = 0.90 \text{ ng / m}^3$$

From Equation 3.5 we obtain $v_{dc} = 4.8$ cm / s. The coarse particle flux 4.3×10^{-6} ng/(cm^2 s) constitutes 62% of the total flux. For $v_{dg} = 0.01$ cm/s and $v_{df} = 0.1$ cm/s one obtains a coarse particle flux of 6.6×10^{-6} ng/(cm^2 s) or 94% of the total flux. Thus for the range of realistic deposition velocities for the gas and fine particle phases, the coarse particle flux is large, even though the gas phase PCB concentration is dominant.

Deposition velocities v_d in various size ranges j can also be calculated from a chemical mass balance deposition model (CMBD) as carried out by Caffrey et al. (1998). This model may be written as,

$$F_i = \sum_{j=1}^{11} (v_d c_i)_j \tag{3.6}$$

where
F_i = flux of element i
c_{ij} = concentration of element i in particles of size class j.

For the measurement of the concentrations c_{ij}, there were 11 size classes based on a micro-orifice uniform deposit impactor (MOUDI), with 50% aerodynamic cutoff diameters of 15, 3.2, 1.8, 1.0, 0.55, 0.29, 0.17, 0.091, and 0.053 μm, and a Noll rotary impactor (NRI) with cutoff diameters of 36.5, 24.7, 11.5, and 6.5 μm, The first three stages of the MOUDI impactor were covered by ~2 mg of dimethylpolysiloxane to reduce particle bounce and re-entrainment. The NRI contained similar coated Gelman Zefluor Teflon filters as impaction substrates.

MOUDI was developed in the 1980s (Marple et al., 1991), and has been the most commonly used collector of particulate samples. It has been used with various cutoff aerodynamic diameters to fractionate airborne particles, as well as many organic and inorganic pollutant groups, such as elemental carbon, vehicle exhaust aerosols, PAHs, dioxins, PBDEs, PCBs, pesticides, metals, sea salt, and other inorganic species, based on particle sizes (Caffrey et al., 1998; Chao et al., 2003; Zhang et al., 2012b).

For measurement of deposition flux F_i in Equation 3.6, 10 cm diameter coated Gelman Zefluor Teflon filters were mounted on symmetric frisbee-shaped airfoils.

Caffrey et al. (1998) determined element concentrations for As, Ca, Mg, Se, Sb, V, and Zn using instrumental neutron activation analysis and for S using X-ray fluorescence. Sets of solutions were obtained using As, Ca, S, Se, Sb, and Zn for one period (July 21–25, 1994) and As, Mg, Sb, V, and Zn for the second period (July 25–28, 1994). Ca and Mg were associated with larger particles of aerodynamic diameter ~10 μm, whereas As, Se, and Sb were associated with fine particles. Thus from Equation 3.6, there are 6 and 5 equations with 11 unknowns (v_{dj}), and a solution can be obtained for the particle velocities v_{dj}.

Results for July 21–25 are shown in Table 3.1 along with concentrations and calculated fluxes for As, Ca, Se, and Zn. Calculated deposition velocities are also shown in Figure 3.1. The total calculated fluxes are in reasonable agreement with the measured fluxes, especially for Ca and Zn. Note that the element concentrations are typically maximal in a small size range, e.g. 0.40 μm for As, while the fluxes have peaks in the larger size range, e.g. 21.2 μm for As.

By imposing constraints it may be possible to determine settling velocities from an underdetermined version of Equation 3.6, i.e. one characterized by fewer equations than unknown velocities.

Experimental determination of deposition velocities in specific size ranges was accomplished by Zufall et al. (1998) based on the equation,

TABLE 3.1. Measured Concentrations and Calculated Fluxes for Four Elements in Each Size Range. The Total Fluxes are Compared with Measured Fluxes

Midpoint diameters		Deposition	Element							
			As		Ca		Se		Zn	
Aerodynamic μm	Physical μm	velocity[a] cm/s	Conc. ng/m^3	Flux μg/m^2/h	Conc. ng/m^3	Flux μg/m^2/h	Conc. ng/m^3	Flux μg/m^2/h	Conc. ng/m^3	Flux μg/m^2/h
60.4	42.7	10.98	0.0161	0.0064	120	47.4	0.00270	0.00107	1.74	0.688
30	21.2	2.67	0.0071	0.00068	47.5	4.57	ND	0	0.61	0.059
21.2	15	1.35	0.0334	0.00162	202	9.82	0.050	0.00243	3.09	0.150
6.92	4.90	0.144	0.0964	0.00050	479	2.48	0.069	0.00036	17.6	0.091
2.40	1.70	0.0228	0.0486	0.00004	59.4	0.05	0.096	0.00008	13.3	0.011
1.34	0.949	0.0180	0.0351	0.00002	21.6	0.01	0.113	0.00007	19.5	0.013
0.74	0.525	0.0073	0.0358	0.00001	8.10	0.002	0.392	0.00010	13.5	0.004
0.40	0.282	0.0043	0.172	0.00003	2.82	0.0004	0.590	0.00009	8.85	0.001
0.22	0.158	0.0052	0.133	0.00002	1.04	0.0002	0.194	0.00004	3.11	0.0006
0.12	0.087	0.0059	0.0968	0.00002	ND	0	0.101	0.00002	1.25	0.0003
0.07	0.049	0.0054	0.0383	0.00001	ND	0	0.026	0.00001	0.51	0.0001
Sum	-	-	-	0.00935	-	64.3	-	0.0043	-	1.018
Measured	-	-	-	0.0048	-	58.7	-	0.0058	-	0.992

[a]Calculated from CMBD model assuming a particle density of 2 g/cm^3. (Source: Caffrey et al., 1998).

$$v_d = \frac{F}{C} \qquad (3.7)$$

where
F = deposition flux
c = concentration of airborne particles.

Both particles and several elements were investigated. The flux F was measured with frisbee-shaped symmetric airfoils. Zefluor Teflon filters coated with dimethyl polysiloxane and upturned pieces of tape with liners were placed on the airfoils. The concentration c was determined using a filter system with 0.8 or 0.4 µm pore size polycarbonate filter in a stainless steel filter holder and a Savillex Teflon filterpack with a 47 mm diameter 1 µm pore size Teflon Zefluor filter.

Tape and polycarbonate filters were analyzed by scanning electron microscopy (SEM) with energy dispersive X-ray detector (EDX), and Teflon filters were analyzed with instrumental neutron activation analysis (INAA). For individual particles, size range in µm, i.e. 0.25–0.5, 0.5–1.0, 1.0–2.0, 2.0–4.0, 4.0–8.0, 8.0–16.0, 16–32, 32–50, 50–100, and greater than 100 µm, was determined by SEM. Cumulative mass was determined from projected area and density calculated as a weighted average based on elemental composition (EDX). The results with error bars are shown for seven size ranges in Figure 3.1. It may be seen that these experimentally determined deposition velocities are consistent with those based on Equation 3.1, and with velocities derived from a mass balance model (Caffrey et al., 1998).

Instrumental neutron activation analysis was used to determine fluxes and concentrations of several elements, i.e. Al, Ba, Br, Ca, Cl, Cu, K, Mg, Mn, Na, Ti, and V. Size ranges were estimated from physical particle mass median aerodynamic diameters (mmd), determined from impactor measurements taken during the same time period (Caffrey et al., 1998). The deposition velocity for the elements vs. physical particle diameter (Zufall et al., 1998) is similar to the deposition velocities plotted in Figure 3.1.

Dry deposition fluxes of Pb, Cu, and Zn over Lake Michigan were measured by Paode et al. (1998). They used a dry deposition plate of polyvinyl chloride (PVC) with a sharp leading edge that was pointed into the wind by a wind vane. The plate was covered with Mylar strips coated with Apezion L grease. In order to obtain calculated fluxes, airborne concentrations of these metals were also measured. Concentrations of metals in the size range from $d = 0.1–100$ µm were measured with a Noll rotary impactor (NRI) for the size range 6.5 – 100 µm, and with an Anderson I ACFM ambient particle-sizing sampler with pre-separator (AMPSS).

Stages of the NRI and Anderson impactor were covered with Mylar coated with Apezion L grease. The total mass collected was determined by weighing media before and after sampling. Calculated metal fluxes were obtained from Equation 3.6 using the measured concentrations and deposition velocities in various size ranges from the Sehmel–Hodgson dry deposition model. The ratio of calculated to measured fluxes was generally between 0.2 and 2, indicating a fairly good agreement between flux model and measurements.

The metal fluxes were mostly in the $d > 2.5$ µm range whereas the atmospheric concentrations were mainly in the $d < 2.5$ µm range. This distinction parallels that found for PCBs (Holsen et al., 1991) and for the elements As, Ca, Se, and Zn (see Table 3.1 and Caffrey et al., 1998).

The fraction ϕ of gaseous material adsorbed on aerosols depends on aerosol surface area and the saturation vapor pressure of the organic compound (Junge, 1977; Andren and Strand, 1981),

$$\phi = \frac{c\theta}{P_0 + c\theta} \tag{3.8}$$

where

ϕ = amount adsorbed on aerosol per volume unit / total concentration
θ = aerosol surface area, $cm^2/(cm^3$ air)
P_0 = saturation vapor pressure, mm Hg
c = constant dependent on molecular weight and heat of condensation
($c \cong 0.13$ for many organic compounds)

Thus, the amount of an organic compound adsorbed to aerosols increases with the aerosol concentration but decreases with higher vapor pressure. This relationship is depicted in Figure 3.2. For example, using saturation vapor pressures with solid phase of 0.0136 Pa for PCB 33 (2′,3,4-trichlorobiphenyl) and 0.001 Pa for the insecticide malathion, along with an aerosol surface area of 3×10^{-6} cm² /(cm³ air) (urban air), it is seen that the aerosol adsorbed fraction is 0.004 for the PCB but 0.05 for malathion. Also, for DDT with a very low saturation vapor pressure with solid phase

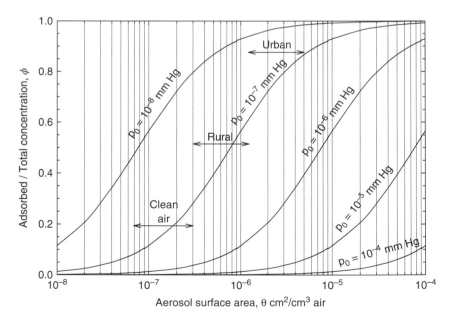

Figure 3.2. Fraction ϕ of compounds attached to aerosols vs. aerosol surface area θ for various saturation pressures p_0. (Source: Adapted from Junge, (1977))

of 2×10^{-5} Pa, the adsorbed fraction is 0.72. However, under clean air conditions, DDT mainly occurs in the gas phase ($\theta \sim 0.15$).

3.2.3 Wet Deposition

Atmospheric pollutants may be associated with dry aerosols, aerosols in cloud water, aerosols in precipitation, or they may be present in the gas phase (Scott, 1981). For the first three forms we may write,

$$\chi = \chi_a + \chi_c + \chi_p \tag{3.9}$$

where

χ = total mass of pollutant associated with aerosols per unit volume of space,
χ_a = concentration of pollutant associated with dry aerosol,
χ_c = concentration of pollutant associated with cloud water, and
χ_p = concentration of pollutant associated with precipitation water.

Washout is a removal mechanism that is normally only an issue with particles. Since HOCs are only slightly soluble in water, little from the vapor phase will be dissolved in rain. The particle phase is usually completely washed out in rain, but since the proportions of PCBs, PAHs, and Hg are low for particles in comparison to vapors, the quantity of these substances contributed to lakes by particle washout is not large (Nelson et al., 1998). For pesticides, e.g. lindane, aldrin, and dieldrin and metals such as Pb and Zn, the washout contribution can be significant (Scott, 1981). Washout increases with aerosol concentration (available surface area), and is inversely related to saturation vapor pressure.

Aerosols can grow in diameter from less than 1 μm to approximately 10 μm within minutes (nucleation scavenging). This was illustrated for aerosols composed of 50% $(NH_4)_2SO_4$ by Scott (1981). Thus, if a pollutant such as lead is attached to a soluble or partially soluble particle, this process can occur. Once the pollutant has been incorporated into 10 μm diameter droplets it may be more easily removed from the atmosphere by precipitation scavenging. The flux or deposition of pollutant with precipitation can be written as,

$$\chi_p \upsilon_p = C \cdot J \tag{3.10}$$

where
υ_p = velocity of rain drops,
C = concentration of pollutant in precipitation water, and
J = precipitation rate.

The surface deposition D is,

$$D = (C \cdot J)_0 \tag{3.11}$$

Defining the washout ratio W as,

$$W = \frac{C}{(\chi_a)_0} \tag{3.12}$$

the surface deposition becomes,

$$D = W \cdot (J\chi_a)_0 \tag{3.13}$$

Therefore, the washout ratio is a concentration factor for pollutants in rain relative to the dry aerosol phase. For many elements including metals, this factor is about 10^6 (Scott, 1981). The physical significance of $W = 10^6$ may be thought of as the concentration increase for a pollutant occurring when 10^6 aerosol droplets of 10 μm diameter contained in 1 L volume of air, i.e. the aerosols are spaced 1 mm apart, are all coalesced into one raindrop of 1 mm³ volume. The overall washout ratio W_T including both particle and gas scavenging may be written as,

$$W_T = W_G(1 - \phi) + W_P(\phi) \tag{3.14}$$

where W_G and W_P are the gas and particle washout ratios respectively, and ϕ is the fraction of aerosol-bound particle persistent organic compound (POP) to the total atmospheric POP concentration, $\chi_a / (\chi_a + \chi_g)$ with χ_g indicating the gas phase concentration (Jurado et al., 2005).

The process of precipitation scavenging of aerosols is irreversible. By contrast, gas scavenging is reversible, meaning that volatile or semi-volatile organics can be transferred from the gas to the aqueous phase and back from the aqueous to the gas phase.

Consider now a raindrop falling in an atmosphere with the concentration χ_g of a gaseous pollutant. The fall distance necessary to establish equilibrium between the gas phase concentration χ_g and the concentration c of the dissolved gas in the aqueous phase is proportional to the square of the diameter of the drop, and to the drop fall speed. This fall distance is also inversely related to the diffusivity of the pollutant in water.

The equilibrium distance and time are of the order of 1 m and 1 s respectively, for most cases (Scott, 1981). Thus, the gas and liquid phases may be considered to be in equilibrium so that the ratio of their concentrations equals the dimensionless Henry's law constant, H:

$$H = \frac{\chi_g}{c_{eq}} \tag{3.15}$$

where

$c_{eq} \cong c$ is the aqueous phase concentration of the pollutant.
Equation 3.15 leads to,

$$c_{eq} = \frac{\chi_g}{H} \tag{3.16}$$

By comparing this equation with Equation 3.12, it is seen that the washout ratio W for the gas phase is

$$W = \frac{1}{H} \tag{3.17}$$

The deposition rate D can be written as

$$D = \frac{1}{H}J\chi_g \tag{3.18}$$

Henry's law constants, gas phase washout ratios, and saturation vapor pressures for several pesticides, PCBs, and PAHs are presented in Table 3.2. Particle washout ratios are significantly higher for PCBs and PAHs ($\sim 10^3 - 10^8$), highlighting the dominant roles of particles in wet deposition fluxes of these pollutants (Jurado et al., 2005). For semi-volatile pesticides, both gas and particulate phases can be important. For PCDD/Fs, gas scavenging makes a significant contribution to wet deposition fluxes as a result, in part, to adsorption to water droplets.

Since washout ratios for all organics, except lindane and malathion, are much lower than the common value for metals of 10^6 (Scott, 1981), it may be expected that for low particle concentrations, wet deposition of most organics would be much lower than for metals. This is true for comparable values of the concentrations of organics in the gas phase (χ_g) and of metals on aerosols (χ_a).

TABLE 3.2. Gas Phase Washout Ratios and Saturation Vapor Pressures for Selected Organic Chemicals at 25°C

Compound	Henry's law constant H, Pa m^3/mol	Washout ratio $W = RT/H$ (dimensionless)	Saturation vapor pressure ρ_s, Pa
Pesticides			
Lindane	0.149	1.66×10^4	0.0274(l)[a]
Aldrin	91.23	27.2	0.0302(l)
Dieldrin	1.120	2.21×10^3	0.016(l)
Malathion	2.28×10^{-3}	1.09×10^6	0.001
PCBs			
16(22′3)	20.22	1.23×10^2	-
32(24′6)	20.22	1.23×10^2	-
44(22′35′)	12.47	1.99×10^2	0.013(l)
70(23′4′5)	10.13	2.45×10^2	-
76(2′345)	10.13	2.45×10^2	-
PAHs			
Fluorene	7.87	3.15×10^2	0.715(l)
Phenanthrene	3.24	7.65×10^2	0.113(l)
Fluoranthene	1.037	2.39×10^2	8.72×10^{-3}(l)
Pyrene	0.92	2.69×10^2	0.0119(l)

Source: For the pesticides and PAHs, the values are from Mackay et al. (1997, Vol V, pp 567–569), and Mackay et al. (1992c, Vol II, p. 251). For PCBs, the Henry's law constants are from Achman et al. (1993). The vapor pressure of PCB44 is listed in Mackay et al. (1992b, Vol I, p. 418).
[a]With liquid phase.

3.2.4   Gas Exchange

Volatile or semi-volatile organic compounds, and certain metals such as mercury that may occur in the gas phase, can be absorbed by or emitted from water bodies. This is dependent on the actual concentration of the compound in water and its saturation concentration in water under the prevailing atmospheric vapor pressure.

The rate of gas exchange is calculated based on simultaneous measurement of the water and air concentration of the chemicals considered (Achman et al., 1993).

Differences in volatility, i.e. vapor pressures of various PCB congeners, can cause changes in the temporal homolog patterns at different latitudes. For example, while total PCB concentrations have declined throughout Norway, the relative concentrations of hexa- and hepta-chlorinated homolog groups have decreased to a larger extent in the south and in the north (Lead et al., 1996).

The vapor pressure p, or the gas phase concentration, increases with increasing temperature and may be expressed as,

$$\ln p = \frac{m}{T} + b; \quad m < 0 \tag{3.19}$$

where
p = atmospheric partial pressure of the compound (atm),
T = atmospheric temperature (K),
m = constant (< 0), and
b = constant

An example of a modified relationship of this kind for PCB-52 (2,2′,5,5′) and γ-HCH (lindane), as measured in Egbert, Ontario, is shown in Figure 3.3 (Wania

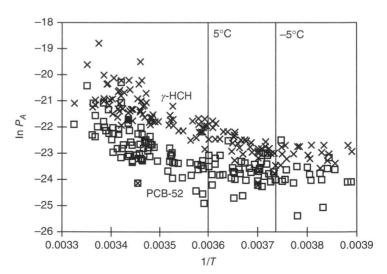

Figure 3.3. Log vapor pressure vs. reciprocal temperature for PCB 52 (2,2′,5,5′)-tetrachlorobiphenyl and γ-HCH (lindane). (Source: Reprinted from Wania et al. (1998) with permission from American Chemical Society.)

et al., 1998). The modification lies in the fact that, in contrast to Equation 3.18, both chemicals display constant vapor pressures at low temperatures. A shallow slope at higher temperatures, i.e. a low temperature dependence, indicates that long-range atmospheric transport controls atmospheric levels at the sampling site, whereas a steeper slope reflects high surface concentrations near the site (Wania et al., 1998). In the latter case, m may be deemed a thermodynamic constant that is given by,

$$m = -\frac{\Delta H}{R} \qquad (3.20)$$

where ΔH is the heat of phase transition (kJ/mole), and R is the gas constant (8.314 J/(mole °K) (Cortes et al., 1999). Equation 3.19 is then the classical Clausius–Clapeyron equation.

Once the vapor and aqueous phase concentrations of a compound are known, the flux F (mole/(m^2 d)) of the compound from water to air is given by,

$$F = K_{ol}(c_w - c^*) \qquad (3.21)$$

where

c_w = concentration of the compound in the water phase (mole/m^3)
c^* = saturation concentration of the compound in water (mole/m^3),
   assuming equilibrium between air and water phases, and
K_{ol} = mass transfor coefficient (m/d).

If $F < 0$, the water is under-saturated, and the transfer takes place from air to water, i.e. absorption of the compound by the water. The saturation concentration $c*$ may be expressed as,

$$c^* = \frac{P}{H} \qquad (3.22)$$

where p (atm) is the vapor pressure of the compound above the water, and H (atm m^3/mole) is Henry's law constant.

Mainly following Achman et al. (1993), the mass transfer coefficient K_{ol} may be determined from

$$\frac{1}{K_{ol}} = \frac{1}{k_w} + \frac{RT}{Hk_a} \qquad (3.23)$$

where

k_w = mass transfer coefficient in water (m/d),
k_a = mass transfer coefficient in air (m/d), and
H = Henry's law constant (atm m^3/mole) at the water temperature.

Equation 3.22 expresses the total resistance to mass transfer as a sum of a liquid film resistance ($1 / k_w$) and a gas film resistance (RT / Hk_a). The gas phase controls for H less than 1.2×10^{-5} atm m^3/mole, and the liquid phase for $H > 4.4 \times 10^{-3}$ atm m^3/mole. The two-film model was reviewed by Weber and Digiano (1996).

The mass transfer coefficient k_w (cm/h) in water increases with wind speed and may be determined from

$$
k_w = \begin{cases}
0.17 u_{10} & \text{for} \quad u_{10} < 3.6 \text{ m/s} \\
2.85 u_{10} - 9.65 & \text{for} \quad 3.6 \text{ m/s} < u_{10} < 13 \text{ m/s} \\
5.9 u_{10} - 49.3 & \text{for} \quad u_{10} > 13 \text{ m/s}
\end{cases}
\tag{3.24}
$$

In these expressions, u_{10} (cm/s) is the wind speed at a reference height of 10 m. The value of k_w is normalized to CO_2 with a Schmidt number of 600 (20°C). The Schmidt number is derived by dividing the kinematic viscosity of the medium by the compound's diffusivity through the medium. To find the mass transfer coefficient of the compound, e.g. PCB in water, we use the following equation,

$$
\frac{k_w(CO_2)}{k_w(PCB)} = \frac{Sc^n(CO_2)}{Sc^n(PCB)}
\tag{3.25}
$$

where

$$
n = -\frac{2}{3} \text{ for } u_{10} < 3.6 \text{ m/s}
$$

$$
n = -\frac{1}{2} \text{ for } u_{10} > 3.6 \text{ m/s}
$$

The molecular diffusivity of the compound in water is determined from the Wilke–Chang equation (Weber and DiGiano, 1996: 162).

The mass transfer coefficient in air, k_a, may be calculated from the evaporation rate of water itself from the water body:

$$
k_a(H_2O) = 0.2 \, u_{10} + 0.3
\tag{3.26}
$$

where k_a (H_2O) is in units of cm/s, and the wind speed at 10 m reference height u_{10} is in m/s. The mass transfer coefficient k_a for the organic chemical in air is then calculated from,

$$
k_a(\text{org chem}) = k_a(H_2O) \left(\frac{D_a(\text{org chem})}{D_a(H_2O)} \right)^{0.61}
\tag{3.27}
$$

where D_a (org chem) and D_a (H_2O) are the molecular diffusivities in air of vapors of the organic chemical and water. These diffusivities are estimated by the method used by Fuller et al. as described by Reid et al. (1987).

Note that there is no apparent temperature dependency of k_a (org chem) since D_a (org chem) and D_a (H_2O) have the same temperature factor, and k_a (H_2O) is not specified in terms of temperature. This contrasts with the analogous expression for the liquid phase, Equation 3.24 where k_w (CO_2) and Sc (CO_2) are to be taken at a reference temperature of 20°C. In accordance herewith it would be physically more meaningful to specify Equation 3.26 at a reference temperature, and then use the same temperature for D_a (H_2O) in Equation 3.27.

Henry's law constants H at 25°C were adopted or calculated from data in Brunner et al. (1990). The following expression proposed by Tataya et al. (1988) and quoted in Nelson et al. (1998) may be used to correct H for water temperature:

$$\ln H_T = \ln H_{298} + 26.39 - \frac{7868}{T} \qquad (3.28)$$

where H_T is Henry's law constant at temperature T (K) and H_{298} is the same constant at 25°C. The Henry's law constant is quite temperature dependent. For example, for a 10°C temperature increase, e.g. from 10°C to 20°C, the Henry's law constant increases by a factor of 2.6 according to Equation 3.28. The following illustrates the calculations using data from two areas of the Laurentian Great Lakes: Green Bay and Lake Superior.

Green Bay. PCB flux calculations for five PCB congeners listed in Table 3.3 are carried out by estimating gas transfer based on Equations 3.21 and 3.28. Volatilization of PCBs from Green Bay, Lake Michigan, using data from Achman et al. (1993) is considered here. The authors made simultaneous measurements of PCB congener concentrations in water (XAD-2) and air (glass fiber filter + polyurethane foam plug, PUF), and used the results to calculate PCB fluxes according to Equation 3.21. The specific PCB concentration, wind speed ($u_{10} = 4$ m/s) and water (15.3°C) and air (26.4°C) temperatures are those that applied to station 4 on August 1, 1989.

The calculations begin with the estimation of temperature-corrected Henry's law constants based on the values in Table 3.2 and Equation 3.28. Note that due to the lower water temperature (15.3°C), the dimensionless Henry's law constants of Table 3.3 only amount to approximately 42% of the equivalent values in Table 3.2.

Transfer through the water film is characterized by the mass transfer coefficient k_w (PCB) which is related to k_w (CO_2) through Equation 3.25. For example, with a

TABLE 3.3. Calculation of Selected Individual PCB Congener Fluxes from Water to Air in Green Bay, Lake Michigan based on Data from Achman et al. (1993)[a]

| PCBs | Henry's law constant H/RT (dimensionless)[b] | PCB concentrations | | Mass transfer coefficients | | | PCB flux ng/m²/d |
		Air ng/m³	Water ng/m³	k_a m/d	k_w m/d	k_{ol} m/d	
16(22'3)	0.00343	0.080	183	368	0.209	0.179	28.6
32(24'6)	0.00343	0.080	183	368	0.209	0.179	28.6
44(22'35')	0.00211	0.121	450	359	0.204	0.161	63.2
70(23'4'5)	0.00172	0.040	199	359	0.204	0.153	26.9
76(2'345)	0.00172	0.040	199	359	0.204	0.153	26.9

[a]Site 4, 8/1/1989, wind speed = 4 m/s, air temperature = 26.4°C.
[b]At a water temperature of 15.3°C.

diffusivity of 4.69×10^{-6} cm^2/s for trichloro PCBs we obtain a Schmidt number of 2429. In addition, k_w (CO$_2$, 20°C) = 1.75 cm / h and Sc (CO$_2$, 20°C) = 600. Thus, from Equation 3.24 we obtain k_w (PCB) = 0.209 m/d (Table 3.3).

To calculate the air-side mass transfer coefficient k_a, we first determine k_a (H$_2$O) = 950 m/d. The diffusivities for H$_2$O and trichloro PCBs are 0.256 and 0.054 cm^2/s respectively. Thus, from Equation 3.27, k_a (PCB) = 368 m/d. The overall mass transfer coefficient K_{ol} is then determined from $1/K_{ol} = 1/0.209 + 1/(0.00343 \times 368) =$ 5.58 d/m, meaning that K_{ol} = 0.179 m/d. Finally, the PCB flux from water to air is $F = 0.179$ (183 − 0.080/0.00343) = 28.6 ng/(m^2 d). For the tetra-chloro PCBs, we obtain 63.2 and 26.9 ng/(m^{2}d) (Table 3.3). With all significant congeners included, Achman et al. (1993) calculate a total PCB flux of 800 ng/(m^{2}d).

Considering that the saturation PCB concentration c^* is only about 12% of the PCB water concentration c_w, and that K_{ol} is mainly determined by the liquid mass transfer coefficient k_w, the temperature dependence of the PCB flux is as $(D_w$ (PCB)$/v_w)^{0.5}$ or $(T/v_w \eta_w)^{0.5}$ where D_w (PCB) is the diffusivity of PCB in water and v_w and μ_w are the kinematic and dynamic viscosities respectively, in water. This gives a rather modest temperature influence on the flux, e.g. an increase by a factor of 1.3 by increasing the temperature from 10°C to 20°C.

The main factors affecting the PCB flux are, therefore, the concentration of PCBs in water c_w and the wind speed u_{10}. This is illustrated in Figure 3.4 where the flux calculation referred to in Table 3.3 (8/1/89) is labeled B. By comparing events A and B it is seen that increasing u_{10} from 1 to 4 m/s at nearly the same c_w dramatically increases the PCB flux from 90 to 800 ng/(m^2 d). Similarly, looking at events C and D, one notes that the flux increases significantly from 70 to 1300 ng/(m^2 d) when c_w

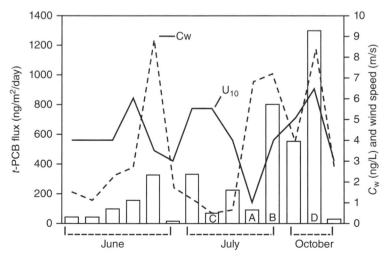

Figure 3.4. Total PCB flux vs. PCB water concentration c_w and wind speed u_{10}. See text for explanation of A-D. (Source: Reprinted from Achman et al. (1993) with permission from American Chemical Society.)

increases from 350 to 6670 ng/m^3 at near constant u_{10}. The water temperatures for events A, B, C, D are 15.6, 15.3, 14.0, and 7.6°C respectively.

Measurements of PCBs in vapors over land and water, and dissolved in the water from the area around Green Bay, support the hypothesis that volatilization from natural waters is an important source of PCBs to the atmosphere (Hornbuckle et al., 1993). The total PCB concentrations over the water were higher in southern Green Bay (670 – 2200 pg/m^3) than over the northern bay (160 – 520 pg/m^3) compared with concentrations over nearby land areas of 70 to 760 pg/m^3. In addition, there was a high correlation between PCB concentrations in the water of southern Green Bay and the overlying air.

Lake Superior. Estimation of air–water PAH diffusive fluxes in Lake Superior differed from the Green Bay work because passive polyethylene samplers for the water phase were used (Ruge et al., 2013). The polyethylene sheets were immersed in water at the sampling site for several weeks. The aqueous concentration of PAH, c_w is then,

$$c_w = c_{PE}/K_{PE_W} \qquad (3.29)$$

where c_{PE} is the PAH concentration of the polyethylene and K_{PE_w} is the polyethylene–water partition coefficient. The PAH flux is calculated as,

$$Flux = k_o * \left[c_w - \frac{c_A}{K_{aw}} \right] \qquad (3.30)$$

where c_A is the PAH air concentration obtained from the Integrated Atmospheric Deposition Network (IADN), k_o is the overall water-phase mass transfer coefficient, and K_{aw} is the air–water partition coefficient.

The results showed that polyethylene samplers are efficient and inexpensive samplers that provide good spatial and temporal resolution. The maximum PAH deposition occurred for phenanthrene (~ 17 pg/m^2 /d) near industrial sources in Sault Saint Marie, and the maximum volatilization for retene (0.21 pg/cm^2/d), a unique product of biomass burning, from the central lake.

All three processes of dry deposition, wet deposition, and air–water diffusive exchanges can play a role in the distribution of pollutants. For example, Jurado et al. (2005) studied PCBs and PCDD/Fs in the Atlantic Ocean and compared their findings with global data. They concluded that wet deposition was dominant during precipitation events in the Intertropical Conversion Zone and in low temperature regions. However, there was a measurable deposition with dry aerosols. Diffusive exchanges, especially deposition, were important over longer time spans. These included both rain and dry conditions, and could be dominant in many cases, for example for octachloro dibenzofuran (OCDF) in equatorial and southern hemisphere regions, and for PCB 180 in most regions, except the Arctic.

3.3   DEPOSITION AND GAS EXCHANGE OF ORGANIC CONTAMINANTS

While atmospheric inputs of pollutants by wet and dry deposition have been measured over the last 25 years, it is only recently that gas exchange of organics such as PCBs and PAHs have been quantified.

PCB Volatilization from Green Bay.   Using the methodology outlined earlier in this chapter, Achman et al. (1993) calculated PCB fluxes from measured concentrations in water and air in Green Bay, Wisconsin, over 14 days during June–October 1989. There was a net transfer of PCBs from the water to the air. The major factors controlling the flux were wind speed and dissolved PCB concentrations in the water (Figure 3.4). PCB volatilization rates for Green Bay range from 15 to 1300 $ng/m^2/d$ for wind speeds u_{10} of 1–6 m/s.

For Lake Michigan, the rate is 240 $ng/m^2/d$ at $u_{10} = 5$ m/s, and for Lake Ontario, 81 $ng/m^2/d$. These rates are quoted in Achman et al. (1993) based on reports by Strachan and Eisenreich (Lake Michigan) and Mackay (Lake Ontario).

The study by Hornbuckle et al. (1993) on atmospheric PCB concentrations confirmed that volatilization from natural waters is an important source of PCBs to the atmosphere. Atmospheric total PCB concentrations were 400 – 500 pg/m^3 over land near Green Bay at the following stations: University of Wisconsin-Green Bay, Peninsula State Park, and Fayette State Park. By contrast, the PCB concentration over southern Green Bay was 800 – 2200 pg/m^3, suggesting that the water is a source of PCBs. Over northern Green Bay, the PCB levels were at or slightly above land-based concentrations.

Among the PCB homologs, tri- and tetra-chlorinated compounds are particularly subject to volatilization, whereas the higher chlorinated homologs show no or little enrichment over the Bay. The importance of the more contaminated southern Green Bay as a source of PCBs is illustrated by the difference between over-water and over-land concentrations of tri- and tetra-chlorobiphenyls in three areas of the Bay. For southern Green Bay, this difference is 100 – 900 pg/m^3, for the central bay it is 50 – 200 pg/m^3, and it is less than 150 pg/m^3 for northern Green Bay.

PCB Volatilization from Contaminated Sediments.   Study of the volatilization of PCBs from Green Bay raises the question of whether PCB-contaminated sediments contribute to the volatilization. Qi et al. (2006) considered this possibility by modeling the fate of PCBs in a sediment–water–air system and comparing the results with those derived from a microcosm experiment. They used PCB congeners 15 (4,4′ DCB) and 153 (2,2′,4,4′,5,5′ HCB). The experimental system was used to measure PCB transfer between the three phases and to calibrate the model. In agreement with the Green Bay results, there was significant volatilization of DCB, but none for HCB. This applied to the first twelve days both with and without overlying water.

Only the upper 0.04 cm of sediment contributed to volatilization within the first 80 days. Increasing the water depth from 1 cm to 80 cm reduced the volatilization from 32 to 16% after 80 days. The experiments did not include mixing in the water column or sediment, both of which are likely to enhance volatilization in a real system.

PCBs and PAHs in Chesapeake Bay. In order to evaluate the impact of gas exchange of volatile organic contaminants on the overall mass balance for a lake or an estuary, one should consider loadings from wet and dry deposition and riverine inputs, in addition to the contribution from diffusive gas exchange. This was undertaken by Nelson et al. (1998) in a study of PCBs and PAHs in Chesapeake Bay.

Nelson et al. (1998) found that volatilization is the dominant removal process for PCBs from the bay, removing an estimated 400 kg/yr. This value is larger than the current external PCB loading (Table 3.4), suggesting that the release of PCBs from contaminated sediments supports volatilization from the bay. The average total PCB flux for Chesapeake Bay is 96 ng/m^2/d which is less than the 240 ng/m^2/d estimated for Lake Michigan. PCBs volatilized from Chesapeake Bay throughout the study period from March to September 1993, with the largest fluxes occurring in September and the smallest in June.

In contrast to PCBs, gas exchange is not a source of volatile PAHs to the air over Chesapeake Bay (Table 3.4; Figure 3.5). Instead, gas exchange appears be the largest external source to the bay of fluorene and phenanthrene, providing up to three times the combined loading from wet and dry deposition and from riverine inputs. For higher molecular weight PAHs, e.g. fluoranthene, pyrene, and benz(a)anthracene,

TABLE 3.4. Annual Organic Contaminant Loadings to and Losses from Chesapeake Bay (Source: Nelson et al., 1998)

Compound	Annual loadings (kg/yr)			Net gas deposition[a] (or volatilization)	
	Susquehanna River	Wet deposition	Dry aerosol deposition	µg/m^2/yr	kg/yr
Fluorene	122	16	12	300-(167)	379
Phenanthrene	450	63	92	70-719	2875
Anthracene	NA	6	6	31-(3.6)	130
Fluoranthene	1130	70	120	157-(13)	679
Pyrene	1030	75	109	76-(9.7)	361
Benz(a)anthracene	376	9	34	0.1-(1.9)	(4.6)
Chrysene	330	29	85	2.1-4.1	29
Total PCBs	165	13	20	(31-112)	(403)

[a]Net annual input to Bay, numbers in parentheses denote net annual losses from Bay.

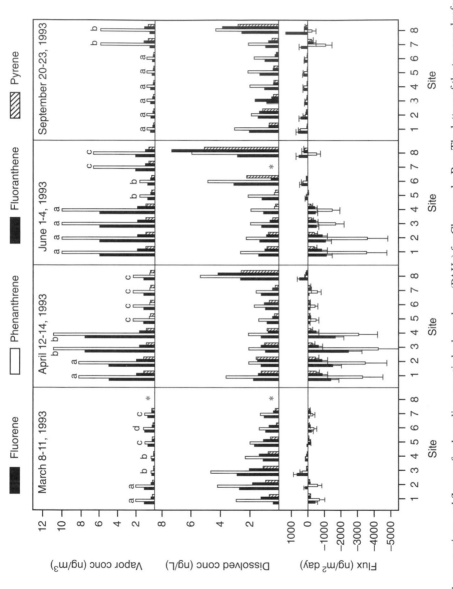

Figure 3.5. Concentrations and fluxes of polycyclic aromatic hydrocarbons (PAHs) for Chesapeake Bay. The letters of the top panel refer to segments of the Bay near relevant sampling sites with a indicating the northernmost segment by Susquehanna River, and subsequent segments b, c, d, and e further south in that order, with segment e by the Atlantic Ocean. (*) no data available. PAH fluxes are calculated from the vapor and water concentrations. A negative flux indicates absorption. (Source: Reprinted from Nelson et al. (1998) with permission from American Chemical Society.)

86

the tributary input is dominant. Absorption from the atmosphere is more important than wet and dry deposition for fluorene, phenanthrene, anthracene, fluoranthene, and pyrene. However, for the high molecular weight compounds benz(a)anthracene and chrysene, dry aerosol deposition is more significant than absorption.

From Figure 3.5 it is seen that dissolved PAH concentrations are high at site 8, near Norfolk, Virginia, which is characterized by contaminated sediments. There is net absorption during April and June and volatilization from warmer waters in September. The high vapor phase concentration of fluoranthene and phenanthrene during April and June may be influenced by winds from the Baltimore–Washington area.

Nelson et al.'s (1998) study demonstrates that gas exchange of organic pollutants can be an important source (PAHs) or sink (PCBs) of organic contaminants to a water body. Dry deposition of selected elements to Chesapeake Bay was considered by Wu et al. (1994).

Organics in Izmir Bay, Turkey and Taihu Lake, China. Cetin and Odabasi (2007) investigated air–water exchange and dry deposition of polybrominated diphenyl ethers (PBDEs) in Izmir Bay, Turkey. BDE-209 was the dominant congener in water samples, followed by BDE-99 and BDE-47. PBDE gas fluxes mainly consisted of deposition fluxes for BDE-47, -100, -153, -154, and -209. Maximum values were recorded for BDE-209 ($22 - 28$ ng/m^2/d) while fluxes for other congeners measured between 1 and 18 ng/m^2/d. Congeners BDE-28, -47, and -99 underwent volatilization during summer. Dry deposition of PBDEs can be quite significant, especially for congeners BDE-99 and -209, with average fluxes in summer of 62 ± 30 and 89 ± 47 ng/m^2/d respectively.

A study of seasonal air–water exchange of the organochlorine pesticides α-hexachloro cyclohexane (α-HCH), chlordane, o,p'-DDT, and α-endosulfan in Taihu Lake, China was conducted by Qiu et al. (2008). They found volatilization of α-HCH ($30 - 100$ ng/m^2/d) and *trans*-chlordane (up to 600 ng/m^2/d), particularly during summer, while there was a net deposition of o,p'-DDT and α-endosulfan with maximum fluxes of ~ 100 ng/m^2/d and ~ 25 ng/m^2/d respectively, in the same season. Because α-HCH had been banned since the late 1980s, Qiu et al. estimated that lake sediments were the source at the time of their study. Chlordane may have originated from upstream manufacturing plants while the deposition of o,p'-DDT and α-endosulfan could be a result of atmospheric transport from land sources.

3.4 MARINE AND FRESHWATER MICROLAYERS

Rivers, lakes, and oceans are characterized by a surface microlayer of thickness between 0.01 and 1 mm in which there is a concentration enrichment of trace metals, pesticides, PCBs, petroleum products, silicones, lipid surfactants, natural surfactants, and microorganisms (Elzerman, 1981). Organic carbon, organic nitrogen, total phosphorus, and particulate matter are also present in higher concentration in the surface microlayer than in the bulk water. Because of the thinness of the microlayer, the mass of contaminants, i.e. surface excess measured in $\mu g/m^2$, is usually very small compared to the total mass in the water column. The importance of the surface microlayer is, therefore, related to microlayer processes and interactions.

The enrichment factor for a substance is the concentration of this substance in the surface microlayer divided by its concentration in the bulk water. Where slicks are present, the film pressure > 1 dyn/cm (10^{-3} N/m) and the enrichment factor is typically between 1.03 and 5.7 for trace metals, organically bound N and P, and for particles. Slicks are areas of capillary wave action resulting from surface active materials. In non-slick areas, the film pressure is less than 1 dyn/cm. The residence time for metals in the surface microlayer is rather short, typically in the range of $1-1100$ min, as for Al, Fe, Mn, Pb, Zn, Cd, and Cr in Lake Superior (Elzerman, 1981).

There can be *in-situ* production in the microlayer of biologically or photochemically derived organic molecules, and higher organisms such as fish may be attracted to the microlayer where they graze on the plankton. Transport of material to or from the microlayer may take place through tributary inputs, sedimentation, atmospheric deposition, diffusion, or bubble flotation. Sorption processes or *in-situ* generation can also play a role.

Hunter and Liss (1981) define the surface active layer as the top $100-300$ μm of the water column. While dissolved organic compounds may be enriched in the microlayer if they are surface active, there must be another mechanism at work for dissolved inorganic species. The enrichment from electrostatic attraction of cations to a negatively charged organic surfactant in the microlayer is negligible compared to the concentrations of major cations in seawater, i.e. Mg^{2+}, Ca^{2+}, Na^+, K^+, and $MgHCO_3^+$. Instead, organic surfactants act as surface ligands for dissolved trace metals. These ligands possess selectivity for trace ions, e.g. Cu and Zn, to lead to enrichment despite high Mg^{2+} and Ca^{2+} concentrations in seawater. Enrichment factors for freshwater microlayers are slightly higher than for seawater microlayers. Adsorption of dissolved inorganic species to particulate matter in the microlayer constitutes an alternative enrichment mechanism.

Sampling devices for surface microlayers include screen ($150 - 300$ μm), glass plate ($60 - 100$ μm), rotating drum (~ 60 μm), and bubble microtome ($0.5 - 10$ μm). The approximate depths sampled with each device are indicated in parentheses. The rotary drum consists of a buoyant, ceramic cylinder that is rolled along the water surface during sample collection.

Meyers et al. (1981) found that transition metals are preferentially enriched in the surface microlayer, largely in the particulate phase. They examined organic carbon, fatty acids, hydrocarbons, and heavy metals in the particulate phase of the surface microlayer and 1 m below the surface in Lake Michigan. The particulate organic carbon is controlled by physical processes such as turbulent mixing and settling. As suggested by Figure 3.6, tributary inputs of particulate organic carbon are important. However, the microlayer enrichment is greatest for the open lake. Particulate fatty acids have a distribution similar to the one shown for particulate organic carbon in Figure 3.6, except that the fatty acid concentrations are ~ 100 times lower, and there is a smaller concentration gradient from the river plume to the open lake. One reason for this is that the particulate fatty acids are more likely to be generated *in-situ* by neustonic and planktonic communities.

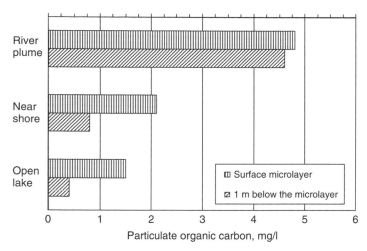

Figure 3.6. Particulate organic carbon in the microlayer vs. bulk water in different aquatic environments. (Source: Adapted from Meyers et al. (1981))

Note that enrichment factors can be less than 1 down plume from a river mouth (St. Joseph River) as a result of particles settling through the water column (Meyers et al., 1981). The top of the water column in this area is relatively free of particles because they are no longer supplied by the river.

Surface enrichment of particulate material was also found to control copper enrichment at the air–sea interface, according to Lion and Leckie (1981). In addition, their results indicate surface enrichment of bacteria. Uptake of materials by bacteria is therefore another mechanism of surface enrichment. There was no surface enrichment of dissolved Cu or dissolved organic carbon, but significant concentration increases of particulate copper. Rising bubbles can contribute to the surface enrichment of particulate trace metals and particulate organic carbon. Trace metals such as Cu can also adsorb onto solid surfaces, which in the presence of surface active organics are carried to the surface by rising bubbles.

In summary, the major mechanisms responsible for surface enrichment of trace metals are trace metal binding with organic surface active ligands, adsorption to solid surfaces, and uptake by microorganisms. For organic compounds, similar mechanisms are relevant, except that binding with surface active ligands may be replaced by complex formation between the organic compounds and surface active organic materials.

3.5   CASE STUDY: EMISSION OF VOCs FROM WASTEWATER TREATMENT PLANTS

Aeration basins of activated sludge plants can be a source of volatile organic compounds (VOCs). Air emission occurs by means of volatilization from the surface or stripping caused by bubble aeration. The degree of stripping of a VOC for a completely mixed continuous flow reactor depends on the mass transfer coefficient K_{ol} of the VOC, the liquid and gas flows, Q_L and Q_G respectively, and on the Henry's law constant H (dimensionless) of the compound (Matter-Müller et al., 1981).

We distinguish between the following cases. Case 1: For low Henry's law constant H and low air flow, there is a greater degree of stripping for higher H. The mass transfer coefficient K_{ol} is not important in this case. Case 2: On the other hand, for high Henry's law constant or high air flow, stripping increases with higher mass transfer coefficient K_{ol} a where a (cm^{-1}) is the interfacial area per unit volume of liquid. In this case, the Henry's law constant is not important.

Surface aeration gives the highest degree of stripping. Here, Q_G/Q_L is undefined but high, and this corresponds to the second case. The concentration of VOCs in the gas phase is similar to that of the ambient air.

For bubble aeration, $Q_G/Q_L > 2$, reflecting an intermediate case between the first and the second. Stripping is smaller in this case than for surface aeration and off gases with small Henry's law constants ($H < 3$) come close to equilibrium with the wastewater. Trickling filters with forced aeration measure $Q_G/Q_L > 30$; therefore they display higher stripping than bubble aeration. This is also an intermediate case between the first and the second.

Roberts et al. (1984) developed a methodology for estimating removal of VOCs, particularly halogenated compounds from aeration basins of wastewater treatment plants. The compounds considered were dichlorodifluoromethane CCl_2F_2, carbon tetrachloride CCl_4, tetrachloroethylene $HClC = CCl_2$, and chloroform $CHCl_3$. These compounds were chosen for their volatility. They display low rates of both adsorption to particles and biodegradation. Only trichloroethylene ($10 - 22$ μg/L), and to some extent chloroform ($2.4 - 4.4$ μg/L), are usually found in actual influents of wastewater treatment plants.

From their model, Roberts et al. (1984) find greater transfer to the atmosphere in surface aeration because of the partial saturation of the gas phase in bubble aeration. This is also clear from Equation 3.20, which shows $c^* \sim 0$ in surface aeration, whereas in bubble aeration $c^* > 0$. For low Henry's law constant H, the gas in the bubbles may be nearly saturated. This result is similar to the findings of Matter-Müller et al. (1981).

For surface aeration, the VOC removal is mainly a function of the specific oxygen transfer requirement (g O_2 transferred/m^3 wastewater treated), whereas for bubble aeration the VOC removal is additionally dependent on the Henry's law constant and the oxygen transfer efficiency. Low oxygen transfer efficiency gives a high rate of VOC removal. The partial saturation of the atmospheric boundary layer above the aeration basin, which would decrease VOC removal, is small ($\leq 5\%$), indicating that VOC removals are not likely to be influenced by this factor (Roberts et al., 1984).

Namkung and Rittmann (1987) consider biodegradation in addition to adsorption and volatilization. Biodegradation rates for benzene, dhlorobenzene, dthylbenzene, and toluene, determined from laboratory experiments, are in the range $0.2 - 0.4$ m^3/(gVSS d), which is an order of magnitude higher than the rates estimated by Zhang et al. (1990) for a high degree of biodegradation, $0.01-0.05$ L/(mg MLVSS d). Their model for the fate of VOCs in aeration basins of treatment plants is based on the assumption of saturation of the exit gas; thus the removal depends on the Henry's law constant. This assumption would appear to be a simplification in view of the above findings of Matter-Müller et al. (1981) and Roberts et al. (1984). Adsorption is found to play an insignificant role ($\leq 1\%$) in the removal of VOCs.

In their evaluation of air emission of VOCs, Namkung and Rittmann (1987) consider two major wastewater treatment plants in Chicago, Illinois: The west-southwest $(3, 164, 300 \ m^3/d$ or 836 mgd) and the Calumet $(866, 800 \ m^3/d$ or 229 mgd) plants. Typical VOCs in the influent of these plants and their concentrations are methylene chloride $(10 - 48 \ \mu g/L)$, toluene $(22 - 48 \ \mu g/L)$, trichloroethylene $(10 - 22 \ \mu g/L)$, benzene $(23 \ \mu g/L)$, and ethylbenzene $(18 \ \mu g/L)$. All of these, except trichloroethylene, are estimated to undergo a significant degree of biodegradation $(\sim 95\%)$.

Zhang et al. (1990) considered air emissions of VOCs, both from stripping and volatilization, in addition to adsorption and biodegradation. Thus, as for Equation 3.21, the air speed above the aeration basin becomes a factor in addition to solute concentration in water. The biodegradation index $I_{bd} = 10^4 \times k$ where k (L/(mg MLVSS d)) is a second order or a pseudo first order degradation constant. Three ranges were defined: $I_{bd} = 1 - 20$ (low, L), $20-100$ (intermediate, I), and $100-500$ (high, H). Biodegradation indices were determined from a correlation of known degradation indices $\log I_{bd}$ with BCF/K_{oc} where BCF = bioconcentration factor, and K_{oc} = organic carbon partition coefficient. The parameters BCF and K_{oc} were determined by numerical estimation methods contained in the graphical exposure modeling system PCGEMS. These methods are available on the internet through the EPA Estimation Programs Interface (EPI) Suite (U.S. EPA, 2013).

Table 3.5 lists several properties of organics of various degrees of volatility. There is qualitative agreement with Namkung and Rittmann (1987) on the relatively high biodegradation potential of ethylbenzene and toluene. However, their prediction of high degradability of chlorobenzene contrasts with Zhang et al.'s (1990) estimate of low biodegradation for this compound. The low biodegradation of trichloroethylene (Table 3.5) is consistent with the assumption of Roberts et al. (1984). The three categories of biodegradation H, I, and L and in general agreement with probabilities of biodegradation obtained with the non-linear model of the Biowin6 program contained in the 4.11 version of the EPI Suite. This model was developed for the Japanese Ministry of International Trade and Industry (MITI) ready biodegradation test (Tunkel et al., 2000).

From Figure 3.7 it is seen that VOCs with a Henry's law constant $> 10^{-3}$ atm m^3/mole have a high potential for air emission, but that high biodegradability will reduce the air emission, typically from 65 to 30% (H = 10^{-3} atm m^3/mole). Other variables that influence the air emission include flow, diffusivity in water, and the exact value of the Henry's law constant. Moreover, plant conditions such as air flow rate play a role in determining the actual emission of VOCs to the atmosphere.

There are few comparisons of calculated VOC removals in aeration basins with actual plant data. However, estimates for toluene and methylene chloride fractions (%) that remain in the effluent of two treatment plants in Milwaukee, Wisconsin, the Jones Island (JI) and South Shore (SS) treatment plants (Zhang et al., 1990), are available. For the JI plant, the calculated percentages of toluene and methylene chloride

TABLE 3.5. Partial Listing of Contaminants in NR-445 with Basic Properties at 25°C Estimated from EPI Suite (Source: U.S. EPA, 2013)[a]

Contaminants	CAS No.	MW	H atm-m^3/mole (10^{-5})	D cm^2/s (10^{-5})	Log K_{ow}	BD
Acrolein	107-02-8	56	12.2	1.25	−0.01	H
1,2,4-Trichloro- benzene	120-82-1	182	142	0.75	4.02	I
1,2-Dichloro-ethylene	540-59-0	97	408	1.12	1.86	I
Chlorobenzene	108-90-7	113	311	0.89	2.84	L
Ethylbenzene	100-41-4	106	788	0.80	3.15	H
Trichloro-ethylene	79-01-6	131	985	0.99	2.42	L
Methylene chloride	75-09-2	85	325	1.26	1.25	H
Toluene	108-88-3	92	664	0.89	2.73	I/L
Vinyl chloride	75-01-4	63	2780	1.26	1.62	I
Acrylonitrile	107-13-1	53	13.8	1.24	0.25	H
Di (2-ethylhexyl) phthalate	117-81-7	391	0.027	0.37	7.6	L
o-Toluidine	95-53-4	107	0.198	0.83	1.32	L
Benz(a)anthracene	56-55-3	228	1.2	0.57	5.76	L
Benz(a,h)acridine	226-36-8	280	0.00182	0.53	5.63	L
Glycidol	556-52-5	74	0.0545	1.08	−0.95	H
Hydroquinone	123-31-9	110	4.73E-06	0.82	0.59	H
Ethyl butyl ketone	106-35-4	114	9.08	0.74	1.73	I
Morpholine	110-91-8	87	0.116	0.99	−0.86	I

[a]H = Henry's law constant; D = Diffusivity in water (Zhang et al., 1990); K_{ow} = Octanol−water partition coefficient; BD = Biodegradability; H = High, I = Intermediate, L = Low

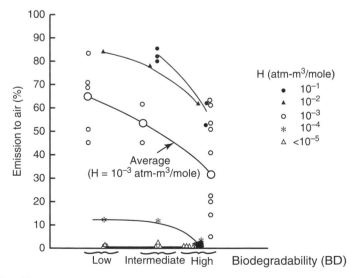

Figure 3.7. Effect of Henry's law constant (H) and biodegradability on the emission (%) to air for Jones Island Wastewater Treatment Plant. (Source: Zhang et al. (1990) with permission from ASCE.)

being removed to the effluent are 11 (2) and 14 (54) respectively, where measured values are given in parenthesis. A better agreement is found for the SS plant, where the corresponding percentage removals to effluent are 20 (13) and 23 (22). Thus, it appears that computer estimation of air emission is feasible as a screening tool.

3.6 THE FUGACITY MODEL

Although the concept of fugacity was introduced in 1901 (Lewis, 1901), it is only since the mid-1970s that it has been effectively applied to the distribution of semi-volatile or volatile organics in the environment (Mackay, 1979; Mackay and Paterson, 1981). Fugacity is the tendency of a substance to escape from one phase to another. The fugacity model is a thermodynamic approach to environmental transport and fate assessment.

The fugacity approach simplifies certain calculation scenarios for fate of chemicals in a multimedia environment. Its strong point is comparison of fate of different chemicals in identical environments. However, these environments or "unit worlds" are abstractions that are rarely achievable. For example, should 1 cm or 10 cm sediment depth be considered, and what is the appropriate height of the atmosphere? What are the details of advective flows? When specific environmental conditions are considered, such as the quantitative estimation of volatilization of HOCs from a wastewater treatment plant or HOC gas exchange in Green Bay or Chesapeake Bay, one is usually left with alternative estimation methods, including parameters such as aeration versus liquid flow in the treatment plant, and wind speed and air–water mass transfer coefficients for the estuary.

The early development of fugacity modeling is summarized in Mackay et al. (1992a). Later, Mackay and co-workers (1996a, 1996b, 1996c) expanded the scope of the model to include chemicals that are insoluble in water, such as long-chain hydrocarbons, silicones, and polymers. In the multimedia equilibrium criterion (EQC) model (2011), fugacity was used as one of the criteria, along with aquivalence, which applies to cations, anions, and involatile organic chemicals. There are four levels of complexity with regard to equilibria and steady-state conditions (see Section 3.6.2), and the level-III module has been incorporated into the widely used Estimation Programs Interface (EPI) Suite developed by the U.S. EPA (2011). More recent advancement, as well as challenges in the applications of the fugacity concept, is discussed in Mackay and Arnot (2011) and Hughes et al. (2012).

Many, but not all, real world settings with defined system boundaries can be approximated by fugacity models with suitable data. For example, aspects of gas exchange between the atmosphere and water may be modeled by a fugacity model but dry or wet deposition of chemicals to water bodies are less easy to model. Fugacity models are also limited in the modeling of turbulent transfer of chemicals between the atmosphere and water as described above.

Although a simplification relative to the real environment, one may argue that an attractive feature of the fugacity model is the use of a well-defined evaluative environment, of which designs with different dimensions and numbers of compartments may be presented (Mackay, 2001). Such a "unit world" allows quick comparisons among chemicals under the same environmental conditions, facilitating the design and screening of chemicals based on expected environmental behavior and impact.

The purpose of this section is to provide an introduction of the basic concept and model process for simple applications. Readers are referred to the book and articles by Mackay and co-workers for more thorough descriptions of the theory and examples of applications.

3.6.1 Fugacity Definitions and Basic Equations

Fugacity represents the tendency of a substance to flee from one phase to another. It is a thermodynamic property analogous to temperature in a thermal system, and can be described by the equation,

$$c_H = T \ Z_H \tag{3.31}$$

where
   c_H = concentration of heat, JL^{-3}
   Z_H = volumetric heat capacity, $JL^{-3}T^{-1}$
   T = absolute temperature, T

For fugacity one can similarly write

$$c = f \ Z_f \tag{3.32}$$

where
   c = solute concentration, ML^{-3}
   Z_f = fugacity capacity, $L^{-2}t^2$
   f = fugacity, $ML^{-1}t^{-2}$

In these equations, J = energy (Joule), L = length, M = moles or kg, and t = time (s).

For a closed system in thermal equilibrium all objects have the same temperature T. Similarly, for a closed environmental system in equilibrium, a chemical would have the same fugacity f in all compartments (air, water, etc.) of the system. Thus for two compartments 1 and 2 we may write

$$f_1 = f_2, \tag{3.33}$$

or

$$\frac{c_1}{Z_{f,1}} = \frac{c_2}{Z_{f,2}}; \ \frac{c_1}{c_2} = \frac{Z_{f,1}}{Z_{f,2}} = K_{1,2} \tag{3.34}$$

where
   $K_{1,2}$ is a partition coefficient for the given chemical between phases 1 and 2.

Fugacities of a solute in the vapor and aqueous phase respectively, f_v and f_a may be expressed as

$$f_v = \alpha_f X_v P_T = P,$$ (3.35)

$$f_a = \alpha_a X_a f_p$$ (3.36)

where

X_v = mole fraction of the solute in the vapor phase
X_a = mole fraction of the solute in the aqueous phase
P_T, P = total and component pressures
f_p = reference fugacity relating to the pure liquid solute
α_f = dimensionless fugacity coefficient accounting for non-ideal behavior.

Except for solutes that associate with one another in vapor phases, its value is close to one at atmospheric pressure (Weber and Digiano (1996))

α_a = aqueous phase solute activity coefficient

Henry's law is concerned with the equilibrium between low mole fractions of organic substances dissolved in water and the corresponding fugacity f_i or vapor pressure P_i,

$$f_i = K_H X_{i,l}$$ (3.37)

where

$f_i \cong$ partial pressure of a compound in a solution mixture $= P_i$
K_H = Henry's law constant
$X_{i,l}$ = mole fraction of a compound in a solution mixture

By contrast, Raoult's law expresses a linear relationship between the fugacity f_i and the generally high mole fraction of organic compounds in a liquid mixture such as gasoline,

$$f_i = f_i^o X_{i,l}$$ (3.38)

where

f_i^o = vapor pressure of compound in the pure form

An application of these laws is shown in Sawyer et al. (2003: 26). In this case the vapor pressure of benzene over gasoline consisting of an equimolar mixture of benzene, toluene, ethylbenzene, and xylene (BTEX) is determined from Raoult's law, with $X_{i,l} = 0.25$ for each chemical. The resulting vapor pressure for benzene 0.0315 atm is used with Henry's law constant for benzene $K_H = 0.0055$ atm m^3/mole (Table 3.6) to determine an aqueous benzene concentration of 5.73 mole/m^3 or 447 mg/L.

TABLE 3.6. Physical-Chemical Properties of Selected Organic Chemicals

Compound	Formula	T_m (°C)	T_b (°C)	C_s (mg/L) at' 25°C	p^v (Pa) at' 25°C	K_H (Pa m³/mole) at' 25°C	Log K_{ow} at' 25°C	Mackay et al. (1992b,1992c, 1997) Vol, p
Carbon tetrachloride	CCl_4	−22.9	-	800	15,250	2989	2.64	III, 621
Methylene chloride	CH_2Cl_2	−95	-	13,200	26,222	300	1.25	III, 621
Vinyl chloride	C_2H_3Cl	−153.8	-	2763	354,600	2685	1.38	III, 622
Trichloroethylene	C_2HCl_3	−73	-	1100	9900	1034	2.53	III, 622
Tetrachloroethylene	C_2Cl_4	−19	-	150	2415	1733	2.88	III, 622
Methyl-t-butyl ether	$C_5H_{12}O$	−109	55.2	42,000	33,500	70.31	0.94	III, 823
Aniline	C_6H_7N	−6.3	184.4	36,070	65,019	0.168	0.90	IV, 898
Benzene	C_6H_6	5.53	80.1	1780	12,700	557	2.13	I, 141
Phenol	C_6H_6O	41	182	88,360	47	0.0539	1.46	IV, 429
Chlorobenzene	C_6H_5Cl	−45.6	132.2	484	1580	368	2.80	I, 269
Pentachlorophenol	C_6HCl_5O	174	-	14	0.00415	0.0790	5.05	IV, 429
Toluene	C_7H_8	−95	110.6	515	3800	680	2.69	I, 141
Ethylbenzene	C_8H_{10}	−95	136.2	152	1270	887	3.13	I, 141
o-Xylene	C_8H_{10}	−25.2	144	220	1170	565	3.15	I, 141
Atrazine	$C_8H_{14}ClN_5$	174	-	30	4.00×10^{-5}	2.88×10^{-4}	2.75	V, 258
Malathion	$C_{10}H_{19}O_6PS_2$	2.9	-	145	1.00×10^{-3}	2.88×10^{-3}	2.8	V, 569
Parathion	$C_{10}H_{14}NO_5PS$	6	-	12.4	6.00×10^{-4}	0.0141	3.8	V, 570
Chlordane (*cis*- or α-)	$C_{10}H_6Cl_8$	107–109	-	0.056	2.65×10^{-3a}	0.342	6.0	V, 567
PCB-18 (25−2)	$C_{12}H_7Cl_3$	44	-	0.4	0.143	92.2	5.6	I, 596
PCB-70 (25−34)	$C_{12}H_6Cl_4$	104	-	0.041	0.006	20.26	5.9	I, 449
Anthracene	$C_{14}H_{10}$	216.2	340	0.045	0.001	3.96	4.54	II, 251
Benzo(a)pyrene	$C_{20}H_{12}$	175	495	0.0038	7.00×10^{-7}	0.046	6.04	II, 251

[a]Subcooled liquid.

The total mass M of the chemical in all compartments (phases) may be written as

$$M = \sum m_i = \sum c_i V_i \tag{3.39}$$

or

$$M = \sum f_i Z_{f,i} V_i \tag{3.40}$$

where m_i is the mass of the chemical in phase i of volume V_i. For a closed system in equilibrium,

$$f_1 = f_2 = \cdots = f_n = f \tag{3.41}$$

Thus, Equation 3.40 may be written,

$$M = f \sum Z_{f,i} V_i,$$

or

$$f = \frac{M}{\sum Z_{f,i} V_i} \tag{3.42}$$

The fugacity of the system is, therefore, equal to the total mass or number of moles of the chemical, divided by the sum of the fugacity capacities, multiplied by the phase volumes over all phases.

Vapor Phase.   For the vapor phase, we obtain from eqs. 3.32 and 3.35 and the ideal gas law,

$$c_v = f_v Z_{f,v} = \frac{n}{V} = \frac{P}{RT} = \frac{f_v}{RT} \tag{3.43}$$

where c_v is the gas phase concentration of a solute, $Z_{f,v}$ is the corresponding fugacity capacity, n is the number of moles, V is the volume, and R is the universal gas constant. Thus the fugacity capacity may be expressed as

$$Z_{f,v} = \frac{1}{RT} \frac{\text{mole}}{\text{atm} \times \text{m}^3 \text{ air}} \tag{3.44}$$

Aqueous Phase.   Considering partitioning of a solute between the aqueous and vapor phase, we may write the aqueous phase fugacity $Z_{f,a}$ as

$$Z_{f,a} = \frac{Z_{f,v}}{K_{v,a}}; \quad K_{v,a} = \frac{Z_{f,v}}{Z_{f,a}} = \frac{c_v}{c_a} \tag{3.45}$$

where $K_{v,a}$ is a partition coefficient for partitioning of a solute between the vapor and aqueous phases. Rewriting this equation we obtain,

$$Z_{f,a} = \frac{Z_{f,v} RT}{K_H} = \frac{1}{K_H} \frac{\text{mole}}{\text{atm} \times \text{m}^3 \text{ water}} \tag{3.46}$$

Pure Solids and Liquids. The fugacity capacity for pure solids and liquids $Z_{f,p}$ is, according to Equation 3.32 and 3.36, given by

$$Z_{f,p} = \frac{c_p}{f_p} = \frac{1}{P^v V_{mo,p}} \qquad \frac{\text{mole}}{\text{atm} \times \text{m}^3 \text{ pure phase}} \tag{3.47}$$

where the subscript p refers to the pure phase. Thus, f_p is the fugacity of the pure compound, and therefore vapor pressure P^v of the compound, and $V_{mo,p}$ is its molar volume. This is based on Raoult's law, which mandates that the activity coefficient $\alpha_p = 1$ for pure solids and liquids.

The partition coefficient for partitioning of a solute between the vapor and its pure state $K_{v,p}$ is given by,

$$K_{v,p} = \frac{Z_{f,v}}{Z_{f,p}} = \frac{P^v V_{mo,p}}{RT} \tag{3.48}$$

In a similar way, the partition coefficient for a solute between the aqueous and its pure state $K_{a,p}$ may be expressed as

$$K_{a,p} = \frac{Z_{f,a}}{Z_{f,p}} = \frac{P^v V_{mo,p}}{K_H} = c_s V_{mo,p} \tag{3.49}$$

where c_s is the water solubility of the solute. For liquid solutes, the aqueous solubility $(c_s)_{\text{liquid solute}}$ may be expressed as,

$$(c_s)_{\text{liquid solute}} = c_{s(l)} = \frac{1}{V_{mo,a}\alpha_a} \tag{3.50}$$

where $V_{mo,a}$ is the aqueous phase molar volume. Henry's law states that the ratio between vapor pressure and solubility is the same for liquid and solid solutes. Thus we have

$$(c_s)_{\text{solid solute}} = c_{s(s)} = \frac{P^v_{(s)}}{P^v_{(l)} V_{mo,a}\alpha_a} \tag{3.51}$$

from which the solubility of solid solutes may be calculated.

Sorbed Phases. For suspended matter, soil, and sediment in the sorbed phase the condition of equal fugacities may be written as

$$f_a = \frac{c_e}{Z_{f,a}} = f_s = \frac{q_e \rho_s}{Z_{f,s}} . \tag{3.52}$$

where the subscript s stands for sorbed phase, c_e is the equilibrium concentration of solute in the aqueous phase, q_e is the mass of solute per mass of sorbent, and ρ_s is the bulk density of solids. The concentration of solute in the sorbed phase c_s may therefore be written as

$$c_s = q_e \rho_s \tag{3.53}$$

$$\frac{\mu g \text{ solute}}{mL \text{ sorbent}} = \frac{\mu g \text{ solute}}{g \text{ sorbent}} \frac{g \text{ sorbent}}{mL \text{ sorbent}}$$

In the linear model for the relationship between q_e and c_e we have,

$$q_e = K_D c_e \tag{3.54}$$

$$\frac{\mu g \text{ solute}}{g \text{ sorbent}} = \frac{mL \text{ water}}{g \text{ sorbent}} \frac{\mu g \text{ solute}}{mL \text{ water}}$$

Thus, from Equation 3.52 we obtain

$$Z_{f,s} = \frac{q_e \rho_s}{f_s} = \frac{q_e \rho_s Z_{f,a}}{c_e} = \frac{K_D \rho_s}{K_H} \qquad \frac{mole}{atm \times mL \text{ sorbent}} \tag{3.55}$$

Biota. Assuming equal fugacities in the aqueous phase and biota, we have

$$f_a = f_b = \frac{q_e \rho_b}{Z_{f,b}} \tag{3.56}$$

where the subscript b refers to biota. The concentration of solute in biota is then,

$$c_b = q_e \rho_b \qquad \frac{\mu g \text{ solute}}{g \text{ biomass}} \frac{g \text{ biomass}}{mL \text{ biomass}} \tag{3.57}$$

The bioconcentration factor BCF is defined by an equation similar to Equation 3.52:

$$q_e = BCF \times c_e \tag{3.58}$$

$$\frac{\mu g \text{ solute}}{g \text{ biomass}} = \frac{mL \text{ water}}{g \text{ biomass}} \frac{\mu g \text{ solute}}{mL \text{ water}}$$

Equations 3.56 and 3.32 allow us now to calculate the fugacity capacity of biota,

$$Z_{f,b} = \frac{q_e \rho_b}{f_b} = \frac{q_e \rho_b}{c_e} Z_{f,a} = \frac{BCF \, \rho_b}{K_H} \qquad \frac{mole}{atm \times mL \text{ biomass}} \tag{3.59}$$

Note that this is similar to the expression for the fugacity capacity for sorbed phases, Equation 3.55.

The partition coefficient K_D and the bioaccumulation factor BCF may be expressed in terms of the organic carbon partition coefficient K_{oc} (mL water/g organic carbon) and the octanol water partition coefficient K_{ow} (mL water/mL octanol) as follows

$$K_D = f_{oc}K_{oc} \tag{3.60}$$

$$\mathrm{BCF} = f_{lipid}K_{ow} \tag{3.61}$$

where f_{oc} (g organic carbon/g sediment) is the organic carbon fraction of sediment and f_{lipid} (g lipid/g biomass) is the lipid fraction of biota. Values of K_{oc} in Equation 3.60 may be determined from the approximate relationship (Karickhoff, 1981),

$$K_{oc} \cong 0.41\,K_{ow} \tag{3.62}$$

In Equation 3.61 it is assumed that the solute has a similar affinity for lipid and octanol and $g_{lipid}/mL_{octanol} \approx 1$.

Octanol. The fugacity capacity of a solute in octanol saturated with water $Z_{f,o}$ is given by

$$Z_{f,o} = \frac{1}{V_{mo,o}\alpha_o P^v} \tag{3.63}$$

where $V_{mo,o}$ is the molar volume of the solute in octanol, α_o is the activity coefficient of the solute in octanol, and P^v is the vapor pressure of liquid solute at system temperature. Considering equilibrium of a solute between the octanol and aqueous phase, K_{ow} may be written as,

$$K_{ow} = \frac{c_o}{c_a} = \frac{Z_{f,o}}{Z_{f,a}} = \frac{V_{mo,a}\alpha_a P^v}{V_{mo,o}\alpha_o P^v} = \frac{V_{mo,a}\alpha_a}{V_{mo,o}\alpha_o} \tag{3.64}$$

where c_o and c_a are the octanol and aqueous phase concentrations respectively of the solute.

Fraction of Solute in Aqueous and Sorbed Phase. Consider a volume V_{tot} of water with suspended solids. The total mass of a solute M_{tot} may be written,

$$M_{tot} = c_w V_w + K_D c_w m_s \tag{3.65}$$

where c_w (μg solute/mL water) is the dissolved concentration of the solute, V_w is the volume of water, and m_s is the total mass of suspended solids. Making the reasonable assumption that $V_w \approx V_{tot}$ and expressing the total mass of suspended solids as $m_s = c_{ss} V_{tot}$ where c_{ss} (g sed/mL water) is the concentration of suspended material, we obtain

$$M_{tot} = c_w V_{tot} + K_D c_w c_{ss} V_{tot} \tag{3.66}$$

The fraction α of solute in the water phase is therefore,

$$\alpha = \frac{c_w V_{tot}}{c_w V_{tot} + K_D c_w c_{ss} V_{tot}} = \frac{1}{1 + K_D c_{ss}} = \frac{1}{1 + f_{oc} K_{oc} c_{ss}} \quad (3.67)$$

The remaining fraction is the sorbed phase

$$1 - \alpha = \frac{f_{oc} K_{oc} c_{ss}}{1 + f_{oc} K_{oc} c_{ss}} \quad (3.68)$$

Fugacity Ratios. In the comparison of solubilities or vapor pressures of organic compounds, the values for the solid phase depend on crystal structure. In order to remove this consideration, the properties of the liquid phase are often used. If necessary, subcooled state is used to obtain a value for the liquid phase if the material is at a temperature at which it is normally in the solid state. As an example, aerosol–air partitioning is often described in terms of the subcooled liquid phase. The relationship between solid and liquid vapor pressures and solubility in water is defined by the fugacity ratio, F (Van Noort, 2006),

$$F = \frac{P_S}{P_L} = \frac{C_S}{C_L} = \exp\left(-\frac{\Delta G_{S \to L}}{RT}\right) \quad (3.69)$$

where P and C indicate vapor pressure and solubility in water and subscripts S and L refer to the solid and liquid state respectively. $\Delta G_{S \to L}$ is the Gibbs free energy change for the solid-to-liquid transition at the given temperature. Note that the Gibbs free energy for melting ΔG_m ($\Delta G_{S \to L}$ at melting point temperature) is zero. Van Noort (2006) reviewed methods for the calculation of fugacity ratios based on fusion thermodynamic data and provided estimated values for all 209 PCB congeners at 298 K. Most F-values are in the range between 0.1 and 0.9 with high values for low molecular weight compounds and low values for highly substituted PCBs.

The equations presented above should be sufficient for level-I calculations. Additional parameters are needed for more complex cases, as briefly introduced in Section 3.6.2.

3.6.2 Levels of Complexity

Four levels of complexity with regards to the equilibrium and steady-state conditions can be pursued in the fugacity model. A system at equilibrium with regard to a specific chemical is one that has the same fugacity of the chemical in all phases. A steady-state condition means that all the state variables of the system do not change with time, and this requires that the input rate equals the output (including internal elimination via transformation reactions) rate for the chemical in each of the compartments.

In reality, any combination of equilibrium and steady-state conditions is possible. In fugacity modeling, the system can be examined under the conditions of equilibrium with no reaction and no inflow or outflow (level I), steady-state and equilibrium (level II), steady-state but non-equilibrium (level III), and non-steady-state and non-equilibrium (level IV).

Level I calculations generate the equilibrium distribution of a chemical in various phases (i) of a closed environment. As it is a closed system without any input and output of the chemical of interest, steady-state consideration is irrelevant. The basic equation is $M = \sum M_i = \sum c_i V_i = f \sum Z_i V_i$ (Equation 3.42), where M could be replaced by percentages (e.g. 100% for M). Level I calculations can be easily carried out in a spreadsheet or by hand, although computer models are readily available.

The Level II model extends the analysis to a system characterized by equilibrium and steady state, in which input of the chemical is balanced by its transformation reactions and outflows via advections. Similar to level I, each phase has the same fugacity. The basic equation is $I = f \sum D_i$, where I is the input rate, and the "D-value" has the dimension of mass/pressure–time for compartment i. For advections, $D_i = G_i Z_i$ with G_i being the steady-state advective flow rate. For internal degradation processes, $D_i = V_i Z_i K_i$, where K_i is the sum of first order rate constants for all degradation reactions in the compartment i. The basic equation for level II calculations may be written as,

$$I + \sum G_i C_{Bi} = I + \sum K_{ai} V_i C_{Bi} = \sum V_i C_i (K_i + K_{ai}) = f \sum V_i Z_i (K_i + K_{ai})$$
(3.70)

where we have assumed that the fugacity $f = C_i/Z_i$ and C_{Bi} represents input concentrations. This equation expresses that under steady-state conditions, the amount of the chemical introduced by emission I or advection $\sum G_i C_{Bi}$ equals the amount leaving the system by reaction $f \sum V_i Z_i K_i$ and advection $f \sum V_i Z_i K_{ai}$. Level II calculations can also be easily performed in a spreadsheet without the use of macros and programming languages.

More complicated scenarios arise when the assumption of equilibrium is dropped at level III. Similar to the case of level II, a chemical continuously enters and is discharged at a constant rate, achieving a steady state condition. However, the fugacities of the chemical from different phases are no longer the same, causing intermedia transport across boundaries between compartments, such as diffusion and evaporation, and those associated with aerosol deposition, sediment re-suspension, and others. A mass balance equation is written for each medium; its terms are dependent on the processes involved. New D-values are defined for various intermedia transport processes (Mackay, 2001). The model-calculated chemical distributions differ depending on the compartment on which the chemical enters the system. Level III calculations require solution of a set of algebraic equations which are facilitated by the computer model "Level III" (Centre for Environmental Modeling and Chemistry at Trent University, Canada, CEMC, 2004).

Non-steady-state conditions (level IV) need a differential equation $dM/dt = $ total input rate – total output rate, in order to obtain time- and media-dependent chemical concentrations and distribution. The time needed for the system to de-contaminate itself to a given extent can be estimated when the chemical input is eliminated or reduced. Sufficient data are often not available for level III and IV calculations on actual environmental systems. However, level III calculations are useful for standard-ized comparison between properties of chemicals and are used for that purpose, for example in the widely used chemical property estimation software, EPI Suite (U.S. EPA, 2013).

Free downloadable Windows-based computer programs are available from CEMC (http://www.trentu.ca/academic/aminss/envmodel/). The multimedia equilibrium cri-terion (EQC) model (2011) relates to the generic environment and consists of levels I to III of the calculations; while the "Fugacity Model" relates to a user-defined real world environment and has three separate programs for levels I, II, and III calcula-tions respectively. In the latest version, which is a template on the Microsoft Excel spreadsheet platform, all chemicals are treated together as a single class (Hughes et al., 2012). These and related models are useful for the design of environmentally benign chemicals (Gama et al., 2012).

3.6.3 Example Calculations – Chlorobenzene

Many example calculations have been presented in Mackay (2001). In this section, we present an example of chlorobenzene using both spreadsheet and the computerized EQC model. The necessary physicochemical properties of chlorobenzene are sum-marized in Table 3.7, along with the compartment parameters for the EQC standard environment.

The results of spreadsheet calculations at level I and level II are presented in Tables 3.8a and 3.8b respectively. In the case of level II calculations we have assumed that the chlorobenzene input concentrations $C_{Bi} = 0$. Thus, from Equation 3.70 we obtain

$$I = f \sum V_i Z_i (K_i + K_{ai}) = f \sum D_i \text{ (total)} \tag{3.71}$$

where $D_i(\text{total}) = D_i(\text{reaction}) + D_i(\text{advection}) = V_i Z_i K_i + G_i Z_i$. Calculations for level-II include the following steps: $f = I/D_i \text{ (total)}$, $C_i = f_i Z_i$, $M_i = C_i V_i$, $R_i = K_i C_i V_i$, $A_i = K_{ai} C_i V_i$ where R_i is the total reaction (mol/h) within com-partment i, and A_i (mol/h) is the advection out of it. The summed values $M = \sum M_i$, $R = \sum R_i$, and $A = \sum A_i$ are used for calculation of residence times for reaction, $\tau_{\text{react}} = M/R$, and advection $\tau_{\text{adv}} = M/A$. The overall residence time is given by $\tau_{\text{res}} = M/I$.

The graphic outputs for level I, II, and III calculations from the EQC model are shown in Figure 3.8. Apparently, in this standard "unit world", chlorobenzene strongly prefers to enter the atmosphere if equilibrium is achieved, with 98% of the mass found in the air based on level I and level II calculations. More than 99% of the chlorobenzene removal (level II) takes place from the atmosphere, with 71% leaving the system with air flow, and 29% being degraded in the air via direct and indirect photolysis.

The environmental behavior of chlorobenzene is simulated at level III using the computer EQC model, with equal amounts of chlorobenzene entering air, water,

TABLE 3.7. Physicochemical Properties of Chlorobenzene and Standard EQC Environment Parameters

Physicochemical properties of chlorobenzene

MW = 112.6	MP (°C) = −45.6	P° (Pa) = 1585	K_H (Pa m^3/mol) = 375
Sw (g/m^3) = 459	log K_{ow} = 2.78	Log K_{oc} = 2.39[a]	Log K_{fw} = 1.48[a]

Standard EQC environment parameters

Medium	Area	Height	Volume, V_i	Organic carbon fraction	Density	Residence times, $1/K_a$	Advection rate, G_i
	m^2	m	m^3	f_{oc}	kg/m^3	h	m^3/h
Air	1E+11	1000	1E+14	-	1.2	100	1.00E+12
Water	1E+10	20	2E+11	-	1000	1000	2.00E+08
Soil	9E+10	0.1	9E+09	0.02	2400	-	-
Sediment	1E+10	0.01	1E+08	0.04	2400	50,000	2000
Susp. Sed.	-	-	1E+06	0.2	1500	-	-
Fish	-	-	2E+05	0.05	1000	-	-

(Source: Data from Schwarzenbach et al. (2003)).

[a]The organic carbon/water partition coefficient K_{oc} is estimated using $K_{oc} = 0.41 \times K_{ow}$, and the fish/water partition coefficient K_{fw} using $K_{fw} = f_{lipid} K_{ow}$. For soil, sediment, and suspended sediment, $K_D = f_{oc} K_{oc}$.

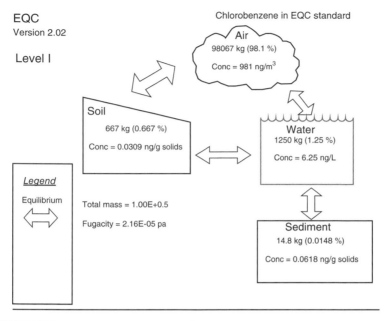

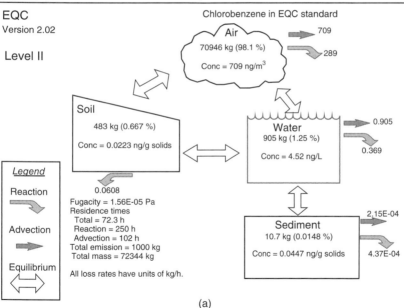

(a)

Figure 3.8. Example graphic outputs of the EQC model (2011): (a) Level I and Level II, (b) Level III.

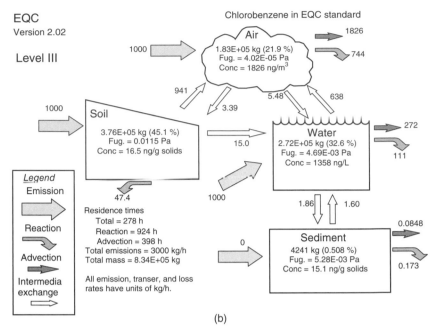

(b)

Figure 3.8. (*Continued*)

TABLE 3.8a. Spreadsheet Calculation of Fugacity Model for Chlorobenzene at 25°C (Level-I, Total mass $= 10^5$ kg)

Medium	Fugacity capacity Z_i mol/(m³ Pa)	V_iZ_i mol/Pa	Fugacity $f_i = M/\Sigma\,(V_iZ_i)$ Pa	Conc. $C_i = f_iZ_i$ mol/m³	Mass $M_i = C_iV_i$ mol	Mass fraction M_i/M %
Air	4.04E-04	4.04E + 10	2.15E-05	8.70E-09	8.70E + 05	98.0
Water	2.67E-03	5.33E + 08	2.15E-05	5.74E-08	1.15E + 04	1.29
Soil	3.16E-02	2.85E + 08	2.15E-05	6.81E-07	6.13E + 03	0.690
Sediment	6.32E-02	6.32E + 06	2.15E-05	1.36E-06	1.36E + 02	0.0153
Susp. Sed.	1.98E-01	1.98E + 05	2.15E-05	4.26E-06	4.26E − 00	0.000
Fish	8.03E-02	1.61E + 04	2.15E-05	1.73E-06	3.46E − 01	0.000
Sum	-	4.12E + 10	-	-	8.88E + 05	100

and soil. Under steady-state conditions, the major intermedia transport takes place from water and soil to air, which is driven by the large difference in fugacity between the phases. The major losses from the system are the advections and degradations in the air, while the advection with water outflow counts for about 9%. A small percentage of the chlorobenzene in water partitions into the sediment, where it is partially degraded and moves out of the system with lost sediment. With a half-life of about 230 days in the soil, 45% of the total chlorobenzene is retained by the soil.

TABLE 3.8b. Spreadsheet Calculation of Fugacity Model for Chlorobenzene at 25°C (Level-II, Total Input = 1000 kg/h)

Medium	D_i (advection) $= G_i Z_i$ mol/(Pa h)	D_i (reaction) $= V_i Z_i K_i$ mol/(Pa h)	Fugacity $f_i = I/\sum D_i$ Pa	Conc. $C_i = f_i Z_i$ mol/m³	Mass $M_i = C_i V_i$ mol	Mass M_i kg	Advection $A_i = K_{ai} C_i V_i$ mol/h	Reaction $R_i = K_i C_i V_i$ mol/h
Air	4.08E+08	1.65E+08	1.56E-05	6.30E-09	6.30E+05	70,938	6.30E+03	2.57E+03
Water	5.33E+05	2.17E+05	1.56E-05	4.16E-08	8.32E+03	936	8.32E+00	3.39E+00
Soil	-	3.59E+04	1.56E-05	4.93E-07	4.44E+03	500	-	5.59E-01
Sediment	1.26E+02	2.58E+02	1.56E-05	9.87E-07	9.86E+01	11.1	1.97E-03	4.02E-03
Susp. Sed.	-	-	1.56E-05	3.08E-06	3.08E+00	-	-	-
Fish	-	-	1.56E-05	1.25E-06	2.51E-01	-	-	-
Sum	4.05E+08	1.65E+08	-	-	6.43E+05	72,385	6.31E+03	2.57E+03

Mass removal rate

Medium	Half-life $t_{0.5}$ (h)	Reaction constant K_i (1/h)	Reaction kg/h	% of emission	Advection kg/h	% of emission	Total kg/h	%
Air	170.00	4.08E-03	289	28.9	709	70.9	998.62	99.86
Water	1700.00	4.08E-04	0.382	-	0.936	-	1.32	0.13
Soil	5.5E+03	1.26E-04	6.30E-02	-	-	-	0.063	0.01
Sediment	17,000	4.08E-05	4.53E-04	-	2.22E-04	-	0.001	6.8E-05
Sum	-	-	290	-	710	-	1000.00	100.00
Sum %	-	-	29.0%	-	71.0%	-	-	-

REFERENCES

Achman, DR, Hornbuckle, KC, Eisenreich, SJ. Volatilization of polychlorinated biphenyls from Green Bay, Lake Michigan. *Environ. Sci. Technol.* 1993;27: 75–87.

Andren, AW, Strand, JW. Atmospheric deposition of particulate organic carbon and polyaromatic hydrocarbons to Lake Michigan. In: Eisenreich, SJ (Editor). *Atmospheric Pollutants in Natural Waters*. Ann Arbor, Michigan: Ann Arbor Science Publishers, Inc.; 1981, pp. 459–479.

Blando, JD, Porcja, RJ, Li, T-H, Bowman, D, Lioy, PJ, and Turpin, BJ. Secondary formation and the Smoky Mountain organic aerosol: An examination of aerosol polarity and functional group composition during SEAVS. *Environ. Sci. Technol.* 1998;32: 604–613.

Brunner, S, Hornung, E, Santi, II, Wolff, E, Piringer, OG, Altschuh, J, Brüggemann, R. Henry's law constants for polychlorinated biphenyls: Experimental determination of structure–property relationships. *Environ. Sci. Technol.* 1990;24: 1751–1754.

Caffrey, PF, Ondov, JM, Zufall, JM, Davidson, CI. Determination of size-dependent dry particle deposition velocities with multiple intrinsic elemental tracers. *Environ. Sci. Technol.* 1998;32: 1615–1622.

Cetin, B, Odabasi, M. Air-water exchange and dry deposition of polybrominated diphenyl ethers at a coastal site in Izmir Bay, *Turkey. Environ. Sci. Technol.* 2007;41(3):785–791.

Chao, M-R, Hu, C-W, Ma, H-W, Gou-Ping, C-C, Lee, W-J, Chang, LW, Wu, K-Y. Size distribution of particle-bound polychlorinated dibenzo-p-dioxins and dibenzofurans in the ambient air of a municipal incinerator. *Atmos. Environ.* 2003;37: 4945–4954.

Cortes, DR, Hoff, RM, Brice, KA, Hites, RA. Evidence of current pesticide use from temporal and Clasius-Clapeyron plots: A case study from the integrated atmospheric deposition network. *Environ. Sci. Technol.* 1999;33: 2145–2150.

Elzerman, AW. Mechanisms of enrichment at the air–water interface. In: Eisenreich, SJ (Editor). *Atmospheric Pollutants in Natural Waters*. Ann Arbor, Michigan: Ann Arbor Science Publishers, Inc.; 1981, pp. 81–97.

Gama, S, Mackay, D, Arnot, JA. Selecting and designing chemicals: application of a mass balance model of chemical fate, exposure and effects in the environment. *Green Chemistry.* 2012;14(4):1094–1102.

Gasic, B, Moeckel, C, Macleod, M, Brunner, J, Scheringer, M, Jones, KC, Hungerbühler, K. Measuring and modeling short-term variability of PCBs in air and characterization of urban source strength in Zürich, Switzerland. *Environ. Sci. Technol.* 2009;43(3):769–776.

Genualdi, S, Lee, SC, Shoeib, M, Gawor, A, Ahrens, L, Harner, T. Global pilot study of legacy and emerging persistent organic pollutants using sorbent-impregnated polyurethane foam disk passive air samplers. *Environ. Sci. Technol.* 2010;44: 5534–5539.

Gouin, T, Jantunen, L, Harner, T, Blanchard, P, Bidleman, T. Spatial and temporal trends of chiral organochlorine signatures in Great Lakes air using passive air samplers. *Environ. Sci. Technol.* 2007;41(11):3877–3883.

Gouin, T, Wania, F, Ruepert, C, Castillo, LE. Field testing passive air samplers for current Use pesticides in a tropical environment. *Environ. Sci. Technol.*, 2008;42(17):6625–6630.

Hicks, BB. Atmospheric deposition and its effects on water quality. In: Christensen, ER, O'Melia, CR. *Workshop on Research Needs for Coastal Pollution in Urban Areas*. Milwaukee, Wisconsin: University of Wisconsin-Milwaukee; 1998.

Holsen, TM, Noll, KE, Liu, S-P, Lee, W-J. Dry deposition of polychlorinated biphenyls in urban areas. *Environ. Sci. Technol.* 1991;25: 1075–1081.

Hornbuckle, KC, Achman, DR, Eisenreich, SJ. Over-water and over-land polychlorinated biphenyls in Green Bay, Lake Michigan. *Environ. Sci. Technol.* 1993;27: 87–98.

Hughes, L, Mackay, D, Powell, DE, Kim, J. An updated state of the science EQC model for evaluating chemical fate in the environment: Application to D5 (decamethylcyclopentasiloxane). *Chemosphere.* 2012;87(2):118–124.

Hunter, KA, Liss, PS. Principles and problems of modeling cation enrichment at natural air–water interfaces. In: Eisenreich, SJ (Editor). *Atmospheric Pollutants in Natural Waters,* 99–127. Ann Arbor, Michigan: Ann Arbor Science Publishers, Inc; 1981.

Jaward, FM, Zhang, G, Nam, JJ, Sweetman, AJ, Obbard, JP, Kobara, Y, Jones, KC. Passive air sampling of polychlorinated biphenyls, organochlorine compounds, and polybrominated diphenyl ethers across Asia. *Environ. Sci. Technol.* 2005;39(22): 8638–8645.

Junge, CE. Basic considerations about trace constituents in the atmosphere as related to the fate of global pollutants. In: *Fate of Pollutants in the Air and Water Environments, Part 1,* Suffet, IH (Editor). *Advances in Environmental Science and Technology,* Vol. 8. New York, NY: John Wiley and Sons; 1977, pp. 7–25.

Jurado, E, Jaward, F, Lohman, R, Jones, KC, Simó, R, Dachs, J. Wet deposition of persistent organic pollutants to the global oceans. *Environ. Sci. Technol.* 2005;39(8): 2426–2435.

Karickhoff, SW. Semiempirical estimation of sorption of hydrophobic pollutants on natural sediments and soils. *Chemosphere* 1981;10: 833–846.

Koblizkova, M, Genualdi, S, Lee, SC, Harner, T. Application of Sorbent Impregnated Polyurethane Foam (SIP) Disk Passive Air Samplers for Investigating Organochlorine Pesticides and Polybrominated Diphenyl Ethers at the Global Scale. *Environ. Sci. Technol.,* 2012;46(1): 391–396

Lead, WA, Steinnes, E, Jones, KC. Atmospheric deposition of PCBs to moss (*Hylocomium splendens*) in Norway between 1977 and 1990. *Environ. Sci. Technol.* 1996;30: 524–530.

Lewis, GN The law of physico-chemical change. *Proceedings of the American Academy of Arts and Sciences.* 1901;37(3): 49–69.

Lion, LW, Leckie, JO. Copper in marine microlayers: Accumulation, speciation and transport. In: Eisenreich, SJ (Editor). *Atmospheric Pollutants in Natural Waters.* Ann Arbor, Michigan: Ann Arbor Science Publishers, Inc.; 1981, pp. 143–163.

Mackay, D. Finding fugacity feasible. *Environ. Sci. Technol.,* 1979;13(10): 1218–1223.

Mackay, D, Paterson, S. Calculating fugacity. *Environ. Sci. Technol.* 1981;15: 1006–1007.

Mackay, D, Paterson, S, Shiu, WY. Generic models for evaluating the regional fate of chemicals. *Chemosphere,* 1992a;24(6): 695–717

Mackay, D, Shiu, W-Y, Ma, K-C. Illustrated Handbook of Physical-Chemical Properties and Environmental Fate for Organic Chemicals, Vol. I, Monoaromatic Hydrocarbons, Chlorobenzenes, and PCBs. Boca Raton, Florida: Lewis Publishers; 1992b.

Mackay, D, Shiu, W-Y, Ma, K-C. *Illustrated Handbook of Physical-Chemical Properties and Environmental Fate for Organic Chemicals,* Vol. II, Polynuclear Aromatic Hydrocarbons, Polychlorinated Dioxins, and Dibenzofurans. Boca Raton, Florida: Lewis Publishers; 1992c.

Mackay, D, Shiu, W-Y, Ma, K-C. Illustrated Handbook of Physical-Chemical Properties and Environmental Fate for Organic Chemicals, *Vol. III, Volatile Organic Chemicals.* Boca Raton, Florida: Lewis Publishers; 1993.

Mackay, D, Shiu, W-Y, Ma, K-C. Illustrated Handbook of Physical-Chemical Properties and Environmental Fate for Organic Chemicals, *Vol. IV, Oxygen, Nitrogen, and Sulfur Containing Compounds.* Boca Raton, Florida: Lewis Publishers; 1995.

Mackay, D, Di Guardo, A, Paterson, S, Kicsi, G, Cowan, CE. Assessing the fate of new and existing chemicals: A five stage process. *Environ. Toxicol. Chem.* 1996a;15: 1618–1626.

Mackay, D, Di Guardo, A, Paterson, S, Cowan, C. Evaluating the environmental fate of a variety of types of chemicals using the EQC model. *Environ. Toxicol. Chem.* 1996b;15: 1627–1637.

Mackay, D, Di Guardo, A, Paterson, S, Kicsi, G, Cowan, C, Kane, DM. Assessment of chemical fate in the environment using evaluative, regional and local-scale models: Illustrative application to chlorobenzene and linear alkylbenzene sulfonates. *Environ. Toxicol. Chem.* 1996c;15: 1638–1648.

Mackay, D, Shiu, W-Y, Ma, K-C. *Illustrated Handbook of Physical-Chemical Properties and Environmental Fate for Organic Chemicals*, Vol. V, Pesticide Chemicals. Boca Raton, Florida: Lewis Publishers; 1997.

Mackay, D. *Multimedia Environmental Models: The Fugacity Approach*, 2nd ed. Boca Raton, FL: Lewis Publishers; 2001, p. 261.

Mackay D, Arnot, JA. The Application of Fugacity and Activity to Simulating the Environmental Fate of Organic Contaminants. *J. Chem. Eng. Data* 2011;56: 1348–1355.

Marple, VA, Rubow, KL, Behm, S. A midro-orifice uniform deposit impactor (MOUDI): Description, calibration and use. *Aerosol Sci. and Technol.* 1991;14: 434–446.

Matter-Müller, C, Gujer, W, Giger, W. Transfer of volatile substances from water to the atmosphere. *Water Res.* 1981;15: 1271–1279.

Meyers, PA, Owen, RM, Mackin, JE. Organic matter and heavy metal concentrations in the particulate phase of Lake Michigan surface microlayers. In: Eisenreich, S.J. (Ed.). *Atmospheric Pollutants in Natural Waters*. Ann Arbor, Michigan: Ann Arbor Science Publishers, Inc.; 1981, pp.129–141.

Moeckel, C, Harner, T, Nizzetto, L, Strandberg, B, Lindroth, A, Jones, KC. Use of depuration compounds in passive air samplers: results from active sampling-supported field deployment, potential uses, and recommendations. *Environ. Sci. Technol.,* 2009;43(9): 3227–3232.

Multimedia equilibrium criterion (EQC) model. http://www.trentu.ca/academic/aminss/envmodel/. *Centre for Environmental Modeling and Chemistry at the Trent University, Canada*. Trent University, Canada; 2011.

Namkung, E, Rittmann, BE. Estimating volatile organic compound emissions from publicly owned treatment works. *J. Water Pollut. Control Fed.* 1987;59: 670–678.

Nelson, ED, McConnell, LL, Baker, JE. Diffusive exchange of gaseous polycyclic aromatic hydrocarbons and polychlorinated biphenyls across the air-water interface of the Chesapeake Bay. *Environ. Sci. Technol.* 1998;32: 912–919.

Paode, RD, Sofuoglu, SC, Sivadechathep, J, Noll, KE, Holsen, TM, Keeler, GJ. Dry deposition fluxes and mass size distributions of Pb, Cu, and Zn measured in southern Lake Michigan during AEOLOS. *Environ. Sci. Technol.* 1998;32: 1629–1635.

Pozo, K, Harner, T, Lee, SC, Wania, F, Muir, DCG, Jones, KC. Seasonally resolved concentrations of persistent organic pollutants in the global atmosphere from the first year of the GAPS study. *Environ. Sci. Technol.* 2009;43(3): 796–803.

Qi, S, Alonso, C, Suidan, MT, Sayles, GD. PCB volatilization from sediments. *ASCE J. Environ. Engrg.* 2006;132(1): 102–111.

Qiu, X, Zhu, T, Wang, F, Hu, J. Air-water gas exchange of organochlorine pesticides in Taihu Lake, *China. Environ. Sci. Technol.* 2008;42(6): 1928–1932.

Reid, RC, Prausnitz, JM, Poling, BE. *The Properties of Gases and Liquids*, Fourth Ed. New York, NY: McGraw Hill Book Co.; 1987, pp. 587–589.

Roberts, PV, Munz, C, Dandliker, P. Modeling volatile organic solute removal by surface and bubble aeration. *J. Water Pollut. Control Fed.* 1984;56: 157–163.

Ruge, Z, Lohman, R, Muir, D, Helm, P. Enhancing polycyclic aromatic hydrocarbon (PAH) monitoring of Lake Superior using polyethylene passive samplers. Poster at SETAC Europe 23rd Annual Meeting, May 12–16, 2013, Glasgow, UK; 2013.

Sawyer, CN, McCarty, PL, Parkin, GF. *Chemistry for Environmental Engineering and Science*, Fifth Ed. Boston, MA: McGrawHill Higher Education; 2003.

Schwarzenbach, RP, Gschwend, PM, Imboden, DM. *Environmental Organic Chemistry.* 2nd Ed. New York, NY: John Wiley & Sons, Inc.; 2003.

Scott, BC. Modeling of atmospheric wet deposition. In: Eisenreich, SJ (Editor). *Atmospheric Pollutants in Natural Waters.* Ann Arbor, Michigan: Ann Arbor Science Publishers, Inc.; 1981, 3–21.

Shoeib, M, Harner, T. Characterization and Comparison of Three Passive Air Samplers for Persistent Organic Pollutants. *Environ. Sci. Technol.,* 2002;36(19): 4142–4151.

Slinn, SA, Slinn, WGN. Modeling of atmospheric particulate deposition to natural waters. In: Eisenreich, SJ (Editor). *Atmospheric Pollutants in Natural Waters.* Ann Arbor, Michigan: Ann Arbor Science Publishers, Inc.; 1981, 23–53.

Tao, S, Liu, Y, Xu, W, Lang, C, Liu, S, Dou, H, Liu, W. Calibration of a passive sampler for both gaseous and particulate phase polycyclic aromatic hydrocarbons. *Environ. Sci. Technol.,* 2007;41(2): 568–573.

Tao, S, Cao, J, Wang, W, Zhao, J, Wang, W, Wang, Z, Cao, H, Xing, B. A passive sampler with improved performance for collecting gaseous and particulate phase polycyclic aromatic hydrocarbons in air. *Environ. Sci. Technol.,* 2009;43(11): 4124–4129.

Tataya, S, Tanabe, S, Tatsukawa, R. In: Schmidtke, N (Editor). *Toxic Contamination in Large Lakes.* Chelsea, Michigan: Lewis Publishers; 1988, 237–281.

Tunkel, J, Howard, PH, Boethling, RS, Stiteler, W, Loonen, H. Predicting ready biodegradability in the Japanese ministry of international trade and industry test. *Environ. Toxicol. Chem.* 2000;19(10): 2478–2485.

U.S. EPA. 2011. *Estimation Program Interface (EPI) Suite.* http://www.epa.gov/oppt/exposure/pubs/episuite.htm (accessed January 5, 2011).

U.S. EPA. 2013. *Estimation Program Interface (EPI) Suite, version 4.11.* http://www.epa.gov/oppt/exposure/pubs/episuite.htm (accessed September 8, 2013)

U.S. EPA. 1999. *Compendium of Methods for the Determination of Toxic Organic Compounds in Ambient Air*, Second Edition. United States Enivronmental Protection Agency, Center for Environmental Research Information. Cincinnati, Ohio. EPA/625/R-96/010b. January 1999. http://www.epa.gov/ttnamti1/files/ambient/airtox/tocomp99.pdf (accessed July 6, 2012).

Van Noort, PCM. Estimation of polychlorinated biphenyl fugacity ratios. *Environ. Toxicol. Chem.* 2006;25(11): 2875–2883.

Venier, M, Ferrario, J, Hites, RA. Polychlorinated dibenzo-*p*-dioxins and dibenzofurans in the atmosphere around the Great Lakes. *Environ. Sci. Technol.* 2009;43(4): 1036–1041.

Venier, M, Hites, RA. Regression model of partial pressures of PCBs, PAHs, and organochlorine pesticides in the Great Lakes' atmosphere. *Environ. Sci. Technol.* 2010;44(2): 618–623.

Wania, F, Haugen, J-E, Lei, YD, Mackay, D. Temperature dependence of atmospheric concentrations of semivolatile organic compounds. *Environ. Sci. Technol.* 1998;32: 1013–1021.

Wania, F, Shen, L, Lei,YD, Teixeira, C, Muir, DCG. Development and calibration of a resin-based passive sampling system for monitoring persistent organic pollutants in the atmosphere. *Environ. Sci. Technol.* 2003;37: 1352–1359.

Weber, WJ Jr, DiGiano, FA. *Process Dynamics in Environmental Systems.* New York, NY: John Wiley and Sons, Inc; 1996.

Wei, H, Li, A. Semi-volatile Organic Pollutants in the Gaseous and Particulate Phases in Urban Air. In Zereini, F, Wiseman, C (Editors). *Urban Airborne Particulate Matter: Origins, Chemistry, Fate and Health Impacts.* Springer Publishing Company; 2010, pp. 339–362.

Westgate, JN, Wania, F. On the construction, comparison, and variability of airsheds for interpreting semivolatile organic compounds in passively sampled air. *Environ. Sci. Technol.* 2011;45(20): 8850–8857.

Wu, ZY, Han, M, Lin, Z-C, Ondov, JM. Chesapeake bay atmospheric deposition study, year 1: Sources and dry deposition of selected elements in aerosol particles Original Research Article. *Atmos. Environ.* 1994;28: 1471–1486.

Zhang, X, Christensen, ER, Ni, S-N. Estimation of the potential emission of toxic air pollutants from sewage treatment plants. *Proceedings of the 1990 ASCE Specialty Conference on Environmental Engineering.* New York, NY: American Society of Civil Engineers; 1990, pp. 621–628.

Zhang, X, Wong, C, Lei, YD, Wania, F. Influence of sampler configuration on the uptake kinetics of a passive air sampler. *Environ. Sci. Technol.* 2012a;46(1): 397–403.

Zhang, B-Z., Zhang, K, Li, S-M, Wong, CS, Zeng, EY. Size-dependent dry deposition of airborne polybrominated diphenyl ethers in urban Guangzhou, China. *Environ. Sci. Technol.* 2012b;46(13): 7207–7214.

Zufall, M, Davidson, CI, Caffrey, PF, Ondov, JM. Airborne concentrations and dry deposition fluxes of particulate species to surrogate surfaces deployed in southern Lake Michigan. *Environ. Sci. Technol.* 1998;32: 1623–1628.

4

WATER CHEMISTRY

4.1 INTRODUCTION

An understanding of the chemistry of natural waters is necessary in order to properly evaluate the behavior and effects of aquatic pollutants. The science of water chemistry can be dated back to the seventeenth century, when experiments were carried out to understand what made seawater salty. Studies on chemical components of natural waters progressed quickly in the nineteenth century, and became a specialty in oceanography and limnology. In the first half of the twentieth century, the need to provide drinking water and treat wastewater boosted studies on the chemical processes in lakes, rivers, and groundwater. As one of the areas of environmental chemistry, water chemistry has progressed into the twenty-first century with strong interactions with other areas of environmental sciences. A review of the history of water chemistry and its interactions with other scientific disciplines is given by Brezonik and Arnold (2012).

The fundamental chemical concepts, emphasizing inorganic chemistry, are covered in the authoritative text by Stumm and Morgan (1996) as well as other textbooks (Benjamin, 2010; Morel and Hering, 1993; Pankow, 1991; Snoeyink and Jenkins, 1980). Extensive information about reaction rates may be found in Brezonik (1994). Weber and DiGiano (1996) focus on the application of chemical engineering principles to process dynamics. Sawyer et al.'s (2003) text on chemistry relating to water quality and treatment is a useful guide for environmental engineers. Traditionally, equilibrium chemistry has been one of the better understood areas, partly because the thermodynamic data for free energies and enthalpies are widely available.

Physical and Chemical Processes in the Aquatic Environment, First Edition. Erik R. Christensen and An Li.
© 2014 John Wiley & Sons, Inc. Published 2014 by John Wiley & Sons, Inc.

By contrast, kinetic data that are necessary for estimating reaction rates are less well established.

This chapter concerns aspects of water chemistry that have received renewed interest because of their importance to the challenges that human society is facing. Extensive releases of carbon dioxide and other greenhouse gases as a result of human activities have not only brought about global climate change, but also profoundly affected natural water sources. With the continuous increase in the partial pressure of carbon dioxide in the atmosphere, acidification of the oceans is accelerating at a rate unprecedented in the last 300 million years (Hönisch et al., 2012). The direct consequences include weakened and inhibited calcification, causing accelerated extinction of coral reefs, mollusks, and some plankton.

In this chapter, the aquatic chemistry of carbon and sulfur is presented. Some important gas–water partitioning and acid/base reactions are summarized in Table 4.1, and redox reactions are summarized in Table 4.2. The geochemical cycles of nitrogen and phosphorus will be discussed in Chapter 5 along with other nutrients.

TABLE 4.1. Major Reactions and their Equilibrium Constants for Oxygen, Carbon, and Sulfur Species

Reaction	$K(25°C)$
Oxygen Species	
$O_2(g) \leftrightarrow O_2(aq)$	1.26×10^{-3} [a]
$O_3(g) \leftrightarrow O_3(aq)$	9.40×10^{-3} [a]
Inorganic Carbon	
$CO(g) \leftrightarrow CO(aq)$	9.55×10^{-4} [a]
$CO_2(g) + H_2O(l) \leftrightarrow H_2CO_3(aq)$	3.39×10^{-2}
$H_2CO_3^* \leftrightarrow H^+ + HCO_3^-$	4.45×10^{-7} [b]
$HCO_3^- \leftrightarrow H^+ + CO_3^{2-}$	4.69×10^{-11}
Sulfur Species	
$SO_2(g) + H_2O(l) \leftrightarrow SO_2 \cdot H_2O(aq)$	1.25 [a]
$SO_2 \cdot H_2O \leftrightarrow H^+ + HSO_3^-$	1.29×10^{-2}
$HSO_3^- \leftrightarrow H^+ + SO_3^{2-}$	6.24×10^{-8}
$HSO_4^- \leftrightarrow SO_4^{2-} + H^+$	1.00×10^{-2}
$H_2S(g) \leftrightarrow H_2S(aq)$	1.05×10^{-1} [a]
$H_2S(aq) \leftrightarrow H^+ + HS^-$	9.77×10^{-8}
$HS^- \leftrightarrow H^+ + S^{2-}$	1.00×10^{-19}
Organic Compounds	
$CH_4(g) \leftrightarrow CH_4(aq)$	1.29×10^{-3} [a]
$CH_3COOH(g) \leftrightarrow CH_3COOH(aq)$	7.66×10^{2} [a]
$CH_3COOH \leftrightarrow H^+ + CH_3COO^-$	1.75×10^{-5}

(Source: Data adapted from Stumm and Morgan (1996)).
[a] Henry's constants at 25°C in mole/(L-atm).
[b] $H_2CO_3^*$ includes dissolved CO_2 and carbonic acid.

TABLE 4.2. Redox Reactions in Natural Waters

Reaction	pE^0 (LogK)	pE^0 (W)^a
$\frac{1}{4}O_2(g) + H^+ + e^- \leftrightarrow \frac{1}{2}H_2O$	20.75	13.75
$\frac{1}{6}HSO_4^- + \frac{7}{6}H^+ + e^- \leftrightarrow \frac{1}{6}S(s) + \frac{2}{3}H_2O$	5.70	−2.47
$\frac{1}{6}SO_4^{2-} + \frac{4}{3}H^+ + e^- \leftrightarrow \frac{1}{6}S(s) + \frac{2}{3}H_2O$	6.03	−3.30
$\frac{1}{8}SO_4^{2-} + \frac{5}{4}H^+ + e^- \leftrightarrow \frac{1}{8}H_2S(g) + \frac{1}{2}H_2O$	5.25	−3.50
$\frac{1}{8}SO_4^{2-} + \frac{9}{8}H^+ + e^- \leftrightarrow \frac{1}{8}HS^- + \frac{1}{2}H_2O$	4.25	−3.75
$\frac{1}{2}S(s) + H^+ + e^- \leftrightarrow \frac{1}{2}H_2S(g)$	2.89	−4.11
$\frac{1}{8}CO_2(g) + H^+ + e^- \leftrightarrow \frac{1}{8}CH_4(g) + \frac{1}{4}H_2O$	2.87	−4.13
$H^+ + e^- \leftrightarrow \frac{1}{2}H_2(g)$	0.00	−7.00
$\frac{1}{4}CO_2(g) + H^+ + e^- \leftrightarrow \frac{1}{4}CH_2O + \frac{1}{4}H_2O$	−1.20	−8.20
$\frac{1}{2}CO_2(g) + \frac{1}{2}H^+ + e^- \leftrightarrow \frac{1}{2}HCOO^-$	−4.83	−8.33
$FeOOH(s) + HCO_3^-(10^{-3}) + 2H^+ + e^- \leftrightarrow FeCO_3(s) + 2H_2O$	—	−0.80

(Source: Data adapted from Stumm and Morgan (1996)).
$^a pE^0$ for pH = 7.0 at 25°C in water of unit activities of oxidant and reductant.

These elements have a wide range of oxidation states, and their speciations and interactions affect all other physical, chemical, and biological processes occurring in water.

4.2 CARBONATE AND ALKALINITY

4.2.1 Dissolved CO_2 and Carbonate Speciation in Water

Carbon dioxide (CO_2) is a minor component of air. In unpolluted air, its concentration varies seasonally as a result of the growth cycle of plants. The average annual concentration of CO_2 in the earth's atmosphere had been around 260–280 ppmv for more than 10,000 years before the 1800s. It has increased from 280 ppmv in the 1830s to about 380 ppmv today, and there is no doubt about its continued increase for the coming decades. The predicted CO_2 levels for the year of 2100 range from the best scenario of 450 ppmv to the worst of near 1000 ppmv (USGCRP, 2009). For the calculations presented in this chapter, the current atmosphere CO_2 concentration, 0.039% by volume or 390 ppmv, is used.

Under a total pressure of 1 atm, the current partial pressure of CO_2 is

$$P_{CO_2} = 0.00039 \text{ atm} = 10^{-3.4} \text{ atm} \tag{4.1}$$

Atmospheric CO_2 partitions into water, and the equilibrium concentration ratio of CO_2 between the gas (g) and the aqueous (aq) phases is described by

$$CO_2(g) \leftrightarrow CO_2(aq) \tag{4.2}$$

The equilibrium constant for this reaction is the Henry's law constant K_H for CO_2:

$$K_H = \frac{\{CO_2 \text{ (aq)}\}}{P_{CO_2}} = 10^{-1.5} \frac{\text{mole}}{\text{atm L}} \tag{4.3}$$

CO_2 is reactive in water. At low pH, a portion of $CO_2(aq)$ reacts with water to form carbonic acid (H_2CO_3). Because it is difficult to measure $CO_2(aq)$ and H_2CO_3 separately, they are often collectively termed as $H_2CO_3^*$, that is

$$\{CO_2(aq)\} = \{H_2CO_3^*\} \tag{4.4}$$

In most natural waters of neutral pH (6 to 9), $H_2CO_3^*$ dissociates into bicarbonate HCO_3^- and H^+, thus increasing the acidity of the water. The dominance of HCO_3^- among carbon species will shift to carbonate $CO_3^=$ as the pH rises to above 10.3. These reactions are discussed below.

Dissolved CO_2, $H_2CO_3^*$, is dissociated as follows,

$$H_2CO_3^* \leftrightarrow H^+ + HCO_3^- \tag{4.5}$$

with an equilibrium constant K_1:

$$K_1 = \frac{\{H^+\} \{HCO_3^-\}}{\{H_2CO_3^*\}} = 4.45 \times 10^{-7} = 10^{-6.35} \tag{4.6}$$

The bicarbonate, HCO_3^-, further dissociates to generate carbonate $CO_3^=$

$$HCO_3^- \leftrightarrow CO_3^= + H^+ \tag{4.7}$$

with an equilibrium constant K_2:

$$K_2 = \frac{\{H^+\} \{CO_3^-\}}{\{HCO_3^{2-}\}} = 4.69 \times 10^{-11} = 10^{-10.33} \tag{4.8}$$

Combined with dissolved hardness ions, such as Ca^{2+}, Mg^{2+}, Fe^{3+}, Mn^{2+}, Zn^{2+}, Cu^{2+}, various carbonate salts precipitate. Limestone is one type of sedimentary rock, in which $CaCO_3$ is the dominant chemical component.

The equilibrium constant of the dissolution/precipitation process is solubility product K_{SP}. For $CaCO_3$,

$$CO_3^= + Ca^{2+} \leftrightarrow CaCO_3(calcite) \quad K_{SP} = \{CO_3^=\}°\{Ca^{2+}\}° = 3.4 \times 10^{-9} \quad (4.9)$$

$$CO_3^= + Ca^{2+} \leftrightarrow CaCO_3(aragonite) \quad K_{SP} = \{CO_3^=\}°\{Ca^{2+}\}° = 6.0 \times 10^{-9} \quad (4.10)$$

where the superscript "°" indicates that the activities take place under saturation conditions.

The saturation state of a water with regard to a specific mineral is evaluated by the ratio of measured activity product to K_{SP}. For calcium carbonates, for example, this ratio is:

$$\Omega = \frac{\{Ca^{2+}\}\{CO_3^=\}}{K_{SP}} \quad (4.11)$$

Theoretically, Ω must be higher than 1 to prevent the dissolution of the mineral, including the coral reefs and the shells of mollusks. Between the two common polymorphs of $CaCO_3$, aragonite is more soluble than calcite and responds more rapidly to changes in water chemistry, thus Ω_{arag} is more often used than $\Omega_{calcite}$ in the assessment of the $CaCO_3$ saturation state of water. The global average Ω_{arag} in the oceans has fallen from a pre-industrial level of 3.4 to 2.9 in recent years (NOAA, 2011).

4.2.2 Solving Equilibrium pH

As is known, water itself is both acid and base with a very low disassociation constant:

$$H_2O \leftrightarrow H^+ + OH^- \quad K_w = \{H^+\}\{OH^-\} = 10^{-14} \quad (4.12)$$

Because dissolved CO_2 acts as an acid, as shown by Equation 4.4, waters respond to increasing P_{CO_2} in the atmosphere by reducing their pH values. Knowledge of pH change is one of the basic pieces of information necessary for water quality and usage management.

The calculation of pH for a water will need tools such as mass balance, charge balance or proton balance, in addition to the equilibrium constants for acid/base reactions. As an example, we calculate pH of a water that is open to the atmosphere and has a bed of limestone, as in some freshwater lakes. As a first approximation, we assume that the water is sufficiently dilute that the activity equals concentration for all constituents. We also assume that equilibria between water and both atmospheric CO_2 and a solid $CaCO_3$ (calcite) are reached, and that the concentrations of all constituents, other than water species (H_2O, H^+, and OH^-), inorganic carbonate species, and calcium ion Ca^{2+}, are negligible. A mass balance for the total dissolved carbonate concentration C_T in this water can be written as

$$C_T = [H_2CO_3] + [HCO_3^-] + [CO_3^{2-}] \quad (4.13)$$

For this aquatic system, the C_T is dependent on the atmospheric pressure of CO_2. The charge balance equation of this water is

$$[H^+] + 2\,[\,Ca^{2+}] = [OH^-] + [HCO_3^-] + 2\,[CO_3^{2-}] \tag{4.14}$$

And the proton balance, with H_2O and HCO_3^- as the reference species, can be written as

$$[H^+] + [H_2CO_3^*] = [OH^-] + [CO_3^{2-}] \tag{4.15}$$

From the equilibrium constant definitions (Equations 4.4 through 4.9),

$$[H_2CO_3^*] = [CO_2\,(aq)] = K_H\,P_{CO_2} \tag{4.16}$$

$$[HCO_3^-] = \frac{[H_2CO_3^*]K_1}{[H^+]} = \frac{K_1 K_H P_{CO_2}}{[H^+]} \tag{4.17}$$

$$[CO_3^{2-}] = \frac{[HCO_3^-]K_2}{[H^+]} = \frac{K_1 K_2 K_H\,P_{CO_2}}{[H^+]^2} \tag{4.18}$$

$$[Ca^{2+}] = \frac{K_{SP}}{[CO_3^{2-}]} = \frac{[H^+]^2 K_{SP}}{K_1 K_2 K_H\,P_{CO_2}} \tag{4.19}$$

By substituting Equations 4.17 through 4.19 as well as Equation 4.10, the charge balance (Equation 4.14) becomes

$$[H^+] + 2\,\frac{[H^+]^2 K_{SP}}{K_1 K_2 K_H P_{CO_2}} = \frac{K_W}{[H^+]} + \frac{K_1 K_H P_{CO_2}}{[H^+]} + 2\frac{K_1 K_2 K_H P_{CO_2}}{[H^+]^2}, \tag{4.20}$$

and the proton balance (Equation 4.15) becomes

$$[H^+] + K_H P_{CO_2} = \frac{K_w}{[H^+]} + \frac{K_1 K_2 K_H P_{CO_2}}{[H^+]^2} \tag{4.21}$$

When the atmospheric pressure of carbon dioxide P_{CO_2} is known, both equations have $[H^+]$ as the only unknown. Therefore, they can be readily solved for the pH of the water in question using tools such as the Solver in MS Excel or other mathematic software. For this system, which is in equilibrium with the atmosphere ($P_{CO_2} = 10^{-3.9}$) and has $CaCO_3$ (calcite) as the bottom mineral, the solution to Equation 4.19 is $[H^+] = 10^{-8.3}$, or pH = 8.3. With this pH, the concentration of all other species can be readily calculated using Equations 4.12 and 4.16 through 4.19. It is worth noting that equilibria with both the atmosphere and the bottom minerals can only be assumed in well mixed waters. In reality, the bulk water below the wind-driven mixing layer may not be in equilibrium with atmospheric CO_2, while the water in the mixing surface layer may not be in equilibrium with the bottom calcite or other minerals. This is especially true during the warm seasons in which stratification of the water column occurs.

The pH of a water changes when other acid/base species are added. For example, acid rain can make a lake acidic; conversely adding lime can increase the pH. The variations of concentrations with pH for the water described above are illustrated in Figure 4.1(A) for H^+, OH^-, all carbonate species, and Ca^{2+}, with $P_{CO_2} = 390$ ppmv (Equation 4.1).

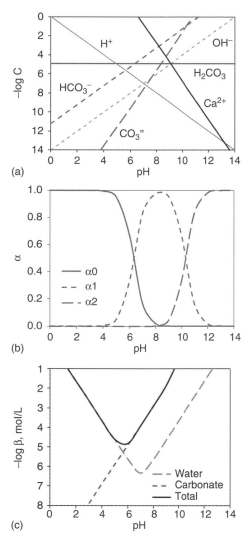

Figure 4.1. (a) pC–pH diagram, (b) species fractions, and (c) buffer intensity, for water in equilibrium with atmospheric CO_2. In panel (a), the calcium line is present when the water is in equilibrium with solid calcite. (*See insert for color representation of the figure.*)

The fractions α_0, α_1, and α_2 of the three carbonate species in C_T can be convenient to use in calculations because, as fractions, they are independent of C_T:

$$\alpha_0 = \frac{[H_2CO_3^*]}{C_T} = \frac{[H^+]^2}{[H^+]^2 + [H^+]K_1 + K_1K_2} \tag{4.22}$$

$$\alpha_1 = \frac{[HCO_3^-]}{C_T} = \frac{[H^+]K_1}{[H^+]^2 + [H^+]K_1 + K_1K_2} \tag{4.23}$$

$$\alpha_2 = \frac{[CO_3^{2-}]}{C_T} = \frac{K_1K_2}{[H^+]^2 + [H^+]K_1 + K_1K_2} \tag{4.24}$$

The α–pH diagram for carbonate system is shown in Figure 4.1(B).

The general approach for pH calculation as demonstrated in the example above is applicable to all natural surface waters. This approach includes identifying the major species (cations, anions, and uncharged species), collecting relevant equilibrium constants, writing the equations for mass balance, charge balance or proton condition, and solving either the charge balance or proton conditions for pH.

This numerical approach to solving pH often needs computer assisted calculations. Another approach using the pC–pH diagram is described in detail by Benjamin (2010) and can be used to quickly estimate the equilibrium composition of specific waters. The software commonly used for calculation of equilibrium or non-equilibrium concentrations of species in complex aqueous systems includes MINEQL+ (ERS, 2007), MINTEQA2 (USEPA, 2006), Visual MINTEQ (KTH, 2010), PHREEQC (USGS, 2014), GeoChemist's WorkBench (GWB, 2014), and others.

4.2.3   Alkalinity

Alkalinity represents the ability of water to resist acidification. Conceptually and numerically, alkalinity equates to acid neutralization capacity (ANC), although some textbooks distinguish the two using slightly different equations. In this book, the definition of Stumm and Morgan (1996) is adopted, and alkalinity (Alk) is theoretically defined as

$$[Alk] = [HCO_3^-] + 2[CO_3^{2-}] + [OH^-] - [H^+] + \text{other bases} \tag{4.25}$$

In natural waters, especially those with carbonate materials such as limestone in the bottom sediment, the dominant components of alkalinity are carbonate species including $CO_3^=$ and HCO_3^-, as shown in Equation 4.25. Alkalinity can also be defined as the difference between the concentrations of positive charges from the conservative cations (e.g., Na^+ and K^+) and negative charges from the conservative anions (e.g. Cl^- and $SO_4^=$). These ions are "conservative" because their concentrations are not affected by the addition of H^+. Apparently, alkalinity is a conservative property; this means the contributions from each species to the total alkalinity can be considered separately and are additive. The commonly used units for alkalinity include eq/L, meq/L, µeq/L, and "mg/L as $CaCO_3$".

Operationally, alkalinity is defined as the concentration sum of all the titratable bases. It is estimated as the amount of H^+ added to a water to lower its pH to 4.5. At this pH, almost 100% of the $CO_3^=$ and HCO_3^- are converted to $H_2CO_3^*$, thus losing their ability to neutralize additional H^+. The standard method of alkalinity measurement is titration using 0.02 N sulfuric acid (H_2SO_4) or hydrochloric acid (HCl) solution. When the final pH of 4.5 is reached, every 1 mL of the H_2SO_4 solution added represents 1 mg/L of $CaCO_3$, or 20 meq/L of alkalinity, if 1.0 L of the water sample is used.

It is worth noting that alkalinity is different from basicity, which is defined by pH. A high pH does not mean high alkalinity, and changing alkalinity does not have to cause pH to change.

For example, a 10^{-3} M NaOH solution in a closed container (no CO_2 gas enters from the air) has a pH of 11 and alkalinity of 1 meq/L. Adding 10^{-3} mole HCl to 1 L of this solution would bring pH to 7 and alkalinity to zero. In contrast, a 10^{-3} M Na_2CO_3 solution in a closed container has a lower basicity with pH = 10.6 but higher alkalinity of 2 meq/L. Adding 10^{-3} mole HCl to 1 L of this solution would drop pH by only 1.3 units to 8.3.

With the use of the fractions α_0, α_1, and α_2, the alkalinity (Equation 4.25) of the water that is defined in the example above can be written as

$$\text{Alk} = \underbrace{\frac{K_H K_1 P_{CO_2}}{[H^+]}}_{HCO_3^-} + \underbrace{\frac{2K_H K_1 K_2 P_{CO_2}}{[H^+]^2}}_{CO_3^{2-}} + \underbrace{\frac{K_w}{[H^+]}}_{OH^-} - [H^+] \qquad (4.26)$$

or

$$\text{Alk} = K_H \left(\frac{K_1}{[H^+]} + 2\frac{K_1 K_2}{[H^+]} \right) P_{CO_2} + \frac{K_w}{[H^+]} - [H^+] \qquad (4.27)$$

or

$$\text{Alk} = K_H(\alpha_1 + 2\alpha_2) \frac{P_{CO_2}}{\alpha_0} + \frac{K_w}{[H^+]} - [H^+] \qquad (4.28)$$

After some acids and/or bases are added to the original water, the pH will change. The discussion below explains the approach to solving for the new pH after equilibrium is reached.

The acid–base reaction for a monoprotic acid HA can be written as

$$HA \leftrightarrow H^+ + A^- \qquad (4.29)$$

with equilibrium constant K_a

$$K_a = \frac{[H^+]\ [A^-]}{[HA]} \qquad (4.30)$$

Defining the fractions $\alpha_{0,\text{acid}}$ and $\alpha_{1,\text{acid}}$ of acid species after Equation 4.29 reaches equilibrium,

$$\alpha_{0,\text{acid}} = \frac{[\text{HA}]}{c_{\text{org}}} = \frac{[\text{H}^+]}{K_a + [\text{H}^+]} \tag{4.31}$$

$$\alpha_{1,\text{acid}} = \frac{[\text{A}^-]}{c_{\text{org}}} = \frac{K_a}{K_a + [\text{H}^+]} \tag{4.32}$$

where the concentration c_{acid} of the added acid is,

$$c_{\text{org}} = c_{\text{acid}} = \quad [\text{HA}] + [\text{A}^-]. \tag{4.33}$$

We obtain

$$[\text{A}^-] = c_{\text{acid}} \frac{K_a}{K_a + [\text{H}^+]} \tag{4.34}$$

Adding $[\text{A}^-]$ to the right-hand side of Equation 4.20 , a new charge balance can be written as:

$$[\text{H}^+] + 2 \frac{[\text{H}^+]^2 K_{SP}}{K_1 K_2 K_H P_{CO_2}} = \frac{K_w}{[\text{H}^+]} + \frac{K_1 K_H P_{CO_2}}{[\text{H}^+]} + 2 \frac{K_1 K_2 K_H P_{CO_2}}{[\text{H}^+]^2} + \frac{c_{\text{acid}} K_a}{K_a + [\text{H}^+]}, \tag{4.35}$$

And a new proton condition, with the added acid HA being a new reference species, can be written by adding $[\text{A}^-]$ to the right-hand side of Equation 4.21:

$$[\text{H}^+] + K_H P_{CO_2} = \frac{K_w}{[\text{H}^+]} + \frac{K_1 K_2 K_H P_{CO_2}}{[\text{H}^+]^2} + \frac{c_{\text{acid}} K_a}{K_a + [\text{H}^+]}. \tag{4.36}$$

When c_{acid} is known, either of Equations 4.35 and 4.36 can be solved for the new pH.

In addition to alkalinity (Equation 4.25), a related concept is acidity. It is the capacity of a water to resist pH change after strong bases are added. For waters open to the atmosphere, acidity (Acy) is defined as

$$[\text{Acy}] = 2[\text{H}_2\text{CO}_3] + [\text{HCO}_3^-] + [\text{H}^+] - [\text{OH}^-] \quad + \text{other acids} \tag{4.37}$$

Operationally, acidity is measured by adding a strong base until the pH reaches about 10.

As an example, let us consider data for Filson Creek (Figure 4.2), collected by Schnoor et al. (1984). In Figure 4.2, pAcy and pAlk represent the negative logarithms of the acidity or alkalinity vs. pH according to Equation 4.26. By contrast, pAcy' and pAlk' are adjusted to fit the actual data points using the concentration of an organic acid $c_{\text{org}} = 10^{-4.16}$ with $pK_a = 4.5$ and $pCO_2 = 10^{-2.8}$ atm. These values reflect the actual conditions under which an acid is present, and bacterial respiration causes the water to become supersaturated with carbon dioxide. The net effect is that the pH is considerably lower for a given alkalinity. For example, for $\text{Alk} = 10^{-4} \mu\text{equiv/L}$ the pH drops from 7.3 to 6.2.

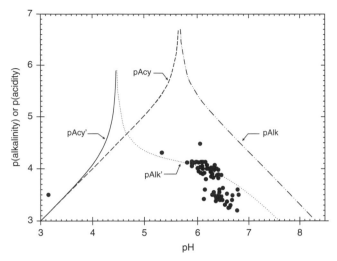

Figure 4.2. Plot of pAlk and pAcy vs. pH for Filson Creek. (Source: Adapted from Schnoor et al. (1984)). Field data are indicated by •. The lines --- and -..-..- indicate Equation 4.26 where the concentration of the anion from an organic acid $[B^-] = 0$ and $pCO_2 = 10^{-3.5}$. The lines ____ and ..... indicate Equation 4.28 with $[B^-]$ on the right hand side from an organic acid of concentration $c_{org} = [HB] + [B^-]$, where $c_{org} = 10^{-4.16}$ mol/L, $pK_a = 4.5$, and $pCO_2 = 10^{-2.8}$ atm.

Since alkalinity can be defined as the difference between the concentrations of conservative cations and anions, it is a conservative property, and we can write a mass balance,

$$V\frac{dA}{dt} = -I L_{acy}(1 - f) - QA + kV \qquad (4.38)$$

where
$A = $ alkalinity of the lake, µeq/L
$V = $ volume of the lake, L
$I = $ precipitation volume, L/yr
$L_{acy} = $ acidity of precipitation, meq/L
$Q = $ outflow (L / yr)
$k = $ zero order neutralization rate by sediments, µeq/(L × yr)
$f = $ fraction of acid neutralized in watershed
$t = $ time (yr)

The solution to this equation may be written as

$$A = A_0 e^{-\frac{Q}{V}t} + \left[\frac{kV}{Q} - \frac{I}{Q}L_{acy}(1 - f)\right] \times \left(1 - e^{-\frac{Q}{V}t}\right) \qquad (4.39)$$

where A_0 is the alkalinity at $t = 0$. When $t \to \infty$, the alkalinity will approach a constant value such that the rate of generation of alkalinity by neutralization in sediments equals the rate of loss by outflow and addition of hydrogen ions from acidic precipitation. Equation 4.39 may be used to estimate the time it takes to reduce the alkalinity to a certain level, for example as a result of a snowmelt event in the spring. The reduction in alkalinity can then be translated into a pH reduction of the lake using a graph such as the one shown in Figure 4.2.

4.2.4   Buffer Index

The buffer index, or buffer intensity, β (equiv/L-pH) is defined as the amount (mol/L) of the strong acid or strong base added to an aqueous solution in order to cause one unit change in pH. It can be estimated from the increase in concentration of cations C_B, such as Na^+, over the corresponding pH increase. Alternatively, the index may be defined as the ratio of increase in concentration of anions C_A, such as Cl^-, over the resulting decrease in pH. Thus we may write,

$$\beta = \frac{dC_B}{dpH} = -\frac{dC_A}{dpH} \tag{4.40}$$

Buffer intensity is an additive property, contributed by water itself and all species of weak acids and weak bases. For solvent water, its buffer index is expressed as

$$\beta_{H_2O} = 2.3([H^+] + [OH^-]) \tag{4.41}$$

As shown by the blue line in Figure 4.1(c), water has relatively high buffer intensity at low and high pHs.

The buffer index attributable to inorganic carbon, β_{CO_2}, may be written as

$$\beta_{CO_2} = \frac{d[HCO_3^-]}{dpH} + 2\frac{d[CO_3^{2-}]}{dpH} \tag{4.42}$$

In waters which are in equilibrium with atmospheric CO_2, this expression becomes,

$$\beta_{CO_2} = 2.3\left([HCO_3^-] + 4[CO_3^{2-}]\right) \tag{4.43}$$

The change of β_{CO_2} with pH is presented by the red line in Figure 4.1(c) with $pCO_2 = 10^{-3.5}$ atm. Therefore, the total buffer intensity for waters dominated by carbonate species is

$$\beta = \beta_{H_2O} + \beta_{CO_2} \tag{4.44}$$

which is represented by the black line in Figure 4.1(c). If the system is also in equilibrium with solid calcite, the carbonate term becomes $\beta_{CO_2,Ca} = 2.3\ ([HCO_3^-] + 4([CO_3^{2-}] + [Ca^{2+}]))$.

In well-buffered lakes, the buffer index is high, meaning that there is a considerable resistance to pH changes, even when there is a large input of acidity from sources such as acid rain. Non-acidic lakes are usually buffered as a result of inorganic carbon, i.e. atmospheric CO_2 dissolved in water, or as a result of equilibrium with calcite in the sediments. Generally, oceans have higher buffer intensity than most fresh waters.

At pH of around 8, the buffer intensity of oceans measures approximately 250 μeq/L (Pytkosoicx and Atlas, 1975).

By contrast, acid lakes, such as the Adirondack lakes during the 1980s, are buffered mainly by aluminum species (Table 4.3) or organic acids. Calculation of individual buffer index contributions for Adirondack water systems (Figure 4.3) proceeds as follows. For a total fluorine concentration of 0.1 mg/L or 5.26×10^{-6} mol/L it may be shown that AlF^{2+} and AlF^{2+} are the major fluorine species using the stability constants of Table 4.3. Similarly, assuming a total, i.e. free and complex, sulfate concentration of 7.2 mg/L or $7.49 \ 10^{-5}$ mol/L, the only sulfate complex contributing significantly to buffering is $AlSO_4^+$. The contribution to the buffer index from aluminum complexed with fluorine or sulfate is

$$\beta_1 = 2.3(6[AlF^{2+}] + 3[AlF_2^+] + 3[AlSO_4^+]) \tag{4.45}$$

The aluminum ion and hydroxyl complexes contribute significantly to the buffering capacity, especially at low pH values (pH less than 5.0). The associated buffer index β_2 is,

TABLE 4.3. Reactions Providing Buffering Capacity in Acidic Lakes

Reaction number	Reaction	K^a
	Hydroxide ligands	
1	$Al^{3+} + H_2O \leftrightarrow Al(OH)^{2+} + H^+$	1.03×10^{-5}
2	$Al^{3+} + 2H_2O \leftrightarrow Al(OH)_2^+ + 2H^+$	7.36×10^{-11}
3	$Al^{3+} + 4H_2O \leftrightarrow Al(OH)_4^- + 4H^+$	6.93×10^{-23}
4	$Al(OH)_3(s) \leftrightarrow Al^{3+} + 3OH^-$	3.09×10^{-34}
	Fluoride ligands	
5	$Al^{3+} + F^- \leftrightarrow AlF^{2+}$	1.05×10^7
6	$Al^{3+} + 2F^- \leftrightarrow AlF_2^+$	5.77×10^{12}
7	$Al^{3+} + 3F^- \leftrightarrow AlF_3$	1.07×10^{17}
8	$Al^{3+} + 4F^- \leftrightarrow AlF_4^-$	5.37×10^{19}
9	$Al^{3+} + 5F^- \leftrightarrow AlF_5^{2-}$	8.33×10^{20}
10	$Al^{3+} + 6F^- \leftrightarrow AlF_6^{3-}$	7.49×10^{20}
	Sulfate ligands	
11	$Al^{3+} + SO_4^{2-} \leftrightarrow AlSO_4^+$	1.63×10^3
12	$Al^{3+} + 2SO_4^{2-} \leftrightarrow Al(SO_4)_2^-$	1.29×10^5
	Calcite	
13	$CaCO_3(s) \leftrightarrow Ca^{2+} + CO_3^{2-}$	3.31×10^{-9}

(Source: Data adapted from Driscoll and Bisogni (1984), and Stumm and Morgan (1996) (Calcite)).
[a]at 25°C.

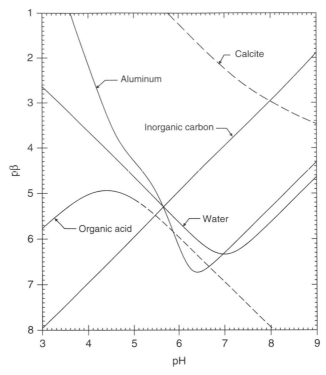

Figure 4.3. Buffer intensity diagram for dilute Adirondack water system. See also Driscoll and Bisogni (1984). Note the high buffering capacity of hypothetical calcite (not present). Inorganic carbon indicates carbonate equilibrium with the atmosphere, $pCO_2 = 10^{-3.5}$ atm. The effect of aluminum sulfate and fluorine complexes is noticeable in the pH range between 4 and 5 ($AlSO_4^+$) and between 5 and 6 (AlF^{2+} and AlF_2^+). The total concentrations of fluorine and sulfate are 0.1 mg/L and 7.2 mg/L respectively.

$$\beta_2 = 2.3(9[Al^{3+}] + 4[Al(OH)^{2+}] + [Al(OH)_2^+] + [Al(OH)_4^-]) \tag{4.46}$$

The main species of aluminum contributing to the buffer index at pH less than 5 is Al^{3+}. For pH greater than 6.4 $Al(OH)_4^-$ is the main contributor. The fluorine complex influences β in the pH range between 5 and 6, and $AlSO_4^+$ has an impact at pH values between 4 and 5 (Figure 4.3). In addition, the buffer intensity contributed by a dissolved organic acid with $K_a = 10^{-4.4}$ mol/L and $C_{acid} = 2 \times 10^{-5}$ mol/L is also shown in Figure 4.3.

The buffer index for an aquatic system in equilibrium with calcite is also shown in Figure 4.3 for comparison, even though calcite is rarely present in significant quantities in acidic lakes. Derivation of the buffer index for calcite begins with a charge balance,

$$[H^+] + 2[Ca^{2+}] + C_B = [HCO_3^-] + 2[CO_3^{2-}] + [OH^-] \tag{4.47}$$

where C_B is the concentration (equiv/L) of a cation such as Na^+. From Equations 5.106 and 5.113 we obtain the following expression for the buffer index of calcite,

$$\beta_{calcite} = \frac{d[HCO_3^-]}{dpH} + 2\frac{d[CO_3^{2-}]}{dpH} - 2\frac{d[Ca^{2+}]}{dpH} \qquad (4.48)$$

Since all inorganic species here are assumed to originate from dissolution of calcite we may write,

$$[Ca^{2+}] = C_T = [CO_3^{2-}] + [HCO_3^-] + [H_2CO_3] \qquad (4.49)$$

where C_T is the total concentration of inorganic carbonate species. Considering the solubility product for calcite (Equation 4.9), we obtain,

$$[Ca^{2+}] = \frac{K_{SP}}{[CO_3^{2-}]} = \frac{K_{SP}}{\alpha_2 C_T} = C_T \qquad (4.50)$$

and,

$$[Ca^{2+}] = \sqrt{\frac{K_{SP}}{\alpha_2}} \qquad (4.51)$$

The coefficient α_2 and the related coefficients α_0 and α_1 are defined in Equation 4.22 through 4.24. Note that α_0, α_1, and α_2 can be expressed in terms of the hydrogen ion concentration $[H^+]$ and the equilibrium constants K_1 and K_2 using Equation 4.6 and 4.8. By differentiation it may be shown that the expression for the buffer index reduces to

$$\beta_{calcite} = 1.15\sqrt{\frac{K_s}{\alpha_2}}[4\alpha_0(1+\alpha_2) + \alpha_1(2-\alpha_1)] \qquad (4.52)$$

This expression is shown as a dashed curve in Figure 4.3 using $K_1 = 4.45 \times 10^{-7}$, $K_2 = 4.69 \times 10^{-11}$, and $K_{SP} = 3.31 \times 10^{-9}$. Note that calcite buffering is quite high compared to other sources of buffering in the pH range between 5 and 8.

4.3   SULFUR CHEMISTRY

Sulfur plays an important role in the aquatic environment. Sources of the sulfur found in natural waters include contaminated water discharge, atmospheric input, and dissolution of sulfide minerals in the bottom bed materials.

Enrichment with sulfur is common in lake sediments where various reduced forms of organic sulfur can form. As a result of restrictions on sulfur dioxide emissions in an effort to reduce acid rain, annual sulfur emissions have been reduced since the late 1980s. The reduced form of sulfur, sulfide, as measured by acid volatile sulfide (AVS), plays an important role in precipitating heavy metals, thereby making them unavailable for aquatic organisms (Di Toro et al., 1990). Thus, it is not just the total concentrations of metals in sediments that determine the aquatic toxicity, but rather the difference between the simultaneously extractable metals and AVS. The relative abundance of stable isotopes of sulfur can be used to trace sources and fate of sulfur in the environment. Average percentages for ^{32}S, ^{33}S and ^{34}S are 95%, 0.8%, and 4.2% respectively.

4.3.1   Sulfur Redox Reactions in Water

The forms of sulfur in water depend on the pH and the redox potential of the water. The form of sulfur depending on oxidation state and pH can conveniently be illustrated using a pe – pH diagram (Stumm and Morgan, 1996). Such a diagram is shown in Figure 4.4 for a total dissolved S species concentration of 10^{-3} mol/L. Boundaries between areas of stability are defined by equal concentration of S in each form, except that the boundaries to solid sulfur are determined by the full dissolved S species concentration of 10^{-3} mol/L. Development of equations for line segments defining these boundaries may be seen as follows: Consider redox reaction

$$SO_4^{2-} + 8H^+ + 6e^- \leftrightarrow S(s) + 4H_2O \quad \log K = 36.2 \tag{4.53}$$

By taking logs and using $pe = - \log(e^-)$ and $pH = - \log[H^+]$ we obtain from the law of mass action,

$$pE = 6.03 + \frac{1}{6}\log[SO_4^{2-}] - \frac{8}{6}pH \tag{4.54}$$

Setting $[SO_4{}^{2-}] = 10^{-3}$ we obtain,

$$pE = 5.53 - 4/3\,pH \tag{4.55}$$

The boundary between $SO_4{}^{2-}$ and $H_2S(aq)$ is determined by reactions

$$SO_4^{2-} + 10H^+ + 8e^- \leftrightarrow H_2S(aq) + 4H_2O \quad \log K = 41.0 \tag{4.56}$$

From this we obtain,

$$pE = 5.13 + \frac{1}{8}\log\frac{[SO_4^{2-}]}{[H_2S(aq)]} - \frac{10}{8}pH \tag{4.57}$$

or since $[SO_4{}^{2-}] = [H_2S(aq)]$, $pe = 5.13-5/4\,pH$.
Considering the line defining the stability areas for $S(s)$ and $H_2S(aq)$

$$S(s) + 2H^+ + 2e^- \leftrightarrow H_2S(aq) \quad \log K = 4.8 \tag{4.58}$$

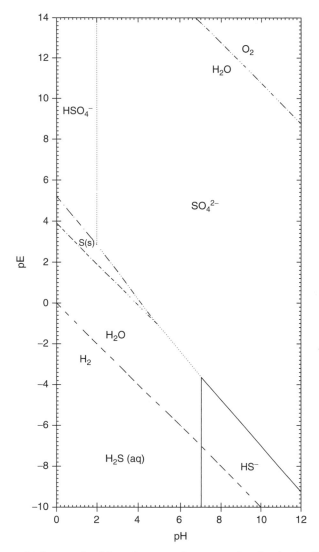

Figure 4.4. pe–pH diagram for $SO_4 - S(s) - H_2S$ system. The dissolved sulfur concentration is 10^{-3} mol/L. Lines between species indicate equal concentrations. Refer to reactions in Tables 4.1 and 4.2.

We have

$$pE = 2.4 - pH - \frac{1}{2}\log[H_2S] \qquad (4.59)$$

which for $[H_2S] = 10^{-3}$ becomes

$$pE = 3.9 - pH \qquad (4.60)$$

The boundary separating stability areas for HSO_4^- and $S(s)$ is obtained from reaction,

$$HSO_4^- + 7H^+ + 6e^- \leftrightarrow S(s) + 4H_2O \quad \log K = 34.2 \tag{4.61}$$

which produces the following equation for pE vs. pH,

$$pE = 5.70 + \frac{1}{6} \log[HSO_4^-] - \frac{7}{6}pH \tag{4.62}$$

or since $[HSO_4^-] = 10^{-3}$ mol/L,

$$pE = 5.20 - 7/6 \, pH \tag{4.63}$$

The line separating the stability areas for SO_4^{2-} and HS^- is derived from reaction,

$$SO_4^{2-} + 9H^+ + 8e^- \leftrightarrow HS^- + 4H_2O \quad \log K = 34.0 \tag{4.64}$$

Or,

$$\frac{1}{8}SO_4^{2-} + \frac{5}{4}H^+ + e^- \leftrightarrow \frac{1}{8}H_2S(g) + \frac{1}{2}H_2O \quad \log K = 4.25 \tag{4.65}$$

This means

$$p\epsilon = 4.25 + \frac{1}{8} \log \frac{[SO_4^{2-}]}{[HS^-]} - \frac{9}{8}pH \tag{4.66}$$

which for equal concentrations of SO_4^{2-} and HS^- becomes

$$pE = 4.25 - 9/8 \, pH \tag{4.67}$$

The boundary, Equation 4.68, between the areas of stability for HSO_4^- and SO_4^{2-} is determined by reaction

$$HSO_4^- \leftrightarrow SO_4^{2-} + H^+ \quad \log K = -2.0 \tag{4.68}$$

yielding

$$\log \frac{[SO_4^{2-}]}{[HSO_4^-]} - pH = -2.0 \tag{4.69}$$

or pH $= 2.0$ for equal concentrations of SO_4^{2-} and HSO_4^-.

Similarly, for the reduced species $H_2S(aq)$ and HS^-, we use Equation 4.70 to find the boundary between these two species,

$$H_2S(aq) \leftrightarrow H^+ + HS^- \quad \log K = -7.0 \tag{4.70}$$

From which

$$\log \frac{[HS^-]}{[H_2S]} - pH = -7.0 \tag{4.71}$$

or pH = 7 when $[HS^-] = [H_2S]$.

The half reactions of Table 4.2 can be combined to yield overall redox reactions. For example, by combining Equations 4.65 and 4.72,

$$\frac{1}{4}CO_2(g) + H^+ + e^- \leftrightarrow \frac{1}{4}CH_2O + \frac{1}{4}H_2O \tag{4.72}$$

the oxidation reaction of organic matter CH_2O by sulfate at pH = 7 is obtained

$$\frac{1}{8}SO_4^{2-} + \frac{1}{4}CH_2O + \frac{1}{4}H^+(W) \leftrightarrow \frac{1}{8}H_2S(g) + \frac{1}{4}CO_2(g) + \frac{1}{4}H_2O \tag{4.73}$$

Considering Equation 4.73 with typical species concentrations and CO_2 pressure, i.e. $P_{CO_2} = 10^{-3.5}$ atm, $[CH_2O] = 10^{-6}$ mole/L, $[SO_4^{2-}] = 10^{-3}$ mole/L and $P_{H_2S} = 10^{-2}$ atm (10^{-3} mole/L aqueous phase), we can prove that this reaction will proceed to the right. Thus sulfate can oxidize organic matter under environmentally relevant conditions (Stumm and Morgan, 1996). The organic matter, CH_2O, donates electrons that are consumed in the sulfate reduction to $H_2S(g)$.

Sulfate-reducing bacteria (SRB) and methanogenic bacteria (MB) can compete for substrate such that either SRB or MB dominate. Under reduced competition they can metabolize available substrates concurrently. In a petroleum-contaminated aquifer with high concentration of hydrocarbons, such as toluene, it was found that SRM dominated over MB above a relatively high sulfate concentration of 1.4 mg/L (Vroblesky et al., 1996). On the other hand, sulfate reduction took place at relatively low sulfate concentration (≤ 1 mg/L) when the electron donor (hydrocarbon) concentration was limited. Both SRB and MB are active when the competition for substrate is reduced as a result of high hydrocarbon concentration and low sulfate concentrations (≤ 1.5 mg/L).

Sulfur emissions from power plants have been reduced significantly since 1970–1987 in the United States and Europe. In the Czech Republic, this decrease has led to decreasing $\delta^{34}S$ values in bulk precipitation (Novak et al., 2001). The decrease measures approximately $1\%_o$ yr^{-1} from 1993 to 1996, and may be explained by the increasing proportion of the light sulfur isotope ^{32}S in biogenic emission of sulfur that was deposited during the years of peak pollution, 1970–1987.

It was also found that there is a yearly cycle of $\delta^{34}S$ in atmospheric $SO_2 - S$ with minimum during winter and maximum in summer. The cycle is contrary to the $\delta^{34}S$ of $SO_2 - S$ found in the United States and Canada and other parts of the world, where a low $\delta^{34}S$ during summer is ascribed to biogenic emissions and the high $\delta^{34}S$ in winter appears to be related to oxidation of SO_2 to sulfate, with a resulting ^{34}S enrichment at low winter temperatures. The atmospheric sulfate in Central Europe is heavier by $4\%_o$ from $SO_2 - S$. The resulting ^{34}S enters the ecosystem through heavy rain in the winter.

Although reemitted SO_2 after bacterial sulfate reduction in soil producing first S_2^- and then SO_x, would be enriched in the lighter isotope ^{32}S, the overall fractionation of atmospheric SO_2 during winter favors the heavier isotope ^{34}S, increasing the $\delta^{34}S$ value. In the Czech Republic this annual cycle is reversed as a result of the burning of light S coal ($\delta^{34}S = 1.6\%o$) during the winter months.

4.3.2   Sulfur in Sediment

Sediments in most eutrophic lakes are rich in organic sulfur. This enrichment depends on the lake eutrophic state but not on the availability of sulfide, organic matter, and reactive iron (Urban et al., 1999). Sulfur enrichment of organic matter is not common in oligotrophic lakes. Measurements by Urban et al. (1999) of C/S ratio and $\delta^{34}S$ of sediment humic acids for the eutrophic Lakes Greifen, Baldegg, and Sempach and the oligotrophic Lake Cadagno are shown in Figure 4.5. High sulfur contents are generally associated with low $\delta^{34}S$ values, indicating microbially reduced sulfur. Thus the high sulfur values reflect diagenetically formed organic sulfur. The reduced sulfur from sulfide is of importance, for example, in lake sediments where it precipitates metals, making them unavailable for toxic action (Christensen, 1998).

In bulk sediments, the organic sulfur is determined as total sulfur minus AVSs and chromium reducible sulfur (CRS). The humic acids were extracted from sediments and analyzed for total C, H, and S. Sediments were extracted under nitrogen with acetone and NaOH.

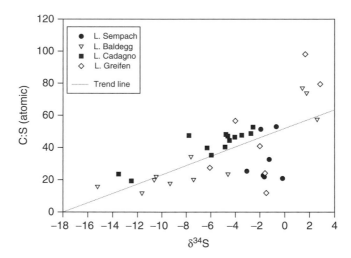

Figure 4.5. Sulfur enrichment of sediment humic acids as measured by C:S ratio and stable isotope ratios $\delta^{34}S$ ($\%o$) for four Swiss lakes. Additional sulfur present in samples with low C:S ratios is derived from microbially reduced sulfur and represents diagenetically formed organic sulfur. (Source: Reprinted from Urban et al. (1999) with permission from Elsevier.)

Speciation of sulfur in the humic acid fraction of the sedimentary organic carbon was examined by means of X-ray photoelectron spectroscopy (XPS). Reduced organic S in humic acids consists of sulfides or thiols. Addition of sulfur to polyunsaturated molecules also occurred. Thiophene was found in 60-year old sediment from Lake Baldegg. In addition, there were a small amount of sulfoxides in the top 2–3 cm of sediment from Lake Cadagno. Sulfoxides may have formed diagenetically in sediments that were exposed to oxygen or those that contain sufficient iron oxides.

4.3.3   Acid Rain

In pure water, pH = 7. Atmospheric CO_2 dissolves into rain water, causing pH lower than 7 due to the reactions shown in Equations 4.2 and 4.5. The electroneutrality condition of normal, unpolluted rain water can be expressed as

$$[H^+] = [OH^-] + [HCO_3^-] \qquad (4.74)$$

In this equation, $[OH^-] \ll [HCO_3^-]$. Next, from Equations 4.1 through 4.6 we obtain $[H^+] = 2.21 \times 10^{-6}$ mol/L or pH = 5.6, which indicates geochemical neutrality.

Acid rain refers to rain with pH lower than 5. Although CO_2 acts as an acid when dissolved in water (Equation 4.5), it does not lower the pH below 5.6 under current partial pressure of 0.00039 atm. Responsible to the formation of acid rain is SO_2, and to a lesser extent NO_x, which have mostly been emitted from human activities.

Acid rain is formed when atmospheric SO_2 reacts with O_2 to form SO_3, which combines with water to produce H_2SO_4

$$2\, SO_2 + O_2 \;\rightarrow\; 2\, SO_3 \qquad (4.75)$$

$$SO_3 + H_2O \;\rightarrow\; H_2SO_4 \qquad (4.76)$$

Similarly, nitric acid, the other major component of acid rain, is formed by oxidation of NO and subsequent reaction with water of the resulting NO_2

$$2NO + O_2 \;\rightarrow\; 2\, NO_2 \qquad (4.77)$$

$$4NO_2 + 2H_2O + O_2 \;\rightarrow\; 4\, HNO_3 \qquad (4.78)$$

In addition, a small amount of HCl is generated in the flue gas of municipal incinerators that burn waste containing chlorinated hydrocarbons. Data from New England during the 1980s indicate that 60–70% of acid rain was caused by sulfuric acid and 30–40% by nitric acid with hydrochloric acid constituting less than 10%. Coal-fired power plants account for 69% of SO_2 emissions and 32% of the NO_x emissions. Another 40% of the NO_x comes from mobile sources.

Acid rain was a major threat to the biodiversity of some lakes in the 1980s. SO_2 emissions caused by coal burning in coal-fired power plants are responsible for most acid rain. SO_2 reacts with elemental oxygen and hydroxyl radicals to form sulfuric acid. NO_x emissions from power plants and transportation sources form nitric acid. About 10% of acidity comes from HCl formed as a result of municipal incinerator emissions. Although now of less significance, acid rain impacts lake ecosystems through pH effect and higher concentrations of metals, such as Al, that affect fish. Acid rain can also have detrimental effects on limestone buildings.

The main reason for the environmental concern about acid rain is that the resulting pH of lakes and streams at 5.0 or less causes the lowering of biodiversity. For example, the lakes in the Adirondack region of New York State have suffered from reduced species richness, with the loss of taxa per unit pH decrease similar across the trophic levels (Nierzwicki-Bauer et al., 2010). The acid rain input, combined with the resulting high concentrations of toxic aluminum compounds, can cause fish kills. Another potentially detrimental effect on aquatic biota is the dissolution of other toxic metals such as mercury, lead, and cadmium. A lake or river with pH less than 5 is deemed to be acidic. Acid rain can damage buildings with exterior limestone surfaces by eroding the limestone and possibly forming porous and less stable gypsum. Buffering of acid rain in lakes by calcite, dissolved carbonate or bicarbonate, aluminum compounds, organic acids, and others is a mechanism that is important in reducing the potential pH decrease from a given amount of acid rain.

Deposition of NO_x, and particularly of sulfate, contributes to the acidity of small, poorly buffered seepage lakes in the eastern United States. However, with the improved sulfur reduction technology for coal-fired power plants, this problem seems to have decreased in importance. Efforts to reduce acid rain in the United States resulted in mean changes in SO_4^{2-} concentrations in the eastern United States from 1983 through 1994 of -7.1 µequiv/L. The mean changes of hydrogen ion concentration and cations measured -4.76 and -4.34 µequiv/L respectively. By contrast, the change in NO_3^- concentration was much smaller (-0.55 µequiv/L) (Lynch et al., 2000). Further reductions in sulfate and hydrogen ion concentrations have been realized through Title IV of the 1990 Clean Air Act Amendments (CAAA), which set limitations on sulfur dioxide emissions from a large number of utility-owned coal-fired power plants. Concentrations of sulfate and free hydrogen ion decreased from 10 to 24% over a large area of the eastern United States from 1995 through 1997 compared to the previous 12-year reference period.

For the lakes of the Adirondack Region of New York this reduced input of acid rain has caused significant changes in the lake chemistry (Driscoll et al., 2003). The mean rates of change for 16 lakes of the Arirondack Long-Term Monitoring (ALTM) program were -2.31 µequiv/L for $SO_4^{2-} + NO^{3-}$, and -2.32 µequiv/L for cations during the period from 1982 to 2000. Note that in a soil-water system, cations are displaced from exchange sites during the addition of sulfate ions, and, similarly, decreases in sulfate in solution are followed by an equivalent decrease of cations in solution.

As discussed previously, sulfur emissions in Central Europe have undergone steep decreases since 1970–1987, which was the period of maximum pollution (Novak et al., 2001). This decrease is accompanied by a decrease, at least from 1993 to 1996, of $\delta^{34}S$ reflecting biologically emitted SO_2, which contains a relatively high proportion of light sulfur ^{32}S from periods of heavy sulfur pollution.

4.4   IMPACT OF GLOBAL WARMING ON NATURAL WATERS

With the continued increase in CO_2 concentration in the atmosphere, its concentration in the oceans and lakes is expected to rise from the current 390 ppmv. Based on the Henry's law constant, the equilibrium activity of CO_2 in waters currently measures approximately 1.2×10^{-5} mol/L. It could reach 1.4×10^{-5} and 3.2×10^{-5} mol/L in 2100, based on the best ($P_{CO_2} = 480$ ppmv) and the worst ($P_{CO_2} = 1000$ ppmv) scenarios of the predicted CO_2 levels in air respectively. When P_{CO_2} reaches 1000 ppmv, the pH in otherwise unpolluted rain water will decrease from the current 5.6 to 5.4, roughly corresponding to a 60% increase in $\{H^+\}$. In the waters that are in equilibrium with limestone, as described in Section 4.2, the pH could drop from 8.3 to 8, meaning doubling of $\{H^+\}$. In less buffered water, the acidification would be more remarkable.

Increased CO_2 input will shift the reaction shown in Equation 4.79, which is the combination of reactions 4.2, 4.5, 4.7, and 4.9, to the right

$$CO_2(g) + CaCO_3(s) + H_2O \leftrightarrow 2\, HCO_3^- + Ca^{2+} \qquad (4.79)$$

causing the enhanced dissolution of $CaCO_3$, which is not only the base material of sedimentary rocks, but also forms the bone structure of coral reefs and is the primary constituent of the shells or skeletons of marine organisms such as coccolithophorids, foraminifera, and pteropods.

Acidification of the oceans due to anthropogenic CO_2 uptake has caused a 0.002 yr^{-1} decrease in pH, with higher rates in coastal waters (NOAA, 2011). With the continued increase of CO_2 in the atmosphere, accelerated disappearance of the corals and shellfish, including mollusks such as oysters and mussels, is expected. In fact it has been occurring for the past few decades.

Freshwater lakes could be acidified more rapidly than the oceans because the buffer intensity is lower in many lakes than in the ocean. In addition, freshwater lakes are affected by the input from tributaries, agricultural runoff, invasive species, and urban pollution from their watersheds. These complicate the assessment of the susceptibility of lakes to accelerated global warming. The calculated average Ω_{arag} is 2.32 in Lake Michigan, which has a limestone/dolomite bottom and thus a relatively high buffer intensity, and only 0.15 for Lake Superior, which sits on the granite-rich Canadian Shield and thus has lower alkalinity (NOAA, 2011). In the worst scenario, that atmospheric CO_2 increases to 1000 ppmv, the pH of Lake Superior water would drop from the current 8.0 to about 7.7, doubling $\{H^+\}$. The impact of water acidification on the plankton and fish in the Great Lakes is yet to be systematically investigated.

For precipitation, the equilibrium pH of otherwise unpolluted rain will be lowered by about 0.1 to 0.2 unit when the atmospheric CO_2 is increased to 500–1000 ppmv respectively. More importantly, the origin, frequency, and patterns of continental rains may be altered by the changes in global climate, causing changes in the precipitation chemistry as demonstrated for the northeastern United States by Kelly et al. (2009).

REFERENCES

Benjamin, MM. *Water Chemistry*. Waveland Press, Inc; 2010.

Brezonik, PL. *Chemical Kinetics and Process Dynamics in Aquatic System*. Boca Raton, Florida: Lewis; 1994.

Brezonik, PL, Arnold, WA. Water chemistry: fifty years of change and progress. *Environ. Sci. Technol.* 2012;46:5650–5657.

Christensen, ER. Metals, acid-volatile sulfides, organics, and particle distributions of contaminated sediments. *Wat. Sci. Technol.* 1998;37(6–7):149–156.

Christensen, ER, O'Melia, CR. *Workshop on research needs for coastal pollution in urban areas*. Milwaukee, Wisconsin: University of Wisconsin-Milwaukee; 1998.

Di Toro, DM, Mahony, JD, Hansen, DJ, Scott, KJ, Hicks, MB, Mayr, SB, Redmond, MS. Toxicity of cadmium in sediments: the role of acid-volatile sulfide. *Environ. Toxicol. Chem.* 1990;9:1487–1502.

Driscoll, CT, Bisogni, JJ. Weak acid/base systems in dilute acidified lakes and streams of the Adirondack Region of New York State. In: Schnoor, JL. *Modeling of Acid Precipitation Impacts*. Boston: Butterworth Publishers, Ann Arbor Science; 1984.

Driscoll, CT, Driscoll, KM, Roy, KM, Mitchell, MJ. Chemical response of lakes in the Adirondack Region of New York to declines in acidic deposition. *Environ. Sci. Technol.* 2003;37:2036–2042.

ERS, 2007. *MINEQL+ version 4.6*. Environmental Research Software, Hallowell, ME. http://www.mineql.com/index.html (accessed March 12, 2014).

GWB, 2014. *GeoChemist's Work Bench version 10*. Aqueous Solutions LLC. Champaign, IL 61820 http://www.gwb.com/ (accessed March 12, 2014).

Hönisch, B, Ridgwell, A, Schmidt, DN, Thomas, E, Gibbs, SJ, Sluijs, A, Zeebe, R, Kump, L, Martindale, RC, Greene, SE, Kiessling, W, Ries, J, Zachos, JC, Royer, DL, Barker, S, Marchitto Jr, TM, Moyer, R, Pelejero, C, Ziveri, P, Foster, GL, Williams, B. The Geological Record of Ocean Acidification. *Science* 2012;335:1058–1063.

Kelly, V, Weathers, KC, Lovett, GM, Likens, GE. Effect of Climate Change between 1984 and 2007 on Precipitation Chemistry at a Site in Northeastern USA. *Environ. Sci. Technol.* 2009;43:3461–3466.

KTH, 2010. *Visual MINTEQ*. Department of Sustainable Development, Environmental Science and Engineering, KTH Royal Institute of Technology, Stockholm, Sweden. http://www.lwr.kth.se/English/OurSoftware/vminteq/ (accessed March 12, 2014).

Lynch, JA, Bowersox, VC, Grimm, JW. Acid rain reduced in Eastern United States. *Environ. Sci. Technol.* 2000;34:940–949.

Morel, FMM, Hering, JG. *Principles and Applications of Aquatic Chemistry*. John Wiley & Sons, Inc; 1993.

Nierzwicki-Bauer, SA, Boylen, CW, Eichler, LW, Harrison, JP, Sutherland, JW, Shaw, W, Daniels, RA, Charles, DF, Acker, FW, Sullivan, TJ, Momen, B, Bukaveckas, P. Acidification in the Adirondacks: Defining the Biota in Trophic Levels of 30 Chemically Diverse Acid-Impacted Lakes. *Environ. Sci. Technol.* 2010;44:5721–5727.

NOAA Ocean Acidification Steering Committee (2010): NOAA Ocean and Great Lakes Acidification Research Plan, NOAA Special Report.

Novak, M, Jackova, I, Prechova, E. Temporal trends in the isotope signature of air-borne sulfur in Central Europe. *Environ. Sci. Technol.* 2001;35:255–260.

Pankow, JF. *Aquatic Chemistry Concepts*. Lewis Publisher; 1991.

Pytkosoicx, RM, Atlas, E. Buffer intensity of seawater. *Limnology and Oceanography* 1975;20(2):222–229.

Saunders, DL Kalff, J. Denitrification rates in the sediments of Lake Memphremagog, Canada – USA. *Wat. Res.* 2001;35:1897–1904.

Sawyer, CN, McCarty, PL, Parkin, GF. *Chemistry for Environmental Engineering and Science*, Fifth Ed. Boston, MA: McGrawHill Higher Education; 2003.

Schnoor, JL, Palmer, WD Jr, Glass, GE. Modeling impacts of acid precipitation of Northeastern Minnesota. In: Schnoor, JL. Modeling of Acid Precipitation Impacts, Boston: Butterworth Publishers, Ann Arbor Science; 1984.

Snoeyink, VL, Jenkins, D. *Water Chemistry*. New York: John Wiley & Sons, Inc; 1980.

Stumm, W, Morgan, JJ. *Aquatic Chemistry - Chemical Equilibria and Rates in Natural Waters*, Third Ed. New York, NY: John Wiley and Sons, Inc.; 1996.

Urban, NR, Ernst, K, Bernasconi, S. Addition of sulfur to organic matter during early diagenesis of lake sediments. *Geochim. Cosmochim. Acta* 1999;63:837–853.

USEPA, 2006. *MINTEQA2. Center for Exposure Assessment Modeling (CEAM)*, U.S. Environmental Protection Agency, Athens, GA. http://www2.epa.gov/exposure-assessment-models/minteqa2 (accessed March 12, 2014).

USGCRP. Global Climate Change Impacts in the United States. T.R. Karl, J.M. Melillo, and T.C. Peterson (editors). *United States Global Change Research Program*. New York, NY: Cambridge University Press; 2009.

USGS, 2014. *PHREEQC version 3.1.2. U.S. Geological Survey*. http://wwwbrr.cr.usgs.gov /projects/GWC_coupled/phreeqc/ (accessed March 12, 2014)

Vroblesky, DA, Bradley, PM, Chapelle, FH. Influence of electron donor on the minimum sulfate concentration required for sulfate reduction in a petroleum hydrocarbon contaminated aquifer. *Environ. Sci. Technol.* 1996;30:1377–1381.

Weber, WJ Jr, DiGiano, FA. *Process Dynamics in Environmental Systems*. New York, NY: John Wiley and Sons, Inc.; 1996.

5

NUTRIENTS

5.1 INTRODUCTION

Aquatic life is dependent on an adequate supply of nutrients. Algae and bacteria exist at the bottom of the food chain. They require inorganic substances, for example, nitrogen and phosphorus, for their growth; trace metals, such as copper and zinc, for enzyme activity; and organic compounds, for example, biotin and other vitamins, for stimulating growth. Nitrogen is used for amino acid and protein synthesis and phosphorus is important for the adenosine triphosphate (ATP) adenosine diphosphate cycle, which provides energy storage for metabolic activity. Heterotrophic bacteria can consume dissolved and particulate organic compounds for carbon supply, whereas autotrophic bacteria and most algae obtain their carbon from inorganic sources, particularly bicarbonate ion. In addition, algae must have sufficient light to sustain photosynthesis of new cell material.

A number of nutrients are essential for microbial activity. The main macronutrients required are carbon, oxygen, nitrogen, phosphorus, and sulfur. Several micronutrients, including copper, manganese, and B vitamins, are also required for biological growth. A listing of nutrient requirements of activated sludge in biological (or secondary) wastewater treatment processes, along with nutrients detected in wastewater, is given in Table 5.1 (Burgess et al., 1999). In some cases, nitrogen and phosphorus must be added to wastewater treatment plants to ensure effective treatment. Nutrient deficiency can promote the growth of undesirable filamentous bacteria that can hinder proper settling in secondary clarifiers. With a sufficient supply of nutrients, floc-forming bacteria providing good settling are allowed to develop, preferably with only a small amount of filamentous bacteria. Thus

Physical and Chemical Processes in the Aquatic Environment, First Edition. Erik R. Christensen and An Li.
© 2014 John Wiley & Sons, Inc. Published 2014 by John Wiley & Sons, Inc.

TABLE 5.1. Nutrient Requirements of Activated Sludge

Nutrients	Theoretical requirement (mg/L)	Detected in wastewater (mg/L)
Macronutrients		
N	15.0	32.0
P	3.0	1.69
S	1.0	100.0
Trace Elements		
Ca	0.4–1.4	0.44
K	0.8 to >3.0	95.0
Fe	0.1–0.4	1.20
Mg	0.5–5.0	10.0
Mn	0.01–0.05	< 1.0
Cu	0.01–0.05	< 1.0
Al	0.01–0.05	0.02
Zn	0.1–1.0	< 1.0
Mo	0.1–0.7	< 1.0
Co	0.1–5.0	< 1.0
Vitamins		
Biotin	0–1.0 µg/L	
Niacin	0–1.0	
Thiamine (B_1)	0–1.0	
Lactoflavin (B_2)	0–1.0	
Pyridoxine (B_6)	0–0.01 µg/L	
Pantothenic acid	0–1.0	

(Source: Burgess et al. (1999)).

phosphorus removal, which otherwise may be desirable to prevent eutrophication of receiving waters, may not be an appropriate treatment; at least not within the activated sludge process itself, in which new bacterial biomass is generated.

An evaluation of regional trends in total phosphorus (TP) and total nitrogen (TN) concentrations in streams and rivers of the United States during 1993–2003 was provided by Sprague and Lorenz (2009). Nutrient data were tested for trend in flow-adjusted concentrations, indicating anthropogenic changes, and observed concentrations, which are indicative of both natural and anthropogenic changes. Using a seasonal Mann Kendall test including spatial correlation they found only one significant trend: TP showed an increase in the central portion of the United States. This trend was significantly linked to increased fertilizer application, although the rank correlation coefficient was low (0.22). Other potential sources, manure and population density, showed positive but insignificant correlation with TP. Omitting spatial correlation causes more regional trends to be displayed, emphasizing the importance of considering spatial correlations in regional trend assessment when site concentrations are not spatially independent.

Most inland lakes and rivers are phosphorus (P) limited. Lake Tahoe, California is an example of an ultraoligotrophic lake with high transparency. Lake Tahoe was previously nitrogen (N) limited but has now become P-limited (Zhang et al., 2002). Atmospheric deposition of N and P appears to have been a major responsible factor for this shift. Measurements made by aircraft of N in the low-Sierras and mid-Sierras showed N concentrations of 660 ± 220 nmol N/m^3-air and 630 ± 350 nmol N/m^3-air respectively, with concentrations over the laboratory ranging from one-half to one-fifth (clear conditions) of these values. There was no measurable P in this part of the Sierras. However, north of Lake Tahoe in an area impacted by forest fires, the atmospheric P concentration was around 26 nmol P/m^3-air and the N concentration 860 nmol N/m^3-air. The measured P concentration over the lake was 2.3 ± 2.9 nmol P/m^3-air (clear air) and 2.8 ± 0.8 nmol P/m^3-air (smoky conditions).

The N/P atomic ratio over the lake of greater than 15:1 suggests P-limitation (Correll, 1998), assuming comparable Henry's law constants for N and P. From the N-measurements, there was a large gradient between the Sierra values of those over the lake. This points to the atmosphere as a source of the nitrogen in the lake, especially when coupled with the fact that the composition of N compounds, e.g. gaseous, particulate, inorganic, and organic, was very similar over the Sierras compared to values over the lake, especially under clear atmospheric conditions.

5.2   INPUT OF NUTRIENTS AND ACIDITY

Throughout the 1980s major concern about the effects of acid rain prompted regulatory agencies, such as the U.S. EPA, to conduct comprehensive studies and to adopt limits on the amount of sulfur in coal and NO_x emissions. These actions have been at least partially successful in curbing and controlling the acidification of lakes. Figure 5.1 shows a map displaying pH of rainfall in the United States in 1982. It is clear that the lowest pH values, i.e. less than 4.4, are reached in the northeastern part of the country.

Acid rain is generated by the oxidation of SO_2 and NO_x emissions from coal-fired power plants and automobiles, e.g. NO and NO_2. The reaction of sulfur with oxygen is enhanced catalytically, by Fe^{3+}, Mn^{2+}, or NH_3, or photochemically. Oxidation products such as NO_2 and SO_3 react with H_2O to form sulfuric and nitric acid. In addition, direct oxidation of NO and SO_2 with ozone (O_3) during daylight hours, when significant ozone concentrations can build up through photochemical reactions, leads to NO_2 and SO_3, which may be hydrolyzed to form nitric and sulfuric acid. A minor portion of acid rain is generated by the reaction of rain water with chlorine from atmospheric emissions. Atmospheric chlorine is present in emissions from municipal incinerators.

Dryfall contains an alkaline component that provides some neutralization to acid rain. This component is most likely to be carbonate and bicarbonate. Dissolved calcium carbonate in precipitation has a similar effect, as has been observed in Florida (Edgerton et al., 1981).

Deposition from the atmosphere may contribute 25–40% of the nutrients that cause eutrophication in Atlantic estuaries (Hicks, 1998). Especially important is NO_x downwind of industrial and urban complexes. Other nitrogen species include

Figure 5.1. Values for pH of wet deposition in the United States, 1982. Precipitation-weighted annual average. (Source: National Acid Precipitation Assessment Program: Annual Report 1983 to the President and Congress, p. 36.)

NO_3^-, NH_3 and NH_4^+. The atmospheric nitrogen species reaches the estuaries through direct deposition and indirectly via deposition to land and subsequent runoff. Sources include cars, power plants, industry, and agriculture. It is now generally recognized that the atmosphere contributes nutrients to the biosphere of rivers, lakes, and oceans.

The turbulent and gravitational transfer of nutrients from air to surfaces makes dry deposition difficult to measure. Dry deposition generally reflects low altitude local sources, whereas the sources of wet deposition originate at higher altitudes and, therefore, tend to transfer material from remote sources. Dry and wet deposition of several pollutants, including nutrients, are of approximately equal significance to deposition to land. However, wet deposition dominates deposition to water surfaces. This may be related to the greater significance of wet scavenging in the water environment.

Note that locations of maximum deposition of NO_3^-, NH_4^+, sulfate, and rain may differ as was demonstrated in a study undertaken near St. Louis, Missouri (Hicks, 1998). Although the atmosphere may have helped to initiate and sustain recent outbreaks of *Pfiesteria piscicida* on the US east coast, it would appear that outbreaks are not directly linked to atmospheric deposition.

For pre-Cambrian lakes in Canada, bulk precipitation (wt + dry) contributed from 5 to 70%, with an average of 32%, of the load of TP (Dillon and Reid, 1981). The biologically available phosphorus (BAP) accounted for 28 ± 27% of TP for bulk precipitation and 40 ± 37% for wet only precipitation. Also, the BAP constituted 105 ± 78% of the filterable reactive phosphorus (FRP) in bulk precipitation and 127 ± 118% of FRP in wet only precipitation. Thus FRP or soluble reactive phosphorus (SRP) may be deemed a surrogate for BAP. Phosphorus is nonvolatile, meaning that it is limited to rock–soil–water phases. Therefore, dryfall of phosphorus is more important (80% of total deposition in Florida) than wet precipitation (Hendry et al., 1981). Levels of SRP are greater than or equal to organic P in order of significance of atmospheric P species.

Regarding atmospheric nitrogen deposition, Hendry et al. (1981) found that the order of significance of nitrogen species in Florida is $NO_3^- $-N $\geq NH_4^+$-N $\geq$ organic N. The wet only mode is the major mechanism for inorganic nitrogen deposition. During the summer contribution of atmospheric nitrogen is greater as a result of the increased biogenic emission of gaseous N because of the higher soil temperature.

The atmospheric loading rate of N to Florida lakes amounts to approximately 37% of the N loading for eutrophic conditions. By contrast, the atmosphere supplies only 12–16% of the P loading required to cause eutrophic conditions. The relatively high N loading is consistent with the fact that nitrogen is seldom the limiting nutrient in lakes.

Ammonia deposition depends on both the NH_3 concentration in the air and the lake water pH, as may be seen from the following equation

$$\frac{p_{NH_3}}{[NH_4^+]} = \frac{K_a}{K_H}\frac{1}{(H^+)}$$
(5.1)

where

p_{NH_3} = partial pressure of NH_3,
K_H = Henry's law constant for NH_3, and
K_a = ionization constant for NH_4^+

For $K_H = 57$ mol/(L atm) (Stumm and Morgan, 1996) and $K_a = 5.56 \times 10^{-10}$ mol/L (Sawyer et al., 2003), we obtain an atmospheric NH_3 concentration of 7.56 $\mu g/m^3$ in equilibrium with lake water at pH = 8.2 and with ammonia-N concentration, NH_4^+–N of 0.1 mg/L. Atmospheric ammonia concentrations of greater than 7.56 $\mu g/m^3$ are not commonly found, e.g. in Florida (Messer and Brezonik, 1981), and absorption of NH_3 from the atmosphere is, therefore, not very likely. Only smaller acidic lakes can absorb ammonia.

Nitrogen fixation by blue-green algae can contribute significantly to the nitrogen budget of some lakes. Atmospheric input of nitrogen to lakes is most significant for smaller lakes that are not heavily impacted by agricultural runoff.

Dry deposition of nitrogen typically accounts for 30–60% of the total atmospheric deposition of nitrogen. Progress in the measurement of dry deposition fluxes of nitrogen has been made by Shahin et al. (1999). They measured dry NO_3^- and NH_3 fluxes in

Chicago during May–October 1997. The total nitric acid (gas) flux measured 3.78 ± 1.24 mg/(m^2 d) and the particulate nitrate flux measured 1.46 ± 0.3 mg/(m^2 d). Similarly, they found that the NH$_3$ gas flux measured 2.64 ± 1.15 mg/(m^2 d) while the NH$_4^+$ particulate flux measured less than 0.06 mg/(m^2 d), which is the detection limit. The nitrate flux depends on time of day of the sample as a result of photo oxidation processes. However, it is independent of wind direction because of the complex chemistry involving both nitrous and nitric acid. By contrast, ammonia fluxes are independent of sampling time but depend on wind direction. Ammonia is emitted primarily from land-based sources, and land surfaces exhibit low reactivity with ammonia gases compared to water surfaces (Lake Michigan).

Surrogate surfaces can be used for the experimental determination of gaseous and particulate fluxes of nitrogen species to water surfaces. The surrogate surface used by Shahin et al. (1999) was a knife leading-edge surrogate sampler (KSS), covered with either a Nylasorb filter for HNO$_3$ (gas) and particulate nitrate, citric acid coated paper filter for NH$_3$ (gas) and particulate ammonia, or a greased disk for particulate species. Nylon in the Nylasorb filter is a perfect sink for HNO$_3$.

For comparison, a water surface sampler (WSS) was used to measure particulate plus gas fluxes. The average mass transfer coefficients for HNO$_3$ and NH$_3$ (gases) measured 1.5 ± 0.22 and 2.46 ± 1 cm/s respectively. An annular denuder system was used to measure ambient concentrations of HNO$_3$ and NH$_3$. Future advances in surrogate surfaces are likely to include samples for dry deposition of phosphorus and other nutrients or pollutants.

5.3 EUTROPHICATION

Eutrophication describes an over enrichment of receiving waters with mineral nutrients (Correll, 1998). Phosphorus is such a nutrient, which in organic forms is hydrolyzed in water by extracellular enzymes to phosphate. In freshwaters, P is frequently the limiting nutrient. Orthophosphate, PO$_4^{3-}$, is the only form of P that autotrophs can assimilate. P is then deemed to be in the pentavalent state. As a result of this enrichment with nutrients, there can be an excessive production of autotrophs, particularly algae, including blue-green algae and cyanobacteria. The high productivity leads to high bacterial activity and respiration. This reduces the oxygen content of the water, leading to hypoxia or anoxia, particularly in the bottom waters that are not well mixed. As a result of this, aquatic animals and fish may not survive. The low oxygen content near the lake bottom can lead to release of P from the bottom sediments, amplifying the effects of eutrophication.

In contrast to lakes, coastal waters are often N-limited. The main reasons for this are the more efficient recycling of P in estuaries, the loss of N through denitrification in coastal sediments, and the role of sulfate in recycling P in coastal sediments. These sediments will reduce the sulfate to sulfide, which will bind ferrous ions to the sediment, thus allowing P to diffuse out from the sediment. In a more oxidizing environment, such as a lake bottom, iron would be precipitated as Fe(OH)$_3$ at the sediment–water interface, preventing release of P. Estuaries can be P-limited at their upstream ends in the spring, and N-limited in the summer and fall.

Under good growth conditions, algae have a fixed elemental atomic ratio between N and P of approximately 15:1 and 16:1. This ratio is known as a Redfield ratio, which can be used to estimate the limiting nutrient. As N/P becomes greater than 16, growth may be P-limited and otherwise N-limited. In case of P-limitation, Correll (1998) and others have found that it is not only the SRP concentration that is important for growth. Under typical conditions there is a growth-rate-dependent transfer from solid or organic forms to SRP. Similar observations have been made for nitrogen; therefore it is the total concentrations of TP and TN that are important for the state of eutrophication.

Environmental costs of freshwater eutrophication in England and Wales were estimated by Pretty et al. (2003). Table 5.2 summarizes the major cost elements consisting of clean water or nonnutrient enriched water (A) and policy response costs, which are costs incurred in responding to eutrophication (B). The damage costs include reduced value of waterside properties amounting to $13.8 million and drinking water treatment costs, DC_1

$$DC_1 = C_0 A_p \times ASP_0 + C_c A_p \times ASP_c + C_r \qquad (5.2)$$

where C_0 and C_c, are annual operating and capital costs respectively, and C_r are reservoir management costs. These costs apply to noneutrophic conditions.

TABLE 5.2. Major Cost Elements of Freshwater Eutrophication in the United Kingdom

Cost categories	Annual costs ($ million)
A. Damage Costs: reduced value of clean or nonnutrient enriched water	
A1. Social damage costs	
Reduced value of waterside properties	13.8
Drinking water treatment costs (remove algal toxins)	26.6
Drinking water treatment costs (remove nitrogen)	28.1
Reduced value of atmosphere (greenhouse and acidifying gas emissions)	7.17–11.2
Reduced recreational value	13.5–47.0
Reduced value for tourist industry	4.12–16.3
A2. Ecological damage cost	
Negative ecological effects on biota (loss of key or sensitive species)	10.3–14.2
Total	104–157
B. Policy response costs: costs involved in responding to eutrophication	
Sewage treatment costs (remove P from large point sources)	70.4
Costs of treatment algal blooms	0.70
Costs of adopting new farm practices that emit fewer nutrients	4.75
Total	76

(Source: Pretty et al. (2003)).

The fraction A_p is the water production suffering from algal proliferation, and ASP_0 and ASP_c are, respectively, fractional operating and capital costs associated with eutrophication. Using $C_0 = 398$ million, $C_c = \$466.1$ million, $C_r = \$5.6$ million, $A_p = 0.33$, $ASP_0 = 0.1$ and $ASP_c = 0.05$ we obtain $DC_1 = \$26.6$ million. The costs DC_2 to remove nitrogen are the sum of operating and capital costs, NC_0 and NC_c respectively,

$$DC_2 = NC_0 + NC_c \qquad (5.3)$$

This cost was estimated to be \$28.1 million. Other damage costs include reduced value of the atmosphere as a result of greenhouse or acidifying gases amounting to \$7.17–11.2 million, reduced recreational value in the order of \$13.5–7.0 million, and reduced value for the tourist industry of \$4.12–16.3 million.

The greenhouse gases are methane CH_4 and nitrous oxide N_2O and the acidifying gas is NH_3. The estimated damage cost for CH_4 is \$109.1 t^{-1}, for N_2O \$4145 t^{-1}, and for NH_3 \$239 t^{-1}. By comparison, the damage cost of carbon dioxide CO_2 is \$41.7 t^{-1}. Other factors involved in the CH_4, N_2O, and NH_3 damage costs are emissions and the fraction of emissions caused by eutrophic conditions of water bodies and water courses. Sewage treatment costs amounting to \$70.4 million, and costs for adopting new farm practices such as no-till plowing that emits fewer nutrients of \$4.75 million constitute the main elements of the policy response expenditure.

While it may be argued that damage costs and policy response costs cannot be added together because of the larger uncertainty of the damage costs, their sum does represent a monetary value of the overall burden of eutrophication. From this point of view, a change in sewage treatment would be justified to the point where there is an overall decrease in the cost of eutrophication.

Dodds et al. (2009) used a slightly different approach in their assessment of potential economic damages of eutrophication of US freshwaters. Rather than categorizing water bodies as eutrophic or noneutrophic on the basis of algal blooms as in Pretty et al. (2003), they provided baseline or reference levels of TN and TP that were compared to current levels of these nutrients. Under this scenario, lakes that were estimated to be hypereutrophic under reference conditions were subtracted from lakes that were hypereutrophic under current conditions.

The overall annual cost of eutrophication of US freshwaters was estimated to be \$2.2 billion, with the greatest economic losses attributed to lakefront property values, (\$0.3–2.8 billion/yr) and recreational use (\$0.37–1.16 billion/yr). Secchi depths of lakes are directly related to property values. Other costs were related to lack of biodiversity and additional drinking water treatment. Agencies from 14 states were surveyed to gather data regarding number of days closed for contact and noncontact use, number of fish kills, restoration efforts, macrophyte removal, and water treatments added because of eutrophication. While all of these categories were represented in the composite data, there was a lack of uniformity in the reporting, which makes accurate cost damage estimating difficult. However, even with these shortcomings the estimate of the economic consequences of eutrophication is useful as a guide to policy setting for nutrient controls.

The Nakdong River system in South Korea is an example of a highly eutrophic water body (Kim et al., 1998). The Nakdong River drains southerly over the southern and northern Kyung Sang Do provinces and empties into the Korea Strait. TN and TP concentrations measured 2.89 ± 0.10 mg/L (206 ± 7.1 μmole/L) and 0.211 ± 0.016 mg/L respectively. The average calculated ratios of these nutrients were TN/TP = 20.8 ± 0.9, based on milligram per liter values, or TN/TP = 46.1 based on atomic ratios. This is significantly greater than 16:1, indicating that P is limiting for algal growth. Tributary regions generally have higher ammonia NH_3-N concentrations (0.3 ± 0.3 mg/L) and lower nitrate NO_3-N concentrations (1.4 ± 0.9) than the main stream regions, in which these values are 0.21 ± 0.2 mg/L NH_3-N and 1.79 ± 0.63 mg/L NO_3-N. This would be expected based on downstream oxidation of organic nitrogen.

Correlation studies demonstrate positive correlation between pH and dissolved oxygen, which is expected during photosynthesis and bicarbonate consumption. Other positive correlations include conductivity with TP and dissolved inorganic phosphorus (DIP); biochemical oxygen demand (BOD) with TN, TP, and DIP; TP with TN; and NH_3-N with dissolved organic nitrogen (DON).

Interstitial waters of the Gulf of Gdansk and of the Pomeranian Bay were examined by Bolalek and Frankowski (2003) for ammonia and phosphates, total iron Fe_{tot} and Fe(II), F(III) concentrations, and organic matter content (Table 5.3). Redox potential indicates conditions for reducing, particularly for the Gulf of Gdansk, which is also reflected in lower concentrations of Fe(III) (0.89 μmole/L) and a lower ratio of Fe(III)/Fe_{tot} (0.11). The Gulf of Gdansk also has the highest organic matter content (17.8%) and lowest sulfate concentration (2.15 mmol/L). Under these conditions, oxidation of organic matter in the sediments may occur as a result of reduction of iron oxides and sulfates and disproportionation

$$(CH_2O)_{106}(NH_3)_{16}(H_3PO_4) + 212Fe_2O_3 + 848H^+ \rightarrow$$

$$424Fe^{2+} + 106CO_2 + 16NH_3 + H_3PO_4 + 530H_2O \qquad (5.4)$$

$$(CH_2O)_{106}(NH_3)_{16}(H_3PO_4) + 53SO_4^{2-} \rightarrow$$

$$106CO_2 + 16NH_3 + 53S^{2-} + H_3PO_4 + 106H_2O \qquad (5.5)$$

$$(CH_2O)_{106}(NH_3)_{16}(H_3PO_4) \rightarrow 53CO_2 + 53CH_4 + 16NH_3 + H_3PO_4 \qquad (5.6)$$

High phosphate concentrations (80.8 μmole/L) are in accordance with the sulfate reduction described in Equation 5.5.

Eutrophication can influence air–water exchange, vertical fluxes, and phytoplankton concentrations of persistent organic pollutants such as polychlorinated biphenyls (PCBs). This was demonstrated by Dachs et al. (2000) in a study of Lake Ontario using a dynamic air–water–phytoplankton exchange model. Under steady-state, near equilibrium, the fugacity ratios f_g/f_w between gas and dissolved phases, and f_w/f_p between dissolved and phytoplankton phases, are close to 1, where f_g, f_w, and f_p are

TABLE 5.3. Concentrations of Substances in Interstitial Waters of Bottom Sediments of the Gulf of Gdansk and the Pomeranian Bay

Substance	Gulf of Gdansk			Pomeranian bay		
	Number of samples	Arithmetic mean	Standard deviation	Number of samples	Arithmetic mean	Standard deviation
Water content (%)	32	66.7	15.1	23	34.5	13.3
Organic matter (%)	32	17.8	5.6	23	3.1	2.3
E_h (mV)	32	−297	46	6	−104	74
pH	32	7.49	0.18	6	7.49	0.15
SO_4^{2-} (mmol/L)	53	2.15	2.17	24	3.26	3.39
PO_4^{3-} (μmol/L)	59	80.8	97.9	30	27.8	18.6
NH_4^+ (μmol/L)	56	258	163	24	94.5	42.8
Alkalinity (mmol/L)	31	2.24	0.92			
Fe_{tot} (μmol/L)	48	7.32	7.50	26	14.8	12.3
Fe(II) (μmol/L)	55	8.93	11.1	27	12.1	11.3
Fe(III) (μmol/L)	47	0.89	1.61	26	3.21	2.68
Fe(III)/Fe_{tot}	47	0.11	0.11	26	0.24	0.17

(Source: Bolalek and Frankowski (2003)).

fugacities (in atm) of the gas, water, and phytoplankton phases. These fugacities may be expressed as

$$f_g = C_g RT \qquad (5.7)$$

$$f_w = C_w H \qquad (5.8)$$

$$f_p = \frac{C_p H}{BCF} \qquad (5.9)$$

where C_g(mole/m^3-air), C_w(mole/m^3-water), and C_p(mole/m^3-phytoplankton) are concentrations of PCBs in air, water, and phytoplankton. R (J/(mol K)) is the gas constant, H ((atm m^3-water)/mol) is the Henry's law constant, and BCF (m^3-phytoplankton/m^3-water) is the bioconcentration factor. The water to phytoplankton fugacity ratio is close to 1 for PCBs such as PCB101; therefore PCBs are in equilibrium between the water and phytoplankton phases. However, there is a lack of equilibrium between the air and water bases as demonstrated by high fugacity ratios f_g/f_w. Phytoplankton uptake of PCBs depletes the dissolved phase and the transfer between the air and water phase is rapid enough to support phytoplankton uptake.

Thus eutrophication with high biomass leads to a disequilibrium between the gas and dissolved phase, resulting in enhanced air–water exchange and vertical sinking fluxes of PCBs. If eutrophication could be stopped, it was projected that PCB concentration in phytoplankton would be increased.

5.3.1 Eutrophication Control

During the 1960s and 1970s many industrialized countries made great efforts to control eutrophication through legislation banning phosphate in detergents. This has helped to bring eutrophication under control in critical waters, for example Lake Erie. However, much remains to be done because nonpoint pollution, such as agricultural runoff, can still create excessive algal growth in lakes or streams. If nutrient inputs are not controlled prior to discharge in a receiving water, one option is to reduce P in inflow, such as in stormwater runoff, by precipitation with ferric chloride (Sherwood and Qualls, 2001). Another option is to remove N and P from wastewater effluent using an ecological wastewater treatment system (Rectenwald and Drenner, 2000). One potential problem with the P reduction method is that the phosphorus is still discharged to the receiving water. However, as precipitate it will adsorb to iron oxy hydroxides in the sediments. The ecological treatment method is not specific to N and P and requires a large treatment area. We consider these two options below. More specific N and P removal processes will be described later.

The removal of P by the addition of ferric chloride to stormwater runoff was investigated by Sherwood and Qualls (2001) in microcosms as an initial step prior to full-scale testing in a demonstration-scale wetland as part of the Everglades Nutrient Removal Project. Microcosms were constructed from polyvinyl chloride (PVC) pipe sections filled with 20 cm of Everglades soil, which was collected from the sampling site. Radiolabeled (^{32}P) precipitate, produced by the reaction of ferric chloride with Everglades water spiked with $^{32}PO_4{}^{3-}$, was placed 0, 3, and 6 cm below the sediment–water interface, and aeration was provided in the water using airstones for 10 h/d, thus simulating the natural diurnal O_2 cycle of Everglades surface water. Precipitation of iron oxyhydroxides, in addition to adsorption of P by ligand exchange, is the main mechanism of P removal.

The results of the experiments demonstrated that sediments became reducing (<120 mV) after 10 d. Most redox potentials 10–20 d after ^{32}P addition measured approximately −150 mV. Under these reducing conditions one could expect Fe^{2+} to form from Fe^{3+} with subsequent release of ^{32}P to the water column. During the 139 d of the experiment, there was some vertical migration via solubilization and diffusion, or via colloid formation, of P in the interstitial water, but less than 1% of ^{32}P was released to the water column. This result suggests that ferrous phosphates, such as vivianite, of low solubility were formed. A certain amount of the Fe^{2+} may be complexed with humic acids.

Overall, the addition of ferric chloride to stormwater entering the system may be an effective method for P reduction in the Florida Everglades. However, the long-term effect of accumulated P precipitate in the sediments and the cost need to be considered. Addition of aluminum sulfate, alum, to stormwater to achieve P removal is a potential alternative; however there are concerns regarding the effects of aluminum and sulfate on plant communities.

The feasibility of using an ecological wastewater effluent treatment system based on fish grazing on periphyton was investigated by Rectenwald and Drenner (2000). Twelve 375-L tanks with *Tilapia mossambica*, an algal grazing chichlid, feeding on periphytic detrital aggregate, i.e. detritus, algae, and bacteria, were arranged in series following the secondary clarifier and a dechlorination tower. Average nutrient removals through fish growth and feces amounted to 27 mg TP/(m^2 d) and 108 mg TN/(m^2 d), with TP and TN removals of 82 and 23% respectively. In order to remove 1 mg/L from a 1 mgd wastewater treatment plant, assuming an optimal P removal rate of 100 mg of P/(m^2 d), a surface area of 3.8 ha would be required. This is a very large area and it would potentially limit the usefulness of this system. Also, because of temperature sensitivity it would have greatest applicability in tropical climates.

5.3.2 Harmful Algal Blooms

There are many examples of harmful algal blooms. Despite reductions in riverine nutrient loading and estuarine nutrient concentrations, the river Neuse in North Carolina has experienced an overall worsening of conditions, including eutrophication with toxic dinoflagellates (Qian et al., 2000). Distinct changes occurred in the watershed, including the impoundment of a large reservoir at the headwaters of the river in 1983 and a phosphate detergent ban in 1988. While nitrogen concentrations had slightly increased over the past 20 yr, those changes were not present in the lower river and estuary. However, the decreases in phosphorus that occurred upstream because of the phosphorus detergent ban persist downstream. As a result, based on the Redfield value, nitrogen to phosphorus ratios have experienced a shift from nitrogen to phosphorus limitation. Phytoplankton species have often preferred dissolved inorganic nitrogen to DIP ratios. The increases in nuisance algal blooms and toxic phytoplankton were therefore thought to have occurred as a result of this shift in the N/P ratio.

Cyanobacterial blooms may be associated with toxins and taste-and-odor compounds. The toxins are of concern for human health and the taste and odor compounds are aesthetically objectionable. Graham et al. (2010) investigated the co-occurrence of toxins and taste-and-odor compounds in several eutrophic lakes with a known history of cyanobacterial blooms in the Midwestern United States. Sampling was carried out during August 2006 in 23 lakes from Iowa, northeastern Kansas, southern Minnesota, and northwestern Missouri. Samples were analyzed for chlorophyll, cyanobacterial community composition, total and dissolved toxins, and dissolved taste-and-odor compounds. Toxins (anatoxins, cylindrospermopsins, lyngbyatoxins, microcystins, nodularins, and saxitoxins) were analyzed by enzyme-linked immunosorbent assay (ELISA) and liquid chromatography followed by tandem mass spectrometry (LC/MS/MS). Taste-and-odor compounds (geosmin and 2-methylisoborneol (MIB)) were determined by solid-phase microextraction gas chromatography/mass spectrometry.

It was found that *Anabaena*, *Aphanizomenon*, and *Microcystis* were present in 96% of the lakes. Microcystins occurred in all blooms. Toxins and taste-and-odor compounds occurred simultaneously in 91% of the blooms. Taste-and-odor compounds were always accompanied by at least one cyanotoxin, but there were cases where toxins were present without taste-and-odor compounds. Thus the presence of earthy or musty odors is a strong indication of aquatic toxins. Even in the absence of such odors, toxins may be present and can present a human health risk.

Blooms of toxic cyanobacteria in the western basin of Lake Erie have been a nuisance since at least the mid-1990s. Rinta-kanto et al. (2005) examined blooms in order to characterize and quantify *Microcystis* cells and measure levels of microcystin toxicity. Samples were collected in August 2003 from *R/V Lake Guardian* and Canada Center for Inland Waters Coast Guard ship (CCGS) *Limnos*. Additional samples were obtained using CCGS *Limnos* in August 2004. Microcystin was detected by conventional polymerase chain reaction (PCR) and quantified by real-time PCR or qPCR. The quantitative polymerase chain reaction (qPCR) test was carried out with specific primer-Taq-man probes targeted on a *Microcystis*-specific 16S rDNA fragment and a microcystin toxin synthetase gene *mcy*D. Microcystin was determined by a protein phosphatase inhibition assay normalized to microcystin-LR (MLR) standards. LANDSAT 7 images taken in August 2003 were used to demonstrate the presence of a major bloom of phytoplankton in the surface waters near the mouth of Maumee River and in Sandusky Bay.

Concentrations of microcystin exceeding the safety limit of 1 µg/L (MLR equivalents) set by the World Health Organization were found in samples from both 2003 and 2004. The abundance of microcystic cells varied between $4 \cdot 10^8$ and $2 \cdot 10^3$ cells/L. Not all strains of *Microcystis* were able to produce toxins. Satellite images showed that the Maumee River and Bay appeared to be a source of the cyanobacteria.

Dreissenid mussels enhance the *Microcystis* biomass at low P concentrations. However, contrary to the expectation without mussels, increased P loading reduces the *Microcystis* abundance (Sarnelle et al., 2012). While the explanation for this phenomenon is not yet clear, it could be that increased filter feeder action of the mussels at high P concentrations suppresses the biomass of cyanobacteria in the water column.

Field control of algal blooms may occur due to loss of aesthetic value of a lake. Nakano et al. (2001) made an attempt to control algal blooms in Lake Senba, a small recreational lake in Mito City, Japan. The lake has a surface area of 33 ha and a volume of $365,000$ m^3. In order to reduce the growth rate of cyanobacteria and provide efficient mixing throughout the lake, several ultrasonic radiation systems (USRS), each consisting of a sonication module and a water jet circulation module, were placed on the surface of the lake during 1998 and 1999.

During most of the project period the Mito municipal government supplied river water to Lake Senba at a flow rate of $75,000$ m^3/d. For perfect mixing, the dilution rate therefore measured 0.207 d^{-1}, corresponding to a detention time of 4.8 d. To wash out cyanobacteria under these conditions, their growth rate must be less than this flushing rate. However, natural *Microcystis aeruginosa*, the most dominant cyanobacteria in Lake Senba have a much higher growth rate than 0.207 d^{-1}. Cyanobacteria form aggregates that are difficult to attack using grazing organisms. The sonication would destroy gas vacuoles and the buoyant ability of cyanobacteria to find optimal levels of illumination in the water column. It also damaged their photosynthetic system. This would slow the growth rate and delay growth recovery. Complete mixing is achieved when the mixing time T_m is less than one tenth of the mean residence time. Thus the number of USRS units N was determined from

$$T_m = \frac{V_L}{Q_e N} < 0.48 \text{ d} \qquad (5.10)$$

where V_L is the volume of the lake and Q_e is the volumetric flow rate of the entrained liquid. Q_e is 8.2 times Q_d where $Q_d = 5.61$ m^3/ min is the volumetric flow rate discharged from the water jet circulator. Equation 5.10 shows that N is greater than 11.5. Thus, while 12 units were necessary, as a result of budget limitations only 10 units were installed. The results of the field experiment showed significant improvement in transparency, and lowered chlorophyll a and suspended solids and chemical oxygen demand (COD) concentrations. Phosphorus concentrations also declined as a partial result of the disappearance of chlorophyll a. However, while the chlorophyll a peak disappeared during 1998, it appeared in 1999 because of reduced flushing during the construction of a canal. One potential limitation of this technology is the impact of mixing on wind-driven currents that could counteract the flushing provided by the USRS units.

Yang et al. (2012) developed a flushing strategy to prevent blooms of the diatom *Stephanodiscus hantzschii* in the lower-middle Hanjiang River, China. River blooms occur often in low flow reaches that contain pools with near stagnant water. The flushing strategy was based on a generalized additive model (GAM), which reflected statistical correlations between air temperature, soluble reactive silicon (SRSi), turbidity, TP, dam release from the upstream Danjiangkou Reservoir, photosynthetically reactive radiation (PAR), pH, and TN. Maximum biomass occurred at 10°C, and there was a strong negative correlation between biomass and dam release for releases measuring less than 2000 m^3/s. There was no time lag between air temperature and phytoplankton, but an 8- to 9-d delay between dam release and biomass. The flushing strategy was used to predict the necessary dam release based on forecasts for the air temperature over the next few days. The dam release estimated from GAM measured less than that which would be obtained directly from measured parameters, thus producing a potential water savings.

The occurrence of toxic blooms in drinking water supplies are quite common and are of concern because of the possible exposure to associated toxins such as hepatotoxins and possible links to human liver cancer (Feitz et al., 1999). One of these cyanotoxins, the hepatotoxin MLR, is present in extracts from *M. aeruginosa* blooms. MLR, the chemical structure that is shown in Figure 5.2, is often identified in cynobacterial blooms and has a high acute toxicity with LD50 values for rodents in the 30–100 μg/kg range. The most common method of controlling cyanobacterial blooms involves application of copper-based algicides. This will destroy the blue-green cells but has unwanted side effects: It could promote blue-green tolerant species, destroy zooplankton, and release toxins upon cell lysis. While chemical coagulation and filtration is ineffective in removing these toxins, other methods, such as ozonation and activated carbon adsorption, have proven effective but could be limited in their application as a result of cost considerations. We describe photocatalytic degradation of MLR using UV or sunlight with TiO$_2$ as a catalyst here (Feitz et al., 1999).

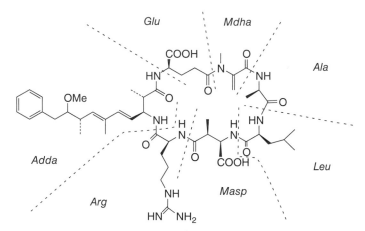

Figure 5.2. Structure of the blue-green algal toxin microcystin-LR (MLR). Adda, 3-amino-9-methoxy-2,6,8-trimethyl-10-phenyldeca-4,6-dienoic acid, Ala, alanine; Arg, arginine; Glu, glutamic acid; Leu, leucine; Masp, methylaspartic acid; Mdha, *N*-methyl-dehydroalanine. (Source: Reprinted from Feitz et al. (1999) with permission from American Chemical Society.)

The adsorption of MLR to TiO_2 (1 g/L) in the dark is maximal at low pH values, e.g. 3.5. The adsorption isotherm follows a Langmuir expression,

$$(MLR)_{ads} = \frac{K_{ads}(MLR)_{ads}^{max}(MLR)_{soln}}{1 + K_{ads}(MLR)_{soln}} \tag{5.11}$$

as can be seen by linearizing the above expression,

$$\frac{(MLR)_{soln}}{(MLR)_{ads}} = \frac{(MLR)_{soln}}{(MLR)_{ads}^{max}} + \frac{1}{K_{ads}(MLR)_{ads}^{max}} \tag{5.12}$$

and plotting data points for MLR_{soln}/MLR_{ads} vs. MLR_{soln} produces a near linear relationship (Figure 5.3). From this regression line, the two parameters K_{ads} and $(MLR)_{ads}^{max}$ can be determined from the slope and intercept. We obtain $(MLR)_{ads}^{max} = 126$ nM and $K_{ads} = 1.00 \times 10^{7.0} M^{-1}$, which then produces the curve shown in Figure 5.4. MLR has two acid groups (erythro-b-methyl aspartic acid and D-glutamic acid, and a basic group [L-arginine]). Complexes of MLR and carbonate may interact with TiO_2 surface hydroxyl groups according to reactions shown in Table 5.4. Reaction 6 of this table was suggested by Feitz et al. (1999) to control the rate of degradation of MLR. The reaction rate also depends on the concentration of oxidizing species but this may rapidly reach a steady-state concentration. Thus at low pH, the degradation rate is controlled by the equation,

$$-\frac{d[MLR]}{dt} = k[> TiMLRH] \tag{5.13}$$

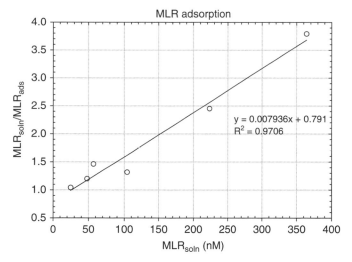

Figure 5.3. Ratio of the concentration of toxin in solution $[MLR]_{soln}$ to that adsorbed to 1 g/L TiO_2 at pH 3.5, $[MLR]_{ads}$, as a function of the concentration of toxin in solution $[MLR]_{soln}$. The straight line represents a Langmuir fit to the data. (Source: Raw data are from Feitz et al. (1999))

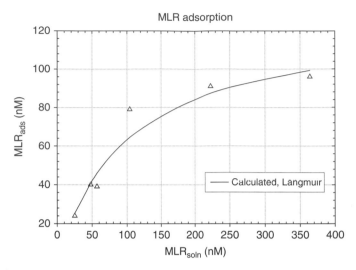

Figure 5.4. Concentration of MLR adsorbed to 1 g/L TiO_2 at pH 3.5, $[MLR]_{ads}$, as a function of the solution concentration of the toxin $[MLR]_{soln}$. The parameters from the straight-line fit in Figure 5.3 are used for the rectangular hyperbola.

TABLE 5.4. Reactions and Associated Constants Used in Surface Complexation Model of Microcystin-LR Interaction with TiO_2 Surface Hydroxy Groups

No.	Reaction	log K
1	$> TiOH_2^+ \Longleftrightarrow > TiOH + H^+$	2.6
2	$> TiOH \Longleftrightarrow > TiO^- + H^+$	9.0
3	$MLRH_3^+ \Longleftrightarrow MLRH_2^0 + H^+$	3.0
4	$MLRH_2^0 \Longleftrightarrow MLRH^- + H^+$	3.0
5	$MLRH^- \Longleftrightarrow MLR^{2-} + H^+$	12.5
6	$> TiOH + MLRH^- + H^+ \Longleftrightarrow > TiMLRH + H_2O$	8.0
7	$> TiOH + CO_3^{2-} + H^+ \Longleftrightarrow > TiCO_3^- + H_2O$	13.0

(Source: Feitz et al. (1999)).

In this case cyanobacterial exudates adsorb to TiO_2 producing long-lived radicals that degrade the adsorbed MLR. At higher pH where MLR adsorption to TiO_2 is insignificant, it is assumed that exudate radicals generated by exudate adsorbed to TiO_2 diffuse out in the solution, where they initiate oxidation of dissolved MLR. In order for degradation to occur UV, or sunlight, and TiO_2 must be present. Also, somewhat surprisingly, sunlight irradiation at a lower level (4.4 mW/cm^2) is more efficient than irradiation from a near-UV lamp at 55 mW/cm^2.

Degradation pathways and intermediates formed during photocatalytic degradation of MLR were examined by Antoniou et al. (2008). During TiO_2 photocatalysis, the aromatic ring is hydroxylated and MLR is demethoxylated. They found that MLR degradation by hydroxyl radicals is initiated at three sites of the Adda amino acid, i.e. at the aromatic ring, the methoxy group, and the conjugated diene bonds; and at one site of the cyclic structure, i.e. the Mdha amino acid. Intermediates were identified by mass spectrometry.

Climate change can affect temperatures of oceans, lakes, and rivers. With increased concentrations of greenhouse gases, such as carbon dioxide CO_2, methane CH_4, and nitrous oxide N_2O, the Earth's ice caps are melting. This has resulted in sea level rise and increased evaporation. Hydrology and frequency of storms and droughts are known to be affected as well. Paerl and Paul (2012) reviewed possible impacts on harmful cyanobacteria blooms as a result of climate change. They considered cyanobacteria blooms in freshwater systems including Lake Erie; Lake Dianchi, Yunan Province, China; Lake Taihu, Jiangsu Province, China; and the lower St. Johns River, Florida. Marine cyanobacterial harmful algal blooms from Fort Lauderdale, Florida; Papua New Guinea; Sanibel Island, coastal Gulf of Mexico; and the Baltic Sea, Gulf of Finland were considered.

Higher nutrient loadings, increased temperatures, and rising atmospheric CO_2 favor larger and more persistent blooms of cyanobacteria. In the Laurentian Great Lakes, dreissenids, e.g. zebra and quagga mussels, can enhance phosphorus cycling and availability and thus increase the significance of blooms of *Cladophora* and other algae (Zou and Christensen, 2012). There are many negative effects of cyanobacterial harmful algal blooms including impacts on drinking water, fisheries, property values, and tourism. Management strategies can focus on a reduction in greenhouse gases, enhanced vertical mixing of inland waters and a more stringent nutrient control.

5.3.3 Cladophora

One filamentous green alga that is often associated with eutrophication is *Cladophora glomerata*. It is a very common weedy river and lake alga in the temperate northern hemisphere (Dodds, 1991), and is deemed a nuisance for chlorophyll a > 100 mg/m^2. Low flow, high NH_4^+, and low epiphytic loads correlate ($P < 0.05$) with high biomass. Possible control parameters include temperature, inorganic phosphorus, inorganic nitrogen, and light. *Cladophora* outbreaks have been a recurring problem in the Great Lakes area, particularly between 1995–2005. One hypothesis to explain this is that filter-feeding zebra mussels improve water clarity, thus increasing the light level and stimulating *Cladophora* growth. However, the exact cause of the proliferation of this alga is not known, making it difficult to control excessive growth. Based on field studies of the East Gallatin River above and below the Bozeman Wastewater Treatment Plant, Montana, Dodds (1991) found significant increased growth of *Cladophora* below the plant, indicating that nutrient levels do have an impact.

Parr et al. (2002) investigated the effect of mats of the duckweeds *Lemna minor* and *Lemna minuscula* floating above *C. glomerata*. The authors determined that *Lemna* mats reduced the photosynthesis system II quantum efficiency of *Cladophora*, thus reducing the dissolved oxygen (DO) levels. The oxygen from *Lemna* is released to the atmosphere, rather than the water. The duckweed mats also reduce the light that is necessary for photosynthesis of *Cladophora*. Optimum temperature of *C. glomerata* can range between 13–17 °C for Lake Huron and 28–31 °C in Lake Michigan.

During high summer temperature and irradiance, shading inhibits oxygen-producing photosynthesis beneath canopies of *Lemna* spp. Thus oxygen electron transport and/or photorespiration will be limited. This inhibition of *Cladophora* results in senescence in slow or standing water and subsequent collapse of DO concentration, which can cause detrimental effects on downstream water quality. Electron transport rate (ETR) for *Cladophora* diminished under *L. minor* mats. In addition, the light levels at which ETR is saturated were reduced from greater than 170 $\mu mole/(m^2$ s) to approximately 70 $\mu mole/(m^2$ s) under the duckweed mat.

5.4   NITROGEN

5.4.1   The Nitrogen Cycle

Nitrogen is an essential nutrient and constituent in amino acids and protein. About 12.4% of cell material, e.g. $C_5H_7NO_2$, is nitrogen. The equivalent percentage for protein is around 16%. Agricultural crops are often nitrogen-limited (Trachtenberg and Ogg, 1994), and excess nitrogenous compounds, including ammonia, organic nitrogen and nitrate, can significantly contribute to eutrophication of receiving waters. Excess nitrogen can also cause acidification of lakes and streams, and toxic algal blooms. There was a renewed interest, especially in the 1980s, in nitrogen in connection with acid rain, in which NO_x compounds from automobile exhaust or coal-fired power plants can be precursors for nitric acid. This can be quite detrimental to fish populations in acid-sensitive lakes. In addition, high concentrations of nitrogen in the form of ammonia can be toxic to aquatic life. In the twentieth century, anthropogenic doubling of the global nitrogen fixation rate has intensified global N_2O greenhouse gas and N_2O induced stratospheric depletion of ozone (Barton and Atwater, 2002).

Ammonia (NH_3) and nitrate (NO_3^-) belong to the category of toxic nitrogen compounds. Ammonia is toxic to fish and the U.S. EPA has, therefore, set a limit of 0.2 mg/L N-NH_3 (Sawyer et al., 2003). Because of the reaction with hydrogen ions,

$$NH_4^+ \leftrightarrow H^+ + NH_3 \tag{5.14}$$

with

$$\frac{[H^+]\ [NH_3]}{[NH_4^+]} = K_A = 5.71 \times 10^{-10} \tag{5.15}$$

the amount of ammonia present depends on the pH of the water. For example, a total concentration of ionized (NH_4^+) and nonionized ammonia (NH_3), e.g. 5 mg/L, violates the EPA standard at pH greater than 7.86, but is in conformance with the standard at lower pH values. Reactions with equilibrium constants for nitrogen species in water are shown in Table 5.5.

TABLE 5.5. Major Reactions and their Equilibrium Constants for Nitrogen Species

Reaction number	Reaction	$K^a(25\,^\circ C)$
(1)	$NH_3(g) \leftrightarrow NH_3(aq)$	57.00
(2)	$NH_3(aq) + H_2O \leftrightarrow NH_4^+ + OH^-$	1.77×10^{-5}
(3)	$N_2(g) \leftrightarrow N_2(aq)$	6.61×10^{-4}
(4)	$NO_2(g) \leftrightarrow NO_2(aq)$	1.00×10^{-2}
(5)	$NO(g) \leftrightarrow NO(aq)$	1.90×10^{-3}
(6)	$N_2O(g) \leftrightarrow N_2O(aq)$	2.57×10^{-2}

(Source: Data adapted from Stumm and Morgan (1996)).
^{a}mol/(L-atm) at 25 °C, Henry's constants. Aqueous concentrations in mol/L.

There is a limit for nitrate of 10 mg/L in drinking water according to the chemical drinking water standards established by the U.S. EPA (Table 11.2). Related limits are 10 mg/L for nitrate + nitrite (as N), and 1 mg/L for nitrite (as N). The reason for these limits is that higher nitrate concentration can cause methemoglobinemia or "blue baby" syndrome in infants. This condition, which is associated with considerable risk, means that the oxygen-carrying capacity of hemoglobin in blood is compromised because of oxidation of the hemoglobin by nitrite. The nitrite is likely to occur as a result of nitrate reduction in the human digestive tract.

Nitrogen pollution of the aquatic environment often occurs in connection with washout of fertilizers that are applied to agricultural fields. This washout takes place either as a result of fertilizers entering streams or lakes, or recharging aquifers. Other sources include untreated or insufficiently treated sewage that is discharged to receiving waters. Note that the ammonium ion tends to adsorb to soil or clay particles that are mainly negatively charged, whereas nitrate NO_3^- stays in solution and can probably leave the soil with washout or leachates.

Atmospheric N_2O levels have increased from a pre-industrial level of 275–311 ppb in 1992 (Barton and Atwater, 2002). Nitrous oxide is 310 times more effective than CO_2 as a greenhouse gas and N_2O destroys atmospheric ozone catalytically. Food production and wastewater treatment operations are significant sources of N_2O.

Beaulieu et al. (2010) measured N_2O emission rates from the water surface and microbial N_2O production in the sediments and water column of the Markland Pool of the Ohio River. The Markland Pool is a 153 km stretch of the Ohio River from the Meldahl Lock & Dam through greater Cincinnati to the Markland Lock & Dam. Emission rates (7–35 (µg N/(m² h)) from the river at 23 km showed a clear seasonal trend with rates 1.2–2.3 times above equilibrium values with the atmosphere during summer and early fall. Pelagic N_2O production made a significant contribution to the overall summertime emission of about 50 kg N_2O-N/d. A wastewater treatment plant in Cincinnati (59 km) was an important source of N_2O. With respect to nitrous oxide generation in biological nutrient removal wastewater treatment plants,

Foley et al. (2010) estimated an average of 0.035 ± 0.027 kg N_2O-N/kg $N_{denitrified}$ based on a study of seven wastewater treatment plants in Australia. Thus, it can be assumed that more than 90% of the N_2O emission from a plant occurs at the plant site. It was found that plants designed and operated for low effluent TN (<10 mgN/L) generally have lower N_2O emissions than plants that only achieve partial denitrification.

It has been estimated that up to 6.8% of nitrogen applied to fields is emitted as N_2O (Barton and Atwater, 2002). In addition, according to the Intergovernmental Panel on Climate Change (IPCC) 9×10^6 t of N_2O-N is naturally emitted to the atmosphere by soil processes and forest and brush fires.

N_2O rises to the stratosphere (> 30 km altitude) where it undergoes the following reactions

$$O_3 + h\nu \rightarrow O(^1D) + O_2 \tag{5.16}$$

$$O(^1D) + N_2O \rightarrow N_2 + O_2 \quad (\text{fate of} \sim 90\% \text{ of the } N_2O) \tag{5.17}$$

$$O(^1D) + N_2O \rightarrow NO + NO \quad (\text{fate of} \sim 10\% \text{ of the } N_2O) \tag{5.18}$$

where $O(^1D)$ is an electronically excited oxygen atom. The nitric oxide destroys ozone catalytically according to the reactions

$$NO + O_3 \rightarrow NO_2 + O_2 \tag{5.19}$$

$$NO_2 + O \rightarrow NO + O_2 \tag{5.20}$$

with the net reaction

$$O + O_3 \rightarrow 2O_2 \tag{5.21}$$

Crutzen and Ehhalt (1977) estimated that N_2O emissions resulting from increased use of nitrogen fertilizer in agriculture could result in a 10% decrease in the ozone layer by the end of the twenty-first century. Another effect of Nitrogen oxides is that they deplete upper atmospheric concentrations of OH radicals that limit the atmospheric lifetime of methane CH_4. This results in CH_4 remaining in the atmosphere for longer, further contributing to the greenhouse effect. Methane is a 21 times more powerful greenhouse gas than CO_2.

To be available for utilization by living organisms, nitrogen must be present in the form of reactive or fixed nitrogen, e.g. N bonded to C, O, H, such as NO_y, NH_x, and organic N. A few species of aquatic and terrestrial bacteria and algae can fix N_2 into NH_4^+ for utilization. Reactive nitrogen is also created by human cultivation of leguminous crops and soybeans, which symbiotically host nitrogen-fixing bacteria in their root nodules. Synthetic fertilizers and fossil fuel combustion are other significant sources of anthropogenic reactive nitrogen (Table 5.6). Combustion converts fuel organic N and atmospheric N_2 into NO and NO_2. From Table 5.6 it is seen that anthropogenic sources (137–164 TgN/yr) are comparable to or larger than natural sources (93–135 TgN/yr). Mineralization describes breakdown of protein and amino acids, assimilation uptake by plants, and immobilization uptake by microorganisms.

TABLE 5.6. Global Reactive Nitrogen Sources

	Global nitrogen fixation (TgN/yr)
Natural biological N fixation	90–130
Lightning	3–5
Natural source subtotal	93–135
Synthetic fertilizers	80–90
Anthropogenic induced N fixation	27–44
Fossil fuel combustion	30
Anthropogenic source subtotal	137–164
Total sources	230–299

(Source: Barton and Atwater (2002) with permission from ASCE).

5.4.2 Nitrification and Denitrification

Nitrification describes the conversion of ammonia to nitrite and nitrate assisted by *Nitrosomonas* and *Nitrobacter* respectively,

$$2NH_3 + 3O_2 \rightarrow 2NO_2^- + 2H_3O^+ \tag{5.22}$$

$$2NO_2^- + O_2 \rightarrow 2NO_3^- \tag{5.23}$$

with the net reaction

$$2NH_3 + 4O_2 \rightarrow 2NO_3^- + 2H_3O^+ \tag{5.24}$$

Denitrification represents conversion of nitrate under anoxic conditions and with the help of bacteria *Pseudomonas* and an organic substrate, methanol, into nitrogen gas N_2

$$6NO_3^- + 5CH_3OH + 6H_3O^+ \rightarrow 3N_2 + 5CO_2 + 19H_2O \tag{5.25}$$

Figure 5.5 schematically illustrates the nitrification and denitrification processes. Removal of ammonia and organic N from wastewater is typically achieved through sequential nitrification–denitrification. Incomplete processes can cause high N_2O production. In the case of denitrification, N_2O production can be avoided by achieving complete denitrification by means of maintaining a high COD/NO_3-N ratio in wastewater N_2O production. Note that conversion of N_2O into N_2 gas is microbially mediated by the enzyme, nitrous oxide reductase. One strong inhibitor of this enzyme is sulfur in the form of H_2S. Since H_2S can be converted to HS^- and S^{2-}, avoidance of sulfur inhibition of nitrous oxide reductase depends on the pH of the wastewater. Nitrous oxide is also emitted from landfilled organic waste, but usually at a slow rate. Chemical reactions achieving denitrification are available (Table 5.7) but are often characterized by slow kinetics, especially for reactions involving nitrogen gas N_2 (Stumm and Morgan, 1996).

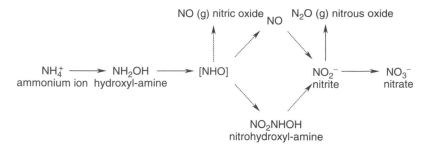

Nitrification

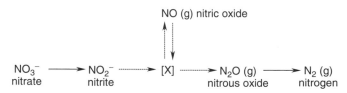

Denitrification

Figure 5.5. Schematic of nitrification and denitrification pathways as described in Barton and Atwater (2002)

TABLE 5.7. Redox Reactions for Nitrogen in Natural Waters

Reaction number	Reaction	pE° (log K)	pE° (W)[a]
(1)	$\frac{1}{5}NO_3^- + \frac{6}{5}H^+ + e^- \leftrightarrow \frac{1}{10}N_2(g) + \frac{3}{5}H_2O$	21.05	12.65
(2)	$\frac{1}{2}NO_3^- + H^+ + e^- \leftrightarrow \frac{1}{2}NO_2^- + \frac{1}{2}H_2O$	14.15	7.15
(3)	$\frac{1}{8}NO_3^- + \frac{5}{4}H^+ + e^- \leftrightarrow \frac{1}{8}NH_3^+ + \frac{3}{8}H_2O$	14.90	6.15
(4)	$\frac{1}{6}NO_2^- + \frac{4}{3}H^+ + e^- \leftrightarrow \frac{1}{6}NH_4^+ + \frac{1}{3}H_2O$	15.14	5.82

(Source: Data adapted from Stumm and Morgan (1996)).
[a]pE° for pH 7.0 at 25 °C in water of unit activities of oxidant and reductant.

More detailed nitrification reactions showing cell mass production and CO_2 consumption are displayed in Equation 5.26. The transformation of ammonia or ammonium ion to nitrate occurs through a microbially mediated two-step nitrification process whereby ammonium ion is first transformed into nitrite (FishDoc, UK, 2013),

$$55\,NH_4^+ + 76\,O_2 + 109\,HCO_3^- \rightarrow C_5H_7O_2N + 54\,NO_2^- + 57\,H_2O + 104\,H_2CO_3$$
$$(5.26)$$

where the active bacteria are typically *Nitrosomonas*, either of the species *Nitrosomonas crytolerans* or *Nitrosomonas europeae* (Biochemical Cycles, 2013). Further oxidation of nitrite can then occur by the action of *Nitrobacter*, *Nitrococcus*, or other related bacteria according to the reaction

$$400\,NO_2^- + NH_4^+ + 4\,H_2CO_3 + HCO_3^- + 195\,O_2 \rightarrow C_5H_7O_2N + 3\,H_2O$$
$$+ 400\,NO_3^- \qquad (5.27)$$

Both of these nitrification reactions are thus aerobic and autotrophic, that is, they consume molecular oxygen, CO_2, and water.

By contrast, most common denitrification reactions involve heterotrophic bacteria, e.g. *Pseudomonas* and an organic substrate, in this case methanol,

$$2\,NO_3^- + CH_3OH + 4\,H_3O^+ \rightarrow N_2O + CO_2 + 8\,H_2O \qquad (5.28)$$

Nitrogen gas (N_2) may also be produced by this reaction, see Equation 5.25. Reaction 5.28 takes place in the absence of oxygen under anoxic conditions ($E_h \leq 400$ mV). By comparison, the redox potential under anaerobic conditions is $E_h \leq 0$ mV (Kielemoes et al., 2000). Note that by combining Equations 5.26–5.28 one can achieve nitrogen removal in that ammonia or organic nitrogen is converted first to nitrate and then to the gaseous forms N_2 or nitrous oxide (N_2O), which escape to the atmosphere.

Anaerobic Ammonium Oxidation—the Anammox Process. The anammox process, using anaerobic ammonium oxidation, is an alternative method of obtaining nitrogen removal in wastewater. Some nitrite must be present either in the wastewater or as a result of nitritation, i.e. partial aeration of ammonia. The anammox stoichiometry is given by (Tokutomi et al., 2011)

$$1.0\,NH_4^+ + 1.32\,NO_2^- + 0.066\,HCO_3^- + 0.13\,H^+ \rightarrow 1.02\,N_2 + 0.26\,NO_3^-$$
$$+ 0.066\,CH_2O_{0.5}N_{0.15} + 2.03\,H_2O$$
$$(5.29)$$

A wastewater treatment plant using this reaction can benefit from reduced operational costs and lower emission of greenhouse gases (Hu et al., 2010). The reduction of operational costs, estimated at 60%, may be achieved because there is no need for external COD, significantly less aeration, and much lower sludge production. Lower greenhouse gas emission, estimated to be 90% lower, is obtained because no NO_2 is emitted and CO_2 is consumed rather than produced. Further details about the use of the anammox process in wastewater treatment are provided in Section 5.4.3 under nitrogen removal.

In the following, we shall examine recent findings regarding nitrification and denitrification processes in the environment, considering water, groundwater, and sediments.

Water. Denitrification can be an important route for removal of nitrogen from water. Usually, an organic substrate is involved as in Equation 5.43. However, it was shown by Till et al. (1998) that Fe(0) can stochiometrically reduce nitrate abiotically, i.e. chemically to ammonium following the reaction

$$4\,Fe(0) + NO_3^- + 7\,H_2O \rightarrow 4\,Fe^{2+} + NH_4^+ + 10\,OH^- \quad \Delta G^o = -196 \ KJ;$$

$$\Delta G^{o'} = -596 \ KJ \qquad (5.30)$$

where $\Delta G^{o'}$ refers to pH = 7. These authors also showed that cathodic hydrogen can sustain microbial denitrification through a two-step reaction,

$$Fe(0) + 2\,H_2O \rightarrow H_2 + Fe^{2+} + 2\,OH^- \qquad (5.31)$$

$$2\,NO_3^- + 5\,H_2 \rightarrow N_2 + 4\,H_2O + 2\,OH^- \qquad (5.32)$$

Combining these reactions yields

$$5\,Fe(0) + 2\,NO_3^- + 6\,H_2O \rightarrow 5\,Fe^{2+} + N_2 + 12\,OH^- \quad \Delta G^o = -636 \ KJ;$$

$$\Delta G^{o'} = -1116 \ KJ \qquad (5.33)$$

Thus the microbially mediated reaction is energetically favored over the abiotic reaction. The reduction of nitrate in Equation 5.33 is achieved using the organism *Parcoccus denitrificans*.

Drinking water denitrification using an ion-exchange membrane bioreactor was demonstrated by Fonseca et al. (2000). Nitrate ion from the polluted water migrates through the membrane into a bioreactor where it is denitrified to N_2 and carbon dioxide using ethanol and nutrients. The bicarbonate can then migrate back through the membrane into the stream of treated water. The system was used to investigate the removal of nitrate from a synthetic groundwater containing 50 mg/L nitrate. The treated water was free of inorganic nitrogen and ethanol and a surface denitrification rate of 7 g N/(m^2 d) was achieved. The permeability of the membrane to oxygen was found to be negligible, thus creating favorable conditions for denitrification in the bioreactor. Alternatively, nitrate ions can be removed from water by ion exchange.

It was demonstrated that Fe(0) can cause biological denitrification through the reactions shown in Equations 5.31 and 5.32, or these reactions combined into Equation 5.33. Formation of N_2 gas in Equation 5.33 can be viewed as a result of the reduction of nitrate to nitrite, and further reduction of nitrite to nitrogen gas,

$$2\,NO_3^- + 2\,H_2 \rightarrow 2\,NO_2 + 2\,H_2O \qquad (5.34)$$

$$2\,NO_2^- + 3\,H_2 \rightarrow N_2 + 2\,H_2O + 2\,OH^- \qquad (5.35)$$

$$2\,NO_3^- + 5\,H_2 \rightarrow N_2 + 4\,H_2O + 2\,OH^- \qquad (5.36)$$

Thus, iron corrosion can be utilized by these hydrogenotrophic, anaerobic organisms to achieve denitrification, i.e. reduction of nitrate and nitrite to gaseous nitrogen products, $NO^{3-} \rightarrow NO^{2-} \rightarrow NO \rightarrow N_2O \rightarrow N_2$. However, high nitrite concentrations (≥ 50 mg NO^{2-}-N/L) have an inhibitory effect on corrosion (Kielemoes et al., 2000). These authors postulated that the inhibitory effect is a result of the formation of high levels of NO, which have a toxic effect on denitrifying organisms. Formation of NO can occur through the following reaction,

$$Fe^{2+} + NO_2^- + H_2O \rightarrow Fe^{3+} + NO + 2OH^- \tag{5.37}$$

The high concentrations of both Fe^{2+} and NO^{2-} are necessary for the generation of significant amounts of NO through this reaction.

Natural chemical denitrification can occur during drying of dew or fog and cloud droplets (Takenaka et al., 1999). It was shown that in the drying process of dew containing high concentrations of ammonium and nitrite, chemical denitrification will occur through the reaction,

$$NH_4NO_2 \rightarrow N_2 + 2H_2O \tag{5.38}$$

Concentrations of NH^{4+} and NO^{2-} in dew obtained in Sakai City, Japan were in fact higher than those of NO^{3-}. Typical values were 6.4 μeq/m^2NO^{2-} and 22.1 μeq/m^2NH^{4+} where the unit is an equivalent value for area of the dew sampler. Under acidic conditions, other possible reactions during drying are

$$2HNO_2 + O_2 \rightarrow 2HNO_3 \tag{5.39}$$

$$2HNO_2 \rightarrow NO + NO_2 + H_2O \tag{5.40}$$

$$HNO_2 (aq) \rightarrow HNO_2(g) \tag{5.41}$$

where the latter two reactions also result in loss of nitrogen from the aqueous phase.

The relative amount of nitrification and denitrification during wastewater treatment can have a significant influence on the emission of N_2O, a known greenhouse gas. Beline et al. (2001) used the ^{15}N technique to investigate the generation of nitrous oxide during biological aerobic treatment of a pig slurry. They ran two sets of experiments with (i) aeration and a high level of DO, which is favorable to nitrification, and (ii) low level of DO, which is favorable to simultaneous nitrification and denitrification. The experiments were conducted in flow-through reactors where a single pulse of NO^{3-}[15]N was added to the reactor after reaching steady-state. It was apparent from the high DO experiment that unlabeled N_2O was produced through the nitrification process, and there was no measurable [15]N_2O generated by denitrification. By contrast, the low DO experiment indicated that 27% of ^{15}N took the N_2O form after 48 h. Based on the decrease of the isotopic excess of the NO^{3-}[15]N, it was estimated that 92% of NO^{2-}N was directly denitrified into gaseous forms (N_2 and N_2O) without any previous oxidation to nitrate. Based on the amount of N_2O produced under high and low DO conditions it was concluded that N_2O was mainly emitted by the denitrification process.

It is well known that PAHs can be degraded under aerobic conditions (Cerneglia, 1992). However, PAHs can also be degraded anaerobically under sulfate- or nitrate-reducing conditions. For example, denitrifying bacteria can biodegrade naphthalene, phenanthrene, and biphenyl (Rockne and Strand, 2001). Typical reaction pathways for naphthalene degradation are as follows:

$$\frac{1}{48}C_{10}H_8 + \frac{5}{12}H_2O \rightarrow \frac{5}{24}CO_2 + H^+ + e^- \tag{5.42}$$

$$\frac{1}{5}NO_3^- + \frac{6}{5}H^+ + e^- \rightarrow \frac{1}{10}N_2 + \frac{3}{5}H_2O \tag{5.43}$$

Part of the CO_2 produced in Equation 5.42 can be used for cell synthesis,

$$\frac{1}{4}CO_2 + \frac{1}{20}NH_3 + H^+ + e^- \rightarrow \frac{1}{20}C_5H_7O_2N + \frac{2}{5}H_2O \tag{5.44}$$

The overall reaction combines the mineralization and the synthesis equations,

$$R_{overall} = f_sR_c + (1 - f_s)R_a + R_d \tag{5.45}$$

where R_d is the electron donor oxidation half-reaction, shown in Equation 5.42, R_a is the electron acceptor half-reaction, shown in Equation 5.43, R_c is the cell synthesis half-reaction, shown in Equation 5.44, and f_s is the fraction of electron donor utilized for cell synthesis. $^{14}CO_2$, produced from PAH degradation, was calculated as the difference between the radioactivity in the alkalized and acidified reaction vials. Nitrous oxide, a specific denitrification product, was also produced from phenanthrene and biphenyl fed cultures in particular. Mineralization was nitrate dependent and was sustainable over several feedings, and the cultures produced N_2O when supplied with PAHs as the only carbon and energy source.

Nitrate that is present in high levels in surface or groundwater can affect arsenic cycling under anoxic conditions (Senn and Hemond, 2002). Nitrate will oxidize As(III) to As(V), which is less toxic and more particle reactive, and oxidizes Fe(II) to hydrous ferric oxide particles that scavenge arsenic.

Groundwater. Nitrogen forms and transformations are important considerations for groundwater plumes. DeSimone and Howes (1996) investigated nitrification and denitrification in a sandy glacial aquiver, which received wastewater from a septic-treatment facility on Cape Cod, MA. They found comparable denitrification rates from three different approaches that maintained the structure of aquifer materials: Acetylene block in intact sediment cores 9.6 ng of $N/(cm^3$ d); in situ N_2 production 3 ng of $N/(cm^3$ d); and in situ NO^{3-} depletion 7.1 ng of $N/(cm^3$ d). These rates are significantly lower than rates determined from the standard slurry method (150 ng of $N/(cm^3$ d)). The higher rates in the slurry method are probably due to enhanced biological activity caused by physical mixing with increased availability of solid-phase organic carbon. In the acetylene block method, the reduction of nitrous oxide N_2O to N_2 is prevented by the introduction of acetylene.

The modest isotopic ^{15}N enrichment of NO^{3-} and enrichment of ^{14}N of the N_2 produced in the groundwater was consistent with the low in situ denitrification rates. From a nitrate and carbon addition experiment it was found that the rate of denitrification was limited by the electron donor; dissolved organic carbon (DOC).

The overall influence of denitrification on the load of nitrogen to the receiving ecosystem was estimated to be less than 10%. Denitrification rates based on laboratory incubations of intact sediment cores, or estimated from in situ N_2 production or NO^{3-} depletion, may be high compared to actual field rates. The reason for this is that microbial processes in the subsurface generally proceed at a slower rate than the groundwater transport. The microbes that are mostly stationary perceive a changing solute environment as a plume moves through an area, resulting in a slower rate of denitrification.

To circumvent this potential problem, Smith et al. (1996) conducted natural-gradient tracer tests using NaBr as tracer and the acetylene block technique for measuring N_2O production rates. The movement of a slug of acetylene through the aquifer was similar to that of bromide, showing that acetylene moved conservatively with the plume. Denitrification rates were based on the fitting of the advection–diffusion equation with a reaction term to measured nitrous oxide distributions. The inferred denitrification rates for two successive tests, 0.6 and 1.51 nmol of $N_2/(cm^3\ d)$, were significantly lower than rates based on tests with intact cores (1.3–4.0 nmol $N_2O/(cm^3\ d)$). Thus, one should exercise caution when extrapolating laboratory results to field situations.

Denitrification in groundwater may be simulated by the injection of formate as electron donor into the subsurface (Smith et al., 2001). Formate can be hydrolyzed to hydrogen and carbon dioxide by several types of organisms and denitrification can proceed via reactions such as Equations 5.48 through 5.50. In these field experiments, Smith et al. (2001) withdrew groundwater from a withdrawal well, mixed it with sodium formate and a NaBr tracer, and reinjected it into an injection well. Concentrations of formate, nitrate, and nitrite were then measured vs. days after injection up to 15 m down gradient using multilevel samplers. The breakthrough curves were also simulated with a two-dimensional site-specific model that includes transport, denitrification, and microbial growth.

The results showed that nitrate concentrations were reduced by 80–100% with a concurrent increase in nitrite concentrations and reduction in formate concentration as a result of reaction with nitrate. The model and laboratory results showed that the rate of formate-enhanced nitrite reduction was almost four times slower than nitrate reduction, but in the laboratory, given sufficient time, nitrate was completely consumed. Thus if nitrite reduction can also be demonstrated in the field within a reasonable time, formate injection may be a viable technology to remediate nitrogen-contaminated groundwater. Note that there are drinking water limits for both nitrate and nitrite (US limits are 10 mg N/L, and 1 mg/L respectively).

Use of methane or natural gas as hydrogen donor to achieve denitrification of groundwaters has been suggested (Eisentraeger et al., 2001). Methane oxidation did not occur in the absence of oxygen. Therefore, denitrification with methane must be performed as a two-stage process with aerobic methanotrophic bacteria producing metabolites, which are used as hydrogen donor by nonmethanotrophic denitrifying bacteria in anoxic areas. For these processes to occur, there may be long lag times. This can result from the low concentration of organic carbon in the groundwater and the common low temperature, e.g. 10 °C, of groundwaters.

Sediment. Sediments can serve as sites for both nitrification and denitrification. Typically, stream sediments are sufficiently aerated to primarily mediate nitrification (Butturini et al., 2000), whereas lake or estuarine sediments are more likely to be sites of denitrification by facultative anaerobic bacteria (Saunders and Kalff, 2001).

In their work on streambed nitrification related to the Mediterranean stream La Solana, Butturini et al. (2000) found that considerable nitrification occurred in stream sediments based on sediment biofilm reactors in which actual stream sediment was subjected to realistic in situ conditions. The effect of various ammonium concentrations and DOC availability on nitrate production and nitrification efficiency was investigated. The one-dimensional advection–diffusion equation with a reaction term was used to quantify distributions and transformations. Nitrification rates stayed between 4 and 20 ng N/(min g) dry mass sediment with 85% efficiency.

Results from successive glucose/ammonium additions suggested that an increased carbon pulse enhances heterotrophic retention of total inorganic carbon. After long-term carbon addition, heterotrophic microorganisms retained four times more ammonium than without glucose enrichment. Thus, long-term addition of highly available organic carbon can shift the microbial community from chemotrophic nitrifying bacteria to primarily nonnitrifying heterotrophic microorganisms.

The importance of denitrification in the littoral and profundal zone of lake Memphremagog, Canada, was investigated by Saunders and Kalff (2001) using the N_2 flux technique on intact sediment cores. Littoral denitrification rates were found to be highly variable with an average rate of 111 µmole N/(m^2 h). Rates were positively related to temperature, percentage of organic matter, and microphyte biomass density. In addition, they negatively correlated with depth. It was concluded that the littoral zone dominates the denitrification of the whole lake.

5.4.3 N Removal

Best Management Practices. Nitrogen reduction strategies in Krka River basin, Slovenia, were described by Drolc et al. (2001). A major incentive for such a reduction is to prevent "blue baby syndrome" or methemoglobinemia in the population. By means of appropriate wastewater management, and agricultural best management plans (BMPs), it was projected that future nitrogen inputs to the Krka River could be curtailed by 20 and 30% respectively. Major N reductions were obtained from upgrading of industrial wastewater treatment plants to include N removal by nitrification–denitrification, reduction of leaching of nitrogen to groundwater, elimination of direct discharge of manure, and reduction of leaching of nitrates to groundwater from agricultural fertilizers.

The most toxic form of nitrogen to aquatic biota is ammonia, followed by nitrite and nitrate. The nitrogen threat to amphibian survival was described by Rouse et al. (1999). Considering ambient concentrations of various nitrogen forms, amphibians are at risk from nitrate in particular. Lethal and sublethal effects on amphibians are present at nitrate concentrations between 2.5 and 100 mg/L. Natural background concentrations of nitrate in groundwater range from 0 to 3.0 mg/L. Environmental nitrate concentrations are highest in late fall, winter, and spring. In the summer, nitrate is assimilated by plants.

Mitigation of nitrate entering surface water can be carried out by establishing vegetated buffer zones around water courses. A 19 m mixed woodland buffer reduced nitrate concentrations in a stream approximately 7 mg/L to less than 0.5 mg/L. Benefits from increased agricultural productivity due to nitrate use must be weighed against the impact of nitrate on wildlife health and survival.

In their analysis of the potential for reducing nitrogen pollution through improved agronomic practices, Trachtenberg and Ogg (1994) concluded that there is insufficient crediting for nutrients, especially nitrogen, coming from manure and legumes. The result is an over application of about 13% of fertilizer N. The nitrogen form most likely to escape from farm fields is nitrate due to its high water solubility.

Wastewater Treatment Processes.   As well as methemoglobinemia, nitrite can form nitrosamines, a group of carcinogens produced from reaction of nitrite with amines. Thus there are several reasons to remove nitrate, a precursor for nitrite. Nitrate can be removed by nitrification–denitrification as described above, or by chemical reduction using iron powder (Huang et al., 1998). Nitrate reduction can also occur with Fe^{2+}. However, this process requires Cu^{2+} as a catalyst, is less effective under oxic conditions, and produces large amounts of sludge. With iron powder, initial reactions include oxidation of Fe^0 by a proton, and reduction of nitrate to nitrite as described in the reactions shown in Equations 5.46 and 5.47 taking place at pH values $\leq$ 4:

$$2H_3O^+ + Fe^0 \Longleftrightarrow H_2(g) + Fe^{2+} + 2H_2O \qquad (5.46)$$

$$NO_3^- + Fe^0 + 2H_3O^+ \Longleftrightarrow Fe^{2+} + NO_2^- + 3H_2O \qquad (5.47)$$

Significant nitrate reduction occurred within minutes. As can be seen from these reactions, an increase in pH would be expected. This was prevented by the use of an automatic pH controller. The overall conversion of nitrate to ammonia was described by the reaction shown in Equation 5.48,

$$NO_3^- + 4Fe^0 + 10H_3O^+ \Longleftrightarrow 4Fe^{2+} + NH_4^+ + 13H_2O \qquad (5.48)$$

The amount of Fe^0 required to remove NO_3 within an hour was 120 m^2 Fe^0/mol NO_3, and, in accordance with Equation 5.48, it was found that equimolar amounts of ammonia were recovered as the final product of nitrate reduction. The apparent reaction order of 1.7 with respect to nitrate was observed.

One disadvantage of chemical nitrate reduction is that there is no removal of nitrogen; it is only converted to ammonia. By contrast, a combined nitrification–denitrification process will oxidize ammonia to nitrate, which is then reduced to nitrous oxide or nitrogen gas that will leave the aqueous phase. If nitrogen gas is the final product this can then be deemed effective nitrogen removal. Van Benthum et al. (1998) demonstrated that denitrification can be integrated in a nitrifying biofilm airlift suspension (BAS) reactor by intermittent aeration. N removal efficiency of 75% at an aerobic ammonia load of 5 kg $N/(m^3$ d) was observed with acetate being added as a carbon source during the anoxic period.

In the BAS reactor, basalt particles of density 3010 kg/m^3 with average diameter of 0.32 mm and initial concentration of 8 kg/m^3 acted as carriers for the biofilms. Under denitrifying operation using acetate as electron donor and carbon source, the COD/N ratio measured 2.6 ± 0.5 kg COD/kg N. This represents the total amount of acetate consumed per total amount of nitrogen compounds removed, including assimilation. The relatively low COD/N ratio indicates that nitrate was an intermediate. The theoretical COD/N ratio will increase when there is a higher biomass yield for $NH_4^+ \rightarrow NO^{3-} \rightarrow N_2$. The yield for denitrification of nitrate using acetate or glucose as carbon source, without consideration of biomass production, is 2.86 g COD/g N, assuming N_2 production. In general, the g oxygen equivalent of nitrate or nitrite reacted equals g COD utilized – g COD of cells.

The acetate must be consumed fully anoxically in order not to limit nitrification during the aerobic cycle. Nitrous oxide was emitted when acetate addition was started. However, during the following 30-d period, nitrous emission decreased to near zero.

Acetate in the form of acetic acid was also the carbon source in a two-step alternate oxic–anoxic process with separate biomass, which was studied by Battistoni et al. (2002). The acetic acid addition took place at the beginning of the anoxic phase. The external carbon source helps to maintain a minimum level of biomass. The nitrified nitrogen in the first step is entirely denitrified without an external carbon source, whereas the same element in the second step needs an external carbon source because normally not enough carbon is left after the first process. In the first reactor, a cycle of 150 min oxidation/45 min denitrification was used, and in the second 105 min oxidation/180 min denitrification was used.

Carrera et al. (2003) used lab scale sequential aerobic and anoxic reactors to remove ammonia from high strength industrial wastewater with 5000 mg/L N-NH_4^+. Separate sludge systems were used for the nitrifying activated sludge and denitrifying activated sludge. The system worked in the presence of high sulfate (15,000–20,000 mg/L) and chloride (500–600 mg/L) concentration. Maximum nitrification rates were approximately 0.21 g N-NH_4^+/(g VSS d) (20 °C). Complete denitrification occurred with two different industrial carbon sources: With mainly ethanol, the maximum denitrification rate (MDR) was 0.64 g N-NO_x^{-1}/(g VSS d)(20 °C), whereas a mixture of 60% methanol, 10% acetone, and 10% isopropylic alcohol (10%) produced a lower MDR of 0.17 g N-NO$_x^-$/(g VSS) (25 °C).

External carbon source addition may serve as a suitable control variable to improve the process, e.g. to compensate for high nitrogen loads or low seasonal temperatures (Isaacs and Henze, 1995). These authors stated that the COD/N ratio should be sufficiently high, around 4.2 g COD/(g N), for TN removal including assimilation when glucose is the carbon source. Because some release of phosphate associated with the carbon source addition occurred, the required COD/N ratio was as high as 7 mg COD/(mg N). Both acetate and hydrolysate derived from biologically hydrolyzed sludge could function as a carbon source during denitrification. The latter form may be attractive because it is derived from the wastewater; therefore no external carbon would need to be added.

Defining N-NO$_x$ as the sum of nitrite and nitrite nitrogen, the MDR determined by Isaacs and Henze (1995) was 0.08 g NO$_x$/(g VSS d) with levels of initial NO$_x$-N of about 7 mg/L NO$_x$-N.

As mentioned previously, the anammox process has many desirable properties for nitrogen removal in wastewater treatment. One potential problem is the high doubling time (11–20 d) of anammox bacteria. To overcome this, Hu et al. (2010) found that a single seeding source could be used as an inoculum of anammox bacteria. This enabled faster start-up of anammox reactors operating under different conditions. Fluorescence in situ hybridization (FISH) confirmed that each community in eight reactors was dominated by just one anammox bacterium. In these reactors at least eight anammox 16S rRNA gene sequences were detected belonging to two genera: *Candidati Brocadia* and *Kuenenia*. Cell densities of around 10^9 cells/mL were determined by qPCR.

For the effective treatment of toxic pharmaceutical wastewater, Tang et al. (2011) addressed the seeding need by using a separate high rate anammox reactor for the production of high activity seed anammox granules, which were added to the main reactors in a process known as sequential biocatalyst addition (SBA). The biotoxicity of the pharmaceutical wastewater was effectively overcome by the addition of the granular anammox seed. The nitrogen removal rate reached up to 9.4 kg N/(m^3 d).

5.5 PHOSPHORUS

5.5.1 The Phosphorus Cycle

Because phosphorus (P) is often a limiting nutrient in rivers and lakes, tracing this nutrient is of interest, for example in runoff from agricultural fields and urban areas to receiving waters, in order to determine P sources for water quality management. Nash and Halliwell (2000) give an overview of P tracing techniques. Most P is transferred from farms in colloidal form (diameter less than 0.45 μm). Surface runoff was found to carry high P loads. Rainfall does not generally contribute significantly to P in runoff. Since natural P is monoisotopic (^{31}P), stable isotope techniques cannot be used for tracing P. In the case of low P levels, or if it is desired to identify and quantify P sources, organic marker compounds (biomarkers) are well suited. Sterols, phospholipids, and alkanes are specific to potential P sources for both agricultural and urban runoff.

Biomarkers are characterized by molecular composition, stereo isomerism, or chirality. For grazing systems, ATP, phospholipids, and possibly nucleic acids are suitable P tracers. Some biomarkers do not contain P. If the ratio between biomarker (B) and P, B/P, is a characteristic for each P source, and the B/P ratio does not change during transport, a chemical mass balance (CMB) model can be used to estimate relative source contributions of P and total P as a weighted sum of P in each source material. To determine n sources, there must be at least m biomarkers where $m \geq n$.

Phosphorus mainly derives from plants, animal wastes, and soils. Plants contribute P by means of decaying vegetation. Feces is a main pathway for return of P to soil, and the soil surface is generally the most organic-rich soil zone.

For management of P in the aquatic environment both concentrations and total maximum daily loads (TMDLs) are used. Previously, concentrations were almost exclusively used because of their role in eutrophication even in very limited areas. However, to consider concentrations and the size of impacted areas, TMDLs are becoming more commonplace, while separate concentration limits are still in effect.

Walker (2003) examined a framework for considering variability and uncertainty in developing lake TMDLs for phosphorus. Variability refers to temporal or spatial variations in water quality conditions, as they relate to the management goal. Uncertainty, on the other hand, refers to random prediction errors resulting from limitation in the data and models to formulate the lake P balance, or the performance of measures to achieve the allocated loads. Along with a recommendation by the National Research Council, he suggested the use of an adaptive approach. This includes measurement of variability and uncertainty over an extended period of time, e.g. 5 yr, and subsequent adjustment of the allocated load. The TMDL is then the sum of allocated load, uncertainty, and variability. In addition, the margin of safety is the sum of uncertainty and variability.

The effects of temporal variations can be expressed, for example, in the correlation of average P concentrations with frequency of algal blooms (chlorophyll a) or frequency of pH or transparency valued linked to algal blooms.

5.5.2   P Removal

Lake Restoration.   The addition of aluminum as alum, $AlSO_4 \cdot 18H_2O$, to lakes has been used in many lakes to remove phosphorus from the water column and reduce the potential for harmful algal blooms. Aluminum hydroxide $[Al(OH)_3]$ adsorbs phosphorus and, in contrast to iron-bound aluminum, the aluminum bound P is not released under reducing conditions. Therefore, Al treatment can offer a long lasting remedy for high P levels in lakes. Reitzel et al. (2005) investigated P reduction in lake water and overall mass balance for P by adding Al to the small hypertrophic Lake Sønderby in the central part of Denmark on the island of Funen. The lake has an area of 8 ha and the average depth is 2.8 m. It is stratified during summer and can be ice-covered during winter.

They determined the mobile phosphorus, P_{mobile}, in sediment as a sum of pore-water P, iron-bound P, and nonreactive polyphosphates and organic P. The authors found a good agreement between winter to summer loss rates from sediments and winter to summer P accumulation in lake water (30 mg of $P/(m^2 \, d)$), suggesting that P_{mobile} was the source of internal P loading in the lake. The aluminum was added as a buffered polyaluminum chloride solution, providing 11 mg/(L of Al m^2) and corresponding to a 4:1 molar ratio between Al and P_{mobile}. The concentration of total P in the lake during summer declined from 1.28 to 1.3 in pretreatment years to between 0.09 and 0.13 mg/L in post-treatment years. Although this result is promising, one should always consider any potential detrimental effects of Al addition. A more conservative approach would be to reduce P input to the lake, for example by restricting agricultural runoff and sources of fertilizer.

Wastewater Treatment Processes. Phosphorus can be removed chemically by means of precipitation with iron or calcium (Tchobanoglous et al., 2003). However, phosphorus can also be removed biologically using an alternating sequence of anaerobic or anoxic reactors and aerobic reactors. Under low oxygen levels, P will be released from bacterial cells, and under aerobic conditions, the bacteria will perform luxury P uptake. As the P-loaded bacteria are settled, there will be a net P removal from the wastewater.

A bench-scale sequencing batch reactor (SBR) for P removal was examined by Kim et al. (2001). The optimization scheme was based on a linear activated sludge model (ASM2) and an optimizer determining the aeration time t_{air}. The objective was to minimize energy consumption by reducing the aeration cycle time while meeting the permit requirement (monthly average of 0.5 mg/L). There were six phases of the SBR with initial cycle times of fill (0.5 h), mix (3 h), aerate (3 h), idle (1 h), settle (0.5 h), and decant (0.5 h), with a total cycle time of 8.5 h. The mix stage represents the anaerobic cycle. On the tenth day and thereafter, the aeration cycle times were determined by the optimizer. A typical model result for t_{air} was 2.8 h with the real system requiring 3.5-h cycle time. With initial concentrations of P, COD, and N of 7.4, 420, and 22.1 mg/L respectively, P removal of 93%, COD removal of 90%, and N removal of 98% were achieved.

Struvite. Struvite is a crystalline mineral ($MgNH_4PO_4 \, 6H_2O$), which can have a negative impact on wastewater treatment plants as a result of scale formation in sludge liquor pipes and centrifuges, but at the same time can be recovered as a valuable byproduct that can be used as a high purity, slow-release P fertilizer (Le Corre et al., 2007; Forrest et al., 2008; Marti et al., 2010). Phosphorus recovery is typically carried out for biological nutrient removal plants in a crystallizer located downstream of anaerobic digesters where NH_4-N and PO_4-P concentrations in supernatants are high. $MgCl_2$ may be added, as well as NaOH, for pH control. Destabilization of struvite fines can be performed with a suitable coagulant such as polydiallyl dimethylammonium chloride (polyDADMAC) (Le Corre et al., 2007). Phosphorus recovery is important in view of limited global phosphorus reserves (Forrest et al., 2008).

5.5.3 Case Study: Phosphorus from Wastewater Treatment, Stormwater, and Rivers in Milwaukee, Wisconsin

Nuisance growth of *C. glomerata* in the nearshore areas of the Laurentian Great Lakes has generated a renewed interest in phosphorus concentrations and loads. Following phosphorus abatement in the late 1970s and early 1980s the *Cladophora* abundance declined significantly. The reason for the current (post 2000) resurgence of *Cladophora* growth may be related to invasion of the zebra mussel *Dreissena polymorpha* and more recently the quagga mussel *Dreissena bugensis* (Ozersky et al., 2009).

The dreissenid mussels are filter feeders that clean the water column so that light becomes more available for algal growth. This increases the likelihood of phosphorus limitation. At the same time, the dreissenids are known to excrete bioavailable SRP, potentially stimulating *Cladophora* growth. Ozersky et al. (2009) suggested that dreissenid recycled phosphorus in the nearshore was a more important P source than local watercourses and wastewater treatment plants. If this were the case, lakewide reductions in TP would be necessary to limit *Cladophora* growth. However, such reductions may not be achievable or desirable because of the effects on feed webs including the fisheries.

Concentrations and loads of TP from rivers and the Jones Island water reclamation facility (WRF) in Milwaukee, Wisconsin, were estimated by Zou and Christensen (2012) in order to evaluate the possible impact on extensive *Cladophora* growth in the nearshore area (Figure 5.6). The average TP value was slightly higher (0.095 ± 0.005 mg/L) at station OH-2 near the outfall of the Jones Island WRF than at the confluence of the three rivers OH-1 (0.073 ± 0.006 mg/L) for the time period 2000–2008. There was a declining trend for TP at OH-1 ($p = 0.76$) and no trend at OH-2 ($p = 0.54$), where probabilities are taken from Mann–Kendall tests (Figure 5.7). The negative trend for OH-1 is thought to reflect the positive impact of phosphorus regulations and possibly a climatic trend as well. Increased TP from the rivers is strongly associated with riverbed erosion during storms ($R^2 = 0.974$), such as occurred in 2004, whereas stormwater runoff alone shows the opposite trend, i.e. dilution of TP during storms. The Jones Island WRF contributes 39,500 kg/yr, or less than 34%, of the total TP load. The remainder originates from the rivers, including stormwater runoff and combined sewer overflows.

There was a significant correlation between TP and total suspended solids (TSS) ($R^2 = 0.360$) at OH-1 but a much lower correlation at OH-2 ($R^2 = 0.012$), indicating that suspended solids from rivers carry a large fraction of TP, whereas the suspended solids contribution to TP at the Jones Island WRF outfall is more variable. Whereas *Cladophora* nuisance growth has increased in the period from 2000 to 2008, TP concentrations from the Jones Island WRF and the rivers have remained nearly the same, suggesting that phosphorus discharges from land-based nearshore coastal sources

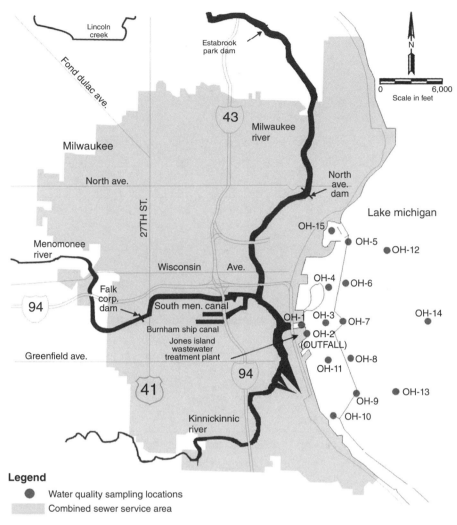

Figure 5.6. Outer Harbor water quality sampling sites. (Source: Zou and Christensen (2012) with permission from MMSD). (*See insert for color representation of the figure.*)

are not the major factor sustaining this growth. Recycling of phosphorus from Lake Michigan by dreissenid mussels may contribute significantly.

5.6 VITAMINS AND TRACE METALS

Microbial growth is dependent on the macronutrients carbon, oxygen, hydrogen, nitrogen, phosphorus, and sulfur. In addition, a number of trace elements and vitamins are required (Burgess et al., 1999). Trace elements of interest include calcium,

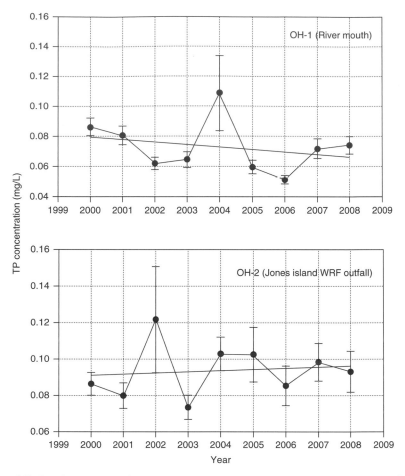

Figure 5.7. Yearly averages of TP concentrations with trend lines and uncertainties of 2012 with permission from ASCE.

potassium, iron, magnesium, manganese, copper, aluminum, zinc, molybdenum, and cobalt, while the required vitamins include biotin, niacin, thiamin (B_1), lactoflavin (B_2), pyridoxine, (B_6), pantothenic acid (Table 5.1), and cyanocobalamin (B_{12}) (Hashsham and Freedman, 2003).

In their investigation of micronutrient requirements of a process for treatment of industrial wastewater using activated sludge, Burgess et al. (1999) found that the addition of the vitamins (1.0 mg/L) or biotin (1 µg/L) to the wastewater stimulated oxygen uptake (biotin) and COD removal (both niacin and biotin). In the case of niacin, 1.0 mg/L added to the wastewater gave an unchanged oxygen uptake rate (0.01 kg O_2/(kg MLSS d)) and increased COD removal from 1.94 to 2.24 kg COD/(kg MLSS d). In some cases there is no requirement of external addition of specific vitamins such as biotin, because several enterobacteria can produce sufficient biotin to supply the sludge.

Several supplements, most of which were added in combination, decreased the COD removal, whereas oxygen uptake can either increase or decrease. The improvement in COD removal by activated sludge dosed with calcium and manganese may be due to the ability of Ca^{2+} ions to increase membrane permeability, allowing Mn better access to the bacterial cells and speeding up metabolism. Decreased COD removal and increased oxygen uptake indicate that the oxygen is not only consumed for biodegradation. Rather, it is also used in cell synthesis and reactions with the added supplements. An added vitamin/metal pair may decrease COD removal as a result of metal adsorption to the cell hindering vitamin utilization.

Vitamin B_{12} (cyanocobalamin, CN-Cbl) is important for stimulating anaerobic bacteria that dechlorinate aliphatic compounds, e.g. chlorinated methanes, ethanes, and ethenes. Hashsham and Freedman (2003) conducted a series of batch and column tests in order to investigate whether the B_{12} vitamin could be distributed without significant loss in an aquifer for the purpose of remediation for these compounds. The Freundlich adsorption isotherm was used in batch adsorption tests

$$q_e = K_f C_e^{1/n} \tag{5.49}$$

where q_e is the equilibrium concentration of solute in the sorbed phase, K_f is the sorption coefficient related to the capacity of the sorbent $((\mu g/g)\,(L/mg)1/n)$, C_e is the equilibrium concentration of the solute (mg/L), and $1/n$ is an experimental constant. For $n = 1$, the relationship is linear. The equilibrium concentration of solute in the sorbed phase q_e is given by

$$q_e = \frac{[(C_o - C_e)V - \beta C_e]}{M} \tag{5.50}$$

where C_o is the initial concentration of the solute (mg/L), V is the volume of solution (L), M is the mass of soil (g), and β is the factor taking into account the mass adsorbed to the bottle surface. Typical results from batch studies with sand indicated values of $K_f = 0.70\ \mu g/g\ (mL/\mu g)^{1/n}$ and $1/n = 0.84$.

A schematic of the column system used is shown in Figure 5.8. Tritium (^3H_2O) tracer without CN-Cbl is used for determining breakthrough curves without adsorption. The ^3H_2O was determined with a liquid scintillation counter, and CN-Cbl by a HP 1090 series II high performance liquid chromatograph. The retardation factor R is the ratio of groundwater velocity to solute velocity which is related to the extent of solute adsorption as follows

$$R = 1 + \frac{\rho_B}{\varepsilon} K_d \tag{5.51}$$

where ρ_B is the bulk density of the sorbent (g/mL), ε is the effective porosity, and K_d the soil partition coefficient (mL/g). K_d is related to K_f when n is near unity.

Using sand with an organic carbon content of 0.14%, $\rho_B = 1.63\ g/cm^3$, $\varepsilon = 0.36$, and a partition coefficient of 0.50 mL/g for a CN-Cbl in distilled water and sandy soil, we obtain $R = 3.3$, which is comparable to the experimentally determined value of

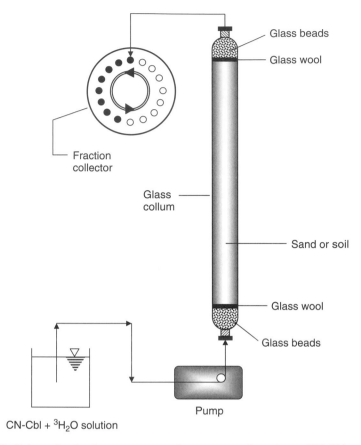

Figure 5.8. Schematic of column system used to measure adsorption to CN-Cbl to sand and sandy soil. (Source: Adapted from Hashsham and Freedman (2003) with permission from ASCE.)

2.1 (Figure 5.9). For groundwater, the slope of the breakthrough curve is steeper, and for a higher organic carbon (2.49%) sandy soil 50% breakthrough occurs at a relative pore volume of $V/V_0 = 2.3$, which is also the retardation factor.

The relatively low value of R indicates that vitamin B_{12} can be distributed throughout a contaminated aquifer fairly rapidly and without experiencing significant losses.

In order to examine the fate of trace metals in lake waters, Hamilton-Taylor et al. (1997) conducted sorption experiments of trace metals (Cu, Pb, Zn) by suspended lake particles in artificial and actual freshwaters. The artificial water was 0.005 M $NaNO_3$ providing an appropriate ionic strength but without reacting significantly with the trace metals. The lake water was collected from Esthwaite Water, Cumbria, UK.

Table 5.8 shows results of experiments where all three metals (Pb, Zn, Cu) were added together and the particulate material was suspended in 0.005 M $NaNO_3$. The metal spikes, added together, measured 1.6 µM Cu (100 µg/L), 0.48 µM Pb

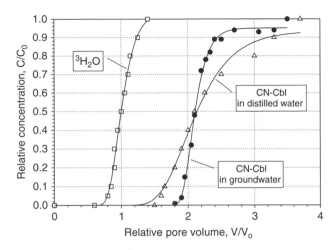

Figure 5.9. Breakthrough curves for tritium and CN-Cbl in distilled water and groundwater passing through a low organic carbon sand. V, volume of water collected; V_0, pore water volume (empty bed volume · porosity). (Source: Sketch based on experimental data of Hashsham and Freedman (2003))

TABLE 5.8. Surface Site Capacities and Formation Constants for Trace Metal Binding by Natural Sediments, Model Solids and Humic Acid

Material	Capacity (meq/g)	pK_{pb}	pK_{Cu}	pK_{Zn}
Esthwaite water particles	0.79 ± 0.07	4.94 ± 0.17	5.57 ± 0.18	6.22 ± 0.17
Hydrous Fe(III) oxide	2.2	4.0	4.3	5.0
Amorphous SiO_2	1.0	4.85	6.15	5.75
Humic acid[a]	3.3	1.7	1.5	2.3

(Source: Hamilton-Taylor et al. (1997)).
[a]Intrinsic binding constants for the reaction involving exchange of a free metal ion for a single proton associated with a carboxyl site.

(100 µg/L), and 0.38 µM Zn (25 µg/L). The results relate to sorption after seven days. The sorption sequence for Esthwaite water particles, Pb > Cu > Zn, is the same as in goethite (Stumm and Morgan, 1996) and for hydrous Fe(III) oxide. The sequence Cu > Pb > Zn has been observed for bacterial surfaces, and is evident in this case for humic acid. If Zn added alone approximately 20% higher adsorption at pH ~ 6–7 occurs, indicating competition for surface sites.

Sorption of Pb, Cu, and Zn increased with time (2 h to 7 d) and pH. In 0.005 M $NaNO_3$ solution, sorption spanned 2–2.5 pH units for Cu and Pb and approximately 4 units for Zn. In natural lake water, the Cu sorption edge was broader, and both Cu and Zn were less strongly sorbed, indicating partial complex formation with substances in lake water.

Particle binding was dominant for Pb. Scavenging or uptake by diatoms was equally important for Zn, whereas complexation with humic substances was an important factor for Cu. Assuming that Pb is associated with lake particles, these features are reflected in relative binding constants for amorphous SiO_2 and humic acid (Table 5.8).

A quantitative treatment of metal adsorption to metal oxides, such as Al_2O_3, Fe_2O_3, FeOOH, or SiO_2, is presented here, expanding on Example 4.36 of Sawyer et al. (2003) to include various concentrations of copper and a graph of the adsorption. The neutral surface of iron oxide is represented by $\equiv$ FeOH. This is a general designation that includes goethite, FeOOH. The relevant equations for the Cu^{2+} ion are

$$\equiv FeOH_2^+ \leftrightarrow \ \equiv FeOH + H^+ \tag{5.52}$$

$$\equiv FeOH \leftrightarrow \ \equiv FeO^- + H^+ \tag{5.53}$$

with equilibrium constants K_{A1} and K_{A2}:

$$K_{A1} = \frac{[\equiv FeOH][H^+]}{[\equiv FeOH_2^+]} \tag{5.54}$$

$$K_{A2} = \frac{[\equiv FeO^-][H^+]}{[\equiv FeOH]} \tag{5.55}$$

Adsorption of Cu^{2+} follows the reaction

$$\equiv FeOH + Cu^{2+} \leftrightarrow \ \equiv FeOCu^+ + H^+ \tag{5.56}$$

with the equilibrium constant K_{SCu}

$$K_{SCu} = \frac{[\equiv FeOCu^+][H^+]}{[\equiv FeOH][Cu^{2+}]} \tag{5.57}$$

which can be developed into a Langmuir expression. Concentrations at surfaces have the meaning mol/kg iron oxide, mol/kg. These concentrations can be converted into M, mol/L as a product of mol/area of solid multiplied by area of solids/kg of solid multiplied by kg of solids/L in solution. The total concentrations of iron Fe_T and copper Cu_T can then be expressed as

$$Fe_T = [\equiv FeOH_2^+] + [\equiv FeOH] + [\equiv FeO^-] + [\equiv FeOCu^+] \tag{5.58}$$

$$Cu_T = [Cu^{2+}] + [\equiv FeOCu^+] \tag{5.59}$$

$$Fe_T = [\equiv FeOH]\left(\frac{[H^+]}{K_{A1}} + 1 + \frac{K_{A2}}{[H^+]} + K_{SCu}\frac{[Cu^{2+}]}{[H^+]}\right) \tag{5.60}$$

$$Cu_T = [Cu^{2+}]\left(1 + K_{SCu}\frac{[\equiv FeOH]}{[H^+]}\right) \tag{5.61}$$

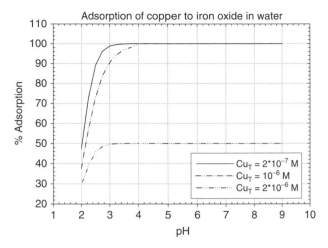

Figure 5.10. Adsorption of copper to iron oxide in water vs. copper concentration and pH.

Equations 5.60 and 5.61 are solved for [≡ FeOH] by inserting [Cu^{2+}] from Equation 5.61 into Equation 5.60. Results of calculations for $K_{A1} = 10^{-6.8}$, $K_{A2} = 10^{-8.8}$ and $Fe_T = 10^{-6}$ M and total copper concentrations of $Cu_T = 10^{-6}$, $2 \cdot 10^{-7}$, or $2 \cdot 10^{-6}$ M, and are shown in Figure 5.10. There is increased metal complex formation and therefore adsorption at higher pH values, as expected from Equation 5.56. Also, as expected from Equation 5.57, a lower concentration of [Cu^{2+}] gives higher adsorption. If the initial copper concentration is twice that of the total iron concentration, only half of the copper can be adsorbed, as seen in Figure 5.10 for $Cu_T = 2 \cdot$ (times) $10^{-6}M$. If several metals are in solution together, adsorption may be modeled as outlined by Hamilton-Taylor et al. (1997).

Recycling nutrients in wastewater sludge via land application is a desirable goal. However, excess micronutrients, singly or in combination, can inhibit growth (Burgess et al., 1999), and can also present a hazard to human health through groundwater contamination. Some metals such as Cd and Cr do not appear to be essential for the growth of organisms or plants and, in the case of Cr (particularly Cr[VI]), are also toxic in high concentrations to humans.

There is a concern that metals can be transferred from land-applied sludge to groundwater, and subsequently to humans with adverse human health effects. With the objective of assessing this risk, Steenhuis et al. (1999) studied trace metal retention in the incorporation zone of land-applied sludge. Below the incorporation zone, there is a vadose zone and below that a saturated aquifer zone. The transfer of metals from the incorporation zone was based on a first-order model,

$$\frac{dM_{Me}}{dt} = -q\frac{M_{Me}}{D_1(\rho K_d + \theta)} \tag{5.62}$$

where M_{Me} is the mass of metals in the zone of incorporation (g/m^2), q the flow rate or recharge rate (m^3/(m^2 s)), D_1 the depth of incorporation (m), ρ the bulk density (g/mL), K_d the partition coefficient (mL/g), and θ the moisture content or porosity.

Integrating Equation 5.62 gives

$$M_{Me} = M_{Me\,0} \exp \frac{-Y}{D_1(\rho K_d + \theta)} \tag{5.63}$$

where Y is the cumulative percolation since the application of sludge ($m^3/m^2 = m$), and $M_{Me\,0}$ the initial amount of metal applied. Considering metals Me and Cr, we obtain from this equation

$$\frac{M_{Me}}{M_{Cr}} = \frac{M_{Me\,0}}{M_{Cr\,0}} \exp \left[\frac{-Y}{\rho D} \left(\frac{1}{K_{Me}} - \frac{1}{K_{Cr}} \right) \right] \tag{5.64}$$

where M_{Cr} and $M_{Cr\,0}$ are actual and initial amount respectively, of Cr in the incorporation zone. We have assumed that $\theta \ll \rho K_d$, which is usually justified. Equation 5.64 can be rearranged slightly to express the ratio of $M_{Me}/M_{Me\,0}$ over $M_{Cr}/M_{Cr\,0}$ the exponential on the right hand side. This ratio is defined as the relative retention.

Steenhuis et al. (1999) used the following assumptions and constants in Equation 5.64 for Cr, Cu, and Cd: There was a 15-yr period after sludge application. Also, $Y = 90$ cm/yr and $\rho = 1.2$ g/mL. Initial metal ratios $M_{Me\,0}/M_{Cr\,0}$ were 1, 0.97, and 0.11 for Cr, Cu, and Cd respectively, and the K_d values were 1229, 216, and 138 mL/g. The observed retentions were 1, 0.67, and 0.58 for Cr, Cu, and Cd respectively, compared with calculated retentions of 1, 0.75, and 0.62. This is a satisfactory agreement between model and measured data, reflecting that the higher K_d gives the highest retention. For Ca, the observed retention was significantly higher than the calculated retention, indicating a violation of the model assumption that the only significant input is from sludge application. There had in fact been other inputs as a result of liming in 1985.

The examples given in this section on vitamins and trace metals demonstrate that the dispersion of organics and metals in the subsurface can be described reasonably using a retardation factor, Equation 5.51 or $R = 1 + \rho K_d/\theta$, using the above symbols, whereas solid–liquid phase hydrous oxide complexation models as given in Equations 5.52–5.61 are more appropriate for dilute systems such as a lake.

REFERENCES

Antoniou, MG, Shoemaker, JA, De La Cruz, AA, Dionysiou, DD. Unveiling new degradation intermediates/pathways from the photocatalytic degradation of microcystin-LR. *Environ. Sci. Technol.* 2008;42(23):8877–8883.

Barton, PK, Atwater, JW. Nitrous oxide emissions and the anthropogenic nitrogen in wastewater and solid waste. *ASCE J. Environ. Eng.* 2002;128(2):137–150.

Battistoni, P, Boccadoro, R, Innocenti, L, Bolzonella, D. Addition of an external carbon source to enhance nitrogen biological removal in the treatment of liquid industrial wastes. *Ind. Eng. Chem. Res.* 2002;41:2805–2811.

Beaulieu, JJ, Shuster, WD, Rebholz, JA. Nitrous oxide emissions from a large, impounded river: The Ohio River. *Environ. Sci. Technol.* 2010;44(19):7527–7533.

Beline, F, Martinez, J, Marol, C, Guiraud, G. Application of ^{15}N technique to determine the contributions of nitrification and denitrification to the flux of nitrous oxide from aerated pig slurry. *Water Res.* 2001;35(11):2774–2778.

Biochemical Cycles, Soil Microbiology. BIOL/CSES 4684. The nitrogen cycle: Nitrification, 2013. Available at: http://filebox.vt.edu/users/chagedor/biol_4684/Cycles/Nitrification .html. Accessed April 23, 2014.

Bolalek, J, Frankowski, L. Selected nutrients and iron in interstitial waters of the estuary of Southern Baltic (Gulf of Gdañsk and the Pomeranian Bay) in relation to redox potential. *Water Air Soil Pollut.* 2003;147:39–50.

Burgess, JE, Quarmby, J, Stephenson, T. Micronutrient supplements for optimisation of the treatment of industrial wastewater using activated sludge. *Water Res.* 1999;33(18): 3707–3714.

Butturini, A, Battin, TJ, Sabater, F. Nitrification in stream sediment biofilms: the role of ammonium concentration and DOC quality. *Water Res.* 2000;34:629–639.

Carrera, J, Baeza, JA, Vicent, T, Lafuente, J. Biological nitrogen removal of high-strength ammonium industrial wastewater with two-sludge system. *Water Res.* 2003;37: 4211–4221.

Cerniglia, CE. Biodegradation of polycyclic aromatic hydrocarbons. *Biodegradation.* 1992;3:351–368.

Correll, DL. The role of phosphorus in the eutrophication of receiving waters: A review. *J. Environ. Qual.* 1998;27:261–266.

Crutzen, PJ, Ehhalt, DH. Effects of nitrogen fertilizers and combustion on the stratospheric ozone layer. *Ambio* 1977;6(2–3):112–117.

Dachs, J, Eisenreich, SJ, Hoff, RM. Influence of eutrophication on air–water exchange, vertical fluxes, and phytoplankton concentrations of persistent organic pollutants. *Environ. Sci. Technol.* 2000;34(6):1095–1102.

DeSimone, LA, Howes, BL. Denitrification and nitrogen transport in a coastal aquifer receiving wastewater discharge. *Environ. Sci. Technol.* 1996;30:1152–1162.

Dillon, PJ, Reid, RA. Input of biologically available phosphorus by precipitation to precambrian lakes. In: Eisenreich, S.J. (Ed.). *Atmospheric Pollutants in Natural Waters.* Ann Arbor, Michigan: Ann Arbor Science Publishers, Inc.; 1981, pp. 183–198.

Dodds, WK. Factors associated with dominance of the filamentous green alga *Cladophora glomerata. Water Res.* 1991;25(11):1325–1332.

Dodds, WK, Bouska, WW, Eitzmann, JL, Pilger, TJ, Pitts, KL, Riley, AJ, Schloesser, JT, Thornbrugh, DJ. Eutrophication of U.S. freshwaters: Analysis of potential economic damages. *Environ. Sci. Technol.* 2009;43(1):12–19.

Drolc, A, Končan, JZ, Cotman, M. Evaluation of total nitrogen pollution reduction strategies in a river basin: A case study. *Water Sci. Technol.* 2001;6:55–62.

Edgerton, ES, Brezonik, PL, Hendry, CD. Atmospheric deposition of acidity and sulfur in Florida. In: Eisenreich, SJ, editor. *Atmospheric Pollutants in Natural Waters.* Ann Arbor, MI: Ann Arbor Science Publishers, Inc.; 1981, pp. 237–258.

Eisentraeger, A, Klag, P, Vansbotter, B, Heymann, E, Dott, W. Denitrification of groundwater with methane as sole hydrogen donor. *Water Res.* 2001;35:2261–2267.

Feitz, AJ, Waite, TD, Jones, GJ, Boyden, BH, Orr, PT. Photocatalytic degradation of the blue green algal toxin microcystin-LR in a natural organic-aqueous matrix. *Environ. Sci. Technol.* 1999;33(2):243–249.

FishDoc, UK, 2013. Available at: www.fishdoc.co.uk/filtration/nitrification. Accessed April 23, 2014.

Foley, J, De Haas, D, Yuan, Z, Lant, P. Nitrous oxide generation in full-scale biological nutrient removal wastewater treatment plants. *Water Res.* 2010;44:831–844.

Fonseca, AD, Crespo, JG, Almeida, JS, Reis, MA. Drinking water denitrification using a novel ion-exchange membrane bioreactor. *Environ. Sci. Technol.* 2000;34:1557–1562.

Forrest, AL, Fattah, KP, Mavinic, DS, Koch, FA. Optimizing struvite production for phosphate recovery in WWTP. *J. Environ. Eng.*, 2008;134(5):395–402.

Graham, JL, Loftin, KA, Meyer, MT, Ziegler, AC. Cyanotoxin mixtures and taste-and-odor compounds in the cyanobacterial blooms from the Midwestern United States. *Environ. Sci. Technol.* 2010;44(19):7361–7368.

Hamilton-Taylor, J, Giusti, L, Davison, W, Tych, W, Hewitt, CN. Sorption of trace metals (Cu, Pb, Zn) by suspended lake particles in artificial (0.005 M $NaNO_3$) and natural (Estwaite Water) freshwater. *Coll. Surf. A Physicochem. Eng. Asp.* 1997;120:205–219.

Hashsham, SA, Freedman, DL. Adsorption of vitamin B_{12} to alumina, kaolinite, sand and sandy soil. *Water Res.* 2003;37:3189–3193.

Hendry, CD, Brezonik, PL, Edgerton, ES. Atmospheric deposition of nitrogen and phosphorus in Florida. In: Eisenreich, SJ, editor. *Atmospheric Pollutants in Natural Waters*. Ann Arbor, MI: Ann Arbor Science Publishers, Inc.; 1981, pp. 199–215.

Hicks, BB. Atmospheric deposition and its effects on water quality. In: Christensen, ER, O'Melia, CR, editors. *Workshop on Research Needs for Coastal Pollution in Urban Areas*. Milwaukee, WI: University of Wisconsin-Milwaukee; 1998.

Hu, B-L, Zheng, P, Tang, CJ, Chen, JW, Vander Biezen, E, Zhang, L, Ni, BJ, Jetten, MSM, Yan, J, Yu, H-Q, Kartal, B. Identification and quantification of anammox bacteria in eight nitrogen removal reactors. *Water Res.* 2010;44:5014–5020.

Huang, C-P, Wang, H-W, Chiu, P-C. Nitrate reduction by metallic iron. *Water Res.* 1998;32(8):2257–2264.

Isaacs, SH, Henze, M. Controlled carbon source addition to an alternating nitrification-denitrification wastewater treatment process including biological P removal. *Water Res.* 1995;29(1):77–89.

Kielemoes, J, De Boever, P, Verstraete, W. Influence of denitrification on the corrosion of iron and stainless steel powder. *Environ. Sci. Technol.* 2000;34:663–671.

Kim, H, Hao, OJ, McAvoy, TJ. SBR system for phosphorus removal: Linear model based on optimization. *J. Environ. Eng.* 2001;127(2):105–111.

Kim, K-H, Heo, W-M, Kim, B. Spatial and temporal variabilities in nitrogen and phosphorus in the Nakddong River System, Koreal. *Water Air Soil Pollut.* 1998;102:37–60.

Le Corre, KS, Valsami-Jones, E, Hobbs, P, Jefferson, B, Parsons, SA. Agglomeration of struvite crystals. *Water Res.* 2007;41:419–425.

Marti, N, Pastor, L, Bouzas, A, Ferrer, J, Seco, A. Phosphorus recovery by struvite crystallization in WWTPs: Influence of the sludge treatment line operation. *Water Res.* 2010;44:2371–2379.

Messer, J, Brezonik, PL. Importance of atmospheric fluxes to the nitrogen balance of lakes in the Florida peninsula. In: Eisenreich, SJ, editor. *Atmospheric Pollutants in Natural Waters*. Ann Arbor, MI: Ann Arbor Science Publishers, Inc.; 1981, pp. 217–236.

Nakano, K, Lee, TJ, Matsumura, M. In situ algal bloom control by the integration of ultrasonic radiation and jet circulation to flushing. *Environ. Sci. Technol.* 2001;35(24):4941–4946.

Nash, DM, Halliwell, DJ. Tracing phosphorus transferred from grazing land to water. *Water Res.* 2000;34(7):1975–1985.

Ozersky, T, Malkin, SY, Barton, DR, Hecky, RE. Dreissenid phosphorus excretion can sustain *C. glomerata* growth along a portion of Lake Ontario shoreline. *J. Great Lakes Res.* 2009;35:321–328.

Paerl, HW, Paul, VJ. Climate change: Links to global expansion of harmful cyanobacteria. *Water Res.* 2012;46:1349–1363.

Parr, LB, Perkins, RG, Mason, CF. Reduction in photosynthetic efficiency of *Cladophora glomerata*, induced by overlying canopies of *Lemna* spp. *Water Res.* 2002;36:1735–1742.

Pretty, JN, Mason, CF, Nedwell, DB, Hine, RE, Leaf, S, Dils, R. Environmental costs of freshwater eutrophication in England and Wales. *Environ. Sci. Technol.* 2003;37(2):201–208.

Qian, SS, Borsuk, ME, Stow, CA. Seasonal and long-term nutrient trend decomposition along a spatial gradient in the Meuse River watershed. *Environ. Sci. Technol.* 2000;34(21): 4474–4482.

Rectenwald, LL, Drenner, RW. Nutrient removal from wastewater effluent using an ecological water treatment system. *Environ. Sci. Technol.* 2000;34(3):522–526.

Reitzel, K, Hansen, J, Andersen, FØ, Hansen, KS, Jensen, HS. Lake restoration by dosing aluminum relative to mobile phosphorus in the sediment. *Environ. Sci. Technol.* 2005;39(11):4134–4140.

Rinta-Kanto, JM, Ouellette, AJA, Boyer, GL, Twiss, MR, Bridgeman, TB, Wilhelm, SW. Quantification of toxic *Microcystis* spp. during the 2003 and 2004 blooms in western Lake Erie using quantitative real-time PCR. *Environ. Sci. Technol.* 2005;39(11):4198–4205.

Rockne, K, Strand, SE. Anaerobic biodegradation of naphthalene, phenanthrene, and biphenyl by a denitrifying enrichment culture. *Water Res.* 2001;35:291–299.

Rouse, JD, Bishop, CA, Struger, J. Nitrogen pollution: An assessment of its threat to amphibian survival. *Environ. Health Perspect.* 1999;107(10):799–803.

Sarnelle, O, White, JD, Horst, GP, Hamilton, SK. Phosphorus addition reverses the positive effect of zebra mussels (*Dreissena polymorpha*) on the toxic cyanobacterium, *Microcystis aeruginosa*. *Water Res.* 2012;46:3471–3478.

Saunders, DL, Kalff, J. Denitrification rates in the sediments of Lake Memphremagog, Canada – USA. *Water Res.* 2001;35:1897–1904.

Sawyer, CN, McCarty, PL, Parkin, GF. *Chemistry for Environmental Engineering and Science*, 5th ed. Boston, MA: McGraw-Hill Higher Education; 2003.

Senn, DB, Hemond, HF. Nitrate controls on iron and arsenic in an urban lake. *Science* 2002;296:2373–2376.

Shahin, UM, Zhu, X, Holsen, TM. Dry deposition of reduced and reactive nitrogen: A surrogate surfaces approach. *Environ. Sci. Technol.* 1999;33:2113–2117.

Sherwood, LJ, Qualls, RG. Stability of phosphorus within a wetland soil following ferric chloride treatment to control eutrophication. *Environ. Sci. Technol.* 2001;35(20):4126–4131.

Smith, RL, Carabadian, SR, Brooks, MH. Comparison of denitrification activity measurements in groundwater using cores and natural-gradient tracer tests. *Environ. Sci. Technol.* 1996;30:3448–3456.

Smith, RL, Miller, DN, Brooks, MH, Widdowson, MA, Killingstad, MW. In situ stimulation of groundwater denitrification with formate to remediate nitrate contamination. *Environ. Sci. Technol.* 2001;35:196–203.

Sprague, LA, Lorenz, DL. Regional nutrient trends in streams and rivers of the United States, 1993–2003. *Environ. Sci. Technol.* 2009;43(10):3430–3435.

Steenhuis, TS, McBride, MB, Richards, BK, Harrison, E. *Environ. Sci. Technol.* 1999;33(8):1171–1174.

Stumm, W, Morgan, JJ. *Aquatic Chemistry Chemical Equilibria and Rates in Natural Waters*, 3rd ed. New York, NY: John Wiley and Sons, Inc.; 1996.

Takenaka, N, Suzue, T, Ohira, K, Morikawa, T, Banelow, H, Maeda, Y. Natural denitrification in drying process of dew. *Environ. Sci. Technol.* 1999;33:1444–1447.

Tang, C-J, Zheng, P, Chen, T-T, Zhang, J-Q, Mahmood, Q, Ding, S, Chen, X-G, Chen, J-W, Wu, D-T. Enhanced nitrogen removal from pharmaceutical wastewater using SBA-ANAMMOX process. *Water Res.* 2011;45:201–210.

Tchobanoglous, G, Burton, FL, Stensel, HD. *Wastewater Engineering: Treatment and Reuse.* New York, NY: Metcalf and Eddy, Inc., McGraw-Hill Education; 2003.

Till, BA, Weathers, LJ, Alvarez, PJJ. Fe(0)-supported autotrophic denitrification. *Environ. Sci. Technol.* 1998;32:634–639.

Tokutomi, T, Yamauchi, H, Nishimura, S, Yoda, M, Abma, W. Application of the nitration and anammox process into inorganic nitrogenous wastewater from semiconductor factory. *J. Environ. Eng.* 2011;137(2).146–154.

Trachtenberg, E, Ogg, C. Potential for reducing nitrogen pollution through improved agronomic practices. *Water Resour. Bull.* 1994;30(6):1109–1118.

Van Benthum, WAJ, Garrido, JM, Mathijssen, JPM, Sunde, J, Van Loosdrecht, MCM, Hijnen, JJ. Nitrogen removal in intermittently aerated biofilm airlift reactor. *J. Environ. Eng.* 1998;124(3):239–248.

Walker, Jr WW. Consideration of variability and uncertainty in phosphorus total maximum daily loads for lakes. *J. Water Resour. Plann. Manag.* 2003;129(4):337–344.

Yang, Q, Xie, P, Shen, H, Xu, J, Wang, P, Zhang, B. A novel flushing strategy for diatom bloom prevention in the lower-middle Hanjiang River. *Water Res.* 2012;46:2525–2534.

Zhang, Q, Carroll, JJ, Dixon, AJ, Anastasio, C. *Environ. Sci. Technol.* 2002;36(23):4981–4989.

Zou, Y, Christensen, ER. Phosphorus loading to Milwaukee harbor from rivers, storm water, and wastewater treatment. *J. Environ. Eng.* 2012;138(2):143–151.

6

METALS

6.1 INTRODUCTION

Of the more than 110 known elements, about 90 are metals plus some metalloids. Some metals, such as sodium, potassium, calcium, and magnesium, are macronutrients to life; some others, including iron, zinc, cobalt, copper, manganese, selenium, and molybdenum, are micronutrients. However, many transition metals are toxic. In natural waters, micronutrient metals, such as Fe and Zn, may become biolimiting to a healthy ecosystem. On the other hand, toxic metals, such as Hg and Cd, even at trace levels, can exert detrimental impact on aquatic organisms, ranging from phytoplankton and bacteria to fish and mammals.

Unlike organic compounds (see Chapter 7), metals do not disappear under the normal conditions in the environment; they only change their forms from one speciation to another. In natural water bodies, a metal can be categorized based on the physical form of its existence, such as "dissolved," "suspended," and "sorbed." Based on chemical forms, each metal may exist as a pure element or inorganic cations with different charges. Metals can also be "embedded" in anions or organic compounds. For example, lead (Pb) can take the form Pb^{2+}, $Pb(OH)_2$, $PbCl^+$, $PbCl_2$, $PbCl_3$, $PbCO_3$, PbO, PbO_2, $(CH_3CH_2)_4Pb$, or one of many others. Speciation—the particular form or forms of a metal—should be clearly recognized when metal is concerned, because both the environmental behavior and the toxicity of ecosystems and humans are very different between different species of the same metal. In response to this

Physical and Chemical Processes in the Aquatic Environment, First Edition. Erik R. Christensen and An Li.
© 2014 John Wiley & Sons, Inc. Published 2014 by John Wiley & Sons, Inc.

recognition environmental analytical chemistry has shifted from analyzing "total dissolved" in the 1970s–1980s toward identifying and quantifying particular species of metals.

This chapter discusses the sources and the behaviors of metals in natural waters, with the emphasis on the toxic metals and metalloids that are of greatest concern. In Section 6.2, air deposition is described because it is a significant but easily neglected source of metals in natural waters. The basic theories of the major reactions of metal ions, including hydration; hydrolysis and other complexation reactions; precipitation and dissolution of minerals; and redox, are briefly reviewed. Detailed calculation procedures are beyond the scope of this book, but can be found in a number of textbooks (Benjamin, 2002; Jensen, 2003; Stumm and Morgan, 1996; Morel and Hering, 1993; Pankow, 1991; Snoeyink and Jenkins, 1980). To solve problems in field practice applications, the calculation efforts can be facilitated by the use of water chemistry and geochemistry softwares such as MINTEQA2 (U.S. EPA, 2006), MINEQL + (ERS, 2007), Visual MINTEQ (KTH, 2013), and Geochemist's Workbench (Aqueous Solutions, 2013). Section 6.3 discusses the behavior of cadmium, mercury, and arsenic in natural waters. Finally, in Section 6.4, the uses of elemental iron in environmental remediation to destroy halogenated organics and reduce toxic heavy metal ions are discussed.

6.2   TRENDS, MEASUREMENT, AND TOXICITY

Table 6.1 provides an overview of metals and the metalloid arsenic that are deemed of interest in terms of environmental impact. Arsenic pollution of groundwaters is prevalent in the southwestern United States and in certain other regions, for example Bangladesh and Southeast Asia (Twarakavi and Kaluarachchi, 2006; Smedley and Kinniburgh, 2002). The use of arsenic as a pesticide is declining due to its persistence and unintended impact on the environment and human health. Chromates (VI) are known for their carcinogenicity, but have several beneficial applications. Cadmium is used as, among other uses, an antifriction agent, and copper has applications for wires and plumbing. Both metals are toxic to biota and humans in high concentrations. Copper may be used as an algaecide, for example in antifouling paint. Copper ions are often more toxic than copper complexed with organic compounds such as ethylenediamine-N,N,N',N'-tetraacetic acid (EDTA). Tributyl tin (Sn) was earlier used in antifouling paints for ships to prevent the growth of marine organisms. However, because of its extreme toxicity to marine life it was banned by the European Union in 2003 (European Parliament and Council, 2003), and its use has been restricted in other countries.

Lithium is used as a tracer in groundwater (Tables 9.2 and 9.4) but is also a toxic pollutant. Another important metal pollutant is lead, the environmental concentration of which has declined due to the phase-out of tetraethyl lead as gasoline engine antiknock compound in the 1970s (Seyferth, 2003). Use of lead in paint is also being phased out. Despite this, concern over lead pollution continues, particularly with regard to its neurotoxic effects on children (Levin et al., 2008). Mercury still retains

TABLE 6.1. Metals and The Metalloid Arsenic with Common Forms at Near-Neutral pH Values

Name	Symbol	Oxidation state	Chemical form(s)	Comments
Arsenic[a]	As	0		Toxic element
Arsenic acid		+5	$AsO(OH)_3$	Used in manufacturing of arsenates
Arsenious acid		+3	$As(OH)_3$	Anhydride form used as herbicide, pesticide, rodenticide
Methane arsonic acid		+3	$CH_3AsO(OH)_2$	The disodium salt used as a herbicide
Cacodylic acid		+1	$(CH_3)_2AsOOH$	Herbicide
Trimethylarsine oxide		−1	$(CH_3)_3AsO$	Produced by spontaneous (pyrophoric) combustion of trimethylarsine
Arsine		−3	AsH_3	Applications in semiconductor industry and for synthesis of organoarsenic compounds[b]
Dimethyl arsine		−3	$(CH_3)_2AsH$	Ignites spontaneously in air
Trimethyl arsine		−3	$(CH_3)_3As$	Similar applications as for arsine; "Garlic"-like smell
Chromium[c]	Cr	+6	CrO_4^{2-} (chromate)	For tanning leather, dying, paints, etc.; Cr(VI) is more toxic than Cr(III)
		+6	$Cr_2O_7^{2-}$ (dichromate)	
		+3	Cr^{3+}, $Cr(Cl)_2^+$ $Cr(Cl)^{2+}$, $Cr(HCO_3)^{2+}$ $Cr(HCO_3)_2^+$ $Cr(OH)_2^+$, $Cr(OH)^{2+}$	
Cadmium	Cd	+2	Cd^{2+} $CdCl^+$ $Cd(OH)^+$ $Cd(HCO_3)^+$	Antifriction agent (bearings). Rust proofer and other uses, e.g. in batteries. Itai-Itai Byo incident (Japan), high Cd content in rice and soya beans. Accumulates in the kidneys. From rat experiments, Cd may cause hypertension, heart enlargement, and reduction in life span[d,e]
Copper[f]	Cu	+2	Cu^{2+} (cupric)	Uses: wires and cables, plumbing, machinery, algacide

(*continued*)

TABLE 6.1. (*Continued*)

Name	Symbol	Oxidation state	Chemical form(s)	Comments
			$Cu(OH)^+$, $Cu(Cl)^+$	
			$Cu(HCO_3)^+$	
		+1	Cu^+ (cuprous)	
Lithium	Li	+1	Li^+	Used for processing silica; in batteries and cell phones.[g] Lithium oxides often a component of ovenware.[g] Found in groundwater. Can cause deformed young in animals. Li^+ may injure the kidneys of humans
Lead[h]	Pb	+2	Pb^{2+} $Pb(OH)^+$ $Pb(Cl)^+$ $Pb(HCO_3)^+$	Used in batteries, construction. May accumulate in liver and kidney. Inhibits enzyme systems. Children are particularly susceptible to lead poisoning[i]
			$Pb(C_2H_5)_4$ $Pb(CH_3)_4$	Tetraethyl form used as antiknock compounds in gasoline in the United States until the 1970s.[j] Methylated form generally more toxic than the nonmethylated form
Mercury	Hg	+2 +1 0 −2	Hg^{2+} (mercuric) Hg^+ (mercurous) ^oHg, CH_3Hg^+ $(CH_3)_2Hg$	Electrical and electronic applications; used in fluorescent lamps, dental amalgams, and metallurgical amalgamation. Emissions to the environment from coal-fired power plants, gold and silver production, and other sources.[k,l] Minamata incident 1953–1961 (Japan); methyl mercury poisoning from a chemical plant using mercuric sulfate catalysts for producing acetaldehyde. Microbial transformation from inorganic to organic form.[m] Most Hg compounds are highly toxic

TABLE 6.1. (*Continued*)

Name	Symbol	Oxidation state	Chemical form(s)	Comments
Nickel	Ni	+2	Ni^{2+}, $Ni(OH)^+$ $Ni(HCO_3)^+$ $NiCl^+$	Used for making nickel steels, nonferrous alloys, and electroplating.[n,o] Can cause dermatitis
Silver[p]	Ag	+1	Ag^+	Jewelry, industrial uses, coins and medals. Use in photography declining. Skin irritant in prolonged exposure. Toxic to microorganisms[q]
Zinc[r]	Zn	+2	Zn^{2+}, $Zn(OH)^+$ $ZnCl^+$, $Zn(HCO_3)^+$	Several uses, e.g. galvanizing, zinc-copper (brass) and other alloys. Toxic to organisms in high concentrations

Sources:
[a]Grund et al. (2005).
[b]Institut National de Recherche et de Sécurité (2000).
[c]Kotaś and Stasicka (2000).
[d]ARL Analytical Research Labs, Inc. (2013).
[e]Balaraman et al. (1989).
[f]Joseph (1999).
[g]U.S. Geological Survey. Mineral commodities – Lithium (2013a).
[h]Sutherland et al. (2005).
[i]Levin et al. (2008).
[j]Seyferth (2003).
[k]Pacyna et al. (2006).
[l]Maprani et al. (2005).
[m]Princeton University, Department of Geosciences (2013).
[n]Kuck (2006).
[o]Kuck (2012).
[p]U.S. Geological Survey, Minerals, Silver statistics and information (2013b).
[q]Maillard and Hartemann (2013).
[r]Tolcin (2011).

many electrical and electronic applications, but is no longer widely used in chlor-alkali plants. Mercury is an unusual case because the main source to the environment is atmospheric deposition, for example from coal-fired power plants, rather than as a direct by-product of Hg applications. The role of microorganisms in methylation of Hg to MeHg, with associated toxic effects, is well known (Princeton University, Department of Geosciences, 2013).

Other metals include nickel, silver, and zinc. Except for special cases, they are usually of lower significance as toxic agents. However, silver is especially toxic to microorganisms. Its use in photography has declined significantly as a result of the increased popularity of digital photography.

Metal concentrations in the aquatic environment can alter as a result of changed industrial practices, changing demographics, and new regulations. Mahler et al. (2006) studied trends of metals (Cd, Cr, Cu, Pb, Hg, Ni, and Zn) in ^{210}Pb-dated sediments from 35 lakes across the United States from 1970 to 2001. Elemental concentrations were determined by acid digestion followed by inductively coupled plasma (ICP) emission spectroscopy or inductively coupled plasma-mass spectrometry (ICP-MS). Because of the volatile nature of many Hg compounds, Hg was determined by cold-vapor atomic absorption spectroscopy. Most metal concentrations showed a decrease, while Cu and Hg remained at approximately the same level, and median Zn concentrations increased in urban watersheds, possibly as a result of Zn from automobile tire wear (Christensen and Guinn 1979). The largest decreases were found for Pb (−46%) and Cr (−34%) with slightly lower decreases for Cd and Ni (both −29%). There was a maximum Pb level between 1970 and 1980 in Lake Washington, Washington, and Newbridge Pond, New York. This finding is in general agreement with the 1970 maximum of Pb found for atmospheric input to the Great Lakes region (Eisenreich et al., 1986; Christensen and Osuna, 1989). The median background concentrations in sediments measured Cd 0.14 ppm, Cr 52 ppm, Cu 16 ppm, Hg 0.040 ppm, Ni 22 ppm, Pb 21 ppm, and Zn 72 ppm. Anthropogenic metal accumulation rates in urban areas were much higher (Cr, Cu, Zn) than that in rural areas, pointing to the significance of fluvial over atmospheric inputs in urban areas.

For the surface waters of the coastal ocean off Los Angeles, California, Smail et al. (2012) found a reduction in particulate metal levels (Cd, Zn, Pb, Ba, Ni, Cu, and V) in 2009 compared with the period of the 1970s prior to implementation of the Clean Water Act. Despite these reductions, the 2009 particulate metal concentrations were still orders of magnitude higher than background crustal levels. The enrichment factor was taken as the ratio ([metal sample]/[Fe sample])/([Metal crust]/[Fe crust]), where the brackets indicate concentrations. By contrast, dissolved trace metal levels in the Los Angeles coastal waters were very low, with values comparable to those found in the pristine waters off Mexico's Baja California peninsula near Punta Banda. The samples from the Mexico waters were obtained biweekly from 2004 to 2005. Ocean samples were subjected to complexation and solvent extraction prior to analysis.

In order to evaluate the potential impact of these metals on the marine food chain, experiments were conducted with phytoplankton that had been exposed to water from the Southern California Bight (SCB). Axenic cultures of *Synechococcus* sp. CC9311 isolated from the California current were grown in filtered and N, P amended SCB water. Trace metal internalization experiments were conducted with bioactive trace metals at concentrations measuring five times the dissolved concentrations during February 2009 at near effluent discharge stations in the Palos Verdes area (0.5 nM Al, 15 nM Ni, 10 nM Cu, 0.05 nM Ag, 1 nM Cd, 0.3 nM Pb, 595 nM Mo, 1 nM Co, 15 nM Zn, and 15 nM Fe). The metals were not strongly chelated by DOC. Within 3 h, there significant increases in the total and intracellular levels of several metals including Pb(+45%), Cu(+29%), and Cd(+16%) were observed, with mortality of

Synechococcus occurring after 48 h exposure. The rapid internalization of the metals in *Synechococcus* suggests that these metals can be incorporated into the marine food webs.

The speciation of trace metals in freshwaters plays an important role for their reactivity, bioavailability, and toxicity to aquatic organisms. For example, Cu^{2+} is generally more toxic than Cu–EDTA complexes. Experimental techniques have been developed to measure concentrations of total metals and complexes with various inorganic and organic ligands (Table 6.2). Dynamic techniques, e.g. DGT (diffusive gradients in thin-film gels), GIME (gel-integrated microelectrode arrays), and SCP (stripping chronopotentiometry), follow the development in time of various complexes, depending on their liability and diffusivity through a diffusion layer. For example, metals bound to colloids or large organic molecules have a lower diffusivity than the metal ions. Thus, sensors that are responsive to metal ions or small

TABLE 6.2. Analytical Techniques for Measurement of Trace Metal Speciation in Natural Freshwaters

Technique	Name	Comments
ICP-MS	Inductively coupled plasma-mass spectrometry	Total metal concentrations
DGT	Diffusive gradients in thin-film gels	Dynamic (days) metal species; Exposed in situ + laboratory analysis
GIME	Gel-integrated microelectrode arrays	Dynamic (minutes) metal species; in situ measurement
SCP	Stripping chronopotentiometry	Dynamic metal species; laboratory analysis
FTPLM	Flow-through permeation liquid membranes	Dynamic (hours to days) metal species or free metal ions. Also equilibrium species; laboratory analysis
HFPLM	Hollow fiber permeation liquid membranes	Dynamic (hours to days) metal species or free metal ions. Also equilibrium species. Exposed in situ + laboratory analysis
DMT	Donnan membrane technique	Free metal ions; dynamic or equilibrium; exposed in situ + laboratory analysis
CLE-SV	Competitive ligand-exchange stripping voltammetry	Free metal ions; equilibrium; laboratory analysis
CLE-AdSV	Competitive ligand-exchange adsorption stripping voltammetry	

(Source: Adapted from Sigg et al. (2006)).

complexes, but not large molecules, can respond rapidly, for example within minutes as for GIME, whereas detectors that respond in a slower manner, for example, within days as for DGT, are more likely to accumulate a signal for all-size complexes and therefore provide a larger response.

The ability of detectors to measure in situ concentrations is another important consideration. There are two categories: in situ measurement, e.g. GIME, and in situ exposure, e.g. DGT, HFPLM (hollow fiber permeation liquid membranes), and DMT (donnan membrane technique). In situ measurements have the advantage that they avoid any artifacts introduced during sampling and sample handling. Sigg et al. (2006) found that the measured concentrations of Cu, Pb, and Cd decreased in the order DGT $\geq$ GIME > FTPLM (flow-through permeation liquid membranes) $\geq$ HFPLM $\approx$ DMT in accordance with the effective measurement time of the methods. For example, for Lake Greifen, Switzerland, measured copper concentrations (in nM) were as follows: ICP-MS average (19 $\pm$ 3), DGT Duebendorf (3.8 $\pm$ 0.6), GIME Geneva 4/9/2003 (1.3 $\pm$ 0.2), SCP Belfast 4/9/2003 (1.3 $\pm$ 0.2), FTPLM Geneva 4/9/2003 (0.05 $\pm$ 0.02), HFPLM Geneva (0.057 $\pm$ 0.009), and DMT Wageningen (0.073 $\pm$ 0.005).

Equilibrium concentrations of the metal species measured by Sigg et al. (2006) were calculated by Unsworth et al. (2006). They used WHAM 6 (Centre for Ecology & Hydrology, 2013) and visual MINTEQ (KTH, 2013). The WHAM 6 incorporated the humic binding model VI (Tipping, 1998) and Visual MINTEQ included the NICA–Donnan model or metal binding to humic substances (Benedetti et al., 1995). DGT and GIME measurements were used to estimate labile and mobile species, whereas measured free ion activities were estimated by DMT and HFPLM. The authors compared measured and calculated metal concentrations for River Wyre, England, and Lake Greifen and Furtbach Stream both near Zürich, Switzerland. Humic substances consisted of 90% fulvic acid and 10% humic acid. For Lake Greifen, there was a good agreement between in situ measurements of Cu using DGT (3.8 $\pm$ 0.6 nM) and model predictions based on WHAM and MINTEQ; however, the GIME value (1.3 $\pm$ 0.2 nM) was lower. The WHAM prediction of free Cu ion concentrations was approximately two orders of magnitude lower than the DMT and HFPLM measured values, and the MINTEQ prediction was about three orders of magnitude lower. Model predictions of free ion concentrations were generally better for Cd, Ni, and Pb. However, they were still only within one to two orders of magnitude of the measured values. The lack of agreement between model and measurement for free ion activities indicates a need for further work, and possibly additional data.

There are many reports on ecotoxicological effects of metals in the aquatic environment. For example, Schmitt et al. (2005) assessed the exposure of fish from the Spring and Neosho Rivers of northeast Oklahoma to lead, zinc, and cadmium from the Tri-States Mining District. Fish from Big River in St. Francois County, Missouri, which is contaminated with mine tailings, were also examined. Reference fish in relatively pristine environments were obtained from four sites in Missouri and Long Branch (LB) Lake in Macon County, Missouri. The fish species included channel catfish, flathead catfish, largemouth bass, spotted bass, carps, and crappies. The results for catfish and bass indicated that concentrations of Pb, Zn, and Cd were highest in fish from contaminated areas. There was a negative correlation between the logarithm of the hemoglobin-adjusted enzyme δ-aminolevulinic acid dehydratase (ALA-D/Hb) activity and the logarithm of the Pb concentration in blood. This shows that Pb is both bioavailable and active biochemically. However, there was no indication that Pb influenced hemoglobin synthesis or iron content of the blood. The authors concluded that further work is necessary to identify higher level end points of Pb, Cd, Zn, and other metals. Other examples of ecotoxicological effects of metals are provided in Chapter 10.

6.3 MAJOR SOURCES AND REACTIONS OF METALS IN WATER

6.3.1 Atmospheric Deposition of Metals

Metals in the atmosphere are derived from wind suspension of continental dust, sea spray, volcanic activity, chemical reactions between trace gases, and the atmospheric emissions from industries such as mining, smelting, and manufacturing. Worldwide monitoring results indicate that "the emission of airborne metallic pollutants has now reached such proportions that long-range atmospheric transport causes contamination, not only in the vicinity of industrialized regions, but also in more remote areas" (Bartram and Ballance, 1996).

Atmospheric deposition through gas/water exchange is a major source of the mercury that is found in natural waters because mercury is the most volatile of all metals. Therefore, it is capable of long-range atmospheric transport (see Section 6.3.1). Inorganic vapor of arsenic and selenium may be emitted from coal burning (Stumm and Morgan, 1996), and many toxic metals are released into the air from forest fires and volcanoes. Organometallic compounds, such as ethylated lead, were also emitted to air from traffic vehicles before the phase-out of leaded gasoline.

Dry deposition with settling particles in the atmosphere is an important source of many metals found in natural waters. Major metals, such as Ca, Mg, Al, Fe, and Mn, have the greatest variability in ion concentrations and are associated with particles of greater than 4.5 μm. On the other hand, trace metals, for example Cr, Ni, Cu, Zn, Cd, and Pb, show lower variability in concentrations and are less abundant in precipitation samples. They are associated with aerosol-size particles of diameter less than 1 μm. Therefore, these metals have lower deposition velocities, and longer atmospheric residence times.

In a study of metal concentrations and deposition in northern Minnesota, Thornton et al. (1981) found high atmospheric concentrations of major metals in the western area and lower levels in the east. In the west, these metals originate in prairie soils that are turned over and exposed to wind, whereas in the east (Hovland), there are fewer soil-derived metals because of the effect of forests to the west and Lake Superior to the east. Atmospheric trace metals showed no consistent concentration patterns.

In northeastern Minnesota, deposition of Fe (4.8 kg/(ha yr)), Mn (0.23 kg/(ha yr)), and Zn (1.15–0.55 kg/(ha yr)) is primarily dry. Wet and dry fractions are more evenly distributed for Cu, Cd, and Cr. The wet deposition rates (kg/(ha yr)) of several metals were: Ca (3.16), Al (0.39), Fe(0.25), Mn(0.02), Pb (0.049), Ni (0.010), Zn (0.65−0.05), Cd (0.0017), Cr (0.0017), and Cu (0.025).

Sievering et al. (1981) demonstrated that atmospheric input of trace metals to Lake Michigan is significant compared to surface runoff. Dry and wet (precipitation) input of Fe and Mn were about equal. However, dry input was greater than precipitation inputs of Pb and Zn. Including lake runoff, atmospheric inputs from dry loading are at least 44% of total Pb input, 21% of total Zn input, and 14% of total Fe input. Based on mean dry inputs, the dry deposition rate of Pb is greater than 72% of the total input, and the dry deposition rate of Zn is greater than 47% of total inputs. These dry loadings are of concern because of the direct contribution to the biologically active surface layers.

Metals input to Lake Michigan depend on meteorological conditions at the macro-, meso-, and microscale. Macroscale describes air masses and mixed layer variations. Transport over the lake as a function of stability, including low-lying inversions that can prevail under spring and summer conditions, is the focus at the mesoscale.

For the micro scale, the focus is on variations in mid-lake concentrations with surface layer stability and wind speed u_z. Surface layer stability depends on the air–lake surface temperature difference $\Delta T = T_a - T_s$, where T_a is the sampling height temperature and T_s is the lake surface temperature. Stable conditions characterized by high temperature differential, e.g. ΔT greater than 8.2 °C or low wind speed, e.g. u_z less than 2.2 m/s are associated with low aerosol deposition velocities (< 0.1 cm/s). Higher deposition velocities (> 0.9 cm/s), on the other hand, are obtained when the temperature differential is small, or even negative, e.g. less than −0.9 °C, and for higher wind speeds (> 8.6 cm/s).

The geometric mean aerosol mass for the 1977 study was 35 ± 23 $\mu g/m^3$. However, during periods of strong stability, the aerosol mass increased to 90 $\mu g/m^3$. The mass median diameter for all metals, except Ca, Mg, and Fe measured less than 1 μm. It was also evident that high concentrations of Al, Ca, Fe, Mg, Mn, Pb, and Zn came from Chicago, Illinois, and Gary, Indiana. In addition, since the large diameter particles settle first, there was an increasing magnitude of small-to-large aerosol ratio with increasing fetch over the lake for Al, Fe, Mn, Pb, and Zn. A mass balance study indicated that most Pb, Zn, Mn, and probably Fe at 40–100 km off-shore may be of anthropogenic origin.

A later investigation of dry deposition of metals, i.e. Pb, Cu, and Zn, to Chicago, southern Lake Michigan, and South Haven, Michigan (non-urban) was conducted by Paode et al. (1998). These authors measured total dry deposition fluxes (Figure 6.1) and concentrations of metals in aerosols of nine mass median diameters (in μm): 0.15, 0.4, 1.2, 2.5, 4.5, 7.0, 18, 30, and 60. This contrasts with Sievering et al.'s (1981) three stage impactor measuring aerosols in only three size ranges: greater than 1 μm, less than 1 μm, and between 0.5 and 8 μm.

From the measured concentrations, Paode et al. (1998) calculated fluxes in different size intervals using deposition velocities obtained from the Sehmel–Hodgson dry deposition model. The results are shown in Figure 6.2. As may be seen from this

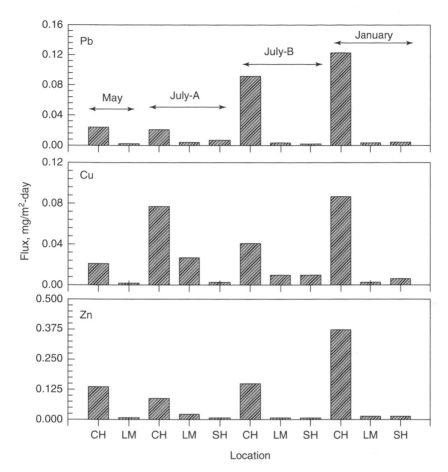

Figure 6.1. Measured fluxes of Pb, Cu, and Zn to Chicago, Illinois (CH), Lake Michigan (LM), and South Haven (SH). May and July samples were taken in 1984, and the January sample was taken in 1995. (Source: Reprinted from Paode et al. (1998) with permission from ASCE.)

figure, fluxes mostly fall within the d greater than 2.5 μm range, whereas the atmospheric concentrations mainly fall within the d less than 2.5 μm range. However, there were some exceptions where the atmospheric size distributions had maxima in the 2–8 μm range for Pb, Cu, and Zn that were sampled in Chicago during May 1994. This is similar to the size distribution for crustal elements.

As may be seen from Figures 6.1 and 6.2, fluxes are substantially higher in Chicago (CH) than either in South Haven (SH) or over Lake Michigan (LM). When the wind over the lake (July A) came from the direction of Chicago, the fluxes and concentrations over the lake (LM) were higher than when the wind came from other directions (May, July B).

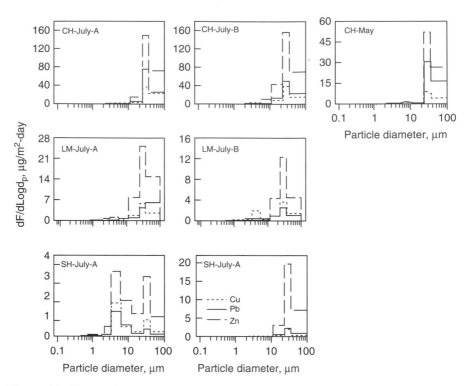

Figure 6.2. Fluxes of Pb, Cu, and Zn vs. particle diameter based on measured size distributions and on deposition velocities calculated from Sehmel and Hodgson's dry deposition model. (Source: Reprinted from Paode et al. (1998) with permission from American Chemical Society.)

The aerosol concentration for d less than 10 μm, PM10, was found to be 33 ± 7 μg/m^3 considering the median value for five sites in Chicago. Metals were determined by an ICP-mass spectrometer.

It is apparent from Figures 6.1 and 6.2 that the ratio between calculated and measured fluxes is mostly in the range of 0.15–2, indicating that most calculated fluxes are underestimated. According to Paode et al. (1998), the probable reason is that the Sehmel–Hodgson dry deposition model underestimates the deposition velocity for aerosols in the 5–80 μm size range.

For the mid-Atlantic bight, less than 50% of Cd, Cu, and Mn are of atmospheric origin, whereas the majority of Ni, Zn, Fe, Al, and Pb come from the atmosphere (Hicks, 1998). The atmosphere is known to be a significant source of Hg to Wisconsin inland lakes. The Hg is present in fly ash from coal-fired power plants.

6.3.2 Hydration, Hydrolysis, and Complex Formation

It is important to realize that bare "free" ions do not exist in water. Hydration occurs to all ions upon entering water. For metal ions, the free pairs of electrons of the oxygen atom in the water molecules are attracted to the positive charge of the metal ion; thus water molecules orient themselves around the metal ions to form "hydrates" or "aquo complexes" $M(H_2O)_n^{z+}$

$$M^{z+} + nH_2O \rightarrow M(H_2O)_n^{z+} \tag{6.1}$$

The water molecules surrounding the metal ion form a "hydration shell" (or "hydration sphere"), losing their positions in the hydrogen-bonded network of the bulk water. Hydrated metal ions $M(H_2O)_n^{z+}$ are often called "free" or "uncomplexed", and customarily abbreviated as M^{z+} for simplicity. This is to say, for example, Fe^{3+} is in fact $Fe(H_2O)_n^{3+}$. In the formula, n is the number of water molecules in the inner hydration sphere. It ranges from 4 to 8, increasing with the charge/radius ratio of the ions and decreasing with increasing temperature. It also varies with solution composition. Secondary or higher outer hydration spheres may form around the inner sphere, mainly as a result of the polar–polar interactions of the water molecules. Such metal–water complex is a different entity and has completely different size and behavior than the bare metal and metal ion that may exist in the gas phase. However, the metal hydrates possess the same charge as the metal ion in the center, because water molecules are neutral.

Coordination, or complex formation, describes the combination of cations with molecules or anions, known as ligands, which contain free electron pairs. The positive charge of the metal ions attracts the electron pairs, thus a special covalent bond is formed in which both electrons shared in the bond come from the same atom in the molecule or anion. In the resulting complex, the metal ion is the central atom, and the atom in the ligand that actually provides the electron pair, for example, N in NH_3 and O in H_2O and OH^-, is the donor atom. Mixed ligand complexes with more than one type of ligands, for example $Pb(H_2O)_2NH_3Cl^+$ in which the central Pb^{2+} is bonded to three types of ligands, can form.

Chelation is a special type of coordination reaction, in which the ligands are "multi-dentate", i.e. they have more than one pair of electrons to donate to the central metal. For example, EDTA is hexa-dentate, glycine (NH_2CH_2COOH) is di-dentate and so, sometimes, are inorganic ligands CO_3^{2-} and SO_4^{2-} (Figure 6.3). Such ligands "grab" the central metal with multiple "teeth", resulting in highly stable chelate complexes, which often have ring structures. On the other side, when the metal-to-ligand concentration ratio is high, polynuclear complexes with more than one central metal atom may form. The growth of polynuclear complexes may eventually lead to the precipitation of the metals; for example, the floc formed after the addition of Fe(III) or Al(III) chloride or sulfate salts during the flocculation process in wastewater treatment.

At appropriate pH, hydroxide ion OH^- may replace the water molecules in the hydration shell. This reaction is known as hydrolysis, in which hydrogen ions (proton) H^+ are lost from the aquo complex $M(H_2O)_n^{z+}$, which is customarily abbreviated as M^{z+}, forming hydroxo-complexes $M(H_2O)_{n-L}(OH)_L^{(z-L)+}$, which is abbreviated as $M(OH)_L^{(z-L)+}$. Further increases in pH may remove all hydrogen ions from the complex, thus metal oxide (oxo-complex) forms.

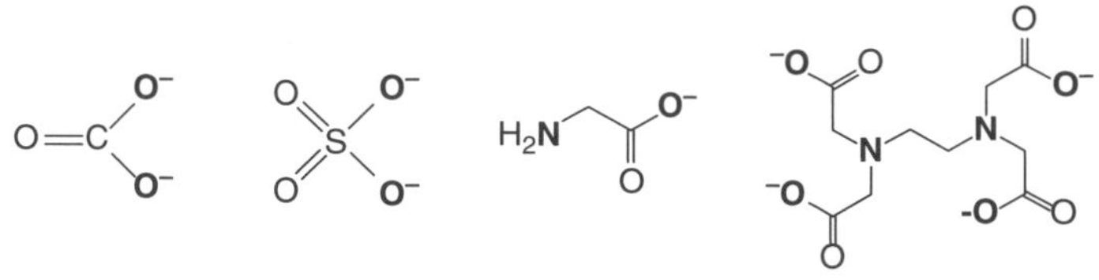

Figure 6.3. Example chelants. The donor atoms are in bold letters. (a) CO_3^{2-} can be di-dentate; (b) SO_4^{2-} can be di-dentate; (c) glycine is di-dentate; (d) EDTA is hexa-dentate.

The reaction can be written as stepwise acid dissociation reactions

$$M^{z+} + H_2O \rightarrow M(OH)^{(z-1)+} + H^+ \quad K_{a1} \tag{6.2}$$

$$M(OH)^{(z-1)+} + H_2O \rightarrow M(OH)_2^{(z-2)+} + H^+ \quad K_{a2} \tag{6.3}$$

$$M(OH)^{(z-2)+} + H_2O \rightarrow M(OH)_3^{(z-3)+} + H^+ \quad K_{a3} \tag{6.4}$$

$$M(OH)^{(z-3)+} + H_2O \rightarrow M(OH)_4^{(z-4)+} + H^+ \quad K_{a4} \tag{6.5}$$

The same reactions can also be expressed as the formation of metal hydroxide ions:

$$M^{z+} + OH^- \rightarrow M(OH)^{(z-1)+} \quad K_1 \tag{6.6}$$

$$M(OH)^{(z-1)+} + OH^- \rightarrow M(OH)_2^{(z-2)+} \quad K_2 \tag{6.7}$$

$$M(OH)_2^{(z-2)+} + OH^- \rightarrow M(OH)_3^{(z-3)+} \quad K_3 \tag{6.8}$$

$$M(OH)_3^{(z-3)+} + OH^- \rightarrow M(OH)_4^{(z-4)+} \quad K_4 \tag{6.9}$$

Apparently, the two types of K's can be linked by the water dissociation constant K_w. That is, $K_{ai} = K_w \times K_i$. Alternatively, the reaction steps can be combined

$$M^{z+} + H_2O \rightarrow M(OH)^{(z-1)+} + H^+ \quad \beta_{a1} = K_{a1} \tag{6.10}$$

$$M^{z+} + 2H_2O \rightarrow M(OH)_2^{(z-2)+} + 2H^+ \quad \beta_{a2} = K_{a1} \times K_{a2} \tag{6.11}$$

$$M^{z+} + 3H_2O \rightarrow M(OH)_3^{(z-3)+} + 3H^+ \quad \beta_{a3} = K_{a1} \times K_{a2} \times K_{a3} \tag{6.12}$$

$$M^{z+} + 4H_2O \rightarrow M(OH)_4^{(z-4)+} + 4H^+ \quad \beta_{a4} = K_{a1} \times K_{a2} \times K_{a3} \times K_{a4} \tag{6.13}$$

Or

$$M^{z+} + OH^- \rightarrow M(OH)^{(z-1)+} \quad \beta_2 = K_1 \tag{6.14}$$

$$M^{z+} + 2OH^- \rightarrow M(OH)_2^{(z-2)+} \quad \beta_2 = K_1 \times K_2 \tag{6.15}$$

$$M^{z+} + 3OH^- \rightarrow M(OH)_3^{(z-3)+} \quad \beta_3 = K_1 \times K_2 \times K_3 \tag{6.16}$$

$$M^{z+} + 4OH^- \rightarrow M(OH)_4^{(z-4)+} \quad \beta_4 = K_1 \times K_2 \times K_3 \times K_4 \tag{6.17}$$

As an example, the sequential deprotonations of $Fe(H_2O)_n^{3+}$ (same as Fe^{3+}) will produce $Fe(OH)^{2+}$, $Fe(OH)_2^{+}$, and $Fe(OH)_3^{0}$, which may precipitate. A further increase in solution pH may result in the dissolution of the metal hydroxide solids as a result of the formation of charged species such as $Fe(OH)_4^{-}$. In this sense, all hydrated metals are potentially polyprotic acids, and hydrolysis is an acid/base reaction.

In an aqueous environment, water is the most important ligand to all metals because of its abundance. Hydration is a special case of complexation, with water being the only ligand; hydrolysis changes some water to hydroxide ions. However, metal ions may combine more strongly with other ligands. In general, complexation reactions can be similarly written as

$$M + L \rightarrow ML \quad K_1 \tag{6.18}$$

$$ML + L \rightarrow ML_2 \quad K_2 \tag{6.19}$$

$$ML_2 + L \rightarrow ML_3 \quad K_3 \tag{6.20}$$

$$ML_3 + L \rightarrow ML_4 \quad K_4 \tag{6.21}$$

Or alternatively,

$$M + L \rightarrow ML \quad \beta_1 = K_1 \tag{6.22}$$

$$M + 2L \rightarrow ML_2 \quad \beta_2 = K_1 \times K_2 \tag{6.23}$$

$$M + 3L \rightarrow ML_3 \quad \beta_3 = K_1 \times K_2 \times K_3 \tag{6.24}$$

$$M + 4L \rightarrow ML_4 \quad \beta_4 = K_1 \times K_2 \times K_3 \times K_4 \tag{6.25}$$

The equilibrium constants presented above are the stepwise (K) or combined (β) stability constant. They are available for numerous metals in chemistry handbooks and textbooks (Stumm and Morgan, 1996; Benjamin, 2002). The values of log β_i for selected metals are summarized in Table 6.3.

Inorganic ligand species Cl^-, HCO_3^- and $CO_3^=$, NH_3 and NH_4^+, S^{2-} and HS^-, are widely present. The bicarbonate HCO_3^- is generally the most abundant anion in freshwaters, and has a high affinity toward transition metals. Phosphates and polyphosphates, other halogen ions (F^- and Br^-), and sulfate (SO_4^{2-}) and thiosulfate ($S_2O_3^{2-}$), can also be important inorganic ligands. Some such ligands come from anthropogenic sources; for instance NH_3 from organic wastes, S^{2-}, SO_4^{2-}, SO_3^{2-} from pulp and paper mill effluents, PO_4^{3-} and polyphosphates from detergents and fertilizers, and CN^- from various industries. Among these, cyanide CN^- forms extremely stable complexes with transition metals, whereas fluoride F^- combines firmly with Al^{3+}.

The most important organic ligands are the humic substances—falvic acids, humic acids, and humin. They form very stable complexes, often chelates, with polyvalent metal ions. Among the major anthropogenic organic ligands, EDTA is a colorless, water-soluble solid, with a wide range of industrial uses. It is deemed a hexa-protic

TABLE 6.3. Stability Constants log β_i for Metal–Ligand Complexes

Metal ion	i	OH⁻	CO₃²⁻		Cl⁻		EDTA	
Ca^{2+}	1	1.40						
Cd^{2+}	1	3.92	CdL	5.4	CdL	1.98	CdL	16.28
	2	7.65	CdL₃	6.22	CdL₂	2.6	CdHL	2.9
	3	8.70	CdHL	12.4	CdL₃	2.4		
	4	8.65						
Cr^{3+}	1	10.00	CrL	1.34	CrL	−0.25		
	2	18.38			CrL₂	−0.96		
	3	25.25						
	4	28.23						
Cu^{2+}	1	6.00	CuL	6.73	CuL	0.43	CuL	18.78
	2	14.32	CuL₂	9.83	CuL₂	0.16	CuHL	11.2
	3	15.10	CuHL	13.6	CuL₃	−2.29		
	4	16.40			CuL₄	−4.59		
Fe^{2+}	1	4.50			FeL	0.9	FeL	16.7
	2	7.43					FeHL	20.1
	3	11.00						
Fe^{3+}	1	11.81			FeL	1.48	FeL	27.8
	2	22.33			FeL₂	2.13	FeHL	29.4
	3	28.40			FeL₃	1.13		
	4	34.40						
Hg^{2+}	1	10.60			HgL	6.75		
	2	21.90			HgL₂	13.12		
	3	20.86			HgL₃	14.02		
					HgL₄	14.43		
Ni^{2+}	1	4.14	NiL	6.87	NiL	0.4	NiL	20.33
	2	9.00	NiL₂	10.11	NiL₂	0.96	NiHL	11.56
	3	12.00	NiHL	14.27				
Pb^{2+}	1	6.29	PbL	7.24	PbL	1.6	PbL	17.86
	2	10.88	PbL₂	10.64	PbL₂	1.8	PbHL	9.68
	3	13.94	PbHL	13.2	PbL₃	1.7	PbH₂L	6.22
	4	16.30			PbL₄	1.38		
Zn^{2+}	1	5.04	ZnL	5.3	ZnL	0.43	ZnL	16.44
	2	11.10	ZnL₂	9.6	ZnL₂	0.45	ZnHL	9
	3	13.60	ZnHL	12.4	ZnL₃	0.5		
	4	14.80			ZnL₄	0.2		

(Source: All values are taken from Benjamin (2002)).

acid with four dispensable hydrogens attached to carboxylic acid groups ($pK_a = 0.0$, 1.5, 2.0, and 2.7 respectively) from the neutral molecule, and the two amine groups are capable of accepting protons ($pK_a = 6.1$ and 10.4 respectively). Based on these pKa values, EDTA in the normal pH range (6–9) of natural waters would be mostly deprotonated. It forms very stable hexa-dentate complexes with transition metals, as can be seen from the stability constants in Table 6.3. Nitrilotriacetic acid (NTA) is a triprotic acid with pKa values 1.66, 2.95, 10.28, and has been used as a builder in detergents to combat the hardness ions, such as Ca^{2+} and Mg^{2+}, thus enhancing the efficiency of the detergents. NTA has been banned in the United States as a result of health concerns regarding certain types of birth defects. It can be mostly removed in secondary wastewater treatment.

Evidently, complex stability depends on the speciation and concentrations of both metal ions and ligands themselves, which in turn are functions of solution temperature, pH, ionic strength, and redox condition. As mentioned above, most metal ions are acidic because of hydrolysis and their speciation is therefore pH dependent. Oxidized metal ions (e.g. Fe^{3+}) tend to hydrolyze at lower pH and form more stable complexes (Table 6.3) than the reduced forms (e.g. Fe^{2+}) of the same ML_n formula. The solution conditions also affect the speciation of most of the ligands discussed above.

Under certain conditions, the most dominant complex can be easily determined from the predominant area diagram (PADs). Examples of PADs are presented in Section 6.4. PADs are independent of total concentration of the dissolved metal. It depends on ratios between species, and the ratios depend on K's, pH, etc. All relevant reaction equilibrium constants, including those in Table 6.3, as well as the mass balance equations, charge balance, proton condition, etc., are necessary for drawing PADs and concentration calculations for all the species. The calculation efforts are facilitated by the use of water chemistry and geochemistry software, as mentioned in the introduction of this chapter. It is worth mentioning that such calculations are meaningful only when all the major constituents, including metal cations and anions, are identified and included in the calculations.

Complex formation has significant impact on the functioning of natural water systems. It stabilizes metal ions in solution by preventing metals from precipitating, thus promoting their bioavailability and facilitating their uptake by organisms. Complexation may reduce the toxicity of some metal ions that are more toxic in the "free" form than if complexed by ligands. For example, the "free" aquo complex copper ion, Cu^{2+}, is very toxic to fish, whereas many of its organic chelates are biologically inert. Thus, much lower total dissolved copper concentration is needed in order to prevent fish kills in relatively pure water than in water containing humic materials. On the other hand, excessive input of ligand species, such as phosphates from fertilizers and detergents, may result in "biolimiting" states of trace nutrient metals, such as iron, manganese, and zinc, as a result of excessive complexations with these ligands. Similarly, complex formation in the human body helps maintain the healthy amounts of various metals. In cases of acute metal poisoning, EDTA could be added into the diet in order to accelerate the excretion of the unwanted metals, although the safety of such "oral chelation" is a concern.

6.3.3 Dissolution of Metals from Minerals

Many elements, such as calcium, magnesium, selenium, and arsenic, exist in water mainly as a result of the geological formation of the bottom bed of lakes and rivers or the soil grains of groundwater. The dissolution of calcite (and other minerals) as a source of calcium (and other major metals) in water and its impact on water hardness and acid neutralization capacity have been discussed in Chapter 4. The ecological impact of some trace metals is included in Chapter 5. The release of arsenic from soil grains into groundwater has caused widespread poisoning among people drinking well water in Southeast Asia; this will be further discussed in Section 6.4.

Most toxic metals mentioned in this chapter exist in the earth's crust in the form of oxides and sulfides, which are fairly insoluble. Strictly speaking, nothing is absolutely "insoluble," but this term is applied when solubility is very low and/or the dissolution is very slow. In natural waters, common bottom solids include sedimentary rocks such as limestone ($CaCO_3$) and gypsum ($CaSO_4 \cdot 2H_2O$), and volcano lava rocks such as basalt and granite, which are composed of various minerals that are rich in silica and alumina, such as feldspar ($KAlSi_3O_8$ – $NaAlSi_3O_8$ – $CaAl_2Si_2O_8$) and quartz (SiO_2). Natural water bodies differ according to their ionic strength, salinity, hardness, alkalinity, etc., because the bottom bed materials differ.

Solid formation from dissolved ions in natural waters is a long and stepwise geo-chemical process, as described by Stumm and Morgan (1996). First, increasing metal concentration leads to the formation of polynuclear complexes, or polymers. When the ion concentration exceeds that allowed by the solubility product, the solution becomes super-saturated, and a three-dimensional network of the polymers builds up, and many ions in the interior lose contact with the bulk water. Tiny nuclei form to become the cores for crystal growth. After nucleation, more ions spontaneously deposit on the nuclei, resulting in increased particle size. The newly formed crystals are "immature" and will undergo a long period of slow agglomeration and a ripening process to become the thermodynamically stable mineral. Therefore, kinetics is often more important, and equilibrium may not always be achieved.

The reverse process is the dissolution of minerals and other solid materials. It is understandably slow for the minerals concerned here, and diffusion of ions in solution is often the barrier. The minerals dissolve to the level allowed by the equilibrium constant of the dissolution reaction K_{si}, where s in the subscript stands for solubility and i is the number of ligands associated with the dissolved metal ion. For simple solids,

$$ML_n(s) \rightarrow M + nL \quad K_{s0} \tag{6.26}$$

$$ML_n(s) \rightarrow ML + (n-1)L \quad K_{s1} = K_{s0}\beta_1 = K_{s0}K_1 \tag{6.27}$$

$$ML_n(s) \rightarrow ML_2 + (n-2)L \quad K_{s2} = K_{s0}\beta_2 = K_{s0}K_1K_2 \tag{6.28}$$

$$ML_n(s) \rightarrow ML_3 + (n-3)L \quad K_{s3} = K_{s0}\beta_3 = K_{s0}K_1K_2K_3 \tag{6.29}$$

where the K's and β's are the stepwise and overall equilibrium constants for the general complexation reactions respectively, as presented in Equations 6.26–6.29. Note that the charges, although not shown, must be balanced.

In general, mineral dissolution reactions can be written as

$$M_xL_y(s) \leftrightarrow xM(aq) + yL(aq), \quad K_{s0} = [M]^x[L]^y \tag{6.30}$$

The equilibrium constant K_{s0} is commonly known as "solubility product" or K_{SP}. In applications of the K_{si}'s, it is often assumed that the invasion of the solvent (e.g. water) and other foreign substances to the solid crystalline structure is negligible; that is, the solid is pure and the activity of the solid phase is unity. The K_{si}'s are temperature dependent, and their unit depends on the formula of the solid. The K_{si}'s work only when the solid is in equilibrium with water, thus they should be used with caution where kinetics dominates and prevents equilibrium from occurring. In addition, reported K_{si} values vary greatly for some minerals. For instance, the K_{s0} for hydroxyapatite $Ca10 \, (PO_4)_6 \, (OH)_2$ was found to vary over a range of 10^{11} (Larsen, 1966).

Above graphs do not consider solid precipitation. Thus, TOTM = constant, and PAD is independent of TOTM. When solids precipitate, TOTMdiss is less than TOTM, although solids may not exist if TOTM is low. Whenever a solid exists, it is in control and dominant.

Metal ions (and the anions) released from solids undergo various further reactions such as hydraton, hydrolysis and other complexations, gas transfer, and redox. Few solids (e.g. $BaSO_4$) release ions for which these reactions are insignificant. Therefore, apparent solubility of the mineral in water should be the sum of its entire dissolved species. For instance, $Cd(OH)_2$ (s) releases Cd^{2+}, which forms complexes with various ligands.

6.4 BEHAVIOR OF SELECTED METALS IN WATER

6.4.1 Mercury

Mercury mainly occurs in nature as the red ore cinnabar HgS, although the free element also exists. Mercury is extracted by heating cinnabar in a stream of air and condensing the vapor

$$HgS(s) + O_2 \quad \rightarrow \quad Hg(s) + SO_2 \tag{6.31}$$

The main source of mercury in the environment is atmospheric deposition, from coal-fired power plants, gold and silver production, and other sources (Pacyna et al., 2006; Maprani et al., 2005). Most mercury compounds, especially methyl mercury MeHg, are highly toxic and considerable efforts are made to limit environmental concentrations and human exposure. The U.S. Environmental Protection Agency has provided a seafood consumption limit for MeHg in fish of 0.95 mg/kg (Evers et al., 2012). Information about mercury fisheries advisories for Wisconsin, USA, is provided in Chapter 10 on Ecotoxicology.

Mercury concentration in newly constructed reservoirs is an active area of research. Relatively low levels of methyl mercury were found in Chinese reservoirs, whereas reservoirs in North America and Europe appear to contain higher levels of mercury and methyl mercury (Yao et al., 2011). The authors studied Hongjiadu and Suofengying Reservoir within the Wujiang River, Guizhou Province, China.

They found low mercury and methyl mercury and an average Hg content in fish tissue of only 0.044 mg/kg wet weight, which is also low compared to results from other reservoirs. They estimated that the low MeHg concentrations result from low dissolved organic carbon (DOC) levels (0.5–4.9 mg/L). Sufficient DOC is required for microbial methylation. The two reservoirs are located in the karstic region of southwest China where limestone is the main bedrock, and this may be the reason for the unusually low MeHg levels.

An example of a reservoir that receives mainly non-atmospheric mercury is the Lahontan Reservoir in Nevada, USA (Gandhi et al., 2007). The main source of mercury is estimated to be Hg that is tightly bound to gold–silver–mercury amalgam from the Carson River. During mining of precious metals in Virginia City, Nevada, from 1860 to 1895, gold and silver was extracted in a mercury amalgamation process. Once extracted, the mercury was heated away from the precious metals and then recondensed and reused. The mercury pollution of the Carson River occurred as a result of Hg losses in this process. Gandhi et al. (2007) developed a mercury speciation, fate, and biotic uptake model and estimated that Hg-DOM, $Hg(OH)_2$ and HgClOH were the major species in the aqueous phase, and more than 90% of the mercury from Carson River was retained in the sediment. Methyl mercury MeHg, accounting for 10% of the total loading to the water column, diffuses from sediment to water column from which it is removed as a result of uptake by fish and demythelation. Fish accumulated 45% of the total MeHg in the water column, of which 9% leaves the system as a result of fishing. Concentrations of total mercury in water (bulk 99–3300 ng/L) and surficial sediments (369–30,400 ng/g) were quite high, and even though MeHg concentrations were moderate, there may be a significant potential for methylation once Hg is taken up by fish.

By contrast, the work of Kirk et al. (2011) demonstrates that atmospheric deposition of mercury in 14 subarctic lakes could be the main mercury source, and that it has been increasing since the start of the industrial revolution in the 1850s. The authors considered the hypothesis that Hg increases in sediments were caused by climate change through increased algal derived organic carbon S2, and changes in the phytoplankton composition (Chrysophytes, diatoms, green algae, and blue-green algae). However, in most cases there was not a significant correlation between Hg and S2, and in the few cases that such a correlation was found, changes in species composition did not take place at times of change in the Hg flux. Based on this, increased Hg fluxes appear to be independent of S2 and algal species composition.

6.4.2   Zinc and Cadmium

The types and concentrations of metal–ligand complexes formed by metal ions in a given aquatic environment are of interest for the study of metal bioavailability and toxicity. Table 6.3 lists stability constants for several metals described in Table 6.1. Note that metals can have more than one oxidation state, e.g. Cu^{2+} and Cu^{+}, and Hg^{2+} and Hg^{+}. For chromium, chromate CrO_4^{2-} or dichromate $Cr_2O_7^{2-}$ where Cr is in oxidation state, six is often of higher interest than Cr^{3+}. Other metals such as tetraethyl lead and MeHg are toxic and can further interact with dissolved organic carbon or organic biomolecules. Metals used in plating applications are often solubilized by cyanide complexation. Less toxic alternatives for this are being developed. Furthermore, kinetic considerations can be important in order to evaluate if a given equilibrium composition is achievable within a reasonable time.

While multiple oxidation states for metals other than iron, chromates, MeHg, and tetraethyl lead are not considered in Table 6.3, the complexes that are included provide a good starting point for quantitative metal complexation. We shall illustrate speciation calculations here by considering zinc for metal removal through precipitation, and cadmium speciation with hydroxyl, chloride, and carbonate with a view to the effects on cadmium toxicity.

For the divalent metal, Zn^{2+}, the solubility product K_{s0} is defined as

$$K_{s0} = [Zn^{2+}][OH^-]^2 = 10^{-15.55} \tag{6.32}$$

which is based on the reaction,

$$\alpha - Zn(OH)_2(s) \leftrightarrow Zn^{2+} + 2OH^- \tag{6.33}$$

Similarly, for carbonate precipitation the reaction is

$$ZnCO_3 \cdot H_2O \ (s) \ \leftrightarrow \ Zn^{2+} + CO_3^{2-} \tag{6.34}$$

with the solubility product

$$K_{s0} = [Zn^{2+}][CO_3^{2-}] = 10^{-10.26} \tag{6.35}$$

where the values for both solubility products are taken from Benjamin (2002). Thus, the zinc ion concentration can be written as

$$[Zn^{2+}] = \frac{K_{s0}}{[OH^-]^2} \tag{6.36}$$

Using the notation introduced in Equations 6.32–6.36, we obtain

$$[Zn(OH)^+] = \frac{K_{s1}}{[OH^-]} = \frac{\beta_1 K_{s0}}{[OH^-]} = \frac{K_{s0}}{[OH^-]^2}\beta_1[OH^-] \tag{6.37}$$

Continuing this procedure for higher order hydroxy complexes, the total soluble zinc S (mol/L) can be written as

$$S = \frac{K_{s0}}{[OH^-]^2}(1 + \beta_1[OH^-] + \beta_2[OH^-]^2 + \beta_3[OH^-]^3 + \beta_4[OH^-]^4) \tag{6.38}$$

In the case where there is no equilibrium with the solid phase, the expression becomes

$$S = [Zn^{2+}] \ (1 + \beta_1[OH^-] + \beta_2[OH^-]^2 + \beta_3[OH^-]^3 + \beta_4[OH^-]^4) \tag{6.39}$$

If there is another ligand such as chloride ion Cl$^-$, the second parenthesis will have similar terms defined by the product of the stability constant β_i and the chloride concentration [Cl$^-$] raised to the power i, β_i[Cl$^-$]^i.

Figure 6.4 displays the soluble zinc concentration vs. pH. The curve may be seen as a sum of contributions from five equilibria between Zn(OH)$_2$ (s) and Zn^{2+}, Zn(OH)$^+$, Zn(OH)$_2$(aq), Zn(OH)$_3^-$, and Zn(OH)$_4^{2-}$. Using the stability constants listed in Table 6.3, these equilibria are described by straight line segments in the log S vs. pH plot (Figure 6.4)

$$\log\ S = 12.45 - 2\text{pH}, \quad \text{Zn}^{2+} \tag{6.40}$$

$$\log\ S = 3.49 - \text{pH}, \quad \text{Zn(OH)}^+ \tag{6.41}$$

$$\log\ S = -4.45, \quad \text{Zn(OH)}_2\ (\text{aq}) \tag{6.42}$$

$$\log\ S = -15.95 + \text{pH}, \quad \text{Zn(OH)}_3^- \tag{6.43}$$

$$\log\ S = -28.75 + 2\text{pH}, \quad \text{Zn(OH)}_4^{2-} \tag{6.44}$$

Consider, for example, Equation 6.43 for Zn(OH)$_3^-$

$$S = [\text{Zn(OH)}_3^-] = \frac{K_{s0}}{[\text{OH}^-]^2}\ \beta_3\ [\text{OH}^-]^3 \tag{6.45}$$

$$K_{s0}\ \beta_3\ [\text{OH}^-] = 10^{-15.55}\ 10^{13.6}\ \frac{10^{-14}}{[\text{H}^+]} = 10^{-15.95}\ [\text{H}^+]^{-1} \tag{6.46}$$

from which we obtain Equation 6.43 by taking log S [metal precipitation]. An important application of metal precipitation is metal removal, for example in rinsewaters from electroplating processes (Christensen and Delwiche, 1982). As may be seen from Figure 6.2, there is a minimum zinc solubility at pH 9–10 resulting in an effluent-dissolved zinc concentration of only 10$^{-4.45}$ M.

Examples of cadmium speciation with hydroxyl, and both hydroxyl and chloride ions, in solution are shown in Figures 6.5(a) and (b) respectively. Ionic concentrations are estimated by first calculating [Cd^{2+}] from expressions such as Equation 6.39 with total Cd concentration $S = 10^{-5}$ M. Next, individual ion concentrations are calculated from the stability constants and hydroxy or chloride concentrations (Tables 6.3 and 6.4). For example, [CdCl$_3^-$] = [Cd^{2+}] β_3 [Cl$^-$]^3. It is clear that Cd^{2+} dominates in freshwater at common pH's between 6.5 and 8.5, but that hydroxy complexes become more important at higher pH values (Figure 6.3a). In sea water, Cd^{2+} is also dominant with increasing concentrations of Cl$^-$ complexes at higher chloride concentrations (Figure 6.3b). The Cd(OH)$^+$ concentration is fairly low, but it is still two orders of magnitude higher than that of the doubly hydroxy complexed species, Cd(OH)$_2^0$.

Carbonate is commonly found in natural waters and it is therefore of interest when considering the speciation as a function of both carbonate and hydroxyl ions (Figure 6.6). Equation 6.39 cannot be used in this case because of the equilibrium

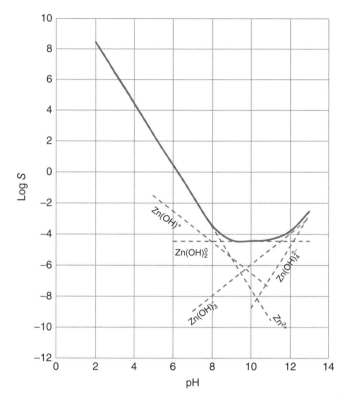

Figure 6.4. Log S, total zinc concentration in solution vs. pH. All zinc species Zn^{2+}, $Zn(OH)^+$, $Zn(OH)_2^0$, $Zn(OH)_3^-$, and $Zn(OH)_4^{2-}$ are in equilibrium with the solid phase $Zn(OH)_2(s)$. (*See color plate for color version*)

between carbonic acid, bicarbonate ion, and carbonate ion. Therefore, MINTEQ (KTH, 2013) was used to estimate the speciation both for a closed system with total carbonate concentration $= 5\ 10^{-4}$ M (Figure 6.6a) and for a system open to the atmosphere (Figure 6.6b). At typical pH values between 6.5 and 8.5, Cd^{2+} dominates in both cases, but $CdCO_3^0$ becomes significant near pH 8.5, followed by $CdHCO_3^+$ and $Cd(OH)^+$. The actual metal speciation in a given case can be a determining factor for the toxicity to aquatic life because the metal ion is often the more reactive and toxic species.

6.4.3 Arsenic

Arsenic is one of the metalloids (Grund et al., 2005) and it is naturally contained in many soils. It may seep into groundwater. Unintended release occurs during mining and smelting of gold, lead, copper, and nickel, in the ores of which arsenic commonly occurs. Combustion of coal: Most coal has As of less than 5 ppm,

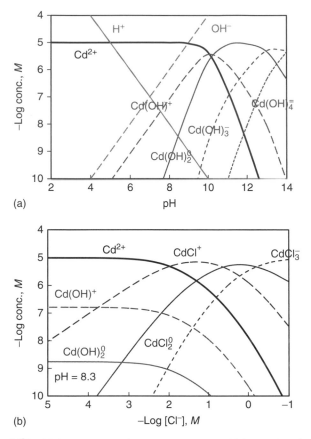

Figure 6.5. (a) Cd^{2+} with hydroxy complexes vs. pH. The solubility product for Cd^{2+} and OH^- is exceeded for $pH \geq 10$ which, depending on kinetics, may lead to precipitation of $Cd(OH)_2(s)$; (b) Hydroxy and chloride complexes of Cd^{2+} vs. log of chloride concentration at $pH = 8.3$. Sea water is typically at $-\log [Cl^-] \approx 0.24$. (*See insert for color representation of the figure.*)

but some coal contains as high as 10,000 ppm (1%) As. Indoor arsenic pollution may occur when coal is used for cooking or heating. Arsenic is released by the iron and steel production processes. Industrial uses for arsenic are detailed below.

Arsenic is used in pesticides. For example, $Pb_3(AsO_4)_2$ is an insecticide and $Ca_3(AsO_4)_2$, Ca_3AsO_3, $Cu_3(AsO_3)_2$, are herbicides. Chromated copper arsenate (CCA) is a wood preserver. Wood ash from burnt CCA lumber has caused fatalities in animals and serious poisoning in humans. An organic arsenic compound is used in chicken feed to stimulate growth and prevent disease. Arsenic has been used in therapeutic medicine for centuries in many countries, and it is also used in green pigment. Lewisite is an arsenic-containing gas and was used in World War I (WWI) as mortar gas.

TABLE 6.4. Reactions and Equilibrium Constants used in Figures 6.5 and 6.6

Reaction	Equilibrium constant
Water and Carbonate Only (See Chapter 4)	

$H_2O = H^+ + OH^-$ $K_w = \{H^+\}\{OH^-\} = 10^{-14}$

$CO_2(g) + H_2O = H_2CO_3$ $H = \{H_2CO_3\}/P_{CO_2} = 10^{-1.5}$ [a]

$H_2CO_3 + H_2O = H^+ + HCO_3^-$ $K_{a1} = \{H^+\}\{HCO_3^-\}/\{H_2CO_3\} = 10^{-6.35}$

$HCO_3^- = H^+ + CO_3^{2-}$ $K_{a1} = \{H^+\}\{CO_3^{2-}\}/\{HCO_3^-\} = 10^{-10.33}$

Hydrolysis of Cd^{2+}

$Cd^{2+} + OH^- = CdOH^+$ $\beta_1 = \{CdOH^+\}/\{Cd^{2+}\}\{OH^-\} = 10^{3.92}$

$Cd^{2+} + 2\,OH^- = Cd(OH)_2^0$ $\beta_2 = \{Cd(OH)_2^0\}/\{Cd^{2+}\}\{OH^-\}^2 = 10^{7.65}$

$Cd^{2+} + 3\,OH^- = Cd(OH)_3^-$ $\beta_3 = \{Cd(OH)_3^-\}/\{Cd^{2+}\}\{OH^-\}^3 = 10^{8.70}$

$Cd^{2+} + 4\,OH^- = Cd(OH)_4^{2-}$ $\beta_4 = \{Cd(OH)_4^-\}/\{Cd^{2+}\}\{OH^-\}^4 = 10^{8.65}$

Complexation of Cd^{2+} with Carbonate

$Cd^{2+} + HCO_3^- = CdHCO_3^+$ $\beta_{HCO_3^-} = \{CdHCO_3^+\}/\{Cd^{2+}\}\{HCO_3^-\} = 10^{12.4}$

$Cd^{2+} + CO_3^- = CdCO_3^0$ $\beta_{CO_3^{2-}} = \{CdCO_3^0\}/\{Cd^{2+}\}\{CO_3^{2-}\} = 10^{5.4}$

Dissolution of $CdCO_3(s)$ and $Cd(OH)_2(s)$

$CdCO_3(s) = Cd^{2+} + CO_3^{2-}$ $K_{s0} = \{Cd^{2+}\}\{CO_3^-\} = 10^{-13.74}$

$Cd(OH)_2(s) = Cd^{2+} + 2OH^-$ $K_{s0} = \{Cd^{2+}\}\{OH^-\}^2 = 10^{-14.27}$

(Source: All values are taken from Benjamin (2002)).
[a] moles/(l atm).

6.5 ZERO-VALENT IRON IN REMEDIATION OF CONTAMINATED WATER

The electrochemistry of zero-valent iron is well known, and its potential for remediating pollutants in the environment, especially the subsurface, has been explored. Zero-valent iron has been successfully used to dechlorinate several organic compounds. Other successful applications include arsenate reduction and removal of UO_2^{2+} and MoO_4^{2-} ions.

6.5.1 Dechlorination of Chlorinated Hydrocarbons

Zero-valent iron can cause reductive dehalogenation of chlorinated and brominated organics in the subsurface, and can serve as a reductant for arsenate and chromate reduction. The reactive iron can take the form of filings and turnings arranged to form permeable reactive barriers in groundwater remediation (Blowes et al., 1997; U.S. EPA, 2013; Gu et al., 1999). Reactive subsurface barriers have proven particularly useful for the remediation of plumes of chlorinated hydrocarbons and Cr(VI) species in groundwater using zero-valent iron as the reactive substrate. We shall in the following review the basis for groundwater remediation with zero-valent iron Fe(0), considering dechlorination of chlorinated hydrocarbons and other substances such as arsenate and chromate.

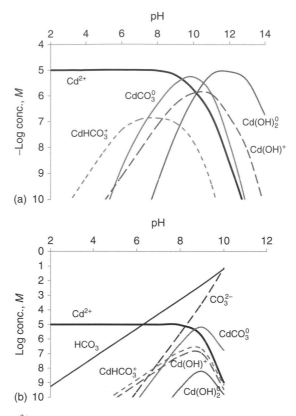

Figure 6.6. (a) Cd^{2+} with hydroxy and carbonate species, closed system, vs. pH. Total carbonate concentration was fixed to $5 \cdot 10^{-4}$ M. The solubility products for Cd^{2+} and OH^- are exceeded at pH $\geq$ 10, and for Cd^{2+} and CO_3^{2-} at $6 \leq$ pH $\leq$ 12, which, depending on kinetics, may lead to precipitation of $Cd(OH)_2(s)$ and $CdCO_3(s)$. (b) Cd^{2+} with hydroxy and carbonate species, system open to the atmosphere, vs. pH. Possible $CdCO_3(s)$ precipitation for pH $\geq$ 7. Calculations performed with Visual MINTEQ version 3.0 (KTH, 2013). (*See color plate for color version*)

The overall dechlorination reaction may be written as (Sayles et al., 1997)

$$Fe(0) + RCl + H^+ \leftrightarrow Fe^{2+} + RH + Cl^- \tag{6.47}$$

In the absence of alkyl chloride, iron can react with water anaerobically to form hydrogen,

$$Fe(0) + 2H_2O \leftrightarrow Fe^{2+} + H_2 + 2OH^- \tag{6.48}$$

or under aerobic conditions to produce just ferrous and hydroxyl ions,

$$2Fe(0) + O_2 + 2H_2O \leftrightarrow 2Fe^{2+} + 4OH^- \tag{6.49}$$

Note that the pH of the solution would be expected to increase for all three reactions, 6.47–6.49.

To evaluate the thermodynamics of dechlorination using acetate rather than Fe(0) as electron donor, consider the transformation of carbon tetrachloride into chloroform at pH 7 (Table 6.5),

$$\frac{1}{2}CCl_4 + \frac{1}{2}H^+ + e^- \leftrightarrow \frac{1}{2}CHCl_3 + \frac{1}{2}Cl^- \quad E^{o'} = 0.58 \text{ V} \tag{6.50}$$

$$\frac{1}{8}CH_3COO^- + \frac{1}{4}H_2O \leftrightarrow \frac{1}{4}CO_2(g) + \frac{7}{8}H^+ + e^- \quad E^{o'} = 0.29 \text{ V} \tag{6.51}$$

Sum $E^{o'} = 0.87$ V

For the reaction shown in Equation 6.51, the calculation of $E^{o'}$ proceeds as in Equations 6.52 and 6.53

$$\Delta G_f^{o'} = \frac{1}{4}(-394.37) + \frac{7}{8}(-39.89) + 0 + \frac{1}{8}(-369.4) + \frac{1}{4}(237.18)$$

$$= -28.026 \text{ kJ/electron-eq} \tag{6.52}$$

$$E_H^{o'} = \frac{\Delta G_f^{o'}}{nF} = \frac{-28.02 \text{ kJ/(electron eq)}}{1 \cdot 96.485 \text{ kJ/(V eq)}} = 0.29 \text{ V} \tag{6.53}$$

This can also be seen from

$$\Delta G_f^{o'} = -RT \ln K \tag{6.54}$$

$$E_H^{o'} = \frac{2.3RT}{nF} \log K = 0.059 \times 4.912 = 0.29 \text{ V} \tag{6.55}$$

If a significant amount of nitrate is present, it will compete with carbon tetrachloride for electrons (Table 5.6),

$$\frac{1}{5}NO_3^- + \frac{6}{5}H^+ + e^- \leftrightarrow \frac{1}{10}N_2(g) + \frac{3}{5}H_2O \quad E^{o'} = 0.748 \text{ V} \tag{6.56}$$

Thus, for unit molar concentrations nitrate is a more important electron acceptor, meaning that carbon tetrachloride may not be fully dechlorinated according to Equation 6.50.

Pentachlorophenol (PCP) may be dechlorinated with zero-valent iron (Kim and Carraway, 2000) and it is important to separate reaction and sorption processes. It was also found that tetrachloroethylene (TCE) can undergo reductive dechlorination in the presence of zero-valent iron, and that an observed decrease in reaction rate, i.e. electron yield over time, can result from anodic control over iron corrosion, i.e. by reaction 3, Table 6.5 (Farrell et al., 2000). Sayles et al. (1997) investigated dichlorodiphenyltrichloroethane (DDT), dichlorodiphenyldichloroethane (DDD), and dichlorodiphenyldichloroethylene (DDE) dechlorination by zero-valent iron.

TABLE 6.5. Half Reactions of Interest for Environmental Remediation

Reaction number	Reaction	pE° (Log K)	E° (V)	$E^{\circ\prime}$ (V)	References
1	$\frac{1}{4}CO_2(g) + \frac{7}{8}H^+ + e^- \leftrightarrow$ $\frac{1}{8}CH_3COO^- + \frac{1}{4}H_2O$	1.20	0.071		Sawyer et al. (2003, p. 22)
2	$\frac{1}{4}CO_2(g) + H^+ + e^- \leftrightarrow$ $\frac{1}{24}C_6H_{12}O_6 + \frac{1}{4}H_2O$	−0.19	−0.011		Sawyer et al. (2003, p. 22)
3	$\frac{1}{2}Fe^{2+} + e^- \leftrightarrow \frac{1}{2}Fe(s)$	−6.91	−0.409		Sawyer et al. (2003, p. 22)
4	$\frac{1}{2}Hg^{2+} + e^- \leftrightarrow \frac{1}{2}Hg(l)$	13.45	0.796		Sawyer et al. (2003, p. 22)
5	$\frac{1}{2}Zn^{2+} + e^- \leftrightarrow \frac{1}{2}Zn(s)$	−12.88	−0.762		Sawyer et al. (2003, p. 22)
6	$Ag^+ + e^- \leftrightarrow Ag(s)$		0.799		Sawyer et al. (2003, p.84)
7	$\frac{1}{2}Cu^{2+} + e^- \leftrightarrow \frac{1}{2}Cu(s)$	5.74	0.339		Sawyer et al. (2003, p.22)
8	$\frac{1}{2}Pb^{2+} + e^- \leftrightarrow \frac{1}{2}Pb(s)$		−0.126		Sawyer et al. (2003, p. 84)
9	$\frac{1}{3}Al^{3+} + e^- \leftrightarrow \frac{1}{3}Al$		−1.66		Sawyer et al. (2003, p. 84)
10	$\frac{1}{2}Mg(OH)_2(s) + e^- \leftrightarrow$ $\frac{1}{2}Mg(s) + OH^-$		−2.69		Sawyer et al. (2003, p. 84)
11	$\frac{1}{8}CO_2(g) + \frac{1}{8}HCO_3^- +$ $H^+ + e^- \leftrightarrow$ $\frac{1}{8}CH_3COO^- + \frac{3}{8}H_2O$	0.13		−0.28	Sawyer et al. (2003, p. 334)
12	$\frac{1}{6}CO_2(g) + H^+ + e^- \leftrightarrow$ $\frac{1}{6}CH_3OH + \frac{1}{6}H_2O$	0.03		−0.38	Sawyer et al. (2003, p. 335)
13	$Fe^{3+} + e^- \leftrightarrow Fe^{2+}$	0.77		0.77	Sawyer et al. (2003, p.335)
14	$\frac{1}{2}CCl_4(aq) + \frac{1}{2}H^+ + e^- \leftrightarrow$ $\frac{1}{2}CHCl_3(aq) + \frac{1}{2}Cl^-$	0.79		0.58	Sawyer et al. (2003, p.344)
15	$\frac{1}{2}CHCl_3(aq) + \frac{1}{2}H^+ + e^- \leftrightarrow$ $\frac{1}{2}CH_2Cl_2(aq) + \frac{1}{2}Cl^-$	0.68		0.47	Sawyer et al. (2003, p.344)

$E^{\circ\prime}$ indicates that reactants and products are at unit activity except $[H^+] = 10^{-7}$ M. E° can be calculated from $E^{\circ} = (-\Delta G^{\circ}/96.485)$ kJ/(V-eq). Also, $pE^{\circ} = 16.9E^{\circ}$. $\Delta G^{\circ\prime}$ is calculated as for ΔG° except that ΔG° for H^+ is taken to equal -39.87 kJ/electron-eq the value it would have at pH $= 7$ when $[H^+] = 10^{-7}$ M.

They developed a mechanistic model indicating that the rate of dechlorination of the solid-phase reactants was limited by the rate of dissolution into the aqueous phase.

Gas-phase halogenated organic compounds can also be removed by zero-valent iron. Scheutz et al. (2000) studied transformation of gaseous CCl_3F and CCl_4 by zero-valent iron in systems that were unsaturated with water under atmospheric conditions, such as occur in landfill top covers containing granular zero-valent iron. Removal rates were found to range between 1100 mL/(m^2 h) (batch) and 200 mL/(m^2 h) (flow through), where rate constants refer to the water phase and are normalized to 1 m^2 iron surface/mL.

Halogenated aromatic compounds are intrinsically more stable than halogenated aliphatic ones. The effort to dechlorinate polychlorinated aromatic hydrocarbons, such as polychlorinated biphenyls (PCBs), by zero-valent iron has achieved only limited success under environmental conditions (Gardner, 2004; Varanasi et al., 2007). Rapid and efficient dechlorination of PCBs by zero-valent iron has been observed at elevated temperatures or when the iron particles are palladized by depositing a small amount of palladium on the iron surface as catalyst (Choi et al., 2008; Noma et al., 2003; Wang and Zhang, 1997; Chuang et al., 1995; Yak et al., 1999, 2000; Kim et al., 2004). Nanoscale (<100 nm id) zero-valent iron was found to be more reactive than submicro- and micro-sized zero-valent iron for the dechlorination of PCBs, probably as a result of the higher reactivity on larger specific surface area (Lowry and Johnson, 2004). Bimetallic and nano-sized particles have also been used for the dechlorinations of polychlorinated benzenes (Zhu and Lim, 2007) and dibenzo-dioxins (Kim et al., 2008).

Polybrominated hydrocarbons have been used in large quantities as flame retardants in various industrial and consumer products from the 1970s to the present. As a result, they have been found to be ubiquitously present in the global environment. Remediation of contaminated groundwater, soils, and sediments calls for effective removal of brominated flame retardants (BFRs), and zero-valent iron is one of the chemical reactants investigated in laboratories since the 2000s (Li et al., 2007; Carvalho-Knighton et al., 2009; Zhuang et al., 2010, 2011; Shih and Tai, 2010). As expected from the weaker C−Br bonds compared with C−Cl bonds, the debromination of polybrominated diphenyl ether (PBDE) congeners is orders of magnitude faster than the counterpart PCB congeners. No catalyst metals are needed, and the reactions proceed rapidly under room temperature in various organic solvents. This is particularly true for heavily brominated compounds. For example, Li et al. (2007) used nanoscale iron powder embedded in a cation exchange resin to compare the dehalogenation kinetics and pathways between decachlorobiphenyl (PCB209) and decabromodiphenyl ether (BDE209) in a water/aceton (1:1) mixture. BDE209 disappeared completely after 8 h, while no dechlorination of PCB209 was noticeable when the experiment ended after 10 d. In general, the dehalogenation appears to be stepwise, with halogen atoms being sequentially substituted by hydrogen; therefore lower congeners were produced with increasing reaction time.

In addition to the halogenated hydrocarbons, nitroaromatic compounds such as nitrobenzene may also be remediated with zero-valent iron. However, the formation of siderite ($FeCO_3$) on the metal inhibits nitro reduction (Agrawal and Tratnyek,

1996). In general, a potential limitation of Fe(0) technology is the deterioration of the Fe(0) metals by corrosion and the subsequent precipitation of minerals that may cause cementation and decreased permeability of the Fe(0) barrier (Gu et al., 1999).

Linear free energy relationships (LFERs) can be used to predict rates of dehalogenation caused by environmental reductants (Scherer et al., 1998). Successful LFERs were reported by these authors based on estimated lowest unoccupied molecular orbital (LUMO) energies calculated from semi-empirical and ab initio methods and one- and two-electron reduction potentials (E_1 and E_2). The best LFER explained 83% of the variability in surface area normalized rate constants with ab initio LUMO energies.

6.5.2   Reduction of Uranium Carbonate, Chromate, and Arsenate

Permeable reactive barriers have been used by the U.S. Department of Energy (DOE), Grand Junction, Colorado, at its Monticello, Utah, Superfund site, a former uranium and ore processing mill. The contaminants at the site included uranium, vanadium, selenium, and arsenic, which were present in groundwater at concentrations of 700, 400, 40, and 10 µg/L respectively. After groundwater passed through the barrier, the concentrations of these contaminants were reduced to nondetectable levels. The reactive barrier was installed in 1999 and the cleanup was completed 1 year later. The site was removed by the EPA from the National Priorities List in 2000 (U.S. EPA, 2013). Zero-valent iron removes uranium by precipitation through the reaction,

$$Fe(0) + UO_2(CO_3)_2^{2-} + 2H^+ \leftrightarrow Fe^{2+} + 2HCO_3^- + UO_2(\text{uraninite}) \qquad (6.57)$$

Thus a uraninite precipitate is formed in the presence of uranium-contaminated carbonate solution. A fraction of the contaminated groundwater was still flowing around the south slurry wall and in 2005 an additional treatment cell containing a mixture of zero-valent iron and gravel was installed at the site.

Chromate containing chromium in the hexavalent state Cr(VI) may be converted to nontoxic Cr(III) by the reaction,

$$Cr(VI)(aq) + 3Fe(II)(aq) \leftrightarrow Cr(III)(aq) + 3Fe(III)(aq) \qquad (6.58)$$

The Cr(III) in solution can then react with hydrogen ions to form $Cr(OH)_3$, which will be removed from solution by precipitation. The divalent iron is derived from zero-valent iron via the reaction,

$$Fe(0) \leftrightarrow Fe^{2+} + 2e^- \qquad (6.59)$$

where ferrous ion can be further oxidized as follows:

$$Fe^{2+} \leftrightarrow Fe^{3+} + e^- \qquad (6.60)$$

Inorganic arsenic species such as arsenates and arsenites may be immobilized by iron filings. One method uses zero-valent iron and sand to reduce inorganic arsenic species to iron co-precipitates, mixed precipitates and, in the presence of sulfates, to arsenopyrite (FeAs) (U.S. EPA, 2002). In this case, water to be processed would first be treated with barite ($BaSO_4$) and then pass through a reactor with iron filings and sand under anaerobic conditions. Sulfate reduction occurs according to the reaction,

$$8e^- + 9H^+ + SO_4^{2-} \leftrightarrow HS^- + 4H_2O \tag{6.61}$$

In addition, hydrolysis of water may take place,

$$2e^- + 2H^+ \leftrightarrow H_2(g) \tag{6.62}$$

along with reduction of arsenate to arsenite,

$$2e^- + 4H^+ + HAsO_4^{2-} \leftrightarrow H_2O + H_3AsO_3 \tag{6.63}$$

As a result of the presence of sulfide and dissolved iron, several precipitates will form such as $Fe(OH)_3$, $FeAsO_4$, and $FeAsS$. For this technology to work effectively, it is important that precipitates do not block pore spaces and that reactor materials can be rejuvenated by appropriate cleaning procedures.

REFERENCES

Agrawal, AA ,Tratnyek, PG. Reduction of nitroaromatic compounds by zero-valent iron metal. *Environ. Sci. Technol.* 1996;33:153–160.

Aqueous Solutions. *Geochemist's Workbench*. Champaign, IL: Aqueous Solutions LLC; 2013. Available at http://www.gwb.com/. Accessed April 26, 2014.

ARL Analytical Research Labs, Inc. Cadmium toxicity; 2013. Available at http://www.arltma.com/Articles/CadmiumToxDoc.htm. Accessed April 26, 2014.

Balaraman, R, Gulati, OD, Bhatt, JD, Rathod, SP, Hemavathi, KG. Cadmium- induced hypertension in rats. *Pharmacology* 1989;38(4):226–234.

Bartram, J, Ballance, R, editors. *Water Quality Monitoring – A Practical Guide to the Design and Implementation of Freshwater Quality Studies and Monitoring Programmes*. UNEP/WHO; 1996. ISBN 0-419-22320-7. Available at http://www.who.int/water_sanitation_health/resourcesquality/waterqualmonitor.pdf. Accessed May 13, 2014.

Benedetti, MF, Milne, CJ, Kinniburgh, DG, Van Riemsdijk, WH, Koopal, LK. Metal ion binding to humic substances: Application of the nonideal competitive adsorption model. *Environ. Sci. Technol.* 1995;29:446–457.

Benjamin, MM. *Water Chemistry*. New York, NY: McGraw-Hill Companies, Inc.; 2002.

Blowes, DW, Ptacek, CJ, Jambor, JL. In-situ remediation of Cr(VI)-contaminated groundwater using permeable reactive walls: Laboratory studies. *Environ. Sci. Technol.* 1997;31:3348–3357.

Carvalho-Knighton, K, Talalaj, L, DeVor, R. PBDE degradation with zero-valent bimetallic systems. *Environmental Applications of Nanoscale and Microscale Reactive Metal Particles*. Washington, DC: American Chemical Society; 2009, pp. 75–87.

Centre for Ecology & Hydrology. Windermere Humic Aqueous Model, version 7; 2013. Available at http://www.ceh.ac.uk/products/software/wham/ Accessed April 26, 2014.

Choi, H, Al-Abed, SR, Agarwal, S, Dionysiou, DD. Synthesis of reactive nano-Fe/Pd bimetallic system-impregnated activated carbon for the simultaneous adsorption and dechlorination of PCBs. *Chem. Mater.* 2008;20(11):3649–3655.

Christensen, ER, Guinn, VP. Zinc from automobile tires in urban runoff. *J. Environ. Eng.* 1979;105: 165–168.

Christensen, ER, Delwiche, JT. Removal of heavy metals from electroplating rinsewaters by precipitation, flocculation, and ultrafiltration. *Water Res.* 1982;16:29–737.

Christensen, ER, Osuna, J. Atmospheric fluxes of Pb, Zn, and Cd to Lake Michigan from frequency domain deconvolution of sedimentary records. *J. Geophys. Res.* 1989;94(C10):14,585–14,597.

Chuang, FW, Larson, RA, Wessman, MS. Zero-valent iron-promoted dechlorination of polychlorinated biphenyls. *Environ. Sci. Technol.* 1995;29:2460–2463.

Eisenreich, SJ, Metzer, NA, Urban, NR, Robbins, JA. Response of atmospheric lead to decreased use of lead in gasolime. *Environ. Sci. Technol.* 1986;20(2):171–174.

ERS. *MINEQL+*. Hallowell, ME: Environmental Research Software; 2007. Available at http://www.mineql.com/. Accessed April 26, 2014.

European Parliament and Council. On the prohibition of organotin compounds on ships. *Off. J. Eur. Union* 2003. Regulation (EC) No.782/2003.

Evers, DC, Turnquist, MA, Buck, DG. *Patterns of Global Seafood Mercury Concentrations and Their Relationship with Human Health.* BRI Science Communications Series 2012-48. Gorham, ME: Biodiversity Research Institute; 2012, p. 16.

Farrell, J, Kason, M, Melitas, N, Li, T. Investigation of the long-term performance of zero-valent iron for reductive dechlorination of trichloroethylene. *Environ. Sci. Technol.* 2000;34:514–521.

Gandhi, N, Bhavsar, SP, Diamond, ML, Kuwabara, JS, Marvin-DiPasquale, M, Krabbenhoft, DP. Development of a mercury speciation, fate, and biotic uptake (Biotranspec) model: Application to Lahontan Reservoir (Nevada, USA). *Environ. Toxicol. Chem.* 2007;26(11):2260–2273.

Gardner, K. In-Situ Treatment of PCBs in Marine and Freshwater Sediments Using Colloidal Zero-Valent Iron. Final Report to the Cooperative Institute for Coastal and Estuarine Environmental Technology (CICEET); 2004. Available at http://rfp.ciceet.unh.edu/display/report.php?chosen=51. Accessed April 26, 2014.

Grund, SC, Hanusch, K, Wolf, HU. Arsenic and arsenic compounds. *Ullmann's Encyclopedia of Industrial Chemistry*. Weinheim: Wiley-VCH verlag GmBH; 2005.

Gu, B, Phelps, TJ, Liang, L, Dickey, MJ, Roh, Y, Kinsall, BL, Palumbo, AV, Jacobs, GK. Biochemical dynamics in zero-valent iron columns: implications for permeable reactive barriers. *Environ. Sci. Technol.* 1999;33:2170–2177.

Hicks, BB. Atmospheric deposition and its effects on water quality. In: Christensen, ER, O'Melia, CR. *Workshop on Research Needs for Coastal Pollution in Urban Areas.* Milwaukee, WI: University of Wisconsin-Milwaukee; 1998.

Institut National de Recherche et de Sécurité. *Fiche toxicologique n° 53: Trihydrure d'arsenic.* 2000. http://www.inrs.fr/accueil/produits/bdd/doc/fichetox.html?refINRS=FT%2053. Accessed October 24, 2014.

Jensen, JN. *A Problem-Solving Approach to Aquatic Chemistry*. New York: John Wiley & Sons, Inc.; 2003.

Joseph, G, Kundig, KJA, editor. *Copper: Its Trade, Manufacture, Use, and Environmental Status*. Materials Park, OH: International Copper Association, ASM International; 1999.

Kim, JH, Tratnyek, PG, Chang, YS. Rapid dechlorination of polychlorinated dibenzo-p-dioxins by bimetallic and nanosized zero-valent iron. *Environ. Sci. Technol.* 2008;42(11):4106–4112.

Kim, Y-H, Carraway, ER. Dechlorination of pentachlorophenol by zero valent iron and modified zero-valent irons. *Environ. Sci. Technol.* 2000;34:2014–2017.

Kim, YH, Shin, WS, Ko, SH. Reductive dechlorination of chlorinated biphenyls by palladized zero-valent metals. *J. Environ. Sci. Health A Tox. Hazard Subst. Environ. Eng.* 2004;A39(5):1177–1188.

Kirk, JL, Muir, DCM, Antoniades, D, Douglas, MSV, Evans, MS, Jackson, TA, Kling, H, Lamoureux, S, Lim, DSS, Pienitz, R, Smol, JP, Stewart, K, Wang, X, Yang, F. Climate change and mercury accumulation in Canadian high and subarctic lakes. *Environ. Sci. Technol.* 2011;45(3):964–970.

Kotaś, J, Stasicka, Z. Chromium occurrence in the environment and methods of its speciation. *Environ. Pollut.* 2000;107(3):263–283.

KTH. Visual MINTEQ. 2013. Available at http://www.lwr.kth.se/English/OurSoftware/vminteq/. Accessed April 26, 2014.

Kuck, PH. Mineral yearbook: Nickel. U.S. Geological Survey; 2006. http://minerals.usgs.gov/minerals/pubs/commodity/nickel/. Accessed October 24, 2013.

Kuck, PH. Mineral commodity summaries: Nickel. U.S. Geological Survey; 2012. http://minerals.usgs.gov/minerals/pubs/commodity/nickel/. Accessed October 24, 2013.

Larsen, S. Solubility of hydroxyapatite. *Nature* 1966;212:605. DOI:10.1038/212605a0.

Levin, R, Brown, MJ, Kashtock, ME, Jacobs, DE, Whelan, EA, Rodman, J, Schock, MR, Padilla, A, Sinks, T. Lead exposures in U.S. children, 2008: Implications for prevention. *Environ. Health Perspect.* 2008;116(10):1285–1293.

Li, A, Tai, C, Zhao, Z, Wang, Y, Zhang, Q, Jiang, G, Hu, J. Debromination of decabrominated diphenyl ether by resin-bound iron nanoparticles. *Environ. Sci. Technol.* 2007;41(19):6841–6846.

Lowry, GV, Johnson, KM. Congener-specific dechlorination of dissolved PCBs by microscale and nanoscale zerovalent iron in a water/methanol solution. *Environ. Sci. Technol.* 2004;38(19):5208–5216.

Mahler, BJ, Van Metre, PC, Callender, E. Trends in metals in urban and reference lake sediment across the United States, 1970 to 2001. *Environ. Toxicol. Chem.* 2006;25(7):1698–1709.

Maillard, J-Y, Hartemann, P. Silver as an antimicrobial: Facts and gaps in knowledge. *Crit. Rev. Microbiol.* 2013;39(4):373–383.

Maprani, AC, Al, TA, MacQuarrie, KT, Dalziel, JA, Shaw, SA, Yeats, PA. Determination of Mercury Evasion in a Contaminated Headwater Stream. *Environ. Sci. Technol.* 2005;39(6):1679–1687.

Morel, FMM, Hering, JG. *Principles and Applications of Aquatic Chemistry.* New York: John Wiley & Sons, Inc.; 1993.

Noma, Y, Ohno, M, Sakai, S. Pathways for the degradation of PCBs by palladium-catalyzed dechlorination. *Fresen. Environ. Bull.* 2003;12(3):302–308.

Pacyna, EG, Pacyna, JM, Steenhuisen, F, Wilson, S. Global anthropogenic mercury emission inventory for 2000. *Atmos. Environ.* 2006;40(22):4048–4063.

Pankow, JF. *Aquatic Chemistry Concepts.* Chelsea, MI: Lewis Publishers; 1991.

Paode, RD, Sofuoglu, SC, Sivadechathep, J, Noll, KE, Holsen, TM, Keeler, GJ. Dry deposition fluxes and mass size distributions of Pb, Cu, and Zn measured in southern Lake Michigan during AEOLOS. *Environ. Sci. Technol.* 1998;32:1629–1635.

Princeton University, Department of Geosciences. Environmental chemistry and microbiology of trace metals: Mercury cycling and methylation. 2013. Available at http://www.princeton.edu/morel/research/mercury-methylation/. Accessed April 26, 2014.

Sawyer, CN, McCarty, PL, Parkin, GF. *Chemistry for Environmental Engineering and Science*, 5th ed. Boston, MA: McGraw-Hill Higher Education; 2003.

Sayles, GD, You, G, Wang, M, Kupferle, MJ. DDT, DDD, and DDE dechlorination by zero-valent iron. *Environ. Sci. Technol.* 1997;31:3448–3454.

Scherer, MM, Balko, BA, Gallagher, DA, Tratnyek, PG. Correlation analysis of rate constants for dechlorination by zero-valent iron. *Environ. Sci. Technol.* 1998;32:3026–3033.

Schmitt, CJ, Whyte, JJ, Brumbaugh, WG, Tillitt, DE. Biochemical effects of lead, zinc, and cadmium from mining on fish in the tri-states district of Northeastern Oklahoma, USA. *Environ. Toxicol. Chem.* 2005;24(6):1483–1495.

Scheutz, C, Winther, K, Kjeldsen, P. Removal of halogenated organic compounds in landfill gas by top covers containing zero-valent iron. *Environ. Sci. Technol.* 2000;34:2557–2563.

Seyferth, D. The rise and fall of tetraethyllead. 2. *Organometallics* 2003;22(25):5154–5178.

Shih, Y, Tai, Y. Reaction of decabrominated diphenyl ether by zero valent iron nanoparticles. *Chemosphere* 2010;78:1200–1206.

Sievering, H, Dave, M, Dolske, DA, McCoy, P. Transport and dry deposition of trace metals over southern Lake Michigan. In: Eisenreich, SJ, editor. *Atmospheric Pollutants in Natural Waters*. Ann Arbor, MI: Ann Arbor Science Publishers, Inc.; 1981, pp. 285–325.

Sigg, L, Black, F, Buffle, J, Cao, J, Cleven, R, Davison, W, Galceran, J, Gunkel, P, Kalis, E, Kistler, D, Martin, M, Noel, S, Nur, Y, Odzak, N, Puy, J, Van Riemsdijk, W, Temminghoff, E, Tercier-Waeber, M-L, Toepperwien, S, Rown, RM, Unsworth, E, Warnken, KW, Weng, L, Xue, H, Zhang, H. Comparison of analytical techniques for dynamic trace metal speciation in natural freshwaters. *Environ. Sci. Technol.* 2006;40(6):1934–1941.

Smail, EA, Webb, EA, Franks, RP, Bruland, KW, Sañudo-Wilhelmy, SA. Status of metal contamination in surface waters of the coastal ocean off Los Angeles, California since the implementation of the Clean Water Act. *Environ. Sci. Technol.* 2012;46:4304–4311.

Smedley, PL, Kinniburgh, DG. A review of the source, behaviour and distribution of arsenic in natural waters. *Appl. Geochem.* 2002;17(5):517–568.

Snoeyink, VL, Jenkins, D. *Water Chemistry*. New York: John Wiley & Sons, Inc.; 1980.

Stumm, W, Morgan, JJ. *Aquatic Chemistry Chemical Equilibria and Rates in Natural Waters*, 3rd ed. New York, NY: John Wiley & Sons Inc.; 1996.

Sutherland, CA, Milner, EF, Kerby, RC, Teindl, H, Melin, A, Bolt, HM. Lead. In: *Ullmann's Encyclopedia of Industrial Chemistry*. Weinheim: Wiley-VCH Verlag GmBH; 2005.

Thornton, JD, Eisenreich, SJ, Munger, JW, Gorham, E. Trace metal and strong acid composition of rain and snow in northern Minnesota. In: Eisenreich, SJ, editor. *Atmospheric Pollutants in Natural Waters*. Ann Arbor, MI: Ann Arbor Science Publishers, Inc.; 1981. pp. 261–284.

Tipping, E. Humic ion binding model VI: An improved description of the interactions of protons and metals ions with humic substances. *Aquat. Geochem.* 1998;4:3–48.

Tolcin, AC. U.S. Geological Survey, Mineral Commodity Summaries 2009: Zinc. 2011. http://minerals.usgs.gov/minerals/pubs/commodity/zinc/myb1-2011-zinc.pdf. Accessed November 1, 2013.

Twarakavi, NKC, Kaluarachchi, JJ. Arsenic in the shallow ground waters of conterminous United States: Assessment, health risks, and costs for MCL compliance. *J. Am. Water Resour. Assoc.* 2006;42(2):275–294.

Unsworth, ER, Warnken, KS, Zhang, H, Davison, W, Black, F, Buffle, J, Cao, J, Cleven, R, Galceran, J, Gunkel, P, Kalis, E, Kistler, D, Van Leeuwen, HP, Martin, M, Noël, S, Nur,

Y, Odzak, N, Puy, J, Van Riemsdijk, W, Sigg, L, Temminghoff, E, Tercier-Waeber, M-L, Toepperwien, S, Town, RM, Weng, L, Xue, H. Model predictions of metal speciation in freshwaters compared to measurements by in situ techniques. *Environ. Sci. Technol.* 2006;40(6):1942–1949.

U.S. Environmental Protection Agency. Arsenic treatment technologies for soil, waste, and water. Solid waste and emergency response. EPA-542-R-02-004. 2002.

U.S. Environmental Protection Agency. MINTEQA2. 2006. Available at http://www2.epa.gov/exposure-assessment-models/minteqa2. Accessed April 26, 2014.

U.S. Environmental Protection Agency. Permeable reactive barriers, permeable treatment zones, and application of zero-valent iron. Technology and Field Services Division. 2013. Available at http://www.clu-in.org/techfocus/default.focus/sec/permeable_reactive_barriers/. Accessed April 26, 2014.

U.S. Geological Survey. Mineral commodity summaries – Lithium, 2013a. Available at http://minerals.usgs.gov/minerals/pubs/commodity/lithium/mcs-2013-lithi.pdf. Accessed April 26, 2014.

U.S. Geological Survey. Silver statistics and information. 2013b. Available at http://minerals.usgs.gov/minerals/pubs/commodity/silver/. Accessed April 26, 2014.

Varanasi, P, Fullana, A, Sidhu, S. Remediation of PCB contaminated soils using iron nanoparticles. *Chemosphere* 2007;66:1031–1038.

Wang, CB, Zhang, WX. Synthesizing nanoscale iron particles for rapid and complete dechlorination of TCE and PCBs. *Environ. Sci. Technol.* 1997;31:2154–2156.

Yak, HK, Wenclawiak, BW, Cheng, IF, Doyle, JG, Wai, CM. Reductive dechlorination of polychlorinated biphenyls by zerovalent iron in subcritical water. *Environ. Sci. Technol.* 1999;33:1307.

Yak, HK, Lang, QY, Wai, CM. Relative resistance of positional isomers of polychlorinated biphenyls toward reductive dechlorination by zerovalent iron in subcritical water. *Environ. Sci. Technol.* 2000;34:2792–2798.

Yao, H, Feng, X, Guo, Y, Yan, H, Fu, X, Li, Z, Meng, B. Mercury and methylmercury concentrations in two newly constructed reservoirs in the Wujing River, Guizhou, China. *Environ. Toxicol. Chem.* 2011;30(3):530–537.

Zhu, BW, Lim, TT. Catalytic reduction of chlorobenzenes with Pd/Fe nanoparticles: reactive sites, catalyst stability, particle aging, and regeneration. *Environ. Sci. Technol.* 2007;41(21):7523–7529.

Zhuang, YA, Ahn, S, Luthy, RG. Debromination of polybrominated diphenyl ethers by nanoscale zerovalent iron: pathways, kinetics, and reactivity. *Environ. Sci. Technol.* 2010;44(21):8236–8242.

Zhuang, Y, et al. Dehalogenation of polybrominated diphenyl ethers and polychlorinated biphenyl by bimetallic, impregnated, and nanoscale zerovalent iron. *Environ. Sci. Technol.* 2011;45:4896–4903.

7

ORGANIC POLLUTANTS

7.1 INTRODUCTION

Millions of organic compounds have been synthesized and produced in significant quantities since industrialization began. The CAS has registered more than 71 million compounds, of which more than 68 million are commercially available (CAS, 2013). The term anthropogenic chemicals describes substances that are manufactured for the benefit of humanity, but some have entered the environment in unforeseen and undesirable ways. Examples of such chemicals are the "legacy" pollutants polychlorinated biphenyls (PCBs) and pesticide DDT, and the "emerging" pollutants, such as the perfluorinated chemicals (PFCs) and numerous flame retardants. Many chemicals exist in nature, but some of them have become widespread as a result of human activities, leading to concern regarding their effects. For example, fossil fuels exist in nature and have served human society for centuries, but their intensive use has caused severe pollution. Therefore, that a chemical is deemed a pollutant does not mean it is in itself harmful. In fact, PCBs were once thought of as "wonders" of the chemical industry, and viewed as chemicals that "not only enhance the quality of life but also save lives" (McLearn, 1999).

Some chemicals, such as the organophosphate pesticide parathion, are acutely toxic to a range of organisms apart from their targeted pests. For others, such as PCBs and DDT, their chronic effects on the ecosystem and human health may outweigh their acute toxicity. The general public are mainly unaware that many benign and beneficial chemical substances are also deemed pollutants by environmental agencies and

Physical and Chemical Processes in the Aquatic Environment, First Edition. Erik R. Christensen and An Li.
© 2014 John Wiley & Sons, Inc. Published 2014 by John Wiley & Sons, Inc.

scientists. For example, various pharmaceuticals and personal care products (PCPPs), ranging from aspirin to perfume, have found ways to enter the water supply system in many cities, causing low dose, long term exposure to the public, with a potentially large but difficult to evaluate impact. As mentioned earlier, that a chemical is called a pollutant does not necessarily mean it is toxic. In fact, many PPCPs keep us healthy and save lives.

Then, what are "pollutants"? They can be deemed "chemicals in the wrong place". PCBs and pesticide, fossil fuel components, PPCPs, and many others evaporate into air, leach into groundwater, and enter food chains. If a chemical alters the environment in a way that is detrimental to human welfare by damaging the ecosystem that humans depend on, it is deemed a pollutant, regardless of where it comes from, whether it is useful or useless, toxic or benign. As such, many compounds and substances that are in commercial use, particularly those in large production quantities, have been deemed pollutants and placed within the radar of environmental monitoring and risk assessment (Howard and Muir, 2010, 2011; Muir and Howard, 2006).

7.2 IMPORTANT ORGANIC POLLUTANT GROUPS

In environmental organic chemistry, organic pollutants are often categorized based on their physicochemical properties. Therefore, terms such as VOCs (volatile organic compounds), SVOCs (semi-volatile organic compounds), HOCs (hydrophobic organic compounds), POPs (persistent organic pollutants), PTS (persistent and toxic substances) are often used in the literature. The definitions of these categories often lack clarity. In this section, selected organic chemical pollutant groups are briefly described. There are many other synthetic chemical groups, such as surfactant (e.g. detergent), organic dyes and pigments, and relatively strong acids and bases, which may be more abundant than the selected chemical groups. However, they exhibit distinctly different behavior in water and are thus excluded from the discussions below.

7.2.1 Petrochemicals and Industrial Solvents

Petroleum is fractionated into gases, gasoline, kerosene and diesel, lubricating and heavy fuel oils, wax, and asphalt, in the order of decreasing volatility. Each fraction is a complex mixture containing hydrocarbons and other compounds with heteroatoms such as sulfur and nitrogen. Much of the environmental pollution that occurred in the twentieth century was related to the use of petroleum products. Their transportation, processing, distribution, and uses have caused tremendous pollution of the natural environment through discharges, leaching, and evaporation.

The chemical structures of selected compounds discussed below are given in Table 7.1. Benzene, toluene, ethylbenzene, and xylenes (BTEX) form a group of particular concern due to their adverse health effect and relatively high concentration in air, water, and soils. Methyl-*tert*-butyl ether (MTBE) was used in large quantities in the United States as a gasoline additive to combat the problem of engine knocking.

TABLE 7.1. Selected Industrial Solvents

Abbr.	Name	Formula	MW	Structure
-	Benzene	C_6H_6	78.1	
-	Toluene (methylbenzene)	C_7H_8	92.1	
-	Ethylbenzene	C_8H_{10}	106.2	
-	o-Xylene (o-dimethylbenzene)	C_8H_{10}	106.2	
-	m-Xylene (m-dimethylbenzene)	C_8H_{10}	106.2	
-	p-Xylene (p-dimethylbenzene)	C_8H_{10}	106.2	
TCE	Trichloroethylene	C_2HCl_3	131.4	
PCE	Tetrachloroethylene	C_2Cl_4	165.8	
1,1,1-TCA	1,1,1-Trichloroethane (methyl chloroform)	$C_2H_3Cl_3$	133.4	
1,1,2-TCA	1,1,2-Trichloroethane (vinyl trichloride)	$C_2H_3Cl_3$	133.4	

As a result of their widespread uses, these petroleum components and solvents are frequently found in air and water, especially at locations near petroleum refineries and gasoline stations, as well as urban areas. After entering groundwater, they form a substance known as LNAPL (light non-aqueous phase liquid), which floats on the water table.

Chlorinated methanes, ethanes, and ethylenes are solvents that are widely used for a variety of industrial purposes. For example, trichloroethylene (TCE) and 1,1,1-trichloroethane are used for degreasing of fabricated metal parts; perchloroethylene or tetrachloroethylene (PCE) is the major solvent used for dry cleaning of cloth; tetrachloromethane or carbon tetrachloride was used as a refrigerant and in fire extinguishers as well as being involved in the production of chlorifluorocarbons (CFCs); dichloromethane or methylene chloride is an ideal solvent for many chemical processes; chloroethylene or vinyl chloride and 1,1-dichlorothylene are the used in the manufacture of polymers such as polyvinyl chloride or PVC. These solvents are denser than water, and are the major components of DNAPL (dense non-aqueous phase liquid), which is found in the groundwater aquifers of the United States, challenging the engineering efforts of environmental remediation.

7.2.2 Polycyclic Aromatic Hydrocarbons (PAHs)

PAHs are generated from incomplete combustion of virtually all organic matter (Harvey, 1991). Molecules of PAHs contain only carbon and hydrogen and feature two or more fused aromatic rings. There is no limit for the number of rings, and the "ultimate" PAHs are graphite and soot, which is essentially amorphous elemental carbon. Sixteen individual PAHs have been identified as priority pollutants by the U.S. EPA. The chemical structures of these PAHs are given in Table 7.2.

With the exception of naphthalene, a major ingredient of mothballs, PAHs are not intentionally manufactured. However, numerous studies have proven the ubiquitous existence of PAHs in the environment. The sources of environmental PAHs include power generation using fossil fuels; industrial processes such as steel and coke manufacturing and asphalt production; exhaust emissions from vehicles that use internal combustion engines; industrial incineration of wastes and uncontrolled open burning; as well as natural sources such as forest fires and volcano eruptions. Residential indoor combustion such as cooking and heating also produces PAHs. Studies have shown that concentration of PAHs in residential indoor air is generally higher than that of outdoor air, particularly for 2- to 4-ring PAHs (Li et al., 2005). Cigarette smoke is the predominant source of PAHs in smokers' homes.

PAHs are considered the most widely distributed class of potent carcinogens that are present in the human environment. Some of them, such as benzo[a]pyrene (BaP), are probably among the strongest chemical carcinogens. In addition to cancer, PAHs have also been found to cause morphological, physiological, and developmental abnormalities in test animals, increase allergic immune responses in humans at low levels, and may act synergistically with other air-borne toxics to cause adverse health effects (Harvey, 1997; ATSDR, 1995).

7.2.3 Polychlorinated Biphenyls (PCBs)

Generalized chemical structure of PCBs is shown in Table 8.3. There are 209 possible PCBs, differing in the number and positions of chlorine substitutions. The common nomenclature is based on the International Union of Pure and Applied Chemistry (IUPAC) system, although the Ballschmiter & Zell (BZ) numbering system has

TABLE 7.2. Selected PAHs

Abbr.	Name	MF	MW	Structure
NaP	Naphthalene	$C_{10}H_8$	128	
AcNP	Acenaphthylene	$C_{12}H_{10}$	152	
can	Acenaphthene	$C_{12}H_{10}$	154	
FL	Fluorene	$C_{13}H_{10}$	165	
PhA	Phenanthrene	$C_{14}H_{10}$	178	
An	Anthracene	$C_{14}H_{10}$	178	
FlA	Fluoranthene	$C_{16}H_{10}$	202	
Py	Pyrene	$C_{16}H_{10}$	202	
BaA	Benz(a)anthracene	$C_{18}H_{12}$	228	
Chy	Chrysene	$C_{18}H_{12}$	228	
BbFLA	Benzo(b)fluoranthene	$C_{20}H_{12}$	252	

TABLE 7.2. (*Continued*)

Abbr.	Name	MF	MW	Structure
BkFLA	Benzo(k)fluoranthene	$C_{20}H_{12}$	252	
BaP	Benzo(a)pyrene	$C_{20}H_{12}$	252	
IP	Indeno(1,2,3-cd)pyrene	$C_{22}H_{12}$	276	
dBahA	Dibenz(ah)anthracene	$C_{22}H_{14}$	278	
BghiP	Benzo(ghi)perylene	$C_{22}H_{12}$	276	

also been widely used (Ballschmiter and Zell, 1980). Both numbering systems have evolved over years (Mills et al., 2007). Among the 209 PCBs, there are 68 "co-planar" congeners that contain either no, or only one, ortho (at substitution position 2) chlorine atom in the molecules. Among the 68 co-planar congeners, 12 have both para (4 and 4') chlorines, 2 to 4 meta (3, 3', 5, or 5') chlorines, and at least 4 chlorines in total. These 12 congeners (PCBs 77, 81, 105, 114, 118, 123, 126, 156, 157, 167, 169, and 189) are "dioxin-like", because they can achieve conformations that are similar to the structure of 2,3,7,8-tetrachlorodibenzo-p-dioxin, and are thus expected to have similar toxicity (U.S. EPA, 2003). Chlorines at ortho (2, 2', 6, 6') positions, especially when two or more are present, sterically prevent such dioxin-like conformation, reducing the toxicity that depends on planarity of the molecules.

All PCB technical products were mixtures of dozens of congeners, with various trade names such as Aroclors (Monsanto Co., St. Louis, MO, United States), Phenoclor (Prodelec, France), and Kanechlor (Kanegafuchi Chemical Industry Co., Ltd., Japan). About 100 PCB trade names are listed in the PCB page of the U.S. EPA website (U.S. EPA, 2013a).

PCBs are completely human-made with no known natural sources. PCBs are excellent electrical insulators, and were used heavily in electrical transformers and capacitors. They were also used as heat transfer fluids, plasticizers, hydraulics and lubricants, flame retardants, etc. in various materials varying from construction materials to printing inks. PCB manufacturing started in 1927 in the United States, and annual production peaked from the 1950s to 1970s. During the period from 1965 to 1974, more than 600 million pounds of PCBs were manufactured by Monsanto alone. PCB manufacturing was banned by federal law in 1979.

PCBs are ubiquitous and persistent in the environment. They are also present in humans, including those living in remote regions. In the human body, PCBs tend to accumulate in tissues and organs that are rich in lipids, and are poorly metabolized. PCBs are classified by the U.S. EPA as probable human carcinogens. They also cause adverse effects on the immune system, reproductive system, nervous system, and endocrine system of humans and animals. Other known toxic effects of PCBs in humans include an acne-like skin eruption (chloracne), pigmentation of the skin and nails, excessive eye discharge, swelling of eyelids, and distinctive hair follicles.

7.2.4   Polyhalogenated Dibenzo-*p*-Dioxins and Dibenzofurans (PXDD/Fs)

The generalized structures of PXDDs and PXDFs are given in Table 7.3. Poly-chlorinated dibenzo-*p*-dioxins and dibenzofurans (PCDD/Fs) are among chemical pollutants of the highest concern as a result of their high toxicity. One PCDD in particular, 2,3,7,8-tetrachlorodibenzo-*p*-dioxin (TCDD), is the criterium chemical for the widely used international toxicity equivalency factor (I-TEF) system. Polybrominated dibenzo-*p*-dioxins and dibenzofurans (PBDD/Fs) are the cause of growing concerns as a result of the large quantities of PBDEs (see below) produced and the possible transformation from PBDEs to PBDD/Fs. The molecules of PXDD/Fs are planar, and planarity is believed to be critical in causing strong interactions with the biological receptors.

No PXDD/Fs have ever been produced intentionally except in small quantities for research purposes. They are generated from combustion as well as several chemical processes involving organics and halogens. For example, manufacturing of pesticide 2,4,5-trichlorophenoxyacetic acid (2,4,5-T) produces PCDDs, and that of PCBs produces PCDFs, as by-products. Bleaching of pulp in pulp and paper mills using chlorination techniques has also resulted in the formation of PCDD/Fs. Other industrial processes producing PXDD/Fs include chlorophenol production, scrap metal melting when polyvinyl chloride (PVC) is present, and the chloralkali processes that produce sodium hydroxide and chlorine from sea salt. Combustion sources of PXDD/Fs include vehicle exhaust, incineration of industrial, medical and domestic wastes, and burning of biomass that has been treated with certain pesticides.

7.2.5   Polybrominated Diphenyl Ethers (PBDEs) and other Flame Retardants

Carbon-bromine bond energy (averaged at 281 kJ/mol) makes many brominated organics ideal flame retardants. In incidences of fire, brominated flame retardants

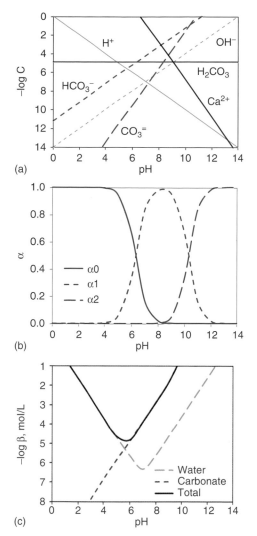

Figure 4.1. (a) pC–pH diagram, (b) species fractions, and (c) buffer intensity, for water in equilibrium with atmospheric CO_2. In panel (a), the calcium line is present when the water is in equilibrium with solid calcite.

Physical and Chemical Processes in the Aquatic Environment, First Edition. Erik R. Christensen and An Li.
© 2014 John Wiley & Sons, Inc. Published 2014 by John Wiley & Sons, Inc.

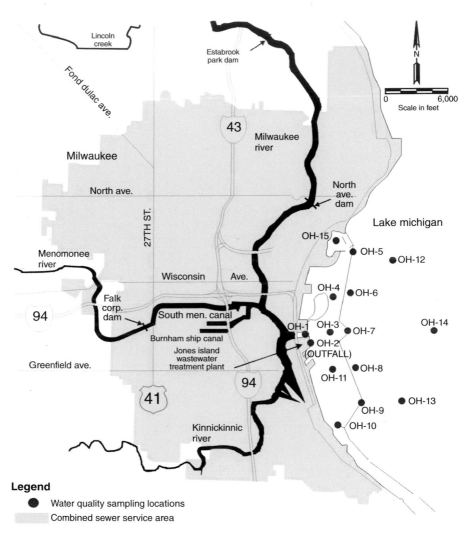

Figure 5.6. Outer Harbor water quality sampling sites. (Source: Reproduced with permission from Zou and Christensen (2012) with permission from MMSD).

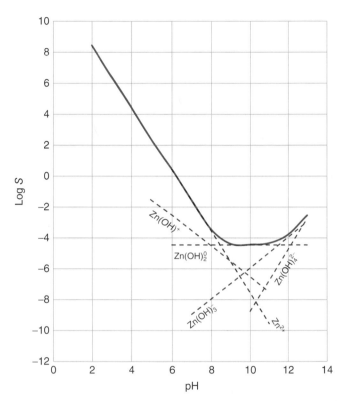

Figure 6.4. Log S, total zinc concentration in solution vs. pH. All zinc species Zn^{2+}, $Zn(OH)^+$, $Zn(OH)_2^0$, $Zn(OH)_3^-$, and $Zn(OH)_4^{2-}$ are in equilibrium with the solid phase $Zn(OH)_2(s)$.

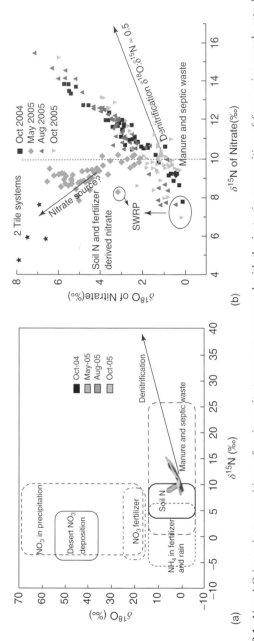

Figure 9.1. N and O isotopic compositions of various nitrate sources compared with the isotopic composition of four major sample sets. Part (a) compares fields of various sources of nitrate. Part (b) shows details of the data from this study. Arrow represents typical denitrification trend. Data from a tile study by (Panno et al., 2006) are also plotted. SWRP = Stickney Water Reclamation Plant (SWRP) in southwest metropolitan Chicago. (Source: With permission from Huang (2006))

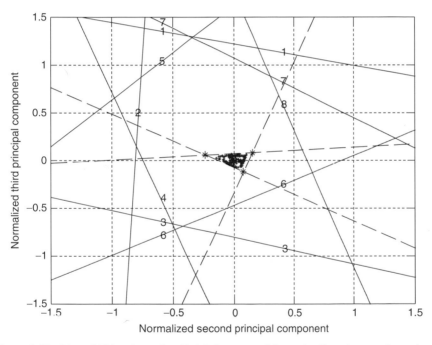

Figure 9.12. Plot of 200 points of artificial data set and 8 species lines in two dimensional plot based on normalized second and third principal components. (Source: Zou (2011) with permission from University of Wisconsin-Milwaukee.)

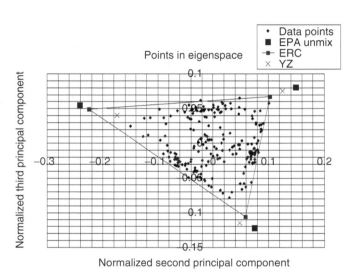

Figure 9.13. Edges of data points and source vertices based on artificial data set. (Source: Zou (2011) with permission from University of Wisconsin-Milwaukee.)

TABLE 7.3. PCBs, PCDDs, PCDFs, and PBDEs

Abbr.	General molecular formula	# Homologs	# Congeners	Generalized structure
PCBs	$C_{12}H_{10-N}Cl_N$	10	209	Cl_X — — Cl_Y
PCDDs	$C_{12}H_{8-N}O_2Cl_N$	8	75	Cl_X — — Cl_Y
PCDFs	$C_{12}H_{8-N}OCl_N$	8	135	Cl_X — — Cl_Y
PBDEs	$C_{12}H_{10-N}OBr_N$	10	209	Br_X — — Br_Y

$N = X + Y$, the total number of chlorine or bromine atoms in a molecule.

(BFRs), which are added to many consumer goods, decompose before the host polymer material. This results in the release of bromine radicals, which interfere with the chain reactions of combustion, thus preventing the fire from spreading quickly. PBDEs are one of approximately 75 different BFR groups that were commercially produced in 2004 (Birnbaum and Staskal, 2004). Stimulated by the increasingly strict law on fire prevention, large quantity production of PBDEs began in the early 1970s, just before the production of PCBs ceased. Although trade names of various chemical manufacturers differ, there are three types of commercially available PBDE products. They are commonly referred to as pentaBDE, octaBDE, and decaBDE. The global production and use of pentaBDE and octaBDE were ended before 2007. However, the use of decaBDE continued in the United States until the end of 2012. At present, decaBDE production may still be continued in some countries although no confirmative information is available.

The general chemical structure of PBDEs is shown in Table 7.3. There are 209 PBDE congeners, and the nomenclature for PBDEs is adopted from that used for PCBs. However, in contrast to PCBs, the two phenyl rings of PBDEs are roughly perpendicular to each other, as a result of the presence of tetrahedron electrons of the oxygen between the rings. In addition, unlike PXDD/Fs that are planar, the two rings of PBDE molecules are capable of rotating relative to each other. These structural characters affect not only their biological activity but also reactivity, and thus persistence in the environment.

After being first reported in soil and sludge from the United States in 1979 (DeCarlo, 1979) and in fish from Sweden in 1981 (Andersson and Blomkvist,

1981), PBDEs largely fell outside of the environmental monitoring radar until the late 1990s. Since then, they have been found in measurable amounts in various samples around the world with concentration generally doubling in as short as 3 to 5 years during the period from the 1970s to 2000 (Hites, 2004). Waters around PBDE manufacturing facilities in southern Arkansas were found to contain as high as 57 $\mu g/g$ in sediments (Wei et al., 2012). The widespread contamination of the environment has been reported worldwide (Wang et al., 2007; Li et al., 2006; Song et al., 2004, 2005b, 2005a; Hites, 2004). PBDE concentrations in the United States population measured around 35 ng/g lipid in the mid-2000s, which is about 10 to 100 times higher than those found in continental Europe and Asia in the same time period (Hites, 2004; Nanes et al., 2014; Sjödin et al., 2004; Schecter et al., 2003; Dassanayake et al., 2009).

Much of the controversy in the environmental research on PBDEs has focused on the bioaccumulation and toxicity of decabromodiphenyl ether (BDE209), the dominant component of decaBDE. Although early studies conducted by PBDE manufacturers reported that BDE209 is neither bioconcentrating nor toxic to aquatic species (Hardy, 2002), evidence of its bioaccumulation in both animals and humans is mounting. In addition, its propensity to debrominate in the environment via chemical, photolytic, and biological transformation reactions is troubling, because the lower brominated PBDE congeners generated from these reactions are much more bioaccumulating and toxic.

In addition to PBDEs, many other brominated or chlorinated organic compounds have been used as flame retardants in various industrial and consumer goods. Selected historically and currently used halogenated flame retardants (XFRs) are shown in Table 7.4. Polybrominated biphenyls (PBBs) were widely used in products such as home electrical appliances, textiles, and plastic foams until an accidental release in west-central Michigan resulted in the culling of many farm animals and elevated levels of PBBs in residents. The dominant congener of PBBs is 2,2',4,4',5,5'-hexabromobiphenyl (BB-153). Under the trade name Dechlorane, the pesticide mirex was also used as an additive flame retardant in plastics, paints, and electrical goods from 1959 to 1972. Its manufacturing was terminated in the United States in 1978.

Among the XFRs that are currently in use, some have a similar production history to PBDEs, while others were developed in recent years to replace PBDEs. These include tetrabromobisphenol A (TBBPA), hexabromocyclododecane (HBCD), dechlorane plus (DP), decabromodiphenylethane (DBDPE), 1,2-Bis(2,4,6-tribromophenoxy)ethane (TBE or BTBPE), 2,3,4,5,6-pentabromoethylbenzene (PEB), halogenated cyclohexanes such as 1,2-dibromo-4-(1,2-dibromoethyl) cyclohexane (TBECH), hexachlorocyclopentadienyl dibromocyclooctane (HCDBCO), 2-ethylhexyl-2,3,4,5-tetrabromobenzoate (EH-TBB or TBB), and bis(2-ethylhexyl)-3,4,5,6-tetrabromophthalate (TBPH or BEH-TeBP) (Table 7.4) and others. Due to their widespread uses and persistent nature, many of these emerging BFRs have been found ubiquitously, even in remote regions of the earth (Meyer et al., 2012; Hermanson et al., 2010; Yang et al., 2011, 2012; Zhu et al., 2012).

TABLE 7.4. Selected Halogenated Flame Retardants

Abbreviation and name		Chemical formula	MW	Structure
Dechlorane	Mirex	$C_{10}Cl_{12}$	545.5	
syn-DP	Dechlorane Plus (*syn*)	$C_{18}H_{12}Cl_{12}$	654	
anti-DP	Dechlorane Plus (*anti*)	$C_{18}H_{12}Cl_{12}$	654	
Dec602	Dechlorane 602	$C_{14}H_4Cl_{12}O$	613.6	
HBB	Hexabromo-benzene	C_6Br_6	551.5	
TBB	1,3,5-Tribromo-benzene	$C_6H_3Br_3$	314.8	
PBT	Pentabromo-toluene	$C_7H_3Br_5$	486.6	

(continued)

TABLE 7.4. (*Continued*)

Abbreviation and name		Chemical formula	MW	Structure
PBEB	Pentabromo-ethylbenzene	$C_8H_5Br_5$	500.7	
TBCT	Tetrabromo-*o*-chlorotoluene	$C_7H_3ClBr_4$	442.5	
*p*TBX	2,3,5,6-Tetrabromo-*p*-xylene	$C_8H_6Br_4$	417.7	
ATE (ATBPE)	Allyl 2,4,6-tribromo phenylether	$C_9H_7Br_3O$	367.8	
PBBA	Pentabromo-benzyl acrylate	$C_{10}H_5Br_5O_2$	551.6	
PBBB	Pentabromobenzyl bromide	$C_7H_2ClBr_5$	522.5	
TBECH	1,2-Dibromo-4-(1,2-dibromoethyl)cyclohexane	$C_8H_{12}Br_4$	427.8	
BB153	2,2',4,4',5,5'-Hexabromo-biphenyl	$C_{12}H_4Br_6$	627.6	

TABLE 7.4. (*Continued*)

Abbreviation and name		Chemical formula	MW	Structure
HCDBCO	Hexachloro-cyclopenta-dienyl-dibro-mocyclooctane	$C_{13}H_{10}Cl_6Br_2$	540.8	
HBCD	Hexabromo-cyclododecane	$C_{12}H_{18}Br_6$	641.7	
BTBPE	1,2-Bis(2,4,6-tribromo-phenoxy)ethane	$C_{14}H_8O_2Br_6$	687.6	
DBDPE	Decabromo-diphenyl ethane	$C_{14}H_4Br_{10}$	971.2	
EHTBB or TBB	2-Ethylhexyl-2,3,4,5-tetrabromo-benzoate	$C_{15}H_{18}O_2Br_4$	549.9	
TBPH	Bis(2-ethylhexyl)-2,3,4,5-tetrabromo-phthalate	$C_{24}H_{34}O_4Br_4$	706.1	

Another group of flame retardants of particular emerging concern are the organophosphate flame retardants (OPFRs). As shown in Table 7.5, most of them are tri-esters of phosphoric acid; some are halogenated. Reviews of their production, environmental occurrence, toxicity, and human exposure are available from the U.S. EPA (1976, 2005) and ATSDR (2012). The findings that TDCPP and TDBPP were absorbed by children through their sleepwear caused public concerns and resulted in the ban of OPFRs for pajama manufacture (Kerst, 1974; Prival et al., 1977; Gold et al., 1978; Blum et al., 1978; Ulsamer et al., 1978, 1980). The environmental detection and evidence of bioaccumulation of phosphate esters were also reported in the early 1980s (Muir et al., 1980, 1981). Nonetheless, OPFRs have returned to replace PBDEs in numerous consumer goods, including various products for children (Stapleton et al., 2011). For instance, Firemaster® 550, a product of Chemtura Corporation (Philadelphia, PA) that contains triphenyl phosphate (TPHP) and two brominated FRs (EH-TBB and TBPH), replaced pentaBDE in polyurethane foams used in furniture. As a result, OPFRs have been detected in house dust, with concentrations often higher than PBDEs (Dodson et al., 2012; Brommer et al., 2012; Stapleton et al., 2009). In natural waters, however, recent information for OPFRs is very limited.

Most environmental fate assessments for OPFRs date from the 1980s. These found generally low water solubilities, Kow range of 10^3 to 10^5, vapor pressures of 10^{-2} – 10^{-7} mm Hg, and relatively strong bioconcentration in fish (Muir, 1984; Muir et al., 1983; Sasaki et al., 1982; Mayer et al., 1981). Hydrolysis was deemed the most important abiotic transformation in water (Boethling and Cooper, 1985). Non-halogenated OPFRs are degradable by microorganisms, with a half-life as short as 3 days in sediment (Muir et al., 1989), but halogenated OPFRs are biorecalcitrant (ATSDR, 2012).

7.2.6 Organochlorine Pesticides (OCPs)

The chemical structures of a few major OCPs are shown in Table 7.6. Among them, DDT (abbreviation for dichloro-diphenyl-trichloroethane, but its IUPAC name is 1,1,1-trichloro-2,2-bis[4-chlorophenyl]ethane) was the first synthesized, and probably the best known, OCP. Its discovery was awarded with a Nobel Prize due to the high efficiency of DDT against harmful arthropods, and therefore its application to crop fields around the world resulted in increased yield. DDT was largely replaced by methoxychlor, which has a similar structure but is much less toxic to birds and less persistent in the environment than DDT. Toxaphene (chlorinated camphene) is a complex but reproducible mixture of at least 177 compounds, with a general formula $C_{10}H_{16-N}Cl_N$ ($N = 6$ to 9) and average molecular mass of 414. The insecticide lindane is the only bioactive among the eight isomers of hexachlorocyclohexane (HCH). Chlorinated cyclodienes are a subgroup of OCPs, including heptachlor, chlordane, aldrin, and dieldrin (degradation product of aldrin). The use of mirex as an insecticide for ant control ended in the late 1970s. A ketone analog of mirex is merex, also known as Chlordecone and Kepone. The manufacturing of merex was also terminated in the late 1970s in the United States when production workers developed serious neurological disorders.

TABLE 7.5. Selected Organophosphate Flame Retardants (OPFRs)

Abbr.	Name	Chemical formula	MW	Structure
TBP	Tributyl phosphate	$C_{12}H_{27}O_4P$	266.3	
TiBP	Triisobutyl phosphate	$C_{12}H_{27}O_4P$	266.3	
TPP	Triphenyl phosphate	$C_{18}H_{15}O_4P$	326.3	
TCP	Tricresyl phosphate	$C_{21}H_{21}O_4P$	368.4	
EHDPP	2-Ethylhexyl diphenyl phosphate	$C_{20}H_{27}O_4P$	362.4	
TBEP	Tris(2-butoxyethyl) phosphate	$C_{18}H_{39}O_7P$	398.48	

(*continued*)

TABLE 7.5. (*Continued*)

Abbr.	Name	Chemical formula	MW	Structure
TCEP	Tris-(2-chloroethyl) phosphate	$C_6H_{12}Cl_3O_4P$	285.50	
TCPP	Tri-(2-chloroisopropyl) phosphate	$C_9H_{18}Cl_3O_4P$	327.57	
TDCPP	Tris(1,3-dichloro-2-propyl) phosphate	$C_9H_{15}Cl_6O_4P$	430.88	
TDBPP	Tris(2,3-dibromopropyl) phosphate	$C_9H_{15}Br_6O_4P$	697.6	

Due to the strength of the C-Cl bond, most OCPs are highly persistent in the environment. For example, more than 20 years after their ban in the United States, toxaphenes are still a great threat to the ecosystem of the Great Lakes. In addition, most OCPs are highly hydrophobic; therefore they dissolve in water to trace amounts but bioaccumulate in the fatty tissues of animals and humans. In addition, many of them biomagnify along the food chain. Because of their persistence and bioaccumulative nature, all of the OCPs discussed above are on the list of the "dirty dozen" identified by the Stockholm Convention for persistent organic pollutants held by the United Nations Environmental Program (UNEP).

TABLE 7.6. Selected Chlorinated Pesticides and Derivatives

Group	Common name	Formula	MW	Molecular structure
DDT and derivatives	*p,p*-DDT	$C_{14}H_9Cl_5$	354.5	
	DDE	$C_{14}H_8Cl_4$	318.0	
	Methoxychlor	$C_{16}H_{15}Cl_3O_2$	345.7	
Toxaphenes	Toxaphenes	$C_{10}H_{18-n}Cl_n$ ($n = 6 - 9$)	-	
-	Mirex	$C_{10}Cl_{12}$	545.5	
-	Lindane	$C_6H_6Cl_6$	290.8	
Chlorinated cyclodienes	Chlordane	$C_{10}H_6Cl_8$	409.8	

(*continued*)

TABLE 7.6. (*Continued*)

Group	Common name	Formula	MW	Molecular structure
	Heptachlor	$C_{10}H_5Cl_7$	373.3	
	Aldrin	$C_{12}H_8Cl_6$	364.91	
	Dieldrin/Endrin	$C_{12}H_8Cl_6O$	380.9	

7.2.7 Other Pesticides

Other major groups of human-made pesticides include organophosphates, carbamates, pyrethrum, chlorinated phenoxyalkyl acids and salts, and triazines. The chemical structures of examples can be found in Table 7.7.

There are more than 100 organophosphate pesticides (OPPs), with most being insecticides. They are characterized by their high acute toxicity on not only targeted insects but also fish, birds, and mammals including humans, and are blamed for an estimated 20,000 deaths per year worldwide. Besides their non-selectivity in toxicity, other important environment-relevant characteristics of OPPs include high water solubility, low bioconcentration and bioaccumulation potential, no biomagnification, and high degradability (low persistence). These are in contrast with the environmental properties of OCPs described above.

Carbamate pesticides share a common structural fragment of $R_1-O-C(=O)-N(R_2)-R_3$, where all R_1, R_2 and R_3 can be either alkyl or aryl groups, and R_2 and R_3 may also be just hydrogen atoms. Carbamates include both insecticides and herbicides. Carbaryl (1-naphthol N-methylcarbamate), as an example, is a general use insecticide with a trade name Sevin in the United States and is commonly applied on various crops as well as livestock, poultry, and pets. Carbamate pesticides share many environmental properties with organophosphate pesticides mentioned above.

TABLE 7.7. Examples of Pesticides

Group	Examples	Formula	MW	Molecular structure
Chlorinated phenoxy-acetic acids	2,4-D	$C_8H_6Cl_2O_3$	221.0	
	2,4,5-T	$C_8H_5Cl_3O_3$	255.5	
	Dichlorprop	$C_9H_8Cl_2O_3$	235.1	
Organo-phosphates	Parathion	$C_{10}H_{14}NO_5PS$	291.3	
	Dichlorvos	$C_4H_7O_4PCl_2$	221.0	
	Dimethoate	$C_5H_{12}NO_3PS_2$	229.3	
Carbamates	Carbaryl	$C_{12}H_{11}NO_2$	201.2	
Triazines	Atrazine	$C_8H_{14}N_5Cl$	216.7	

(*continued*)

TABLE 7.7. (*Continued*)

Group	Examples	Formula	MW	Molecular structure
Pyrethrum	Pyrethrins	-	-	
	Pyrethriods	-	-	

Pyrethrum pesticides include both natural pyrethrins and synthetic pyrethriods. Pyrethrins are often referred to as the mixture extracted from the flowers of some Chrysanthemum plants. They contain six active components (cinerin I, cinerin II, jasmolin I, jasmolin II, pyrethrin I and pyrethrin II) differing in the R and R' groups in the general structure shown in Table 7.7. Pyrethriods are simply man-made compounds mimicking pyrethrin. They keep the core structure of the pyrethrin (Table 7.7), but have various substitutions, including halogens in some of them, on both sides of the core structural moiety. Pyrethriods are used in many commercial household insect repellents and insecticides. Although they contain polar function groups, pyrethrins and pyrethriods are much more hydrophobic than OPPs and carbamates. Natural pyrethrins degradate in water through hydrolysis of the ester group and photolysis on conjugated group $C = C - C = O$ in the 5-member ring. Although possessing low toxicity to mammals and birds, pyrethrins are highly toxic to fish and tadpoles.

"Agent orange", a defoliant used to clear the visual path to the ground from helicopters during the Vietnam War, is a mixture of 2,4-dichlorophenoxyacetic acid (2,4-D) and 2,4,5-trichlorophenoxyacetic acid (2,4,5-T). Whereas 2,4-D is still the most widely used herbicide for weed control, 2,4,5-T has been phased out due to the presence of the by-product 2,3,7,8-TCDD (see above). Other chlorinated phenoxy alkyl acid herbicides include 2-methyl-4-chlorophenoxyacetic acid (MCPA), 2-(2-methyl-4-chlorophenoxy) propionic acids (mecoprop, MCPP), 2-(2,4-dichlorophenoxy)propionic acid (dichloroprop, 2,4-DP), and (2,4-dichlorophenoxy)butyric acid (2,4-DB).

Another major type of herbicides is the triazines. Triazine herbicides include amitrole (aminotriazole), atrazine, cyanazine, simazine, and trietazine (IPCS, 1998). These and other triazines, as well as their derivatives, are also used for other industrial purposes such as the manufacturing of resins, dyes, and explosives. The chemical structure of atrazine, one of the most widely used herbicides worldwide, is shown in Table 7.7. The human toxicity of triazine herbicides is generally considered to be low. In 2006, an assessment of cumulative exposures to these pesticides was conducted by the U.S. EPA, and showed that the levels of atrazine and simazine that

Americans are exposed to in their food and drinking water combined are below the level that would potentially cause health effects (U.S. EPA, 2012a).

7.2.8 Perfluorinated Compounds (PFCs)

Fluorine forms the strongest single covalent bond with carbon (bond energy ≈ 486 kJ/mol), due to the large difference between the electronegativity of these two elements. The strong C-F bond renders exceptionally high stability of PFCs, which are therefore extremely resistant to breakdown in the environment. The chemical structures of selected PFCs are given in Table 7.8. As can be seen, they are surfactants with a polar "head" and a non-polar "tail".

Perfluoroalkane surfactants, such as POSF, form the starting material in the manufacture of polymers, which are used in large quantities in various consumer products such as stain-resistant fabrics, non-stick cookware, and fire-fighting foam. Perfluoroalkyl sulfonates (PFAS), such as perfluorooctanyl sulfonate (PFOS), are the degradation product of POSF-based compounds. PFOS was also used directly in numerous commercial goods. For instance, it is the key ingredient in many commercial stain repellents, and is impregnated in carpets, textiles, paper, leather, wax, polishes, paints,

TABLE 7.8. Selected Perfluorochemicals

Abbr.	Name	Structural formula	MW	Structural formula
POSF	Perfluoro-octane-sulfonyl-fluoride	$C_8F_{18}SO_2$	502.1	C_8F_{17}—S(=O)(=O)—F
PFOS	Perfluoro-octane-sulfonate	$C_8F_{17}SO_3^-$	499.1	C_8F_{17}—S(=O)(=O)—O^-
PFOSA	Perfluoro-octane-sulfonamide	$C_8F_{17}SO_2NH_2$	499.1	C_8F_{17}—S(=O)(=O)—NH_2
PFOA	Perfluoro-octanoic acid	$C_8F_{15}O_2H$	414.1	C_7F_{15}—C(=O)—OH
PFBS	Perfluorobutane sulfonate	$C_4F_9SO_3^-$	299.1	C_4F_9—S(=O)(=O)—O^-
PFBA	Perfluorobutanoic acid	$C_4F_9O_2H$	252.0	C_4F_9—C(=O)—OH

varnishes, and cleaning products for general use. Perfluorinated fatty acids (PFFAs), such as perfluorooctanoic acid (PFOA), are also commonly used in industries and for consumer products. One of the most common uses of PFOA is for processing poly-tetrafluoroethylene (PTFE), most widely recognized under the brand name Teflon, which is manufactured by DuPont.

PFCs were first manufactured in the late 1940s. Production soared decades later, with a global annual production of about $500\,t$ in the 1970s to as high as $4500\,t$ by the 1990s (Paul et al., 2009). PFCs are now being widely detected including in wildlife in polar areas (Houde et al., 2011; Giesy and Kannan, 2001; Kannan et al., 2001). PFOS and PFOA are the most common PFCs found in the environment because they are exceptionally persistent. In humans, PFCs accumulate with age. They are believed to be capable of endocrine disruption, and are potentially carcinogenic. They were also found to cause birth defects in test animals.

7.2.9   Pharmaceuticals and Personal Care Products (PPCPs) and other Endocrine Disrupting Compounds (EDCs)

The terms pharmaceuticals and personal care products (PPCPs) refer, in general, "to any product used by individuals for personal health or cosmetic reasons or used by agribusiness to enhance growth or health of livestock" (U.S. EPA, 2000). PPCPs include medicinal drugs for humans and animals, skin care and sun-screen agents, cosmetic products, chemical agents used for medical diagnosis, bioactive food supplements, and fragrances (e.g. musks). The acronym "PPCPs" was originally used in a critical review by Daughton and Ternes (1999). PPCPs are highly diverse in their chemical structures and applications. Examples of PPCPs are presented in Table 7.9. Reviews on environmental PPCPs are available (Boxall et al., 2012; Rosi-Marshall and Royer, 2012; U.S. EPA, 2000; Daughton and Ternes, 1999; Daughton and Jones-Lepp, 2001; Kümmerer, 2004).

PPCPs enter natural water bodies, including the groundwater from mainly untreated and insufficiently treated wastewaters. Many of them have been found in natural waters and water supplies in the United States (Benotti et al., 2009; Kolpin et al., 2002). Although the concentrations of most PPCPs in waters are low, their inputs tend to be continuous. The presence of PPCPs in the environment is of great concern because these chemicals are intentionally designed to generate biological impacts. Their effects on the ecosystem, even at the low concentration level, can be significant (Mason, 2003). For example, some synthetic hormones in birth control pills in some rivers are believed to contribute to the feminization of some male fish (Raloff, 1998). For human health, environmental PPCPs are of great concern because some PPCPs have been found in drinking water supplies. This is primarily a result of PPCPs' high aqueous solubilities, which make their removal at water purification facilities difficult. The problem can be significant for communities for which water supplies contain recycled or re-used wastewater (Daughton, 2004). The long-term, continuous, and low dose uptake of these chemicals from drinking water may pose risks that are difficult to assess.

TABLE 7.9. Examples of Pharmaceuticals and Personal Care Products (PPCPs) and other Endocrine Disrupting Compounds (EDCs)

Abbr.	Major use	Common name	Structural formula	Structural formula
TCS	Antiseptic	Triclosan	$C_{12}H_7Cl_3O_2$	
-	Synthetic oral contraceptive	17_7-Ethinyl estradiol	$C_{20}H_{24}O_2$	
MX	Fragrances	Musk xylene	$C_{12}H_{15}N_3O_6$	
MBC	Sun-screen agent	Methylbenzy-lidene camphor	$C_{18}H_{22}O$	
-	X-ray contrast agent	Diatrizoate (sodium salt)	$C_{11}H_8I_3N_2O_4Na$	
BPA	For making plastics and epoxy resins	Bisphenol A	$C_{15}H_{16}O_2$	

(continued)

TABLE 7.9. (*Continued*)

Abbr.	Major use	Common name	Structural formula	Structural formula
-	Plasticizer	Phthalates	-	(phthalate ester structure with OR and OR' groups)
NP	Surfactant	Nonylphenol	$C_{15}H_{24}O$	OH—(benzene ring)—C_9H_{19}

(Source: Some structural formulas are taken from Daughton and Ternes (1999)).

PPCPs may serve as "probes" in environmental research. Their uses as "tracers" or "sentinels" of human or animal wastes have been reported. For example, caffeine was used as a tracer for human sewage in Boston Harbor, because it is exclusively consumed by humans in the forms of coffee, tea, and soft drinks, in the same way that fecal e-coli have been used (Twombly, 2001). PPCP compounds with fluorescence characteristics, such as brighteners and whitening agents used in many personal care products, are particularly attractive "indicators" (Hayashi et al., 2002). In addition, many PPCPs have well-defined dates of debut, and thus can be used as marker compounds for "dating" various environmental samples. Compared with a single PPCP, combinations of well selected PPCPs may provide more accurate information on sources and chronologies of pollution by various chemicals.

Also included in Table 7.9 are bis-phenol A, nonylphenols, and phthalates. All these have caused high concern in recent years, due to their ubiquitous presence and their potential impact on human health. These "everywhere" chemicals are capable of endocrine disruption, i.e. interfering with the hormone system in living organisms including humans. The results of such interference are brain development problems, various cancers, sexual problems, etc., observed in animals and humans.

Bisphenol A (BPA) is used in the manufacture of epoxy resins and polycarbonate plastics. BPA has been found in various products, including infant milk bottles and food storage containers. As a result, it has been found in human breast milk and other samples (NIEHS, 2012).

Nonylphenols (NPs) belong to the alkylphenol chemical family, and possess many isomers differing in the carbon arrangement of the nonyl chain, which are highly branched in most NP molecules. NPs are surfactants involved in the manufacture of detergents, rubber, vinyl, polyolefins, and polystyrenics. NPs are persistent in the aquatic environment, moderately bioaccumulative, and extremely toxic to aquatic organisms (U.S. EPA, 2013b).

Phthalate is a group of numerous esters of phthalic acid, with different R groups. The major examples include di-(2-ethylhexyl)phthalate (DEHP), di-*n*-butyl phthalate, di-*n*-octylphthalate (DNOP), and diethyl phthalate. Phthalates are added to plastics to enhance flexibility, and can be found in numerous consumer products, ranging from toothbrushes and toys to food packaging and cosmetics. Phthalate is the most commonly found laboratory contaminant, and often interferes with targeted analyses of chemicals present at trace levels in environmental samples.

7.3 DESCRIPTORS OF ORGANIC MOLECULES

Numerous molecular constitutional, topological, geometrical, and electrostatic descriptors were calculated for environmentally concerned chemicals using quantum chemical computation and semi-empirical methods. The simplest of these include the number of carbons or other elements (e.g. the number of halogens in PCBs, PBDEs, etc.). A few that are relatively frequently used in the literature are briefly introduced in this section.

Molecular volume (MV) and total surface area (TSA) are related to various physicochemical properties that determine the partitioning and transport behavior of molecules. Van der Waal's volume and surface area are the most widely used. The McGowan characteristic volume (McGowan and Mellors, 1986), which can be easily calculated from the chemical formula, was recommended by Schwarzenbach et al. (2003). The Le Bas molar volume has also been used to correlate with various environmentally important physicochemical properties (Reid et al., 1987). Both Le Bas and McGowan volumes do not distinguish structural isomers, because of their atomic contribution approach. It can be important to divide TSA into "hydrophobic" and "hydrophilic" components when dealing with process parameters, e.g. aqueous solubility (see Section 7.4), which are determined by the interactions between organic solute and polar solvents.

Polarity of a molecule depends on bond polarity and molecular geometry. It determines the strength and types of intermolecular forces. Molecular polarity is mostly represented by its permanent dipole moment, which can be estimated from the vector sum of the dipole moment of individual bonds of the molecule. The bond dipole moment results from the difference in electronegativity between the two bonding atoms. Molecules such as CH_4 and benzene have zero permanent dipole moment thus they are "non-polar"; while CH_3Cl and chlorobenzene are "polar". All molecules, however, may possess instantaneous dipole moment, as a result of the random motions of the electrons causing alteration of the permanent dipoles. Meanwhile, dipole moment can be "induced" when the molecule gets close to a charged species (ions) or other molecules with dipole moment. The most commonly used unit for dipole moment is debye (D). The SI unit is coulomb-meter ($1\,C\,m = 2.9979 \times 10^{29}$ D). Molecular polarity is different from bulk polarity, which is often represented by Kow (see Section 7.4), dielectric constant, solubility parameter, etc. (Li et al., 1996), and those based on proton donating and accepting capacities (Schwarzenbach et al., 2003).

Polarizability (α) of a molecule represents its electric susceptibility. It is the ability to develop uneven electron distribution in response to an imposed electric field, and thus acquire an electric dipole moment on femtosecond time scales (Staikova et al., 2004; Schwarzenbach et al., 2003). Polarizability increases with atomic or molecular size, particularly the number of electrons of conjugated bond. For instance, α value is in the order pyrene > naphthalene > benzene > hexane. The experimental values of α can be derived from measured refractive index (Schwarzenbach et al., 2003).

In thermodynamic calculations, the reaction from which one mole of a pure substance is formed from the most stable forms of its elements under standard conditions (e.g. 1.0 atm of pressure, a specified temperature) is a fundamental reaction for other reactions in which this substance is involved. The standard enthalpy, entropy, and free energy of formation reactions (ΔH_f°, S_f°, and ΔG_f° respectively), summarized in handbooks for many environmentally relevant substances, are commonly used to estimate those of other reactions (ΔH_r, ΔS_r and ΔG_r respectively) under various conditions when the reference states for all reactants and product are defined. For substances which do not have established standard values, ΔG_r° can be obtained by experimentally measuring the equilibrium constant (K) of the reaction or process using $\Delta G_r^\circ = -RT \ln K$ (where R is the gas constant and T is the absolute temperature in Kelvin). A widely adopted simplification in describing reactions occurring in the natural environment is to assume the temperature independence of ΔH_r and ΔS_r, based on the fact that the temperature range is not wide (e.g. −40 °C to 40 °C in most part of the world). The assumption holds true if the regression of ln K against the reverse temperature (1/T) is linear. Then, ΔH_r° can be estimated from the slope of the linear regression, based on the equation ln $K = -\Delta G_r^\circ/T = -\Delta H_r^\circ/RT + \Delta S_r^\circ/R$. However, estimation of ΔS_r° by extrapolation based on the intercept of this regression can result in significantly erroneous values.

Molecular characteristics that are important to reactivity include the energy of the highest occupied molecular orbital (E_{HOMO}) and the energy of the lowest unoccupied molecular orbital (E_{LUMO}). E_{HOMO} provides an estimate of the first ionization potential, and the gap between E_{HOMO} and E_{LUMO} corresponds to the minimum excitation energy of the molecule.

Topological descriptors have been extensively used in chemistry. The first of these was the Wiener index (Wiener, 1947), which is the sum of distances between all the pairs of vertices in a hydrogen-suppressed molecular graph. Molecular connectivity index (χ) is a topological descriptor based on bonding and branching patterns. It was developed by Randic (1975) and refined by Kier and Hall (1976). The calculation of χ starts with the drawing of a hydrogen-suppressed molecular structure and designation of a δ value (the number of adjacent non-hydrogen atoms) for each atom. Next, different types of χ with various orders can be calculated.

Assisted by the advancements in computer technology, theoretical and computational chemistry have now been extensively involved in environmental chemistry research. Various molecular descriptors, including those discussed in this section, can be obtained by either ab initio calculations using computer programs, such as GAUSSIAN, or by semi-empirical calculation methods, such as those built in MOPAC and other computer programs. The calculated molecular descriptors are then used to predict the physicochemical properties and to explore reaction pathways and mechanisms in environmental chemistry.

7.4   BASIC PHYSICOCHEMICAL PROPERTIES

The nature of organic chemicals is described by a wide range of physicochemical properties. The simplest of these include melting point and boiling point temperatures, and molar volume, which is often derived from the molecular mass and liquid density. Various partition coefficients between bulk phases are discussed in this section. They are different from those concerned in Section 7.5 where "real world" distribution ratios are the discussion foci. For all of the physicochemical properties dealt with in Sections 7.4 and 7.5, the reference state for thermodynamic expressions is the pure (mole fraction = 1) chemical in liquid state at 1.0 atmosphere pressure and a specified temperature. The standard state is the reference state under standard conditions (e.g. 1.0 atm total pressure and specified temperature).

The relationships among a few "core" properties discussed in this section are depicted in Figure 7.1. To obtain the values of these physicochemical properties, laboratory measurements are always preferred, and many of the procedures are described by OECD (2007). Over the past decades, various quantitative structure property relationships (QSPRs), quantitative property property relationships (QPPRs), and quantitative property retention relationships (QPRRs) have been developed, enabling estimation from chemical structure alone and with fair to good accuracy. Many such quantitative relationships are discussed in Schwarzenbach et al. (2003), Boethling and Mackay (2000), Baum (1998), Lyman et al. (1982), and others. Databases that compile large amounts of property data are also available (Mackay et al., 1991–1997, 2006). The computerized Estimation Programs Interface (EPIWIN) suite, promulgated by the U.S. EPA and readily downloadable from the Internet (U.S. EPA, 2012b), is a convenient set of tools for orders of magnitude estimations of many of the properties included in Sections 7.4 and 7.5. A final point to note is that the discussions in these two sections are limited to neutral organic compounds in bulk phases. For the phase partition coefficients and distribution ratios of the ionizable organics, refer to relevant chapters in Schwarzenbach et al. (2003).

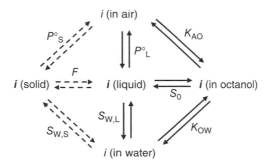

Figure 7.1. Relationship between the key physicochemical properties.

7.4.1 Vapor Pressure

Vapor pressure, $P^°$, is the partial pressure of a chemical in air when at equilibrium with its pure phase. $P^°$ is 1.0 atm for all gases and less than 1.0 atm for liquid and solid chemicals under normal environmental conditions at sea level. Vapor pressure is the upper limit of partial pressure, which is a fraction of the chemical of interest when it is in a gas mixture, of the total pressure exerted by all the molecules in the mixture.

Vapor pressure can be viewed as coefficients of partitioning between air and the pure chemical, or the equilibrium constant of the evaporation process for liquids. It is related to the free energy of the process, $\Delta_{vap}G^°$:

$$\Delta_{vap}G^° = -RT \ln P^° \tag{7.1}$$

Vapor pressure depends strongly on temperature. It increases with increasing pressure because the enthalpy of evaporation, $\Delta_{vap}H$, is greater than zero for all organic compounds. The integrated Clausius–Clapeyron equation describes the change in vapor pressure with temperature

$$\ln P_L = -\Delta_{vap}H/RT + \text{constant} \tag{7.2}$$

For a particular molecule to evaporate from a pure liquid to vapor phase, it needs to break its "links" with other molecules in the pure phase of the chemical. Therefore, the magnitude of $\Delta_{vap}H$ reflects the strength of the "links", or the cohesive energy of intermolecular interactions in the pure chemical, which, in turn, depends on the chemical structure of the pure chemical.

If the chemical of interest is a solid, it sublimates to pose its vapor pressure. The sublimation process can be viewed as the sum of processes of a hypothetical fusion (melting) to liquid and vaporization from liquid to gas at given temperature and pressure. Such liquid, of course, is hypothetical because it may not exist at the given condition. The vapor pressure of this hypothetical liquid, $P_L^°$, rather than the true vapor pressure of the solid, $P_S^°$, is often the one generated by various QSPR models. $P_L^°$ can be converted to solid vapor pressure $P_S^°$ using the fugacity ratio F

$$P_S^° = F \times P_L^° \tag{7.3}$$

$$F = \frac{f_S}{f_L} \approx \exp\left[\left(\frac{-\Delta_{fus}S}{R}\right) \times \frac{T_m}{T-1}\right] \tag{7.4}$$

where f_S and f_L are the fugacity for solid and its hypothetical liquid respectively; T_m and T are the melting point temperature and the temperature of interest respectively, in Kelvin; and $\Delta_{fus}S$ is the entropy of fusion, which reflects the increases in randomness when solids melt. For organic chemicals that are deemed "rigid", $\Delta_{fus}S$ value is approximately 56 J/K/mol. This value may be applied to PAHs, PCBs, PCDD/Fs, PBDEs, and many others, to generate order of magnitude estimates for their fugacity ratio and vapor pressure. For flexible molecules, such as those with long aliphatic chains, corrections must be made.

7.4.2 Aqueous Solubility

Aqueous solubility (S_W) is commonly defined as the abundance of the chemical per unit volume in the aqueous phase when its solution is in equilibrium with the pure compound in its actual aggregation state (gas, liquid, solid) at specific temperature, pressure, and the aqueous phase composition. Aqueous solubility is the upper limit of aqueous concentration. Aqueous solubility is connected to the free energy of dissolution in water, $\Delta_{diss}G°$, by

$$\Delta_{diss}G° = -RT \ln x_w \qquad (7.5)$$

where x_w is the water solubility expressed in mole fraction, with the standard state being the pure liquid chemical as mentioned above. For organic compounds that are sufficiently hydrophobic, the conversion between molar and mole fraction-based water solubilities can be approximated by

$$x_w = S_W \times V_W \qquad (7.6)$$

where V_W is the molar volume of the aqueous solution and can be approximated by the molar volume of water, 0.018 L/mol.

Large databases of aqueous solubility are available (IUPAC, 2007; Dannenfelser and Yalkowsky, 1991). The experimental measurement tends to be more difficult for compounds that are highly hydrophobic. For example, reported experimental S_W values for decachlorobiphenyl (PCB209) span several orders of magnitude. Various QSPRs and QPPRs have been developed to estimate S_W. Simple correlations with, for example molecular mass, were established for compounds within a chemical family, and others use structural parameters such as the number of carbon or halogen atoms, molecular surface area, and molecular connectivity indices (Nirmalakhandan and Speece, 1988, 1989). Models based on the concept of group contribution, such as AQUAFAC (Myrdal et al., 1993), have also been developed. The most widely used relies on the correlation of S_W with the octanol–water partition coefficient K_{OW} (discussed Section 7.4.4). Many such correlation equations have been published and compiled (Boethling and Mackay, 2000). The applicability of correlations between S_W and K_{OW} is often limited to particular chemical groups. The accuracy of such an approach depends on several factors including the numbers of compounds in the "training set" for model development, the range of the independence variable, and statistics of the equations. For compounds that are solids at the temperature of interest, the fugacity ratio (Equation 7–4) is needed to convert the estimated $S_{W,L}$, which pertains to the hypothetical liquid, to that for solids.

The dependence of water solubility on temperature varies. The solubility of organic gases decreases with increasing temperature, and the opposite tends to be true for organic solids. Some compounds, such as halogenated methanes and ethanes, have minimal aqueous solubility at ambient temperatures. The complex pattern of the temperature effects results from the complexity of the dissolution process, which is viewed as consisting of multiple steps with each having their own enthalpy and entropy changes. These steps include the breaking of interactions of a molecule with other molecules in the pure chemical, cavity formation in water, and establishing interactions between the solute molecules with water. In general, the solubility of all organic compounds increases with decreasing salinity and increasing content of dissolved organic matter (DOM) in water. Compounds that exhibit high S_W tend to become quickly dispersed in the environment via the hydrological cycle, are associated with sediment and biological phases to a lesser extent, and often have higher degradation rates relative to sparsely soluble compounds (Lyman et al., 1982).

7.4.3   Henry's Law Constant

Transfer between the atmosphere and water is one of the key processes affecting the transport of many organic compounds in the environment. Henry's law constant, K_H, is an equilibrium partition coefficient of chemicals between air and water phases.

$$K_H = \frac{P}{C_W} \tag{7.7}$$

where P is the partial pressure of the solute, and C_W is its aqueous concentration. K_H can be converted using the ideal gas law $P = C_A RT$ to the air–water partition coefficient K_{AW}

$$K_{AW} = \frac{C_A}{C_W} = \frac{K_H}{RT} \tag{7.8}$$

For many hydrophobic and semi-volatile organic pollutants such as PCBs, PBDEs, PXDD/Fs, their K_H values can be approximated by the ratio of vapor pressure and water solubility:

$$K_H \approx \frac{P^\circ}{S_W} \tag{7.9}$$

7.4.4   Octanol–Water Partition Coefficient

Due partly to historical reasons, n-octanol has become the most popular solvent in assessing the partitioning behavior of organic solutes between organic and aqueous phases. The presence of both non-polar and polar fragments in the molecules of n-octanol makes it a reasonable surrogate for many kinds of environmental and physiological organic substances. Octanol–water partition coefficient, K_{OW}, is defined as ratio of the concentrations of a chemical in n-octanol (o) to that in water (w) at equilibrium:

$$K_{OW} = C_O/C_W \tag{7.10}$$

K_{OW} indicates the lipophilicity or hydrophobicity of organic chemicals. It is widely used in environmental fate studies, predicting physical and biological phenomena, such as soil sorption and bioconcentration in fish (see Section 7.5 below).

Experimental values of K_{OW} have been compiled for large numbers of organic compounds (Hansch et al., 1995; Mackay et al., 1991–1997, 2006). For those where measured values are not available, prediction of K_{OW} from chemical structure is one of the most developed QSPRs. The best known are the two developed by Hansch and Leo (1979) and by Meylan and Howard (1995), both of which are fragment (or group) contribution methods although their approaches are very different. The Internet online program CLogP (Daylight, 2006) and the U.S. EPA EPI Suite software KowWin (U.S. EPA, 2012b) are readily available.

A related property is the solubility of organic compounds in pure n-octanol, or S_O. It should be an appropriate surrogate for solubility in lipid thus has practical applications, and is an important link in the relationships among the "core" properties shown in Figure 7.1 (Li et al., 1995; Yalkowsky et al., 1983). The ratio of solubilities in octanol and in water, S_O/S_W, can be used to estimate K_{OW} (Pinsuwan et al., 1995). However, the former does not consider the mutual saturation of the two solvents and the latter does, despite the often-negligible differences in their values. On another note, partition coefficients between water and other organic solvents, such as hexane, ethers, and halogenated methanes, have also been investigated. Linear relationships between different partition coefficients often exist when the types of solute–solvent interactions are similar (see Chapter 7 in Schwarzenbach et al., 2003).

7.4.5 Air–Octanol Partition Coefficient

Considering the vast amount of organic substances (tree leaves and soil organic matter, for example) in contact with air, the importance of understanding the partitioning of pollutant chemicals between organic matter and air is apparent. This process significantly affects their long range transport, degradation, bioavailability, etc. The air–octanol partition coefficient, K_{AO}, is defined as the ratio of concentrations in air and in n-octanol

$$K_{AO} = C_A/C_O \tag{7.11}$$

For most organic chemicals, K_{AO} is less than 1, indicating that organic chemicals prefer to stay in organic solvent over air. This is true even for apolar compounds with polar solvents. The choice of n-octanol as surrogate organic phase was largely a result of, understandably, the popularity of K_{OW}. However, partitions between other organic liquids and air have also been investigated (Chapter 6 in Schwarzenbach et al., 2003). The reversely defined octanol–air partition coefficient, K_{OA}, has also been widely used.

Experimental data for K_{AO} and K_{OA} are rare. They can be easily obtained from the ratio of K_{AW} and K_{OW}

$$K_{AO} \approx \frac{K_{AW}}{K_{OW}} \tag{7.12}$$

7.5 DISTRIBUTION OF ORGANIC CHEMICALS IN AQUATIC ENVIRONMENT

Natural aquatic systems – rivers, lakes, estuaries, oceans, and groundwater – are vast, complex, and dynamic compartments of the ecosystem. Each of them can be further divided into sub-compartments such as water (with dissolved species), aquatic plants (e.g. algae) and animals (e.g. fish and shell fish), suspended and settling particles in water, surface sediment (with mixing depth), and buried sediment. Of course, each of these sub-compartments can be divided again, and includes different constituents. Not only do these sub-compartments constantly interact with each other, but the entire aquatic system also exchanges both energy and matter with other major environmental compartments such as air, soil, and terrestrial animals and plants.

Once organic chemical pollutants enter an aquatic environment, their behavior is determined by (i) the intrinsic properties of the chemical, many of which have been discussed above, and (ii) the characteristics of the aquatic system, which can be described by numerous parameters, such as location (latitude and longitude), water body dimension, water recharge and discharge rates, temperature, water chemistry (e.g. alkalinity, dissolved oxygen, ionic strength and ion types), depth of sunlight penetration, composition and concentration of the suspended or settling particles in water, composition of the sediments, and biota species and population. These external factors act on the intrinsic nature of the chemicals, determining their fate.

Therefore, in contrast to the partition coefficients discussed in Section 7.4, most phase distribution ratios described in this section are "operationally defined" between "real world" phases that are not pure and are likely to be dynamic. In these cases, equilibrium partitioning in a thermodynamic sense may or may not be reached, but steady-states could be established within spatial and temporal boundaries.

7.5.1   Air–Water

The physicochemical property that is most relevant to air–water distribution of an organic compound is Henry's law constant (K_H), which largely depends on the vapor pressure ($P°$) and aqueous solubility (S_W) as discussed in Section 7.4.3. In reality, equilibrium between the atmosphere and a natural water body may not be achieved, due to constant advections of the chemical with the flowing air and water, the relatively slow diffusion of the chemical in both bulk phases, the barriers of transfer across the air–water interface, etc. In addition, Henry's law concerns only gaseous chemicals; whereas wet depositions of the pollutant chemical with rain and snow, and the dry deposition with settling particles, into water can be significant. Most of the theoretical and modeling concepts regarding these processes are discussed in Chapter 3 of this book.

7.5.2   Water–Sediment

The organic chemicals distribute between solid materials and water, and the concentration ratio at equilibrium is the solid–water distribution coefficient:

$$K_d = \frac{C_S}{C_W} \tag{7.13}$$

The unit of K_d is usually L/kg. Here, the word solid refers to particles or grains of soils, sediments, and suspended solids in water. These are not well-defined homogeneous "phases", rather, they are aggregates with complex compositions and structures.

The complex nature of K_d is discussed in detail by Schwarzenbach et al. (2003). First, the distribution involves both absorption (3-dimensional sorption into solid aggregates) and adsorption (2-dimensional sorption onto solid surface). Composition of the solid plays a critical role, as well as the nature of the chemical. For hydrophobic organic compounds, the organic matter of the solid is the determinative factor, and the K_d is related to the organic matter (OM) or organic carbon (OC) contents as

$$K_d \approx K_{OM} \times f_{OM} \approx K_{OC} \times f_{OC} \qquad (7.14)$$

where $K_{OM} = C_{OM}/C_W$ and $K_{OC} = C_{OC}/C_W$, and $f(< 1)$ is the mass fraction of the OM or OC in the solid. In natural organic matter, carbon is about 50% by weight; therefore, $f_{OM} \approx 2 \times f_{OC}$ and $K_{OM} \approx 0.5 \times K_{OC}$ for many soils and sediments. As a special type of organic carbon, soot or black carbon, which is formed from many combustion processes, has stronger sorption than organic carbon of natural origin, especially toward organic chemicals possessing planar molecular conformation such as PAHs and coplanar PCBs.

The experimental determination of K_{OM} and K_{OC} involves the measurements of C_S and C_W at equilibrium as well as f_{OM} or f_{OC} of the solid. The long time needed to reach equilibrium, and incomplete phase separation caused by colloidal particles and pore water, are among the difficulties of such experiments. Estimation of K_{OM} and K_{OC} depends almost exclusively on the linear correlation with other properties, primarily K_{OW}.

7.5.3   Water–Biota and Sediment–Biota

Bioconcentration factor (BCF) and bioaccumulation factor (BAF) are the ratios of chemical concentration in biota, which can be either animal or vegetation, to the concentration of the media the biota are in contact with, which can be either air or water

$$\text{BCF} = \frac{C_{biota}}{C_{media}} \qquad (7.15)$$

$$\text{BAF} = \frac{C_{biota}}{C_{media}} \qquad (7.16)$$

The difference between BCF and BAF is that bioconcentration refers to the process of true partitioning in which uptake of chemicals by the biota only takes place from the dissolved phase in the media; while bioaccumulation includes all routes of chemical uptake by biota. BCF can be used for biota for which diet is not an issue, such as phytoplankton and zooplankton. BAF, on the other hand, applies to all biota from fish to grass.

A similar parameter, biota sediment accumulation factor (BSAF), is used for biota living at the water–sediment interface, and is defined as

$$BSAF = \frac{C_{biota}}{C_{sediment}} \qquad (7.17)$$

A related parameter, biomagnification factor, deals with the changes in chemical concentration in various biota along a food chain. The "diet" in Equation 7.18 refers to the biota lower by one tropical level, and thus is the food for the "biota" under concern.

$$BMF = \frac{C_{biota}}{C_{diet}} \qquad (7.18)$$

All of these are "factors" rather than equilibrium constants, and do not require equilibrium condition to apply. In fact, reaching equilibrium between biota and media can be difficult or impossible due to slow kinetics, metabolism, and other practical factors. Theoretically valuable equilibrium constant between biota and water, K_{bio}, requires the knowledge of chemical partitioning between water and well-defined biological phases, such as lipid, proteins, and carbohydrates. Because all biota are complex structured "mixtures" of many different "phases" (e.g. a fish is a "mixture" of skin, liver, lipid, muscle, etc.), the term C_{biota} in these definitions (Equations 7.15 through 7.18) is usually the average concentration of the whole organism (e.g. a fish). In some cases, the major chemical depository "phase" of the biota (e.g. lipid of a fish) can also be used. For example, lipid-based BCF and BAF are related to the whole fish BCF and BAF by

$$BCF(lipid\ based) = BCF(whole\ fish)/f_{lipid} \qquad (7.19)$$

$$BAF(lipid\ based) = BAF(whole\ fish)/f_{lipid} \qquad (7.20)$$

where f_{lipid} is the mass fraction of lipid in the fish.

Correlation between these factors and K_{OW} is the most widely used approach for their estimation if experimental values are not available. Correlation equations are applicable only within the specified range of the independent variable, K_{OW}. They fail when K_{OW} is above the upper limit of the range, resulting in parabolic shaped curves. Such observations are likely a result of the hindrance of the biomembranes toward molecules that are large in size. For organic compounds that are relatively easily metabolized by organisms, dependence of these factors on the hydrophobicity of the chemicals is weak.

7.6   TRANSFORMATIONS IN WATER

All chemical transformation reactions concern two aspects: mechanisms and kinetics. While the two are related, the former emphasizes reaction pathways, which show what and how the products or a sequence of products are formed. Understanding the reaction mechanism is important for assessing the fate, ecological impact, and health effect of the chemical. On the other hand, kinetic studies investigate the rates with which a chemical "disappears", revealing its environmental persistence.

7.6.1 Hydrolysis

Hydrolysis is a "baseline" reaction for many inorganic and organic compounds in water. It occurs when the chemicals react with water (H_2O and OH^-). However, not all reactions with water take the form of hydrolysis. For example, we do not deem acid/base reactions initiated by contacting with water as hydrolysis. Addition of H_2O to unsaturated bonds to form alcohols is also not deemed hydrolysis.

Hydrolytic reactions are largely irreversible, in contrast to hydration. In general, hydrolysis is highly temperature dependent. The products are more polar and generally less persistent and harmful than the parent compound. From a wider perspective, hydrolysis is a specific type of nucleophilic substitution reaction, in which electron-rich nucleophiles (H_2O and OH^- in hydrolysis) attack an electron-deficit site in molecules of the chemical of interest, forming a new bond and breaking another; therefore causing an atom or a group of atoms to leave. There are numerous nucleophiles apart from water, some have stronger nucleophilicity but much lower abundance than water in the environment. Commonly found examples include NO_3^-, $SO_4^=$, Cl^-, HCO_3^-, NO_2^-, Br^-, HS^-, and $S_2O_3^=$, in the order of increasing nucleophilicity toward a saturated carbon (Schwarzenbach et al., 2003).

Not all organic compounds subject to hydrolysis. Pure hydrocarbons, including alkanes, alkenes, alkynes, benzene and biphenyl, and PAHs, are essentially inert under environmental conditions. Chlorinated aromatic hydrocarbons, such as chlorinated pesticides, PCBs, and PCDD/Fs, are also resistant to hydrolysis. Polar functional groups that are difficult to hydrolyze include alcohols, aromatic amines and nitro groups, phenols, glycols, ethers, aldehydes, ketones, carboxylic acids, and sulfonic acids (Lyman et al., 1990).

By comparison, compounds with functional groups, including amide, amines, epoxides, carbamates, nitriles, and alkyl halogens, are vulnerable toward hydrolysis under natural environmental conditions. Some of the structures are:

Ester   Amide   Carbonate   Carbamate   Urea

For example, pesticide carbamates may undergo a complex sequence of reactions to eventually form an alcohol and an amine:

Organophosphates can be attacked by water or other nucleophiles at either the phosphorus atom or the alkyl substitute group, forming different products:

$$
O=\overset{\overset{\textstyle O-R_1}{|}}{\underset{\underset{\textstyle O-R_2}{|}}{P}}-O-CH_2-R_3 \ + \ H_2O
\ \overset{\nearrow}{\underset{\searrow}{}}\
\begin{array}{l}
O=\overset{\overset{\textstyle O-R_1}{|}}{\underset{\underset{\textstyle O-H}{|}}{P}}-O-CH_2-R_3 \ + \ R_2OH \\[2em]
O=\overset{\overset{\textstyle O-R_1}{|}}{\underset{\underset{\textstyle O-R_2}{|}}{P}}-O-H \ + \ R_3CH_2OH
\end{array}
$$

Aliphatic halides, such as halogenated methane and ethane, are another group that is subject to hydrolysis and other nucleophilic substitution reactions. For example,

$$CH_3Br + H_2O \longrightarrow CH_3OH + H^+ + Br^-$$

The reaction rate is often independent of pH. When pH is greater than 10, however, OH^- may compete effectively with H_2O, thus base-catalyzed reaction becomes significant. The reaction rate constant is in an order that $R-Br > R-Cl > R-F$, as a result of increasing bond energy.

Due probably to steric hindrance, the hydrolysis of polyhalogenated methanes is often very slow, and thus not environmentally important. However, a non-hydrolytic reaction known as elimination or dehydrohalogenation, in which a halogen leaves the molecule taking a hydrogen with it, can be significant. For example,

$$CCl_3-CH_3 \longrightarrow CCl_2=CH_2 + H^+ + Cl^-$$

Observed rate of "disappearance" is often the sum of various reactions. The relative contribution of different reactions to the total removal is complex and depends not only on the chemical identity, but also the characteristics of the water in which it resides. Together, these reactions explain the dwindling of pollutants in groundwater as the contaminated water flows away from the input sources.

The rate constant for hydrolysis $k = k_H[H^+] + k_N + k_{OH}[OH^-]$, where k_H, k_N and k_{OH} are the rate constants for acid-catalyzed, neutral, and base-catalyzed hydrolytic reactions. Compiled experimental data of k_H, k_N and k_{OH} are available (Lyman et al., 1990; Wolfe and Jeffers, 2000; Schwarzenbach et al., 2003: Chapter 13). QSARs have been developed to predict ks. Well established QSARs include the Taft, Hammett and Bronsted Relationships. These models take a substitute-effect approach, predicting k values using empirical correlation equations relating to observed ks relative to that of a reference compound to the established substitution constants. Refer to Lyman et al. (1990), Wolfe and Jeffers (2000), Harris (1990), and Exner (1988) for more detailed discussions.

7.6.2 Photochemical Degradation

A portion of the sunlight that reaches the earth's surface has wavelength ranging from about 290 to 600 nm. The corresponding photo energy ranges from 400 to 200 kJ/mol, which is of the same order of magnitude as the energy of many covalent bonds in organic chemicals, including C–C, C–H, C–O, C–Cl, and C–Br. Therefore, sunlight provides energy sufficient to transform a wide range of organic compounds that are present in the atmosphere, on solid surfaces, and in the shallow regions of natural waters.

Photolysis of organic chemicals is a non-reversible reaction. The products are often more biodegradable and hydrolysable, but exceptions exist. Organic compounds can be decomposed via either direct or indirect photolysis.

Direct photolysis is a process initiated by the direct sorption of light by the chemical of interest, followed by the excitation of the molecules and transformation of their identity. According to the Stark–Einstein photochemical equivalence law, for every quantum of radiation that is absorbed, one molecule of the substance A is excited to become A*, an "excited" molecule, with some of its electrons being punched into antibonding orbitals. That is,

$$A + h\nu = A^*$$

Therefore, one mole of photon excites one mole of substances absorbing the light, and involves energy equal to $N_A h\nu$, where N_A is the Avogadro constant (6.022×10^{23}), h is the Planck constant (6.626×10^{-34} J s), and $\nu = c/\lambda$ is the frequency of the light with c and λ being the speed (3×10^8 m/s) and the wavelength of light respectively.

However, not all of the absorbed energy is used for chemical transformation. A large portion of the energy may be (i) released when the excited A* molecules deactivate to ground state via fluorescence or phosphorescence, (ii) consumed on providing additional kinetic energy for the molecule, or (iii) wasted as heat during collisions with other molecules. The efficiency of light sorption regarding photolysis is defined by reaction quantum yield, Φ, which is defined as the ratio of the moles that are transformed (lost) to the total moles absorbing the light.

Φ = number of molecules transformed/photons absorbed

= rate of chemical transformation (dC/dt)/rate of light absorption (I_a)

All the terms are functions of light wavelength λ. Both dC/dt and I_a can be determined experimentally. Values of Φ are typically less than 0.01 for organic compounds in water (Lyman et al., 1990). A Φ greater than 1 is possible and indicates the induction of chain reactions.

The chemical structure of organic compounds determines the energy intervals of its electron orbitals, and thus the wavelength of light it absorbs. Photolysis of most saturated aliphatic hydrocarbons and simple oxygenated compounds, such as alcohols, ethers, acids, and esters, are ignored because they do not absorb near UV and visible lights. Light with wavelength 290 to 600 nm can be absorbed by compounds with conjugated π bonds, which may interact with the non-bonded electrons of heteroatoms such as oxygen and nitrogen. Many of the pollutant groups discussed in this chapter are able to be photodegradated, as can be seen from their structures. The extent of π bond conjugation largely determines the light sorption spectrum. For example, adding an additional double bond in a straight chain of double bonds (polyenes) increases the wavelength of light with the strongest absorption (λ_{max}) by about 30 nm (so called "red shift" or bathochromic shift). Compared with single ring benzene, PAHs have longer λ_{max}. Among PAHs, the range of absorption spectra extends to longer λ regions with the increases in the number of rings; therefore, in general, larger PAHs are more vulnerable to photolysis. Similarly, attaching a functional group, such as $-CH = CH_2, -N = N-, -NO_2, -NH_2, -OH$, which are capable of providing additional π electrons participating and strengthening the conjugated bonds in aromatic ring systems, also causes red shift. The dependence of light absorption on the conjugation of π bonds may be explained by the reduced energy intervals or gaps between the ground and excited states (the difference between the lowest unoccupied and the highest occupied molecular orbitals, $E_{LUMO} - E_{HOMO}$); therefore the molecule can be excited with energy provided by light with longer wavelength. Light absorption spectra and values of λ_{max} can be found in the literature for a variety of organic pollutants.

Indirect, or sensitized, photolysis starts with a non-target compound (sensitizer), which absorbs light energy, gets excited, and passes the extra energy to the chemical of interest causing it to be transformed. Many types of natural organic matter as well as inorganic species such as NO_3^-, NO_2^-, and metal complexes in water are such sensitizers. They absorb light and produce highly reactive oxygen-containing photo-oxidant radicals, including HO, HOO, RO, ROO, and singlet oxygen $^1O_2$. These photo-oxidants will then react with organic compounds in water causing indirect photolytic transformation of these compounds. However, the light-excited sensitizers may also pass the extra energy to other species in solution in a process called quenching. Quenching is a non-photolytic deactivation of excited target molecules; thus reducing their reactivity. In aerobic aqueous solutions, molecular oxygen, O_2, can be an effective quencher.

The intensity of solar radiation reaching the earth's surface is affected by time of the year and time of the day (which determine the incidence angle), location (latitude and altitude), air pressure and temperature, cloud coverage, wind, humidity and precipitation, etc. The photon fluxes of the radiation are tabled in the literature, for example, at noon on a midsummer's day at 40 °N latitude. In addition to the light intensity, the status of the water bodies, such as depth of light penetration, types and concentrations of NOM species, temperature and pH, also affect the rate of photolysis of organic pollutants in water. Molecules of dissolved organic matter have various chromophores, making them strong absorbers of light with $\lambda = 290 - 600$ nm, especially the UV lights (λ less than 400 nm). Such absorption causes attenuation of such lights with increasing depth of water and reduces the direct photolysis of organic pollutants of interest dissolved in water. For quantitative estimation of the reaction rate constants for direct and indirect photolysis, we refer to other literature sources (Schwarzenbach et al., 2003). It is worthy of note that many published studies on photolysis of hydrophobic compounds, many of which are discussed in this chapter, were performed in organic solvent(s) or solvent:water mixtures as a result of restrictions of aqueous solubilities. Caution is needed when extrapolating the reaction rate data obtained in laboratories to natural waters.

7.6.3 Biological Degradation

Biological degradation of organic compounds refers to transformations mediated by living organisms, animals, or plants. The pathways of biodegradation include hydrolysis, oxidation, reduction, and addition. Complete degradation of organic compounds, resulting in final products such as stable inorganic forms of carbon, oxygen, nitrogen, or hydrogen, is known as mineralization and often needs multiple-stage and/or multiple-step degradations. Biological mineralization of organic substances is usually accomplished by microbes, because higher animals tend to excrete most organic compounds they take up, with only limited transformations.

In aquatic systems, degradation mediated by microbial organisms is the major pathway of natural elimination for many organic pollutants, including those that are highly persistent. Microorganisms include bacteria, fungi, protozoa, and microalgae. They exist ubiquitously in the environment even under harsh conditions. The type, diversity, abundance, metabolic activities, and growth rate of microorganisms are strongly affected by environmental conditions such as temperature, pH, oxygen content, salinity, and substrate and nutrient availability. Biological activity is generally high in swamps, wetlands, and sediment–water interfaces of lakes and rivers. In water columns, microbes are often attached to suspended or settling particles, and flourish if the organic matter content of the particles is high. The flow rate of water in rivers may negatively affect microbial growth. By comparison, microbial activities are much lower in oceans, although some microbes and other animal species prosper in the surface water where sunlight penetrates and at certain depths, including the very dark and cold deep water.

Virtually all activities inside cells take place through the mediation of enzymes. Enzymes are a special type of protein. They combine with the substrate molecules to form a complex, from which the degradation products form and are released. As a result, enzymes play a role of catalyst, reducing the activation energy and thus increasing the reaction rate by as much as 10^9 times, and making many thermodynamically favored, but otherwise kinetically inhibited, reactions occur. Enzymes are either extracellular (functions outside the cell from which it originates; for example, digestive enzymes) or intracellular (remains and functions within the cell from which it originates). In the context of biotransformation of substrate chemicals, enzymes are either constitutive (always exist) or inducible (triggered by chemicals). They are highly specific to substrate, and thus often take time to develop their ability in handling new substrate chemicals. They are also highly selective towards reactions, and thus catalyze only a few among many possible degradation pathways. Some enzymes need assistance from cofactors, non-protein molecules that may be tightly bound to the enzyme, to fully function. These cofactors can be inorganic or organic, and they are not altered or released during the reaction. Coenzymes are small molecules that are chemically altered during the reaction, but are able to regenerate to maintain steady-state levels in the cell. The function of coenzymes is to transport chemical groups (e.g. methyl or acetyl group) from one enzyme to another. Enzymatic activities can be prohibited by various toxic inhibitors, which act either via competing with the substrate by blocking the reaction site in the enzyme or via alteration of the enzyme through reversible or irreversible binding to the enzyme. Enzymes are also temperature sensitive, and organisms that thrive at temperature ranges $0 - 10\,°C$, $10 - 40\,°C$, and greater than $40\,°C$ are known as psychrophilic, mesophilic, and thermophilic respectively.

Some organic compounds are chiral, meaning that they have two optical isomers known as enantiomers, which have mirror images to each other and rotate polarized light in opposite directions but are otherwise identical. Numerous pesticides are chiral compounds; for example 2-(2,4-dichlorophenoxy)propanoic acid has two enantiomers and the R-isomer is the herbicide dichlorprop (Table 7.6) while the S-isomer has no weed-killing ability. Among 209 PCB congeners, 78 are possibly chiral, whereas only 19 have stable conformations (Kaiser, 1974). Almost all chiral pesticides and PCBs have been manufactured with equal amounts of their two enantiomers, i.e. racemic mixtures with an enantiomer fraction (EF) of 0.5 or enantiomer ratio (ER) of 1, although the recent advancement in technology has made asymmetric synthesis, and thus enantiopure production, possible. The racemic EF or ER is altered in biological systems because the interactions of chiral compounds with the macromolecules involved in the metabolism of organic substances, such as cytochrome P-450 enzymes, are often enantioselective. With the chromatographic separation of enantiomers, non-racemic EF values (> 0.5 or < 0.5) of many chiral compounds have been found in various environmental samples, including soil and sediment, natural and treated waters, and plant and animals, as well as human samples such as tissues, milk, and feces, as reviewed by Lehmler et al. (2010). The chemical chirality provides a powerful tool not only in toxicological studies, but also in research on biological transformation pathways, foodweb alterations, pollution source identification and apportionment, effective bioremediation of contaminated sites, and many others. For example, Robson and Harrad (2004) reported non-racemic EFs for three chiral PCBs (95, 136, and 149) in the topsoils in the Midlands, UK, while the PCBs in the air above the soil were essentially racemic; suggesting the PCBs in the air came from racemic sources rather than evaporated from the contaminated soil. The ERs measured for PCBs 91, 95, 132, 136, 149, 174, and 176 in Lake Hartwell, SC, not only confirmed previous inconclusive reports of reductive dechlorination of PCBs but also assisted in proposing the dechlorination pathway (Wong et al., 2001).

The rate of biotransformation is therefore limited by various processes. First, the chemical of interest must meet the microbes, and this often involves the transport of both in the aqueous media, as well as other reactant chemicals (e.g. oxygen in oxidation reactions) involved in the reaction. Bioavailability may explain the observation that highly hydrophobic aromatic hydrocarbons are in general oxidized more slowly than less hydrophobic ones, as a result of their relatively strong sorption to solids and slow diffusion in media. Once the substrate molecule and the microbes coexist, the process of microbial transformation may begin in a sequence containing (i) uptake, (ii) binding by enzymes, (iii) transformation reactions, and (iv) product(s) release. The uptake is the approaching and sorption of the substrate chemical to the cell wall and subsequent diffusion through the cell membranes to the reaction site. The membranes are lipid-rich, separating the charged substances within the cell from the polar aqueous medium, and allowing non-polar chemicals to diffuse through into the interior of the cell. To the microbes, the uptake is often passive as the cells do not usually have the ability to pick up and deliver the substrate actively and rapidly. Once in contact, the substrate molecule and the enzyme interact, inducing genetic mutation or activating existing enzymes. The transformation process starts when the enzyme and the substrate react to form a complex, which then dissociates to yield product and free enzyme. Finally, the metabolism of substrate stimulates microbial population growth, which eventually determines the overall rate of biodegradation.

It is a difficult task to predict the rates of biodegradation among various chemicals or chemical groups based on their chemical structures and/or individual physic-ochemical properties. In general, hydrocarbons, in which the carbon atoms have oxidation states of less than zero, are vulnerable in an oxidative environment but resistant to reduction. Highly halogenated compounds, on the other hand, tend to be reductively dehalogenated. Assuming uptake is kinetically limiting, structural parameters or properties that affect diffusive transport and represent hydrophobicity can be important. For example, more hydrophobic chemicals (with higher K_{ow}) may be easier to attach (adsorb) to the lipid-rich biomembrane, and thus have a higher degradation rate when the adsorption is the rate-determining step. On the other hand, large molecules may have difficulties penetrating the cell membrane, and thus cannot compete where diffusion through membrane is the critical uptake step. Sometimes, the rate of mineralization can be experimentally measured and stoichiometrically quantified by the generation of CO_2, but the rate determined in this way can be lower than the overall rate of substrate transformation.

7.6.4   Case Study: Transformation of PBDEs in the Environment

Take polybrominated aromatic compounds as an example, many of which are brominated flame retardants that are still in use and of high emerging concern as discussed earlier in this chapter. The C-Br bond (mean bond energy = 276 kJ/mol) is much weaker than C-Cl bond (mean bond energy = 338 kJ/mol), suggesting organobromine compounds such as PBBs and PBDEs are more vulnerable to dehalogenation compared with their counterparts, PCBs and PCDEs, under the same conditions. In fact, experiments of PCB degradation often use polybromobiphenyls (PBBs) as dehaloprimers, because PBBs are readily debrominated at high rates in sediment enrichments contaminated with PCBs (Wu et al., 1999; Bedard et al., 1998).

Our knowledge and experience with PCBs have indeed been valuable in guiding research on the fate of PBDEs in the environment. Extensive in situ dechlorination has occurred for PCBs in heavily contaminated sediment. In the Hudson River, for example, dechlorination of PCBs was found to be significant in the 1980s, less than three decades after PCB releases began from GE facilities into the river

(Brown et al., 1987, 1988; Quensen et al., 1988). Since then, PCB dechlorination in sediment has been observed in many other waters. Apparently, under the anoxic conditions prevalent in organic-rich sediments, PCBs dechlorinate over time. As high concentrations stimulate microbial growth, PCB dechlorination is often observed at "hot spots" (Brown et al., 1988). The reductive dechlorination of highly chlorinated PCBs is primarily observed under anaerobic conditions.

Research to date has provided strong evidence of PBDE debromination via chemical, photochemical, and biological processes. We have learned: (i) although only about twelve congeners are present as major components in PBDE technical mixtures, numerous congeners may form as debromination products, resulting in complex congener patterns; (ii) PBDE debromination appears to be stepwise: BDE209 debrominates into three nona-BDEs which in turn produce five major and a couple of minor (out of 12 theoretically possible) octa-BDEs. Then, many lower congeners may form; (iii) kinetically, PBDE debromination appears to be first order, and the reaction rate is higher for more highly brominated congeners; (iv) debromination into lower PBDE congeners may occur concurrently with other types of transformation reactions for a particular "mother" congener or a mixture of mother congeners, resulting in a complicated set of reaction products which may include polybrominated dibenzofurans, phenols, and other compounds; and (v) the reaction pathway, which is defined by Wei et al. (2013) as the bromine substitution patterns of the "daughter" congener that is mostly likely to form during a sequential debromination of a PBDE congener, could be similar among various abiotic debromination reactions, but is generally different from that of PCBs.

Reductive dehalogenation is widely recognized as the initial and rate-limiting step towards mineralization of halogenated organics. In the context of this book, the in situ transformations of PBDEs in aquatic systems, especially in the sediments, are particularly important. Based on the previous observations on PCBs, it is reasonable to expect that PBDEs, which have been accumulated in aquatic sediments for over four decades, have debrominated in situ in heavily contaminated sediments. Although microbial debromination has been reported under laboratory and controlled (e.g. in wastewater treatment facilities) conditions (La Guardia et al., 2007), little is known regarding whether in situ PBDE transformation has occurred, and if so, to what extent and how fast. The efforts of Li and her co-workers (Wei et al., 2012; Aziz-Schwanbeck et al., in preparation; Zou et al., 2013) have provided some insights. In a sediment core collected from a previously used wastewater sludge retention pond, the researchers examined the relative abundance of PBDE homolog groups as well as some congener ratios, and found interesting trends suggesting two zones over the sediment depth. In the upper sediment, the increase in the relative abundance of the lower congeners could not be explained by the change in input, but likely resulted from a complex combination of reactions, including direct and indirect photolysis and microbial mediated redox reactions. In the deeper sediment of the same water, the fractions of the octa and nona homologs rose relative to BDE209, and a peak of the penta-to-hepta homolog group was observed at depth of 30–50 cm, in agreement with the expectation that in situ debromination should be more extensive under the anaerobic conditions in deeper sediment (Wei et al., 2012).

However, based on the current research, microbially mediated in situ debromination of PBDEs in contaminated sediments appears not to be as extensive as predicted based solely on the high levels of contamination, the long time span since the input started, and the weak C-Br bond relative to the C-Cl bond in PCBs. Biodegradation reactions can be thermodynamically feasible but kinetically hindered. The low bioavailability of the strongly sorbed PBDE molecules, the lack of sufficient electron donors, as well as the competition among various substances capable of accepting electrons may also hinder the transformation of PBDEs in sediment. It is also possible that the relevant microorganisms are not yet sufficiently adapted and activated, resulting in slow rates even at high contamination levels (Wei et al., 2012).

REFERENCES

Andersson, O., Blomkvist, G. Polybrominated Aromatic Pollutants Found in Fish in Sweden. *Chemosphere* 1981;10:1051–1060.

ATSDR, 2012. *Toxicological Profile for Phosphate Ester Flame Retardants. Agency for Toxic Substances and Disease Registry*, U.S. Department of Health and Human Services. September 2009.

ATSDR, 1995. *Toxicological Profile for Polycyclic Aromatic Hydrocarbons (PAHs).* U.S. Department of Health and Human Services. August 1995.

Aziz-Schwanbeck, AC., Wei, H, Zou, Y, Li, A, Rockne, KJ., Christensen, ER. Distribution of Polybromodiphenyl Ethers and Polychlorinated Biphenyls in Aquatic Sediments from Chicago, Illinois and the West Branch of The Grand Calumet River, Indiana. in preparation.

Ballschmiter, K, Zell, M.. Analysis of polychlorinated biphenyls (PCB) by glass capillary gas chromatography. *Fresenius Z. Anal. Chem.* 1980;302:20–31.

Baum, EJ. *Chemical Property Estimation: Theory and Application.* Lewis Publishers; 1998.

Bedard, DL, Van Dort, HM, DeWeerd, KA. Brominated Biphenyls Prime Extensive Microbial Reductive Dehalogenation of Aroclor 1260 in Housatonic River Sediment. *Appl. Environ. Microb.* 1998;64:1786–1795.

Benotti, MJ., Trenholm, RA, Vanderford, BJ, Holady, JC, Stanford, BD, Snyder, SA. Pharmaceuticals and Endocrine Disrupting Compounds in U.S. Drinking Water. *Environ. Sci. Technol.* 2009;43(3):597–603.

Birnbaum, LS, Staskal, DF. Brominated flame retardants: cause for concern? *Environ. Health Perspect.* 2004;112:9–17.

Blum, A, Gold, MD, Ames, BN, Kenyon, C, Jones, FR, Hett, EA, Dougherty, RC, Horning, EC, Dzidic, I, Carroll, DI, Stillwell, RN, Thenot, J-P. Children Absorb Tris-BP Flame Retardant from Sleepwear: Urine Contains the Mutagenic Metabolite, 2,3-Dibromopropanol. *Science.* 1978;201(4360):1020–1023.

Boethling, RS, Cooper, JC. Environmental fate and effects of triaryl and tri-alkyl/aryl phosphate esters. *Res Rev.* 1985;94:49–99.

Boethling, RS, Mackay, D, *Handbook of Property Estimation Methods for Chemicals*, Lewis Publisher; 2000.

Boxall, ABA, Rudd, MA, Brooks, BW, Caldwell, DJ, Choi, K, Hickmann, S, Innes, E, Ostapyk, K, Staveley, JP, Verslycke, T, et al. Pharmaceuticals and personal care products in the environment: what are the big questions? *Environ. Health Perspectives* 2012;120(9):1221–1229.

Brommer, S, Harrad, S, Van den Eede, N, Covaci, A. Concentrations of organophosphate esters and brominated flame retardants in German indoor dust samples. *J Environ. Monitoring* 2012;14(9):2482–2487.

Brown Jr, JF, Bedard, DL, Brennan, MJ, Carnahan, JC, Feng, H, Wagner, RE. Polychlorinated biphenyl dechlorination in aquatic sediments. *Science.* 1987;236:709–712.

Brown, MP, Bush, B, Rhee, GY, Shane, L. PCB Dechlorination in Hudson River Sediment. *Science.* 1998; New Series, 240:1674–1675.

CAS. 2013. *Organic and Inorganic Substances* To Date. http://www.cas.org/ (accessed April 15, 2013).

Dannenfelser, R-M, Yalkowsky, SH. Database of Aqueous Solubility for Organic Non-electrolytes. *Sci. Total Environ.* 1991;109/110:625–628.

Dassanayake, RMA PS, Wei, H, Chen, RC, Li, A. Optimization of the matrix solid phase dispersion extraction procedure for the analysis of polybrominated diphenyl ethers in human placenta. *Anal. Chem.* 2009;81(23):9795–9801.

Daughton, CG. Groundwater Recharge and Chemical Contaminants: Challenges in Communicating the Connections and Collisions of Two Disparate Worlds. *Ground Water Monitoring & Remediation.* 2004;24(2):127–138.

Daughton, CG, Jones-Lepp, TL. *Pharmaceuticals and Personal Care Products in the Environment: Scientific and Regulatory Issues.* ACS/Oxford University Press; 2001.

Daughton, CG, Ternes, TA. Pharmaceuticals and Personal Care Products in the Environment: Agents of Subtle Change? *Environmental Health Perspectives.* 1999;107, Supplement 6:907–938.

Daylight. 2006. *CLogP Online.* Daylight Chemical Information Systems, Inc., Aliso Viejo, CA. http://www.daylight.com/daycgi/clogp. Accessed in January 2006.

DeCarlo, VJ. Studies on Brominated Chemicals in the Environment. *Ann. N. Y. Acad. Sci.* 1979;320:678–681.

Dodson, RE, Perovich, LJ, Covaci, A, Van den Eede, N, Ionas, AC, Dirtu, AC, Brody, JG, Rudel, RA. After the PBDE Phase-Out: A Broad Suite of Flame Retardants in Repeat House Dust Samples from California. *Environ. Sci. Technol.* 2012;46:13056–13066.

Exner, O. *Correlation Analysis of Chemical Data.* Plenum Press; 1988.

Giesy, JP, Kannan, K. Global Distribution of Perfluorooctane Sulfonate in Wildlife. *Environ. Sci. Technol.* 2001;35(7):1339–1342.

Gold, MD, Blum, A, Ames, BN. Another Flame Retardant, Tris-(1,3-Dichloro-2-Propyl)-Phosphate, and Its Expected Metabolites Are Mutagens. *Science.* 1978;200(4343): 785–787.

Hansch, C, Leo, AJ. *Substituent Constants for Correlation Analysis in Chemistry and Biology.* New York: Wiley; 1979.

Hansch, C, Leo, A, Heller, SR. *Fundamentals and Applications in Chemistry and Biology.* ACS, Washington, DC; 1995.

Hardy, ML. The toxicology of the three commercial polybrominated diphenyl oxide (ether) flame retardants. *Chemosphere* 2002;46:757–777.

Harris, JC. Rate of Hydrolysis. In: *Handbook of Chemical Property Estimation Methods.* Lyman, WJ, Reehl, WF, Rosenblatt, DH. Washington, DC: American Chemical Society; 1990.

Harvey, RG. *Polycyclic Aromatic Hydrocarbons: Chemistry and Carcinogenicity.* Cambridge: Cambridge University Press; 1991.

Harvey, RG. *Polycyclic Aromatic Hydrocarbons.* New York: Wiley-VCH; 1997.

Hayashi, Y, Managaki, S, Takada, H. Fluorescent Whitening Agents in Tokyo Bay and Adjacent Rivers: Their Application as Anthropogenic Molecular Markers in Coastal Environments. *Environ. Sci. Technol.* 2002;36(16):3556–3563.

Hermanson, MH, Isaksson, E, Forsström, MS, Teixeira, C, Muir, DCG, Pohjola, VA, van de Wal, RSV. Deposition history of brominated flame retardant compounds in an ice core from Holtedahlfonna, Svalbard, Norway. *Environ. Sci. Technol.* 2010;44:7405–7410.

Hites, RA. Polybrominated Diphenyl Ethers in the Environment and in People: A Meta-Analysis of Concentrations. *Environ. Sci. Technol.* 2004;38(4):945–956.

Houde, M, De Silva, AO, Muir, DCG, Letcher, RJ. Monitoring of Perfluorinated Compounds in Aquatic Biota: An Updated Review. *Environ. Sci. Technol.* 2011;45(19):7962–7973.

Howard, PH, Muir, DCG. Identifying New Persistent and Bioaccumulative Organics Among Chemicals in Commerce. *Environ. Sci. Technol.* 2010;44(7):2277–2285.

Howard, PH, Muir, DCG. Identifying New Persistent and Bioaccumulative Organics Among Chemicals in Commerce II: Pharmaceuticals. *Environ. Sci. Technol.* 2011;45(16):6938–6946.

IPCS (International Programme on Chemical Safety). InChem. 1998. *Triazine Herbicides*. http://www.inchem.org/documents/pims/chemical/pimg013.htm (accessed on Jan. 21, 2007).

IUPAC 2007. *IUPAC-NIST Solubility Database, Version 1.0: NIST Standard Reference Database 106*. http://srdata.nist.gov/solubility/ (accessed May 10, 2007).

Kaiser, K. On the optical activity of polychlorinated biphenyls. *Environ. Pollut.* 1974;7:93–101.

Kannan, K, Koistinen, J, Beckmen, K, Evans, T, Gorzelany, JF, Hansen, KJ, Jones, PD, Helle, E, Nyman, M, Giesy, JP. Accumulation of Perfluorooctane Sulfonate in Marine Mammals. *Environ. Sci. Technol.* 2001;35(8):1593–1598.

Kerst AF. Toxicology of tris(2,3-dibromopropyl) phosphate. *J. Fire Flammabil. Fire Retard Chem.* 1974;1:205–217.

Kier, LB, Hall, LH. *Molecular Connectivity in Chemistry and Drug Research*. New York: Academic Press; 1976.

Kolpin, DW, Furlong, ET, Meyer, MT, Thurman, EM, Zaugg, SD, Barber, LB, Buxton, HT. Pharmaceuticals, Hormones, and Other Organic Wastewater Contaminants in U.S. Streams, 1999–2000: A National Reconnaissance. *Environ. Sci. Technol.* 2002;36:1202–1211.

Kümmerer, K. *Pharmaceuticals in the Environment - Sources, Fate, Effects and Risks*. 2nd Edition, United Kingdom: CPL Press; 2004.

La Guradia, MJ, Hale, RC, Harvey, E. Evidence of Debromination of Decabromodiphenyl Ether (BDE-209) in Biota from a Wastewater Receiving Stream. *Environ. Sci. Technol.* 2007;41(19):6663–6670.

Lehmler, H-J, Harrad, SJ, Hühnerfuss, H, Kania-Korwel, I, Lee, CM, Lu, Z, Wong, CS. Chiral Polychlorinated Biphenyl Transport, Metabolism, and Distribution: A Review. *Environ. Sci. Technol.* 2010;44(8):2757–2766.

Li, A, Andren, AW, Yalkowsky, SH. Choosing a Cosolvent: Solubilization of Naphthalene and Cosolvent Property. *Environ. Toxicol. Chem.* 1996;15:2233–2239.

Li, A, Pinsuwan, S, Yalkowsky, SH. Estimation of Solubility of Organic Compounds in 1-Octanol. *Ind. Eng. Chem. Res.* 1995;34(3):915–920.

Li, A, Rockne, KJ, Sturchio, N, Song, W, Ford, JC, Buckley, DR, Mills, WJ. Polybrominated diphenyl ethers in the sediments of the Great Lakes. 4. Influencing factors, trends, and implications. *Environ. Sci. Technol.* 2006;40:7528–7534.

Li, A, Schoonover, TM, Zou, Q, Norlock, F, Conroy, LM, Scheff, PA. Polycyclic aromatic hydrocarbons in residential air of ten Chicago area homes: concentrations and influencing factors. *Atmos. Environ.* 2005;39(19):3491–3501.

Lyman, WJ, Reehl, WF, Rosenblatt, DH. *Handbook of Chemical Property Estimation Methods.* American Chemical Society; 1990.

Lyman, WJ, Reehl, WF, Rosenblatt, DH, (Ed.). *Handbook of Chemical Property Estimation Methods: Environmental Behavior of Organic Compounds.* McGraw-Hill; 1982.

Mackay, D, Shiu, WY, Ma, KC, Lee, SC. *Handbook of Physical-Chemical Properties and Environmental Fate for Organic Chemicals*, Vol I–IV. Second Edition. CRC Press; 2006.

Mackay, D, Shiu, WY, Ma, KC. *Illustrated Handbook of Physical-Chemical Properties and Environmental Fate for Organic Chemicals*, Vol I–IV. Boca Raton, Florida: Lewis Publishers/CRC Press, Inc.; 1991–1997.

Mason, B. River fish accumulate human drugs. *Nature.* 2003. http://www.nature.com/nsu/031103/031103-8.html and http://www.bioedonline.org/news/news.cfm?art=671 (accessed on Jan. 8, 2007).

Mayer FL, Adams WJ, Finley MT, et al. Phosphate ester hydraulic fluids: An aquatic environmental assessment of Pydrauls 50E and 115E. In: Branson DR, Dickson KL, editors. *Aquatic toxicology and hazard assessment.* American Society for Testing Materials (ASTM). Vol. STP 737; 1981.

McGowan, JC, Mellors, A. *Molecular Volumes in Chemistry and Biology.* Ellis Horwood; 1986.

McLearn, M. *The PCB Information Manual Volume 1: Production, Uses, Characteristics, and Toxicity of PCBs. Report TR-114091-V1.* Vol. 1. Pleasant Hill, CA: Electric Power Research Institute, Inc.; 1999.

Meyer, T, Muir, DCG, Teixeira, C, Wang, X, Young, T, Wania, F. Deposition of brominated flame retardants to the Devon Ice Cap, Nunavut, *Canada. Environ. Sci. Technol.* 2012;46(2):826–833.

Meylan, WM, Howard, PH. Atom/fragment contribution method for estimating octanol–water partition coefficients. *J. Pharm. Sci.* 1995;84:83–92.

Mills III, SA, Thal DI, Barney, J. A summary of the 209 PCB congener nomenclature. *Chemosphere.* 2007;68:1603–1612.

Muir, DCG, Howard, PH. Are There Other Persistent Organic Pollutants? A Challenge for Environmental Chemists. *Environ. Sci. Technol.* 2006;40(23):7157–7166.

Muir, DCG. Phosphate esters. In: Hutzinger, O (editor). *The Handbook of environmental chemistry. Vol. 3. Part C. Anthropogenic compounds.* New York, NY: Springer-Verlag; 1984, pp. 41–66.

Muir, DCG, Grift, NP. Environmental dynamics of phosphate esters. II. Uptake and bioaccumulation of 2-ethylhexyl diphenyl phosphate and diphenyl phosphate by fish. *Chemosphere* 1981;10(8):847–855.

Muir, DCG, Grift, NP, Blouw, AP, Lockhart, WL. Environmental dynamics of phosphate esters. I. Uptake and bioaccumulation of triphenyl phosphate by rainbow trout. *Chemosphere* 1980;9(9):525–532.

Muir, DCG, Yarechewski, AL, Grift, NP. Environmental dynamics of phosphate esters. III. Comparison of the bioconcentration of four triaryl phosphates by fish. *Chemosphere* 1983;12(2):155–166.

Muir, DCG, Yarechewski, AL, Grift, NP. Biodegradation of four triaryl/alkyl phosphate esters in sediment under various temperature and redox conditions. *Toxicol. Environ. Chem.* 1989;18:269–286.

Myrdal, PB, Ward, GH, Simamora, P, Yalkowsky, SH. AQUAFAC: Aqueous Functional Group Activity Coefficients. *SAR QSAR Environ. Res.* 1993;1:53–61.

Nanes, JA, Xia, Y, Dassanayake, RMAPS, Li, A, Jones, RM, Stodgell, CJ, Walker, CK, Szabo, S, Leuthner, S, Durkin, MS, Moye, J, Miller, RK. Selected Persistent Organic Pollutants in Human Placental Tissue from the United States. *Chemosphere*, 2014;106:20–27.

NIEHS, 2012. Bisphenol A (BPA). http://www.niehs.nih.gov/health/topics/agents/sya-bpa/. Nov. 5, 2012 (accessed on April 28, 2013).

Nirmalakhandan, NN, Speece, RE. Prediction of Aqueous Solubility of Organic Chemicals based on Molecular Structure. *Environ. Sci. Technol.* 1988;22:325–338.

Nirmalakhandan, NN, Speece, RE. Prediction of Aqueous Solubility of Organic Chemicals based on Molecular Structure. 2. Applications to PNAs, PCBs, PCDDs, etc. *Environ. Sci. Technol.* 1989;23:708–713.

OECD, 2007. *OECD Guidelines for the Testing of Chemicals. Section 1: Physical Chemical Properties.* http://www.oecd.org/document/40/0,2340,en_2649_37465_37051368_1_1_1_37465,00.html (accessed on March 8, 2007).

Palm, A, Cousins, IT, Mackay, D, Tysklind, M, Metcalfe, C, Alaee, M. Assessing the environmental fate of chemicals of emerging concern: a case study of the polybrominated diphenyl ethers, *Environ. Pollut.* 2002;117(2):195–213.

Paul, AG, Jones, KC, Sweetman, AJ. A First Global Production, Emission, And Environmental Inventory For Perfluorooctane Sulfonate. *Environ. Sci. Technol.* 2009;43(2):386–392.

Pinsuwan, S, Li, A, Yalkowsky, SH. Correlation of octanol/water solubility ratios and partition coefficients. *J. Chem. Eng. Data* 1995;40(3):623–626.

Prival, MJ, McCoy, EC, Gutter, B, Rosenkranz, HS. Tris(2,3-Dibromopropyl) Phosphate: Mutagenicity of a Widely Used Flame Retardant. *Science* 1997;195(4273):76–78.

Quensen, JF, Tiedje, JM, Boyd, SA. Reductive dechlorination of polychlorinated biphenyls by anaerobic microorganisms from sediments. *Science* 1988;242:752–754.

Raloff, J. Drugged Waters: Does it matter that pharmaceuticals are turning up in water supplies? *Science News Online.* March 21, 1998. (accessed on Jan. 8, 2007).

Randic, M, 1975. On Characterization of Molecular Branching. *J. Am. Chem. Soc.* 1975;97:6609–6615.

Reid, RC, Pransnitz, JM, Poling, BE. *The Properties of Gases and Liquids*, 3rd ed. New York: McGraw Hill; 1987.

Robson, M, Harrad, S. Chiral PCB signatures in air and soil: implications for atmospheric source apportionment. *Environ. Sci. Technol.* 2004;38:1662–1666.

Rosi-Marshall, EJ, Royer, TV. Pharmaceutical Compounds and Ecosystem Function: An Emerging Research Challenge for Aquatic Ecologists. *Ecosystems* 2012;15(6):867–880.

Sasaki, K, Suzuki, T, Takeda, M, et al. Bioconcentration and excretion of phosphoric-acid trimesters by killifish Oryzias-latipes. *Bull. Environ. Contam. Toxicol.* 1982;28(6):752–759.

Schecter, A, Pavuk, M, Papke, O, Ryan, JJ, Birnbaum, L, Rosen, R. Polybrominated diphenyl ethers (PBDEs) in U.S. mothers' milk. *Environ. Health Perspect.* 2003;111:1723–1729.

Schwarzenbach, RP, Gschwend, PM, Imboden, DM. *Environmental Organic Chemistry*. 2nd Ed. New York: John Wiley & Sons, Inc.; 2003.

Sjödin, A, Jones RS, Focant JF, Lapeza C, Wang RY, McGahee EE III, et al. Retrospective time-trend study of polybrominated diphenyl ether and polybrominated and polychlorinated biphenyl levels in human serum from the United States. *Environ Health Perspect.* 2004;112(6):654–658.

Song, W, Ford, JC, Li, A, Mills, WJ, Buckley, DR, Rockne, KJ. Polybrominated diphenyl ethers in the sediment of the Great Lakes. 1. Lake Superior. *Environ. Sci. Technol.* 2004;38:3286–3293.

Song, W, Ford, JC, Li, A, Sturchio, NC, Rockne, KJ, Buckley, DR, Mills, WJ. Polybrominated diphenyl ethers in the sediment of the Great Lakes. 3 - Lakes Erie and Ontario. *Environ. Sci. Technol.* 2005b;39(15):5600–5605.

Song, W, Li, A, Ford, JC, Sturchio, NC, Rockne, KJ, Buckley, DR, Mills, WJ. Polybrominated diphenyl ethers in the sediment of the Great Lakes. 2. Lakes Michigan and Huron. *Environ. Sci. Technol.* 2005a;39(10):3474–3479.

Staikova, M, Wania, F, Donaldson, DJ. Molecular polarizability as a single-parameter predictor of vapour pressures and octanol–air partitioning coefficients of non-polar compounds: a priori approach and results. *Atmos. Environ.* 2004;38:13–225.

Stapleton, HM, Klosterhaus, S, Eagle, S, Fuh, J, Meeker, JD, Blum, A, Webster, TF. Detection of Organophosphate Flame Retardants in Furniture Foam and U.S. House Dust. *Environ. Sci. Technol.* 2009;43(19):7490–7495.

Stapleton, HM, Klosterhaus, S, Keller, A, Ferguson, P.L, van Bergen, S, Cooper, E, Webster, TF, Blum, A. Identification of Flame Retardants in Polyurethane Foam Collected from Baby Products. *Environ. Sci. Technol.* 2011;45(12):5323–5331.

Twombly, R. 2001. Boston Pee Party. http://www.ehponline.org/docs/2001/109-5/forum.html#party (accessed on Jan. 8, 2007).

Ulsamer, AG, Osterberg, RE, McLaughlin Jr, J. Flame retardant chemicals in textiles. *Clin. Toxicol.* 1980;17(1):101–131.

Ulsamer, AG, Porter, WK, Osterberg, RE. Percutaneous absorption of radiolabeled TRIS from flame-retarded fabric. *J. Environ. Pathol. Toxicol.* 1978;1(5):543–9.

U.S. EPA, 1976. The manufacture and use of selected aryl and alkyl aryl phosphate esters. EPA 560/6-76-008. Environmental Protection Agency, Office of Toxic Substances, February 1976.

U.S. EPA, 2000. Pharmaceuticals and Personal Care Products (PPCPs) as Environmental Pollutants. http://www.epa.gov/esd/chemistry/pharma/ (accessed on Jan. 8, 2007).

U.S. EPA, 2003. Table of PCB Species by Congener Number. http://www.epa.gov/epawaste/hazard/tsd/pcbs/pubs/congenertable.pdf (accessed on March 8, 2013).

U.S. EPA, 2012a. Atrazine Updates. May 2012. http://www.epa.gov/oppsrrd1/reregistration/atrazine/atrazine_update.htm (accessed on August 31, 2012).

U.S. EPA, 2012b. Estimation Program Interface (EPI) Suite. November 2012. http://www.epa.gov/oppt/exposure/pubs/episuite.htm (accessed on January 31, 2013).

U.S. EPA, 2013a. Aroclor and Other PCB Mixtures. http://www.epa.gov/epawaste/hazard/tsd/pcbs/pubs/aroclor.htm (accessed on March 8, 2013).

U.S. EPA, 2013b. Nonylphenol and Nonylphenol Ethoxylates Action Plan Summary. http://www.epa.gov/oppt/existingchemicals/pubs/actionplans/np-npe.html (accessed on April 28, 2013).

Wang, Y, Jiang, G, Lam, PKS, Li, A. Polybrominated diphenyl ether in the East Asian environment: a critical review. *Environ. Internat.* 2007;3(7):963–973.

Wei, H, Aziz-Schwanbeck, AC, Zou, Y-H, Corcoran, MB, Poghosyan, A, Li, A, Rockne, KJ, Christensen, ER, Sturchio, N. Polybromodiphenyl Ethers and Decabromodiphenyl Ethane in Aquatic Sediments from Southern Arkansas, USA, *Environ. Sci. Technol.* 2012;46(15):8017–8024.

Wei, H, Zou, Y-H, Li, A, Christensen, ER, Rockne, KJ. Photolytic debromination of polybrominated diphenyl ethers in hexane by sunlight. *Environ. Pollut.* 2013;174:194–200.

Wiener, H. Structural determination of paraffin boiling points. *J. Am. Chem. Soc.* 1947;69:17.

Wolfe, NL, Jeffers, PM. Hydrolysis. In: *Handbook of Property Estimation Methods for Chemicals*. Boethling, RS, Mackay, D (Editors). Lewis Publishers; 2000.

Wong, CS, Garrison, AW, Foreman, WT. Enantiomeric composition of chiral polychlorinated biphenyl atropisomers in aquatic bed sediment. *Environ. Sci. Technol.* 2001;35:33–39.

Wu, QZ, Bedard, D, Wiegel, J. 2,6-Dibromobiphenyl Primes Extensive Dechlorination of Aroclor 1260 in Contaminated Sediment at 8–30 °C by Stimulating Growth of PCB-Dehalogenating Microorganisms. *Environ. Sci. Technol.* 1999;33:595–602 and 33(7):1148–1148.

Yalkowsky, SH, Valvani, SC, Roseman, TJ. Solubility and Partitioning VI: Octanol Solubility and Octanol/Water Partition Coefficients. *J. Pharm. Sci.* 1983;72:866–870.

Yang, R, Wei, H, Guo, J-H, Li, A. Emerging Brominated Flame Retardants in the Sediments of the Great Lakes. *Environ. Sci. Technol.* 2012;46:3119–3126.

Yang, R, Wei, H, Guo, J-H, McLeod, C, Li, A, Sturchio, N. Historical and Currently Used Dechloranes in the Sediments of the Great Lakes. *Environ. Sci. Technol.* 2011;45:5156–5163.

Zhu, N-L, Li, A, Wang, T, Wang, P, Qu, G-B, Ruan, T, Fu, J-J, Yuan, B, Zeng, L-X, Wang, Y-W, Jiang, G. Tris(2,3-dibromopropyl) Isocyanurate, Hexabromocyclododecanes, and Polybrominated Diphenyl Ethers in Mollusks from Chinese Bohai Sea. *Environ. Sci. Technol.* 2012;46:7174–7181.

Zou, Y, Christensen, ER, Li, A. Characteristic Pattern Analysis of PBDEs in Great Lakes Sediments: A Combination of Eigenspace Projection and PMF Analysis. *Environmatrics* 2013;24(1):41–50.

8

PATHOGENS

8.1 INTRODUCTION

Pathogens are of concern both in drinking water and recreational waters (Ford 1999; Wade et al. 2003). Pathogenic microorganisms are classified as bacteria, protozoa, or viruses (Table 8.1). From Table 8.1 it is seen that the infectious dose for bacteria, for example for toxigenic *Escherichia coli* or *campylobacter*, can be very large, up to 10^6 or higher. By contrast, the protozoa *Giardia lambia* or *Cryptosporidium*, which sometimes occur in water treatment plants, and viruses need only to be present in very low numbers (1–10) to generate infection. It is clear that in order to properly assess the risks from waterborne disease, it is necessary to understand survival strategies within the water distribution system. A viable but not culturable (VNC) organism has a lesser chance of spreading disease than one that is both viable and culturable.

Of the bacteria listed, *salmonella*, *shigella*, toxigenic *Escherichia coli*, and *campylobacter* are the most common. Among protozoa, *Giardia lambia* and *Cryptosporidium parvum* are particularly prevalent. Viruses are in many ways the most poorly understood group of organisms carrying waterborne disease. The number of people affected is very large (6,500,000), even though the infectious dose is low (1–10). Some of the most common viruses in water include Norwalk or Norwalk-like viruses, hepatitis A (HAV), and rotaviruses. Other waterborne viruses include various types of polioviruses, coxsachievirus, echovirus, reovirus, adenovirus, and hepatitis E virus (HEV).

In addition to pathogenic microorganisms, there is at least one abiotic molecular pathogen of interest, the pathogenic prion protein (Jacobson et al. 2009). A misfolded

Physical and Chemical Processes in the Aquatic Environment, First Edition. Erik R. Christensen and An Li.
© 2014 John Wiley & Sons, Inc. Published 2014 by John Wiley & Sons, Inc.

TABLE 8.1. Pathogens in Drinking Water, Infectious Dose, Estimated Incidence Through Consumption of Drinking Water in the United States, Survival in Drinking Water, and Potential Survival Strategies

	Infectious dose[a]	Estimated incidence[b]	Survival in drinking water (d)	Survival strategies[c]
Bacteria				
Vibrio cholera	10^8	Very few[c]	30	VNC, IC
Salmonella spp.	10^{6-7}	59,000	60–90	VNC,IC
Shigella spp.	10^2	35,000	30	VNC,IC
Toxigenic *Escherichia Coli*	10^{2-9}	150,000	90	VNC,IC
Campylobacter spp.	10^6	320,000	7	VNC,IC
Leptospira spp.	3	–	–	–
Francisella fularensis	10	–	–	–
Aeromonas spp.	10^8	–	90	–
Helicobacter pylori	–	High	–	–
Legionella pneumophila	> 10	13,000	Long	VNC,IC
Mycobacterium avium	–	–	Long	IC
Protozoa				
Giardia lambia	1–10	260,000	25	Cyst
Cryptosporidium parvum	1–30	420,000	–	Oocyst
Naegleria fowleri	–	–	–	Cyst
Acanthamoeba spp.	–	–	–	Cyst
Enfamoeba histolica	10–100	–	25	Cyst
Cyclospora cayefanensis		–	–	Oocyst
Isospora belli	–	–	–	Oocyst
The microsporidia	–	–	–	Spore, IC
Ballantidium coli	25–100	–	–	Oocyst
Toixoplasma gandii	–	–	–	Oocyst
Viruses				
Total estimates	1–10	6,500,000	5–27	Adsorption/ absorption

(Source: Adapted from Ford (1999)); abbreviations: -, unknown; IC, intracellular survival and/or growth; VNC, viable but not culturable.
[a]Infectious dose is number of infectious agents that produce symptons in 50% of tested volunteers.
[b]U.S. point estimates.
[c]Very few outbreaks of cholera occur in the United States, and these are usually attributable to imported foods.

form of the prion protein (PrP^{TSE}) is an infectious agent causing the fatal neurodegenerative disease transmissible spongiform encephalphalopathies (TSE), which affects many mammals, including humans. Forms of the disease include bovine spongiform encephalopathy (BSE or "mad cow" disease) and chronic wasting disease (CWD) of deer, elk, and moose. Interspecies transmission of TSEs have been documented in humans consuming BSE-infected beef. Following the first case of BSE in the United

**TABLE 8.2. Waterborne Disease Outbreaks Associated with Drinking
Water by Organisms Causing the Outbreak, United States, 1993–1994**

Organism	Outbreaks	Cases
Cryptosporidium parvum	5	403,271
AGI[a]	5	495
Giardia lamblia	5	385
Campylobacter jejuni	3	223
Salmonella serotype	1	625
Typhimurium		
Shigella sonnei	1	230
Shigella flexneri	1	33
Non-01 *Vibrio cholerae*	1	11
Chemical exposures	8	93

(Source: Adapted from Ford (1999)).
[a]AGI, acute gastrointestinal disease (of unknown cause(s)).

States in 2003 many cattle were slaughtered, and in several states CWD-infected herds were depopulated, presenting a challenge for the safe disposal of large numbers of infected carcasses. Landfilling may be a cost-effective option that is subject to a low risk of transmission of these diseases, for example via groundwater, to humans from infected carcasses.

Waterborne disease outbreaks associated with drinking water, listed by etiologic agent, are listed in Table 8.2. These data relate to 1993–1994 in the United States. For this time period, the dominant organism was *Cryptosporidium parvum*, which was responsible for five outbreaks, one of which was the well-known Milwaukee case that affected more than 400,000 people.

Recreational waters, such as lakes, rivers, and nearshore ocean areas, can also be important carriers of pathogens. Indicator organisms include *Enterococci*, Fecal coliforms, *E. coli*, and total coliforms (Table 8.3). Note that the slopes here are 0.30 and 0.75, for *Enterococci* and *E. coli* in marine and fresh water respectively. By contrast, considering estimated rotavirus risk based on number of *E. coli* indicator organisms, López-Pila and Szewzyk (2000) found a 1:1 slope corresponding to a slope of 2.30 of the natural log of risk vs. number of indicator organisms. These organisms are observed in lieu of actual pathogens because they are usually easier to detect, and they indicate a probability that pathogens are present.

The U.S. Environmental Protection Agency has recommended use of *Enterococci* in marine water and *Enterococci* and/or *Escherichia coli* in freshwater as indicator organisms (Wade et al. 2003). The recommended regulatory levels based on geometric means of at least five samples over a 30-d period were 35 colony-forming units (cfu)/100 mL and 33 cfu/100 mL for *Enterococci* in marine and fresh water respectively. Recommended levels for *E. coli* in fresh water were 126 cfu/100 mL. Further details of pathogenic bacteria and protozoa along with indicator organisms for bacteria are given in this chapter.

TABLE 8.3. Model Parameters from Weighted Linear Regression of the Natural Log Relative Risk as a Function of Indicator Density (Log Base 10)[a]

Indicator	Intercept	Coefficient	p-value	r[b]
Marine water				
Enterocci	0.099	0.30	0.05	0.37
Fecal coliform	0.86	−0.024	0.94	−0.017
E. coli	−0.071	0.25	0.24	0.37
Total coliform	−0.096	0.42	0.25	0.36
Fresh water				
Enterococci	0.54	0.0078	0.97	0.016
Fecal coliform	0.53	0.0058	0.98	0.0083
E. coli	−1.07	0.75	0.063	0.86
Total coliform[c]				

(Source: Adapted from Wade et al. (2003)).

[a]Weights for the model were the inverse of the standard error of the natural log of the relative risk.
[b]Correlation coefficient.
[c]not enough data.

8.2 BACTERIA

Most strains of *Escherichia coli* are not pathogenic. These bacteria were thought to be a vanishing cause of diarrhea until 1982 with the recognition of *E. coli* O157:H7 (Arvanitidou et al., 1996). This serotype is now recognized as a common cause of bloody, and other milder non-bloody, forms of diarrhea in many parts of the world. Although most outbreaks have had food or bovine origin, there are documented cases of transmission from person to person via the fecal-oral route. There has been a single case of a water supply that was contaminated following freeze damage and repairs. The original source remained unknown although leaks in water and sewer pipes and lack of chlorination may have contributed to the contamination of the water.

Arvanitidou et al. (1996) surveyed more than 1200 drinking water and recreational water samples from Northern Greece for the presence of *E. coli* O157:H7. A membrane filtration method was used, initially with m-Fc agar (Difco). After 24-hr incubation one or more colonies from each sample were re-cultured on sorbitol-MacConkey agar, and non-sorbitol fermenting colonies of *E. coli* were tested for the somatic O157 antigen. Of the re-cultured colonies, 121 were found to be non-sorbitol fermenting but none of them contained the O157 antigen. This finding supports the hypothesis that drinking and recreational waters are not likely routes of *E. coli* O157:H7 transmission in Northern Greece or other similar geographical areas.

Salmonella is a bacterial pathogen that is associated with fecal matter, and can therefore enter drinking and recreational waters through sewage pollution. *Salmonella* may be enumerated by a traditional MPN method or, more rapidly, by an MPN method in conjunction with an immunoassay technique to screen for positive samples, that is samples containing *Salmonella*. Using these methods, Ho and Tam (2000) studied 52 samples from various recreational and drinking water sources. *Salmonella* concentrations ranged from 1 to 4.6×10^3 per liter. The specificity was 100% on a per sample basis. When individual tubes of lowest dilution were compared with results of the immunoassay technique, it was found that 2 out of 88 MPN culture positive tubes were immunoassay negative, giving 97.7% sensitivity. Similarly, out of 70 MPN culture negative tubes there were 4 immuno-positive tubes, indicating 94.3% specificity. These numbers show that the shortened MPN test including immuno-assay screening for positive samples works quite satisfactorily.

The principle of the VIDAS™ immuno-analyzer is that antigens of *Salmonella* that are present react with *Salmonella*-specific antibodies on the pre-coated wall of a solid receptacle inside the analyzer. The antigen–antibody complex will in turn react with a second antibody conjugated with an enzyme to form a large complex. This larger complex reacts with a specific substrate impregnated on a VIDAS *Salmonella* reagent strip to produce a fluorescent product that can easily be detected (Ho and Tam 2000).

8.3 PROTOZOA

The parasitic protozoa *Cryptosporidium* and *Giardia* are pathogens that have the potential for major impact on humans in the United States (Tables 8.1 and 8.2) and throughout the world. Both of these can be transmitted by animals or humans. There is genetic variability among those infective to mammals. *G. lamblia* falls into two major categories; assemblages A and B. Similarly, *C. parvum* can be classified into genotype 1 (*Cryptosporidium hominis* 20) and genotype 2 (*C. parvum*). Molecular methods, such as quantitative polymerase chain reaction (qPCR), allow sensitive detection of pathogens as well as genetic variability among isolates (Guy et al. 2003). By contrast, the traditional detection method immunofluorescence microscopy detects the species only. The *cryptosporidium* outbreak in Milwaukee, Wisconsin showed that *Cryptosporidium* came both from upstream agricultural runoff and combined sewer overflows. This was demonstrated in frozen samples sent to the U.S. Center for Disease Control, Atlanta, which used molecular techniques to distinguish the two genotypes. Thus both cattle and probably humans could have been carriers of the *Cryptosporidium* entering the Howard Avenue filtration plant. Confirming statistical correlation evidence for the significant contribution of combined sewer overflows to *Cryptosporidium* was provided by Christensen et al. (1997).

Medema and Schijven (2001) modeled the emission and dispersion of parasitic protozoa in the Netherlands from the point of discharge to the point of abstraction for drinking water or bathing sites. This was performed with a PROMISE version 2.5.0 to estimate concentrations in treated and untreated wastewater, and WATNAT version 2.05 for modeling dispersion. WATNAT is a hydrological model that consists of a hydrodynamic transport module for the network of waterways in the Netherlands, and a second module that describes processes that effect water quality such as die-off in each part of the network. Results showed that the load in 1994 from the rivers Rhine and Meusen carried a larger load of *Cryptosporidium* oocysts and *Giardia* cysts (3.2×10^{14} and 2.1×10^{15} respectively) than the load from human waste (3.2×10^{13} and 3.8×10^{14} respectively) within the Netherlands. This international river load is a source of uncertainty that made the comparison between measured and calculated loads difficult as it was not practical to include the measured external load in all calculations. Thus international control strategies are important for effective abatement of the threat to human health from protozoa.

The removal of *Giardia* in wastewater treatment plant was 99% and of *Cryptosporidium* only 75%. As a result, the main (85%) source of *Cryptosporidium* oocysts was wastewater effluent, whereas for *Giardia* the majority (82%) of cysts came from untreated wastewater. Agricultural runoff was not included in the emission model, as reflected in calculated values that were 10 to 37 times lower than measured values of both of these protozoa upstream of a wastewater effluent discharge point, but downstream of an agricultural runoff area with cattle, sheep and application of manure for soil fertilization.

Testing on human volunteers demonstrated that only a dose of a single (oo)cyst results in a 0.4% probability of infection for *Cryptosporidium* and a 2% probability for *Giardia* (Medema and Schijven 2001); therefore it is important to use sensitive and genotype-specific detection methods. Such a method is the qPCR methodology for quantification of these protozoa (Guy et al. 2003). The method relies on a suitable primer and TaqMan probe. The principle is illustrated in Figure 8.1, including an application plot and a standard curve. Figure 8.1(a) shows tenfold serial dilutions of *C. parvum* oocyst DNA ranging from 5.7 ng to 5.7 pg and detection using the *Cryptosporidium* oocyst wall protein (COWP) primer-probe set P702. The COWP probe was labeled with 5'-hexachlorofluorescein (HEX; emission wavelength $\lambda_{em} = 553$ nm). Figure 8.1(b) shows how the standard curve can be derived from the amplification plot using the HEX threshold. The number of cycles necessary to induce the HEX response indicates the initial quantity of *C. parvum* DNA.

8.3.1 Cryptosporidium

Determining the number of *Cryptosporidium* oocysts or other pathogens in water is important in order to take appropriate action or precautions against a potential threat from these organisms. The Milwaukee *Cryptosporidium* outbreak in 1993, where more than 400,000 people were affected, served as a reminder that occurrences can happen with little or no warning, and that the impact from pathogens can be considerable. One consequence of this in the United States is the promulgation of the Information Collection Rule (ICR), which requires large-scale water suppliers to test their waters for certain pathogens in order to arrive at meaningful nation-wide standards for those organisms.

Samples collected for *Cryptosporidium* testing typically contain 11.4 hL (300 gal). This volume (V) is first filtered and then analyzed for *Cryptosporidium*. An estimate with upper confidence limit is desired for the mean bulk water concentration μ_c of pathogens over a certain period of time. The number N_1 of oocysts in the sampled

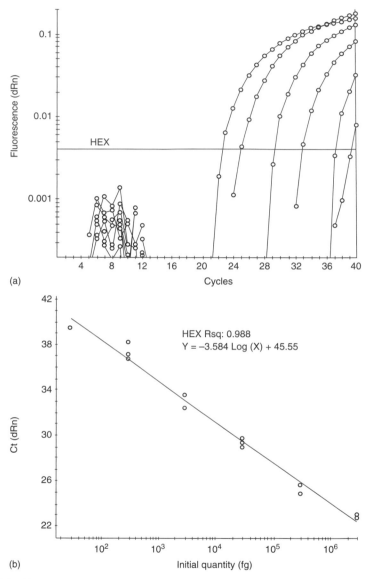

Figure 8.1. Simplex and multiplex detection of *C. Parvum* using qPCR. The amplification plot (a) is shown with corresponding standard curve (b). The gene for the *Cryptosporidium* oocyst wall protein (COWP) was selected for the primer-probe set for detection of *C. Parvum*. The COPW probe was labeled with 5'-hexachlorofluorescein (HEX). The amplification plot (a) shows tenfold serial dilutions of *C. parvum* oocyst DNA ranging from 3 ng to 30 fg and detection using the COWP primer-probe set. The horizontal line shows the threshold. dRn = baseline − corrected normalized fluorescence, Ct = cycle threshold. (Source: Adapted from Guy et al. (2003))

volume follows a Poisson distribution (Parkhurst and Stern 1998). Typically, only a fraction (f) is subsampled. In addition, *Cryptosporidium* oocysts can be lost as a result of adsorption or the transfer process. If the loss fraction is λ, then the retained fraction is $\rho = 1 - \lambda$, and the proportion counted is $\Phi = f\ \rho$.

The number counted (K) in the subsample has a binomial distribution with a probability Φ that each of the N_1 filtered organisms will be counted. The mean number N_1 of oocysts in the filtered volume V is $\mu_N = V\mu_c$ which follows a Poisson distribution. The mean number (K), $\mu_K = \Phi V \mu_c$ in the binomially distributed sub-sample drawn from a Poisson distributed sample also follows, as shown by Parkhurst and Stern (1998), a Poisson distribution. The effective volume EV is defined as $EV = \Phi V = f\rho V$. The number of oocysts counted will, on average, be the same as if the effective volume had been sampled and all organisms in that volume were counted.

Mean concentrations of pathogens can be estimated using at least four different methods. These are: (1) estimates equal to the total number of oocysts observed, divided by the total effective volume of the water investigated, which is the method preferred by Parkhurst and Stern (1998); (2) simple averages of the sample concentration estimates; (3) the "no-zeros" approach where zeros are replaced with one or one half *cryptosporidium*; (4) the "positives-only" approach where only the positive results are averaged; (5) the upper 95% percentile of the data. Only the first method will give unbiased estimates of the true concentrations. Consider, for example, that a given volume was sampled for each of 52 weeks and only two samples had non-zero values; one with a single and one with two *Cryptosporidium*. Method 2 only gives the correct answer if all sample volumes are identical, the no-zeros approach gives a too high estimate, the positives only also gives an inflated estimate, and the percentile method gives a concentration that has no simple relation to the mean bulk concentration of *Cryptosporidium* in the water.

The weighted mean μ_{cw} of the *Cryptosporidium* concentration in water, where the weighting equals the inverse of the estimated variance of each observation, can be expressed as (Parkhurst and Stern 1998)

$$\mu_{cw} = \frac{\sum_i EV_i(K_i/EV_i)}{\sum_i EV_i} = \frac{\sum K_i}{\sum EV_i} \tag{8.1}$$

where K_i and EV_i are actual oocyst counts and effective volumes in sample i respectively. Thus, this is the number determined in the first of the above methods. The concentration is the total number of oocysts divided by the total effective volume.

Masago et al. (2002) assessed the risk of infection of *Cryptosporidium* using a Monte Carlo method with the field data of the Sagami watershed. The daily infection risk was estimated from,

$$DRisk = I(x) = 1 - \exp(-\beta x) \tag{8.2}$$

Where $\beta = 0.0042$ and x is the number of oocysts ingested per day. The annual risk of infection Y was calculated from the equation:

$$Y = 1 - \prod_{i=1}^{365}(1 - I_i(x)) \tag{8-3}$$

The 95% percentile of the annual risk of infection was found to be $10^{-2.6}$. They further estimated that to reduce the 95% percentile value of the annual risk to below 10^{-4}, the risk of days with levels of *Cryptosporidium* in treated water exceeding one oocyst per 80 L should be eliminated.

8.4 MOLECULAR TECHNIQUES FOR DETECTION OF PATHOGENS

8.4.1 Water

Molecular techniques may be used to detect bacterial, protozoan, and viral pathogens, and to track biomarkers for pathogens and pollution sources in the environment (Girones et al. 2010). Many species and strains of microorganisms cannot be detected by conventional cell-culture techniques, whereas molecular techniques make this possible. The molecular techniques offer several advantages over conventional methods, including higher speed, specificity, and possibility of automation. The indicator organisms *E. coli* and *Enterococci* can provide probabilistic guidance on the presence of bacterial pathogens, but cannot, in contrast to molecular methods, provide information on specific pathogens in drinking or bathing water. Pathogens in waters that have been disinfected by treatment or sunlight may be underestimated by indicator bacteria due to the lower sensitivity of these pathogens to inactivation. One advantage of traditional culture-based methods for bacteria is that they, in contrast to molecular techniques, distinguish between viable and non-viable organisms. Note that microscope-based methods, as in U.S. EPA method 1623, for *Cryptosporidium* produce total counts of live and dead cells without distinguishing species or genotypes that can infect humans from those that cannot.

Most molecular methods of microbial detection rely on nucleic acid amplification in polymerase chain reaction (PCR) using temperature cycling with suitable primers and probes. If each cycle is monitored, the method becomes quantitative PCR or qPCR. Quantitative PCR analysis of aquatic DNA provides the number of genomes of a given pathogen per volume unit of water. This is related to, but not identical to, the cell density as determined by fluorescent in situ hybridization (FISH), immunofluorescence, or colony forming units (CFUs) obtained by plate counting. The FISH technique is often used to distinguish between microorganisms in mixed bacterial populations. To obtain sufficient sensitivity, molecular techniques are usually preceded by a concentration step to increase the analyzed biomass.

Simultaneous detection of *Giardia* cysts and *Cryptosporidium* oocysts in water is presently carried out with U.S. EPA method 1623. The inability of this method to distinguish between species and genotype, and to measure viability were addressed by

Baque et al. (2011). They designed a real time or quantitative reverse transcription method, qRT-PCR, with a single primer set and two separate probes. The method is able to detect viable *G. Lamblia* cysts which can be infectious to humans. The primer and probe sets were able to differentiate between different *Giardia* assemblages (genotypes). Viability was indicated by mRNA synthesis after heat induction.

Various human Pathogens, for example *Salmonella* and *Shigella* and enteric viruses such as adenoviruses and *Cryptosporidium*, have been found in surface waters due to human fecal pollution. In addition, domestic and farm animals are known to contribute pathogens such as *E. coli* O157:H7, *Salmonella*, *Cryptosporidium*, and *Giardia*. Ahmed et al. (2009) investigated prevalence and concentrations of *Campylobacter jejuni*, *Salmonella*, and pathogenic *E. Coli* in Brisbane Botanic Gardens and two urban tidal creeks of Brisbane, Australia, using quantitative PCR (qPCR) methods. High levels of pathogens and fecal indicators, especially after rainfall events, can indicate a significant human health risk. There was only a poor correlation with fecal indicators and potential zoonotic (animal-human transmitted) pathogens, emphasizing the need for caution in relying solely on *E. coli* and *Enterococci* indicators for estimating pathogens.

8.4.2 Biosolids

PCR is typically conducted with a few primers and probes in order to look for a limited number of pathogens. However, when considering land application of sludge from sewage treatment or anaerobic digesters a wider overview of many genera and species of pathogenic microorganisms must be obtained in order to estimate the risk to humans of the application. One such technique is pyrosequencing of the 16S rRNA gene, which is performed by first amplifying the 16S rRNA gene from genomic DNA extracted from municipal biosolids (Bibby et al., 2010) and then sequencing the amplicons using parallel pyrosequencing technology.

Pyrosequencing relies on synthesis of complementary strands one base pair at a time and identification of each base added. Sequences are compared to libraries of 16S rDNA sequences with average lengths of about 390 bases. Based on this technology, Bibby et al. (2010) found mainly pathogens of the genera *Clostridium* and *Mycobacterium* in biosolids collected from three US municipal wastewater treatment plants. Pyrosequencing of biosolids is expected to be helpful for regulatory agencies as a guidance for the application of specific culture-based or quantitative methods.

While the advent of molecular methods represents a significant step forward in estimating the threat of aquatic pathogens, they are resource-demanding and will need to undergo further development in automation and standardization to become more generally available. With increased usage, the cost of these techniques is also likely to be further reduced.

8.5 PATHOGEN INDICATOR ORGANISMS AND SURROGATES

Pathogens are usually quite difficult to measure in the environment because of their low concentrations and specialized techniques required for their measurement. This is the main reason that non-pathogens, e.g. *E. coli*, are often used as indicator organisms. The assumption is that there is a fairly constant ratio between the pathogen and *E. coli* such that the pathogen concentration can be inferred from the measured concentration of *E. coli*.

However, there are cases where the exposure probability and risk of pathogens is of sufficient magnitude to warrant direct measurement of the pathogen, for example for *Cryptosporidium* in water supplies. Also, molecular techniques, such as PCR, have improved significantly over the last decade and have become more cost-effective, rendering these techniques more likely to be used than in the past. Surrogates are non-pathogenic organisms that have properties that are similar to those of certain pathogens. Surrogates can therefore be used to study fate and transport of pathogens without unnecessarily jeopardizing personnel or materials involved in testing procedures.

Differences in antimicrobial resistance between indicator bacteria and microbial pathogens is a factor that can limit the ability of indicator organisms to predict the number and persistence of pathogens in the environment. In an effort to address this question, Da Costa et al. (2006) carried out a survey of *Enterococci* bacteria collected from inflow, effluent, and sludge from 14 municipal wastewater treatment plants in Portugal. Antimicrobial resistance to 10 different antimicrobial drugs was measured by the diffusion agar method. They found significant resistance against, for example rifampicin (51.5%), tetracycline (34.6%), erythromycin (24.8%), and nitrofurantoin (22.5%). Even effluent *Enterococci* showed significant resistance, indicating that the use of antimicrobials had created a pool of resistance genes and that sewage treatment is insufficient to avoid dispersion of resistant *Enterococci* in the environment. It may be expected that microbial pathogens could similarly develop antibiotic resistance increasing the risk to human health.

The risk of human infection from pathogens can be estimated from epidemiological studies that consider fecal indicator organism concentrations and the number of bathing-related illnesses. Such studies are costly, however, and as an alternative one can estimate the disease load from known distributions of *E. coli* and known ratios between pathogens and *E. coli*. Such a study was carried out by López-Pila and Szewzyk (2000) who considered the risk from rotaviruses.

Two models have been used to estimate the number of persons infected. According to the exponential model, the probability that each ingested pathogen reaches and infects the target organ has a constant value r, and the probability P_{exp} of becoming infected after the ingestion of n pathogens can be expressed as

$$P_{exp} = 1 - e^{-nr} \tag{8.4}$$

The second dose–response relationship, the Beta–Poisson model, is based on an assumption that r is distributed as a beta probability distribution, and the probability of becoming infected is then,

$$P_{BP} = 1 - \left(1 + \frac{n}{\beta}\right)^{-\alpha} \tag{8.5}$$

Where α and β are specific for the pathogen considered. The exponential model is often selected because of its adequate description of the dose–response function in most cases, and its simple form.

For both equations, the number of ingested pathogens n is specified from a Poisson probability distribution. If the amount of water swallowed while bathing is constant, e.g. 100 mL, the probability of ingesting one organism is,

$$O_1 = Ce^{-C}, \tag{8.6}$$

where C is the concentration of rotaviruses in the water. For two organisms, the expression is,

$$O_2 = \frac{C^2}{2}e^{-C}, \tag{8.7}$$

and for n pathogens,

$$O_n = \frac{C^n}{n!}e^{-C} \tag{8.8}$$

The chance of ingesting n pathogens and becoming infected is then

$$P_n = (1 - e^{-nr})\frac{C^n}{n!}e^{-C}, \tag{8.9}$$

whereas the probability of ingesting at least one pathogen and becoming infected is a sum of probabilities for $i = 1, 2, \ldots, \infty$:

$$P_g = \sum_{i=1}^{\infty} (1 - e^{-ir})\,\frac{C^i}{i!}e^{-C} \tag{8.10}$$

The effect of microorganisms being lognormally distributed was considered by López-Pila and Szewzyk (2000). As a result, these authors were able to calculate disease load from rotaviruses vs. the ratio of rotaviruses/*E-coli* with geometric means of *E. coli* as a parameter. The standard deviation (log10 of numbers/100 mL) and amount of water ingested were assumed to be 0.8 and 100 mL respectively.

As an example of these calculations, the maximum allowed by the European Union of 5% of the analysis gives 2000 *E. coli* per 100 mL. With a 0.8 log standard deviation, this corresponds to $3.301 - 0.8 \times 1.6445 = 1.99 \sim 2$ for the log geometric mean, that is 100 *E. coli*/100 mL. According to López-Pila and Szewzyk (2000), this corresponds to a disease load of 5×10^{-4}, assuming a geometric mean of the rotavirus/*E. coli* ratio of $10^{-5.6}$ found in Germany. Assuming that the above ratio of rotavirus/*E. coli* applies not only to Germany but also to Europe in general, the *E. coli* guidelines thus permit that up to 5 out of 10,000 persons would become infected with rotavirus. Note that this risk assessment would most likely have to be modified for other pathogens such as the bacteria *Salmonella*, *Shigella*, and *Campylobacter*, parasites like *Giardia* and *Cryptosporidium*, and other enteric viruses.

8.5.1 Bacillus Subtilis

Cost-effective disinfection of surfaces, building air, and drinking water supplies that contain spores of the pathogenic bacteria *Bacillus anthracis* Sterne can be carried out by means of ultraviolet (UV) irradiation at a 254-nm wavelength (Nicholson and Galeano 2003). As dosimetry is worked out and appropriate disinfection techniques are investigated, it is helpful to have surrogate spores that are non-pathogenic but otherwise similar to the virulent spores. Spores of *Bacillus subtilis* have often been used as surrogates. However, one report (Knudson 1986) indicated that *B. anthracis* Sterne spores may be three to four times more resistant to 254 nm UV radiation than commonly used indicator strains of *Bacillus subtilis*.

Nicholson and Galeano (2003) decided to test this report and found that the apparent UV inactivation response of a spore population can vary considerably depending on the growth and sporulation environment and the methods used for spore purification, irradiation, and UV dosimetry. They found that UV absorption by the buffer and by the turbid suspension for *B. anthracis* Sterne leads to an overestimate of the spore resistance to UV. In order to eliminate these factors, spores of *B. anthracis* Sterne were purified and their UV inactivation kinetics were determined simultaneously with two indicator strains of *B. subtilis*, WN624 and ATCC 6633, under identical conditions. Under these circumstances, the spores were found to exhibit almost identical UV inactivation kinetics. The lethal UV dose for 90% of the population (LD_{90}) and the UV dose reducing spore viability with a factor 10 for the exponential portion of the inactivation curve were determined to be 260 and 120 J/m^2 respectively. Thus, *B. subtilis* spores can in fact function as a valid surrogate for *B. anthracis* spores.

8.5.2 E. Coli and Fecal Coliforms

Fecal contamination of the wave-washed zone of public bathing beaches can be significant, and this fact is often overlooked (Alm et al., 2003). On a unit weight basis, these authors found that the mean summer abundance of the fecal indicator bacteria *Enterococci* and *E. coli* was 3–38 times higher in the top 20 cm of wet sand cores than in water column at six freshwater bathing beaches. *Enterococci* were more numerous in the 5–10 cm sand layer and *E. coli* in the 0–5 cm layer. The contamination can come from agricultural runoff, sewage, combined sewer overflows, birds, wild and domestic animals, and recreational users. Fecal contamination can be hazardous to human health due to bacteria, viruses, and protozoa that, if ingested, can cause gastrointestinal disease. At present, monitoring of bacterial contamination along the US coastline of the Great Lakes is not mandated. However, in the interest of public health many local health authorities monitor bathing water for the presence of *E. coli* and *Enterococci*. Geometric means of at least five samples over a 30-d period of higher than 126 *E. coli* and 33 *Enterococci* per 100 mL in fresh water have been determined to present a health risk (Wade et al. 2003).

In view of the significant indicator concentrations at the beaches, and the associated apparent risk to children or bathers in general who may be subject to exposure, monitoring programs should be expanded to include the swash-zone of public beaches.

8.6 BACTERIAL CONTAMINATION OF RECREATIONAL WATERS

Between 1988 and 1994 there were over 12,000 coastal beach closings and advisories in the United States, which represented a 4-fold increase over that period. Over 75% of the closings were due to microbial contamination (Henrickson et al. 2001). The types of illnesses developed include gastrointestinal respiratory, dermatologic, and ear, nose, and throat infection. Absorption of pathogenic microorganisms in the blood stream (sepsis) can also occur. Increased bacterial, viral, and toxic contamination appears to be related to watershed pollution, lack of wetlands, and overfishing. Harmful algal blooms, for example the *Pfiesteria* outbreak in 1997, lead to toxic and anoxic conditions. Marine-associated infections increase with exposure duration and level of pollution. Viable but not culturable (VNC) pathogens have virulent potential and may be present in marine systems. Inert materials, such as bottles and plastics, may provide protective niches for bacteria and viruses, which can prolong their survival.

In their review of swimming-related illness, Henrickson et al. (2001) conclude that there is a clear need for increased surveillance and monitoring of coastal ecosystem health. As mentioned previously, with molecular techniques becoming more cost-effective there may be a case to be made for monitoring coastal waters and sediments for pathogens. Source tracking using molecular techniques is increasing. Both public health concerns and coastal zone management need to be addressed, and there are many examples that illustrate potential benefits. For example, beaches near Sydney were dramatically improved by extending outfalls from three sewage treatment plants. To implement these programs and techniques effectively, it seems desirable to develop international agreements on waste release, monitoring, and pathogen standards.

With respect to monitoring, Rathbone et al. (1998) found, perhaps not unexpectedly, that there was a marked improvement in coastal water quality near Durban, South Africa, after commissioning submarine outfalls in 1969. There was also a persistent problem of poor surf water quality in the surf zone surrounding mouths of inland waterways. High bacteria counts were correlated with low salinity due to mixing of polluted freshwater with ocean water. A classification system using *E. coli* I, helminthic parasitic ova, pathogenic *staphylococci*, *salmonella*, and *shigella* was found to be useful in detecting problem areas that would have escaped detection based on *E. coli* alone.

While *E. coli* is the preferred indicator organism for freshwater in the United States and much of the world, Canada uses a limit of 200 fecal coliforms per 100 mL. To give some background for this standard, Lévesque et al. (2000) investigated bird droppings from ring-billed gulls whose populations have grown dramatically in the past 30 yr. Gull droppings were collected at three colonies and examined for fecal coliforms, *Aeromonas* spp., *Campylobacter* spp., *Pseudomonas aeruginosa*, *Salmonella* spp., and *Staphylococcus aureus*. It was suggested that the limit of 200 fecal coliforms per 100 mL provided an acceptable protection with respect to contamination by gulls. This conclusion was reached based on numbers of pathogens found per 200 fecal coliforms, ranging from 5.05 *S. aureus* to 0.13 *Campylobacter*. It was assumed that swimmers swallow 50 mL water and that the infectious dose is 500 microorganisms for *Campylobacter* spp. and 10^5 for most serotypes of *Salmonella* spp.

The role of bed sediment and suspended sediment flocs in providing a reservoir for bacteria is rarely discussed. In their investigation of the South Nation River near Ottawa, Ontario, Canada, Droppo et al. (2009) found that the fate and transport of *E. Coli* and *Salmonella* were highly influenced by suspended sediment flocs and bed sediment particles. Bacterial counts were consistently higher within sediment compartments than for water alone. The incorporation of pathogens within the surficial bed sediments is facilitated by the excretion of extracellular polymeric substances (EPS) from bacteria and algae. A proper evaluation of public health risks from both pelagic and sediment-associated particles will necessitate sampling strategies that include the sediment phase. Note that current sampling protocols for regulatory purposes consider only the water phase. The impact of sediment-associated pathogens may be particularly large during storms, during which sediment re-suspension is likely to occur.

Grant et al (2001) investigated the generation of bacteria in a coastal saltwater marsh and the possible impact on surf zone water quality. Flow of water and *Enterococci* (ENT) into the ocean from the Talbert Marsh and the Talbert watershed was quantified. Forebay water at pump stations had a typically high conductivity (30 mS/cm), indicating that a fraction of the forebay water is ocean water that traveled up into the forebays through leaking flap gates during flood tides. Thus, the runoff fraction of water that was discharged was calculated from,

$$F = 1 - \frac{(C - C_R)}{(C_0 - C_R)} \tag{8.11}$$

Where C_0 (53.5 mS/cm) and C_R (3 mS/cm) are the conductivity of ocean water and runoff, and c is the measured conductivity of samples from the pump stations. The integrated volume of runoff exiting through the Pacific Coastal Highway (PCH) bridge is then,

$$I = \int F(t) Q(t) \, dt \tag{8.12}$$

Where $F(t)$ is the fraction calculated in Equation 8.11, and $Q(t)$, the volumetric flow rate.

Considering the Talbert Marsh during ebb tides, in situ measurements of flow velocity and water elevation at the inlet to the Marsh, Brookhurst station and the outlet, PCH, indicate that these flows Q_{in} and Q_{out} (m³/sec) respectively, are approximately equal. Thus the generation G (ENT/sec) of *Enterococci* within the marsh may be written

$$G = C_{out} Q_{out} - C_{in} Q_{in} \tag{8.13}$$

Or,

$$G = Q \Delta C \tag{8.14}$$

Where C_{in} and C_{out} are concentrations of ENT at the inlet and outlet respectively. Using ebb-tide values of $\Delta C = 29$ MPN/100 mL and $Q = 8.37$ m^3/ sec, a generation rate for ENT in the marsh was estimated at $\approx 10^{10}$ MPN/h. With a max estimated at 10^8 MPN/dropping and an average number of birds in the marsh during the day of 228, this corresponds to about one dropping per 2.3 h. This was comparable to the dropping rate for the same bird species in captivity, and supports, therefore, that the marsh is the source of the bacteria.

There may be other possible sources of ENT in the area, as seen by the fact that even with very low bacteria counts at the outlet, a surf zone station 9N had high bacteria concentrations. One complicating factor is that the possible human health risk from ENT generated in marshland is not well known. However, assuming that ENT is a good pathogen indicator, one should treat any ENT counts with caution. Restoration of coastal wetlands may, therefore, have the unwanted side effect of increasing the bacteria count in the coastal area.

The previously discussed algal blooms of *Cladophora* along coastlines in the Great Lakes area appear to have another undesired effect in addition to the unsightliness of these blooms: They can be associated with bacterial pathogens, especially *Salmonella* (Byappanahalli et al. 2009). The authors found these and other bacterial pathogens *Shigella*, *Campylobacter*, and shiga-toxin producing *E. Coli* to be associated with *Cladophora* growth in southern Lake Michigan. While the *Salmonella* isolates from different areas and time periods exhibited a high degree of similarities they could be differentiated by location and season, indicating that they came from sources such as wastewater effluent, runoff, and birds. The association of bacterial pathogens with *Cladophora* may be a threat to human health and could impact regulatory guidelines.

Fecal coliform bacteria were formally the indicators of choice for routine monitoring in both Milwaukee and Racine. However, because of the direct correlation between *E. coli* and swimmer-associated gastroenteritis and the fact that other fecal organisms, such as *Klebsiella* spp., can be present in the environment regardless of recent fecal contamination, both communities have switched to an *E. coli* standard.

8.6.1   Modeling

In order to evaluate the impact of various sources on the pathogen or indicator count in a given area, modeling can be a very useful tool. Connolly et al. (1999) modeled fate of pathogenic organisms in Mamala Bay, Oahu, Hawaii, United States.

Mamala Bay is at the southern end of the island of Oahu, which includes Pearl Harbor and Honolulu Harbor. There were two parts to the model; hydrodynamic and pathogen fate. The hydrodynamic model provided the circulation and mixing field within Mamala Bay. Results from the hydrodynamic model were then used in the fate model to predict the distribution of indicator organisms and pathogens. The model was calibrated for three indicator organisms, fecal coliforms, *Enterococci*, and *Clostridium perfringens,* and was also used for four pathogens, *Salmonella*, *Cryptosporidium*, *Giardia lambilia*, and enterovirus.

The fate model described by Connolly et al. (1999) was based on mass conservation considering specific loss processes, that is, phytotoxicity, predation, and settling. Phytotoxicity was modeled using the following second order expression,

$$\frac{dB}{dt} = -\alpha E B \qquad (8.15)$$

Where B = organism abundance (CFU/100 mL), α = rate constant (m^2/MJ), E = scalar irradiance (MJ/m^{2}d^{-1}), and t = time (sec). Integration of this equation gives,

$$B = B_0 e^{-\alpha S} \qquad (8.16)$$

The inactivation rate α was measured in bottles inoculated with *E. coli*, *enterococci*, and *Clostridium perfringens*, incubated in the dark or with filtered sunlight simulating different depths of Mamala Bay. Results for the dark bottle were used to determine blank values that were subtracted from the results for filtered sunlight in order to account for losses other than from phytotoxicity. The integrated irradiance or insolation S was computed from hourly measurements of the photosynthetically active radiation (PAR) where the scalar irradiance was determined from

$$E = E_0 e^{-K_{par}z} \qquad (8.17)$$

where z is the depth (m) below the water surface and K_{par} = diffuse attenuation coefficient for PAR (m^{-1}). Attenuation coefficients have been developed for clear water, dissolved organic matter (DOM), chlorophyll pigments, and non-chlorophyll particulates. The DOM component decreases exponentially with increasing wavelength λ (nm) and can be described as,

$$K_d(\lambda) = f(D) e^{-k_d(\lambda-\lambda_0)} \qquad (8.18)$$

where λ (nm) is a reference wavelength, $f(D)$ is an empirical function of D = DOM concentration (mg/L), and k_d(nm^{-1}) is an empirical exponent. The approximation in Equation 8.19 has been used (Connolly et al., 1999)

$$K_d(\lambda) = 0.565D e^{-0.014(\lambda-380)} \qquad (8.19)$$

The chlorophyll component depends on chlorophyll a and b concentrations. Baker and Smith (1982) developed an approximate expression for the attenuation coefficient in the chlorophyll range 0.04–10 mg/m^3.

Attenuation from non-chlorophyll particulates was negligible in Mamala Bay except in nearshore areas under storm conditions. Connolly et al. (1999) used an expression developed by Kirk (1984) to describe the attenuation coefficient K_p,

$$K_p = \frac{1}{\cos(\theta)}[a^2 + (g_1 \cos(\theta) - g_2) a b]^{1/2} \qquad (8.20)$$

where θ is the solar zenith angle at the air–water interface, a and b are absorption and scattering coefficients respectively, and g_1 and g_2 are constants. From this equation it is apparent that radiation coming from directly above the surface ($\theta = 0$) gives minimum, and incident radiation parallel to the air–water surface ($\theta = 90°$) maximum, K_p. In the latter case, light must undergo a very large path length to penetrate a given distance z below the water surface (Equation 8.17). Gallegos et al. (1990) give expressions 8.21 and 8.22 for the constants a and b for a turbid estuary over a TSS range of 2 to 30 mg/L

$$a = (0.153\, m - 0.13)\, e^{-0.0104(\lambda - 400)} \tag{8.21}$$

$$b = 2.41(1 - f_{om})\, m + 0.32 \tag{8.22}$$

where m = TSS concentration (mg/L) and f_{om} is the fraction of organic matter of the particulates. Removal of bacteria by predation was considered by a first order expression with a removal constant of 0.25 d^{-1}. The removal of bacteria and particulate matter by settling was modeled using the equation,

$$\frac{dm}{dt} = -\frac{w_s}{h} m \tag{8.23}$$

Where m = TSS concentration (mg/L), w_s = settling velocity (cm/sec), h = depth of each segment. Re-suspension was not considered. Discharges of organisms (CFU/d) and TSS from each watershed were estimated from discharge records and a runoff model.

The hydrodynamic model was validated by comparing currents and dye concentrations with measured water levels and salinities vs. time over a 25 h period. Mean errors between predicted and observed fecal coliform counts were generally less than 4 CFU/100 mL. The Sand Island wastewater treatment plant was found to be a primary contributor of fecal coliform at eastern recreational beaches. It was also a source of *Enterococci* and *Clostridium* at beaches west of Honolulu Harbor. *Enterococci* originate mostly from shoreline sources. For example, the Ala Wai Canal was a major source of *Enterococci* and *Clostridium* at eastern beaches, especially during storm events.

Model simulations showed that chemically enhanced primary treatment (CEPT) mostly influenced fecal coliform counts and to a lesser extent the *Enterococci* counts. Upgrade of the Sand Island wastewater treatment plant to CEPT reduces fecal coliform counts by a factor of 2. Further reductions could be obtained by adding disinfection and by upgrade to secondary treatment. Upgrade of the Honouliuli wastewater treatment plant, 17 km west of Honolulu Harbor, to CEPT reduced fecal coliform counts at the Oneula beach by a factor of 2, and disinfection and secondary treatment would reduce the counts by an approximate factor of 5. Smaller reductions were achieved for the *Enterococci* counts. Treatment options at the Honouliuli plant had no effect on bacteria levels at the eastern bay stations as a result of the westward flowing currents.

8.6.2 Beaches

Several investigators have studied bacterial contamination of beaches. Haack et al. (2003) collected samples from five beaches of the Grand Traverse Bay, Michigan, in order to determine correlations with ambient conditions and concentrations relative to U.S. EPA limits. As stated previously, the geometric mean limits are 126/100 mL for *E. coli* and 33/100 mL for *enterococci*. Single samples should be lower than 235/100 mL for *E. coli* and 61/100 mL for *Enterococci*. Bacteria were characterized by genomic (EC) or biochemical (ENT) profiling. The genomic profiling used repetitive sequences (rep-PCR), and the phenotype analysis physiological assays, colony color, and hemolysis on sheep blood agar. Cluster analysis was used to check if clusters of base pair signatures came from single or multiple sources.

Generally, ENT criteria were exceeded to a larger extent than EC criteria. In addition, beach water mostly contained multiple sources of bacteria, typically bird droppings, storm drains, and river water, but also beach sands, shallow groundwater, detritus, and effluent from wastewater treatment plants. Some ENT isolates had phenotypes close to those of human pathogens. Not surprisingly, concentrations of these bacteria depend on time of day, currents, and wind patterns. Concentrations tend to be higher later in the day. In some cases, there is a delay of 48–72 h between rainfall and EC concentrations. Interestingly, resistance to antibiotics such as streptomycin, gentamicin, and tetracycline, used for treating human *Enterococci* infection, was higher for *enterococcus* isolates from birds than from other sources.

E. coli was compared with *Enterococci* as indicators of pollution at five designated swimming beaches in southeast Wisconsin by Kinzelman et al. (2003). The U.S. EPA recommendations for single-sample allowable-maximum-density guidelines of 235 CFU/100 mL for *E. coli* and 61 CFU/100 mL for *Enterococci*, as well as the geometric mean of no less than five 1-d samples within a 30-d period equal to 126 CFU/100 mL for *E. coli* and 33 CFU/100 mL for *Enterococci*, were tested on more than 600 samples. It was found that there were 56 incidents where the single-sample limit for *Enterococci* exceeded that for *E. coli*. The additional unsafe-water advisories may or may not be indicative of increased protection of public health because *Enterococci* may be of fecal or plant origin. The authors concluded that these additional advisories would have a negative impact on public perception and the economy.

Boehm et al (2003) used a three-tiered approach to study sources of fecal indicator bacteria (FIB) at beaches near Avalon, Catalina Island, California. The bacteria included EC, ENT, and total coliform (TC). The first tier comprised spatiotemporal distributions of the pollution signal taking into account the influence of sunlight and tides on the FIB concentration. The second tier entailed studies of sources and hot spots. The first and second tier used standard FIB tests. The third tier involved selective sampling of FIB sources and hot spots for the enteric bacteria Bacteroides/Prevotella and enterovirus using nucleic acid techniques to determine whether the FIB was of human origin. The technique applied entailed amplification of Bacteroides/Prevotella using PCR primers that amplify rDNA, which encodes 16S rRNA from the human fecal-specific group.

The FIBs were found to come from multiple sources; bird droppings, contaminated subsurface water, leaking drains, and street runoff. Some shoreline, and two subsurface, samples tested positive for human-specific bacteria and enterovirus. Thus, as was found in the Michigan study, part of the contamination came from domestic (human) sewage.

Predictive models for indicator bacteria at beaches are of value to regulatory agencies issuing beach closures. With the time lag of 24–48h for U.S. EPA culture methods for *Enterococci* analysis, there is a chance of unnecessary closings when conditions improve, and a risk of missing closures when conditions worsen. Predictive models can estimate bacteria levels at the time of observation (nowcast) and forecast future levels. The regulatory background is provided by the U.S. BEACH Act, which requires that swimming advisories be issued when high bacterial concentrations are present.

Zhang et al. (2012) prepared an artificial neural network (ANN) code from MATLAB Neural Network Toolbox (V. 7) to make such predictions, and compared results from linear and non-linear Virtual Beach (VB) regressions of log *Enterococci* concentrations. They considered six sites at Holly Beach, Louisiana and a 6-year period during the swimming season from May 2005 to October 2010. A total of 944 data sets of bacteriological and environmental parameters were used. The ANN model included 15 variables, such as salinity, water temperature, wind speed and direction, tide level and type, and combinations of antecedent rainfalls, while the VB models relied on only five or six variables. Results were measured by linear correlation coefficient (LCC) and root mean square error (RMSE). The ANN model performed better than the VB models during a trial period in 2007, 2008, and 2009 (LCC = 0.857, RMSE = 0.336). The forecasting ability of models was tested by hindcasting for 2005, 2006, and 2010, and in this case the non-linear VB model had the highest LCC value, 0.521 vs. 0.320 for the ANN model, while the RMSE was higher (1.96) than for the ANN model (0.803).

Overall, the ANN model has a good potential for nowcasting and forecasting *Enterococci* levels at Holly Beach and at other beaches. One should realize, however, that these models are not truly predictive but are instead regression models of various degrees of sophistication. The fact that important variables are considered, such as temperature and antecedent rainfall, is a good feature, and as long as easy to handle mechanistic models are not available, they can be useful tools for managers.

8.6.3 Recreational Pools

Disinfection of pool waters is necessary in order to prevent outbreak of infectious diseases. Disinfectants and total organic carbon (TOC) from humic substances and inputs of sweat, urine, and cosmetics from bathers constitute a risk for human health through the formation of harmful disinfection byproducts (DBPs). The risk is amplified by recirculation of pool waters. A class of DBPs of particular concern is nitrogenous DPBs, which include organic chloramines, nitrosamines, and halonitriles. For example, as noted by Liviac et al. (2010), nitrosamines have been shown to be associated with bladder cancer. The authors conducted a study to examine the effect of disinfectant and illumination on the genotoxicity of poolwaters. Chinese hamster ovary cells were used as an indicator for a single cell gel electrophoresis assay for DNA damage to nuclei of cells. The results showed that brominated disinfectants, such as bromochlorodimethylhydantoin, should be avoided if possible, and that a combination of chlorine and ultraviolet (UV) radiation yielded a lower toxic response than chlorine alone. Outdoor pools under illumination of sunlight had a lower genotoxic response than indoor pools, and an indoor warmer pool is less toxic than an indoor cold pool. In order to minimize TOC precursors to DBPs it would be useful to remove TOCs during recycling of poolwaters. Precautions taken by bathers, such as showering before entering poolwaters and abstaining from urinating in the pool, are important as well.

8.7 PATHOGEN REMOVAL IN WATER AND WASTEWATER TREATMENT

The 1993 outbreak of cryptosporidiosis in Milwaukee, Wisconsin, was a pivotal event. About 400,000 people were affected and approximately 100 with weakened immune system perished due to the outbreak. At the time it was thought that *cryptosporidium* oocyst with a diameter of 7 μm came from runoff that resulted from cattle farming upstream the Milwaukee River, but subsequent analysis involving both molecular biological methods and statistical source analysis (Christensen et al. 1997) pointed to urban runoff, possibly including human sewage as an important source. *Cryptosporidium parvum* and *Giardia lamblia* are parasitic protozoa. The event took place in late March 1993 at a time of high runoff and associated turbidity. In addition, a new and apparently less effective coagulant, poly aluminum sulfate, was introduced on a trial basis as a possible replacement for alum at the Howard Avenue filtration plant. No treatment plant standards in place at the time were violated. However, as a result of this occurrence new turbidity standards have been promulgated by the U.S. EPA.

Other pathogens that may be present in water or wastewater are *Giardia*, strains of *E. coli* O157:H7, *Salmonella*, enterovirus, and coliphages that are found, for example, in harmful algal blooms. Effective disinfection is an important consideration in preventing pathogens from exerting harmful effects. For both water and wastewater treatment, chlorine has been replaced in many cases by hypochlorous acid (bleach), which is safer for plant personnel. There is a trend towards using UV for disinfection in small (< 5 mgd) wastewater treatment plants and medium-sized and larger water filtration plants. One clear advantage of UV is that no chemical is introduced into the environment as a result of its use.

8.7.1 Water

As a result of the crypto event, three major actions were taken: Ozone was introduced as a disinfectant at both the Linnwood and the Howard Avenue water filtration plants, the intake for the southern Howard Avenue plant was extended into Lake Michigan by

0.8 mile, and dual-media filtration using both anthracite and sand was put into place. *Cryptosporidium* is better controlled with ozone than chlorine. Previously, chlorine was used for disinfection and single media sand filters for filtration. This solution is typical of the multi-barrier approach used, for example, for *Cryptosporidium* control (Edzwald and Kelley 1998). An effective coagulant can remove both turbidity and natural organic matter (NOM), which may both be associated with *Cryptosporidium* oocyst. Monochloramine has been shown to be a more effective disinfectant than chlorine, ozone, or UV in preventing the attachment of *E. coli* in developing biofilm in drinking water distribution systems (Momba et al., 1999).

The feasibility of using turbidity and particle counts to monitor removal of *Cryptosporidium* by filters was investigated by Huck et al. (2002). They found that both parameters can be used to check filter performance but that particle counts are more suitable than turbidity to monitor deterioration of *Cryptosporidium* removal with time. Experiments were conducted at a plant in Ottawa, Canada, and at Metropolitan Water District of Southern California's (MWDSC's) treatment plant in La Verne, California. The baseline operation involved conventional treatment and dual-media filtration. Coagulants used were alum and polymer at the MWDSC plant, and activated silica at the Ottawa plant. *Cryptosporidium* removal was studied under suboptimal conditions, that is, filter ripening, breakthrough, and for suboptimal coagulation.

The results indicated that *Cryptosporidium* removal during ripening decreased less than for particles (≥ 2 μm), but that the removal decreased more than for particles during breakthrough. Because of the design, only near breakthrough could be achieved at the MWDSC plant. Using 40–65% from optimum amounts of coagulants, particle and *Cryptosporidium* removals decreased by equal amounts, 2 logs. In all three cases there was a better correlation of *Cryptosporidium* with particle counts than turbidity. Although *Cryptosporidium* removal under stable operating conditions was better for the Ottawa plant than the MWDSC plant, the removal decreases for the latter were smaller than for the Ottowa plants, especially during breakthrough.

8.7.2 Wastewater and Solid Waste

Wastewater treatment will typically decrease pathogen levels, either in the physical or biological processes or during disinfection. For example, progressive decrease in phage concentration with treatment steps was found by Tanji et al. (2003). There were no phages detected in the supernatant from the secondary clarifier. Phages infected to strains of *E. coli* O157:H7 produce two toxins, Stx1 and Stx2. The chlorine inactivation levels of the bacteriophages differ from those of non-infected bacteria (Tanji et al., 2003), which should be a consideration for appropriate wastewater treatment.

Simpler wastewater treatment systems, such as anoxic ponds (Almasi and Pescod 1996) or waste stabilization ponds (Awuah et al., 2001), also possess significant capability to reduce pathogen concentrations. Anoxic stabilization ponds can reduce land requirements compared with facultative ponds, and can avoid the environmental nuisances of odor and corrosion that are associated with anaerobic ponds. Bacterial removal rates in anoxic ponds were higher at 25 °C compared with 10 °C, and decreased with higher volumetric organic loading (Almasi and Pescod 1996). Waste stabilization ponds using algae or macrophytes (water lettuce, *Pistia stratiotes*; and duckweed, *Lemna paucicostata*) exposed to sunlight were successful in reducing the fecal *Enterococci* population from 1.18×10^5/mL to values below 100/mL under tropical conditions over a period of 29 d (Awuah et al. 2001). The wastewater was previously treated anaerobically in the laboratory for 2 d.

Composting is one option for the treatment of biosolids. It brings about stabilization of organic matter and can reduce pathogen levels to very low levels. The indigenous microflora inhibits the regrowth potential of *Salmonella typhimurium*. However, long-term storage is not recommended because this can increase the *Salmonella* regrowth potential (Sidhu et al. 2001).

Removal or inactivation of pathogens is subject to the reaction kinetics and the presence of natural organic matter. One important parameter determining the survival rate of most microbes is the product of disinfectant concentration and time of reaction. We shall consider the inactivation kinetics of *E. coli* with ozone and of *Bacillus subtilis* spores with ozone and monochloramine in Section 8.7.3.

8.7.3 Inactivation Kinetics

In order to obtain a given efficiency of inactivation of microorganisms, the kinetics of inactivation must be known. We consider inactivation of *E. coli* with ozone (Hunt and Marinas 1999) in this section. These authors investigated the inactivation of *E. coli* considering different initial ozone concentrations, organism densities, and humic acid concentrations. Experimental results were obtained using a continuous flow tubular reactor. Organism densities were modeled by means of the common Chick–Watson model,

$$\frac{dN}{dt} = -k_i c N \tag{8.24}$$

where N is the density of viable microorganisms, k_i is a rate constant, and c is the concentration of ozone. This is a homogeneous second order rate equation. The rate constant k_i may be expressed by the Arrhenius equation

$$k_i = A \exp\left(-\frac{E_a}{RT}\right) \tag{8.25}$$

where $A = 5.03 \times 10^8$ is a frequency factor, $E_a = 37, 100$ J/mole is the activation energy, $R = 8.314$ J/(mole °K) is the universal gas constant, and T is the absolute temperature in °K.

Decomposition kinetics of ozone was considered by means of the second order rate equation,

$$\frac{dc}{dt} = -k_x c x \tag{8.26}$$

where c is the ozone concentration, mg O_3/L, k_x is a second order rate constant, L/(mg sec), and x is the concentration of fast ozone demand constituents in *E. coli* cells in mg O_3/L. The fast ozone demand x may be expressed as,

$$x = c - c_0 + x_0 \tag{8.27}$$

with c_0 = initial ozone concentration and x_0 = initial concentration of fast oxygen demand constituents. The initial fast oxygen demand x_0 is proportional with the cell density through this equation,

$$x_0 = \alpha_0 N_0 \tag{8.28}$$

where Hunt and Marinas (1999) found $\alpha_0 = 1.29_0 \times 10^{-11}$ mg O_3/CFU by fitting the solution for c to experimental data. Integration of Equation 8.26 using Equation 8.27 gives the expression for c

$$c = \frac{c_0(c_0 - \alpha_0 N_0)}{c_0 - \alpha_0 N_0 \exp(-k_x(c_0 - \alpha_0 N_0)t)} \tag{8.29}$$

Solutions for c as a function of time (t) are shown in Figure 8.2. For large initial ozone concentrations, e.g. $c_0 = 150$ µg/L and moderate organism concentrations for example $N_0 = 3.6 \times 10^9$ CFU/L we obtain $x_0 = 1.29 \times 10^{-11}$ mg O_3/CFU $\times 3.6 \times 10^9$ CFU/L $= 46.4$ µg/L. Thus c will approach $c_0 - x_0 = 150 - 46.4 = 103.6$ µg/L for large times $(t\}$, while the ozone demand x vanishes (middle panel, solid squares). On the other hand, if the initial ozone demand x_0 is smaller than the initial ozone concentration, the ozone concentration will approach zero while the ozone demand x approaches $x_0 - c_0$. Consider $c_0 = 66.5$ µg/L, $N_0 = 7.2 \times 108$ CFU/L, $x_0 = 92.9$ µg/l. The ozone demand becomes $92.9 - 66.5 = 26.4$ µg/L and the ozone concentration 0 for large times (bottom panel, solid circles).

Regarding the influence of humic acid on the degree of inactivation, we may expect that humic acid presents an ozone demand and therefore reduces the inactivation efficiency. This is in fact seen in Figure 8.3, in which the straight line is an average prediction from Equation 8.24, $k_i = 123$ L/(mg sec). Semi-solid symbols reflect higher humic acid TOC_0 levels, and therefore higher N/N_0 values, whereas open symbols, indicating lower humic acid concentrations, show more effective inactivation through the lower N/N_0 values compared to the modeled line.

Larson and Marinas (2003) investigated inactivation kinetics of *Bacillus subtilis* spores with ozone and monochloramine. The kinetics was characterized by a lag phase followed by a pseudo first-order rate law as a function of the product of concentration C and time T, CT, as described by the delayed Chick–Watson expression

$$\frac{N}{N_0} = \begin{cases} 1 & \text{if } CT \le CT_{lag} \\ \exp\left(-k\left(CT - CT_{lag}\right)\right) & \text{if } CT > CT_{lag} \end{cases} \tag{8.30}$$

where k is the rate constant, see Figure 8.4.

The lag phase decreased and the post-lag phase increased with increasing temperature over the ranges of $1-30\,°C$ with ozone and $1-20\,°C$ with monochloramine. The

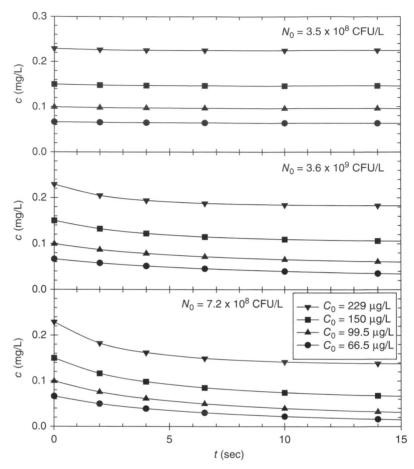

Figure 8.2. Decomposition of ozone in the presence of *E. coli* for different initial ozone concentrations and microorganism densities. (Source: Experimental conditions are described in Hunt and Mariñas (1999))

concept that inactivation is determined by the *CT* product is valid for ozone between 0.44 and 4.8 mg/L and monochloramine in the 3.8–7.7 mg/L range.

The main influence of pH on the activation kinetics of *B. subtilis* with ozone is the increased lag period for low pH = 6–7, which causes the inactivation to be fast at pH = 10 when the lag period is short. For monochloramine, the effect on lag time is small, but the disinfection rate is much higher at low pH values, e.g. pH = 6.

A conservative surrogate organism is one that has a lower inactivation than the target organism. From Figure 8.4 it is apparent that *B. subtilis* spores are conservative surrogates for *C. parvum* for inactivation efficiencies up to approximately 90% using ozone at 1 °C. At higher temperatures, the surrogate range extends to 99.5%. When monochloramine is disinfectant at 1 °C, the lag period for *B. subtilis* is very large,

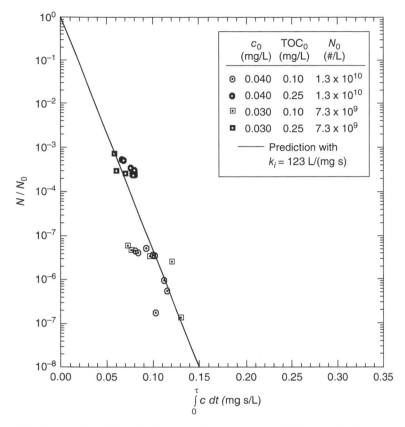

c_0 (mg/L)	TOC_0 (mg/L)	N_0 (#/L)
0.040	0.10	1.3×10^{10}
0.040	0.25	1.3×10^{10}
0.030	0.10	7.3×10^{9}
0.030	0.25	7.3×10^{9}

—— Prediction with $k_i = 123$ L/(mg s)

Figure 8.3. Inactivation of *E. coli* with ozone in the presence of different initial concentrations of humic acid (TOC_0) and ozone (c_0), and microorganism densities (N_0). (Source: Experimental conditions are described in Hunt and Mariñas (1999))

making *B. subtilis* an over conservative surrogate. At higher temperatures (20 °C, pH = 8), lag phases are similar for the two organisms, but the rate vs. *CT* is slightly higher for *C. parvum*, making *B. subtilis* a good surrogate for *C. parvum*. Inactivation of *Cryptosporidium parvum* with ozone and chloramine was described by Driedger et al. (2001).

Human viruses also present a threat. Adenoviruses are transmitted by the fecal-oral route and serotypes 40 and 41 cause enteric diseases, whereas others affect human respiration. Free chlorine is a very effective disinfectant, but it is being phased out by the U.S. EPA for treatment of surface waters (Sirikanchana et al. 2008) because of the risk of handling it and possible formation of toxic DBPs. An alternative treatment process that is being considered involves ultraviolet (UV) radiation at the treatment plant, followed by the longer lasting combined chlorine for the water distribution system.

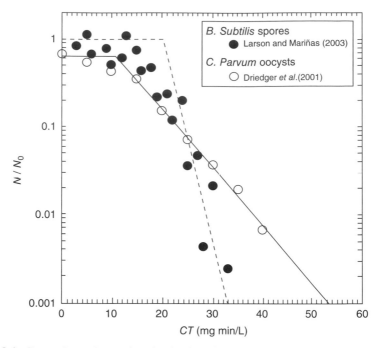

Figure 8.4. Comparison of ozone inactivation kinetics for *B. subtilis* spores (Larson and Marinas 2003) and *C. parvum* oocysts (Driedger et al. 2001). (Source: Adapted from Larson and Mariñas (2003))

Sirikanchana et al. (2008) investigated the disinfection kinetics of adenosine viruses using serotype 2 as a model because of its simpler culture methods and similar UV resistance. The kinetics followed a pseudo Chick–Watson expression at pH 6 and 8, and a Chick–Watson equation with a lag period at pH 10. The rate of inactivation decreased with increasing pH and increased with temperature. The rate constant followed an Arrhenius expression, Equation 8.25. The disinfection kinetics is significantly slower with monochloramine than with chlorine, meaning that monochloramine may not provide adequate adenosine control at low temperature and high pH.

A potential problem with chemical disinfectants is that bacteria and protozoa can develop resistance to these reagents. Examples include *Mycobacterium avium*, which can develop resistance to chlorination. Another mechanism for avoiding the effect of disinfection is the ability of *Legionella* spp. to hide or replicate within host organisms such as amoeba or protozoa. Viruses can resist UV treatment by rearranging DNA after treatment. In order to overcome these obstacles, Morse et al. (2010) suggested the application of gene silencing for microbial disinfection. They developed a proof of concept for microbial inactivation of *Pseudomonas putida* and *Pichia pastoris*. In one version of this technology, antisense deoxyoligonucleotide (ASO) gene silencing is used as a disinfection tool through binding of the ASO with its reverse complementary mRNA; thereby preventing the RNA from making protein. While gene silencing shows promise for microbial disinfection applications, there are still challenges to overcome, such as increasing the efficiency, and development of an effective and targeted delivery system for oligonucletides along with a comprehensive study of possible unwanted toxic effects.

REFERENCES

Ahmed, W, Sawant, S, Huyggens, F, Goonetilleke, A, Gardner, T. Prevalence and occurrence of zoonotic bacterial pathogens in surface waters determined by quantitative PCR. *Wat. Res.* 2009;43:4918–4928.

Alm, EW, Burke, J, Spain, A. Fecal indicator bacteria are abundant in wet sand at freshwater beaches. *Wat. Res.* 2003;37:3978–3982.

Almasi, A, Pescod, MB. Pathogen removal mechanisms in anoxic wastewater stabilization ponds. *Wat. Sci. Technol.* 1996;33(7):133–140.

Arvanitidou, M, Constantinidis, TC, Katsouyannopoulos, V. Searching for Eschirichia Coli 0157 in drinking and recreational waters in northern Greece. *Wat. Res.* 1996;30(2): 493–494.

Awuah, E, Anohene, F, Asante, K, Lubberding, H, Gijzen, H. Environmental conditions and pathogen removal in macrophyte- and algal-based domestic wastewater treatment systems. *Wat. Sci. Technol.* 2001;44(6):11–18.

Baque, RH, Gilliam, AO, Robles, LD, Jakubowski, W, Slifko, TR. A real-time RT-PCR method to detect viable *Giardia lamblia* cysts in environmental waters. *Wat. Res.* 2011;45:3175–3184.

Baker, KS, Smith, RS. Bio-optical classification and model of natural waters. *Limnol. Oceanogr.* 1982;27:500–509.

Bibby, K, Viau, E, Peccia, J. Pyrosequencing of the 16S rRNA gene to reveal bacterial pathogen diversity in biosolids. *Wat. Res.* 2010;44:4252–4260.

Boehm, AB, Fuhrman, JD, Mrše, RD, Grant, SB. Tiered approach for identification of a human fecal pollution source at recreational beach: Case study at Avalon Bay, Catalina Island, California. *Environ. Sci. Technol.* 2003;37(4):673–680.

Byappanahalli, MN, Sawdey, R, Ishii, S, Shively, DA, Ferguson, JA, Whitman, RL, Sadowsky, MJ. Seasonal stability of Cladophora-associated *Salmonella* in Lake Michigan watersheds. *Wat. Res.* 2009;43:806–814.

Christensen, ER, Ab Razak, IA, Phoomipakdeephan, W. Water quality in Milwaukee, Wisconsin versus intake crib location. *J. Environ. Engr. ASCE*, 1997;123(5):492–498.

Connolly, JP, Blumberg, AF, Quadrini, JD. Modeling fate of pathogenic organisms in coastal waters of Oahu, Hawaii. *J. Environ. Engr. ASCE*, 1999;125(5):398–406.

Da Costa, PM, Vaz-Pires, P, Bernando, F. Antimicrobial resistance in *Enterococcus* spp. isolated in inflow, effluent and sludge from municipal sewage water treatment plants. *Wat. Res.* 2006;40:1735–1740.

Driedger, AM, Rennecker, JL, Mariñas, BJ. Inactivation of *Cryptosporidium parvum* oocysts with ozone and monochloramine at low temperature. *Wat. Res.* 2001;35(1):41–48.

Droppo, IG, Liss, SN, Williams, D, Nelson, T, Jaskot, C, Trapp, B. Dynamic Existence of waterborne pathogens within river sediment compartments. Implications for regulatory affairs. *Environ. Sci. Technol.* 2009;43(6):1737–1743.

Edzwald, JK, Kelley, MB. Control of *cryptosporidium*: From reservoirs to clarifiers to filters. *Wat. Sci. Technol.* 1998;37(2):1–8.

Ford, TE. Microbial safety of drinking water: United States and Global perspectives. *Environ. Health Persp.* 1999;107:Suppl. 191–206.

Gallegos, CL, Correll, DL, Pierce, JW. Modeling spectral diffuse attenuation, absorption, and scattering coefficients in a turbid estuary. *Limnol. Oceanogr.* 1990;35:1486–1502.

Girones, R, Ferrús, MA, Alonso, JL, Rodriguez-Manzano, J, Calgua, B, De Abreu Corrêa, A, Hundesa, A, Carratala, A, Bofill-Mas, S. Molecular detection of pathogens in water - The pros and cons of molecular techniques. *Wat. Res.* 2010;4325–4339.

Grant, SB, Sanders, BF, Boehm, AB, Redman, JA, Kim, JH, Mrše, RD, Chu, AK, Gouldin, M, McGee, CD, Gardiner, NA, Jones, BH, Svejkovsky, J, Leipzig, GV, and Brown, A. Generation of Enterococci bacteria in a coastal saltwater marsh and its impact on surf zone water quality. *Environ. Sci. Technol.* 2001;35(12):2407–2416.

Guy, RA, Payment, P, Krull, UJ, Horgen, PA. Real-time PCR for quantification of Giardia and *Cryptosporidium* in Environmental water samples and sewage. *Appl. Environ. Microbiol.* 2003;69(9):5178–5185.

Haack, SK, Fogarty, LR, Wright, C. *Escherichia coli* and Enterococci at beaches in the Grand Traverse Bay, Lake Michigan: Sources, characteristics, and environmental pathways. *Environ. Sci. Technol.* 2003;37(15):3275–3282.

Henrickson, SE, Wong, T, Allen, P, Ford, T, Epstein, PR. Marine swimming-related illness; implications for monitoring and environmental policy. *Environ. Health Persp.* 2001;109(7):645–650.

Hunt, NK, Mariñas, BJ. Inactivation of *Escherichia coli* with ozone: Chemical and inactivation kinetics. *Wat. Res.* 1999;33(11):2633–2641.

Ho, BSW, Tam, T-Y. Rapid enumeration of *Salmonella* in environmental waters and wastewater. *Wat. Res.* 2000;34(8):2397–2399.

Huck, PM, Coffey, BM, Anderson, WB, Emelko, MB, Maufizio, DD, Slawson, RM, Douglas, IP, Jasim, SY, O'Melia, CR. *Wat. Sci. Technol.* 2002;2(3):65–71.

Jacobson, KH, Lee, S, McKenzie, D, Benson, C, Pedersen, JA. Transport of the pathogenic prion protein through landfill materials. *Environ. Sci. Technol.* 2009;43(6):2022–2028.

Kinzelman, J, Ng, C, Jackson, E, Gradus, S, Bagley, R. Enterococci as indicators of Lake Michigan recreational water quality: Comparison of two methodologies and their impact on public health regulatory events. *Appl. Environ. Microbiol.* 2003;69(1):92–96.

Kirk, JTO. Dependence of relationship between inherent and apparent optical properties of water on solar altitude. *Limnol. Oceanogr.* 1984;29:350–356.

Knudson, GB. Photoreactivation of ultraviolet-irradiated, plasmid-bearing, and plasmid-free strains of *Bacillus anthracis*. *Appl. Environ. Microbiol.* 1986;52:544–449.

Larson, MA, Marinas, BJ. Inactivation of *Bacillus subtilis* spores with ozone and monochloramine. *Wat. Res.* 2003;37:833–844.

Lévesque, B, Brousseau, P, Bernier, F, Dwailly, É, Joly, J. Study of the bacterial content of ring-billed gull droppings in relation to recreational water quality. *Wat. Res.* 2000;34(4):1089–1096.

Liviac, D, Wagner, ED, Mitch, WA, Altonji, MJ, Plewa, MJ. Genotoxicity of water concentrates from recreational pools after various disinfection methods. *Environ. Sci. Technol.* 2010;44(9):3527–3532.

López-Pila, JM, Szewzyk, R. Estimating the infection risk in recreational waters from the faecal indicator concentration and from the ratio between pathogens and indicators. *Wat. Res.* 2000;34(17):4195–4200.

Masago, Y, Katayama, H, Hashimoto, A, Hirata, T, Ohgaki, S. *Wat. Sci. Technol.* 2002;45 (11–12):319–324.

Medema, GJ, Schijven, JF. Modelling the sewage discharge and dispersion of *Cryptosporidium* and Giardia in surface water. *Wat. Res.* 2001;35(18):4307–4316.

Momba, MNB, Cloete, TE, Venter, SN, Kfir, R. Examination of the behaviour of *Escherichia coli* in biofilms established in laboratory-scale units receiving chlorinated and chloraminated water. *Wat. Res.* 1999;33(13):2937–2940.

Morse, TO, Morey, SJ, Gunsch, CK. Microbial inactivation of *Pseudomonas putida* and *Pichia pastoris* using gene silencing. *Environ. Sci. Technol.* 2010;44(9):3293–3297.

Nicholson, WL, Galeano, B. UV resistance of Bacillus anthracis spores revisited: Validation of Bacillus subtilis spores as UV surrogates for spores of B. anthracis Sterne. *Appl. Environ. Microbiol.* 2003;69(2):1327–1330.

Parkhurst, DF, Stern, DA. Determining average concentrations of *Cryptosporidium* and other pathogens in water. *Environ. Sci. Technol.* 1998;32(21):3424–3429.

Rathbone, P-A, Livingstone, DJ, Calder, MM. Surveys monitoring the sea and beaches in the vicinity of Durban, South Africa: A case study. *Wat. Sci. Technol.* 1998;38(12):163–170.

Sidhu, J, Gibbs, RA, Ho, GE, Unkovich, I. The role of indigenous microorganisms in suppression of *salmonella* regrowth in composted biosolids. *Wat. Res.* 2001;35(4):913–920.

Sirikanchana, K, Shisler, JL, Mariñas, BJ. Inactivation kinetics of adenovirus serotype 2 with monochloramine. *Wat. Res.* 2008;42:1467–1474.

Tanji, Y, Mizoguchi, K, Yoichi, M, Morita, M, Kijima, N, Kator, H, Unno, H. Seasonal change and fate of coliphages infected to *Escherichia coli* O157:H7 in a wastewater treatment plant. *Wat. Res.* 2003;37:1136–1142.

Wade, TJ, Pai, N, Eisenberg, JNS, Colford, Jr, JM. Do U.S. Environmental Protection Agency water quality guidelines for recreational waters prevent gastrointestinal illness? A systematic review and meta-analysis. *Environ. Health Persp.* 2003;111(8):1102–1109.

Zhang, Z, Deng, Z, Rusch, KA. Development of predictive models for determining enterococci levels at Gulf Coast beaches. *Wat. Res.* 2012;46:465–474.

9

TRACERS

9.1 INTRODUCTION

Tracers are useful to determine pathways and dispersion patterns of pollutants in natural waters. For example, pathogens in the effluent from a sewage treatment plant can cause authorities to close beaches. In a case where there are several possible contributing outfalls, it is desirable to determine the most likely source for pathogens and the potential severity of impact. For this purpose, measuring the levels of pathogens in the outfall plume may not be the best approach because of the intermittent nature of their appearance and their difficulty of measurement.

Outfall plumes may instead be determined from another distinct constituent of the effluent such as turbidity or chloride. Alternatively, an artificial tracer, such as a conventional or fluorescent dye, e.g. rhodamine, may be added to the effluent. The level of this tracer is subsequently measured either by field sampling or remote sensing. This will provide a transient or steady-state picture of the plume and the potential impact of any substances in the effluent, including pathogens, the transport of which is similar to that of the tracer. Desirable properties of a tracer are: The tracer should be (a) easily detectable, even in small concentrations, (b) inert chemically with no precipitation or chemical reactions, and (c) non-hazardous with low levels of radioactivity, if any, and short-lived.

An example of a pathogen in sewage effluent or urban runoff entering the drinking water supply is the Milwaukee *Cryptosporidium* event (Fox and Lytle 1994; Christensen et al., 1997a). *Cryptosporidium*, a parasitic protozoan, caused a gastrointestinal illness, cryptosporidiosis, in a large population segment. More than 400,000 persons were affected with around 100 fatalities in people with weakened immune

Physical and Chemical Processes in the Aquatic Environment, First Edition. Erik R. Christensen and An Li.
© 2014 John Wiley & Sons, Inc. Published 2014 by John Wiley & Sons, Inc.

systems. In this case, the outfall plume from the river flows and the Jones Island wastewater treatment plant exhibit some overlap with the two water intakes, particularly the southern one for the Howard Avenue Filtration Plant. To improve the quality of the intake water, the Howard Avenue intake pipe has been extended by 1.3 km (0.8 miles), sand filters are being replaced with dual media filters, and ozone disinfection is being installed to supplement chlorination. For the design of the intake pipe extension, the pollution plume was estimated based on tracer studies and 2-dimensional hydrodynamic and water quality modeling.

Computer modeling of outfall plumes and near-coastal ocean or lake circulation is an important tool to estimate the impacts of pathogens and other pollutants, such as nutrients, on the receiving waters. Eutrophication in nearshore ocean water is often caused by excess nitrogen and in lakes by excess phosphorus. Tracers serve both to verify computer models and as an independent means of characterizing pollutant transport and dispersal.

9.2   NATURAL VS. ARTIFICIAL TRACERS

A natural tracer is a tracing feature that is already present in the aquatic system that is under investigation. Examples are turbidity, chlorophyll a, color, chloride, and temperature. These parameters can be determined by field sampling, or in some cases by remote sensing, e.g. by aerial photography, coastal zone scanning, AVHHR, or thematic mapping (Mortimer 1988; Lathrop et al., 1990). For example, the effluent plume from Fox River into Green Bay, Wisconsin, can be estimated by the thematic mapper using temperature-dependent infrared emissions from the water surface. Such satellite photos clearly show that the plume follows the west coast of the Door County peninsula under the influence of currents and the earth's rotation through the Coriolis force. Effluent plumes from river mouths or sewage treatment plant outfalls can often be followed using aerial photographs to identify their color or turbidity. If high resolution and low detection limits are desired, addition of an artificial tracer, such as the fluorescent dye rhodamine, may be warranted.

Complex organic contaminants, such as PCBs, PAHs, and dioxins/furans that are already present in the environment, can function as tracers for their sources using their composition profiles or time records in dated sediments (Rachdawong and Christensen 1997; Christensen and Zhang 1993; Su and Christensen 1997). Care must be taken to select marker compounds that are not subject to excessive differential degradation or partitioning.

Similarly, radionuclides that are already in the environment, such as ^{210}Pb, ^{232}Th, and ^{7}Be, are useful tracers for sedimentation processes. Other radionuclides, e.g. ^{137}Cs and tritium, which were introduced with nuclear weapons testing in the 1950s and 1960s, have also found important applications in limnology and hydrology.

Stable isotopes, e.g. ^{18}O, ^{16}O, H, and D, are excellent tracers of the transport of water masses both for surface and ground waters. Other stable isotopes, e.g. ^{14}N, ^{15}N, ^{13}C, and ^{12}C, are useful for checking physical/chemical processes or transfer of materials, including organic contaminants, through the aquatic food chain. Stable isotope analysis has become more widely available with the advent of accelerator mass spectrometry.

9.3   RADIOISOTOPES

Table 9.1 shows a list of radioisotopes that are commonly used as radioactive tracers in limnology, oceanography, and environmental engineering. Radioactive decay is governed by the equation,

$$\frac{dN}{dt} = -\lambda N;$$ (9.1)

where
N = No. of radionuclides at time t
λ = decay constant

Thus, the number of nuclides N_2 at time t_2 is given in terms of the number of nuclides N_1 at time t_1:

$$N_2 = N_1 e^{-\lambda(t_2 - t_1)}$$ (9.2)

The half-life $T_{1/2}$ is then,

$$T_{1/2} = \frac{\ln 2}{\lambda};$$ (9.3)

Goldberg (1963) first demonstrated how ^{210}Pb could be used to date snowfields in Greenland. Following this research, Krishnaswami et al. (1971) and Koide et al. (1972) dated lake and nearshore ocean sediments respectively, based on this isotope.

TABLE 9.1. Radioisotopes Used as Tracers in the Study of Aquatic Systems

Radioisotope	Half life	Origin	Application
^{210}Pb	22.26 y	^{238}U series	Sediment dating
^{222}Rn	3.83 d	^{238}U series	Mixing in lakes; Geothermal reservoirs; Seismic activity; Earthquake prediction
^{228}Th	1.90 y	^{232}Th series	Sediment dating
^{234}Th	24.1 d	^{238}U series	Sediment mixing
^{7}Be	53 d	Cosmogenic	Sediment mixing
^{137}Cs	30.15 y	Bomb fallout	Sediment dating
^{3}H (T-tritium)	12.33 y	Bomb fallout	Water mass ages in lakes and groundwater; Also injected as groundwater tracer
^{99m}Tc	6 h	Reactor produced ^{98}Mo $(\eta, \gamma, \beta^-)^{99m}$Tc	Residence time distribution in chlorination contact reactor in water or sewage treatment plant

Dating of nearshore ocean sediments with ^{228}Th was carried out by Koide et al. (1973), and of lake sediments with ^{137}Cs by Pennington et al. (1973).

Both ^{210}Pb and ^{137}Cs dating of sediments has been widely used since then by many investigators, e.g. Edgington and Robbins (1976) and Hermanson and Christensen (1991). The mobility of ^{137}Cs in marine sediments restricts the use of this radionuclide mainly to freshwater environments. Because of its short half-life (53 d), ^{7}Be is well suited for the measurement of sediment mixing (Krishnaswami and Lal 1978; Li et al., 1995). Bomb-generated tritium ($T_{1/2} = 12.33$ year) may be useful for another 50 year or so for determination of water mass ages of surface and groundwater, and vertical mixing in lakes (Torgerson et al., 1977). For measurements of residence time distributions in various reactor applications, ^{99m}Tc may be a good choice because of its short half-life of 6 h. However, this also means that supplies of this tracer don't last long; therefore fresh supplies may be needed for each application.

There is a special category of tracers that are stable when applied and sampled, but are made radioactive either as a result of neutron activation analysis (NAA), x-ray or ultraviolet/visible fluorescence when measured. Examples of tracers that contain elements that are non-radioactive when injected, but are made radioactive when measured by NAA are $Au(CN)_2^-$, $Au(CNS)_4^-$ (Au); Dy, In, and Br, and $MnNO_3$ (Mn). Similarly, compounds containing Y can be measured conveniently by x-ray fluorescence. Fluorescence spectroscopy obtained by combining emission spectra of organic matter (344–522 nm) at several excitation wavelengths (< 240–370 nm) can be used to obtain qualitative information about organic matter in a sample (Osburn et al., 2012).

In this way, very low detection limits may be obtained, enabling precise determination of the plume even at long distances from the point of application. If the choice is between injecting a hazardous tracer, such as tritium, and an activable tracer, e.g. bromide ion, for tracing groundwater movements (Jester et al., 1977), then bromide ion should be selected in order to minimize the risk of contaminating the subsurface. The need for activation is a complication though, and if possible it is preferable to use a distinct tracer that is already present in the water to be traced. Natural fluorescence with emission wavelengths in the near ultraviolet or visible range (346–500 nm), for example from animal wastes exhibiting a high ratio of protein-like to fulvic/humic-like fluorescence, has also been used as a tracer of diffuse agricultural pollution (Naden et al., 2010). Natural fluorescence is also emitted from chlorophyll a of algae.

9.4   STABLE ISOTOPES

Basic environmental processes can be followed by means of stable isotopic tracers (Hoefs 1987; Rundel et al., 1989). A partial list of common stable isotopes of interest for ecological research is given in Table 9.2a, 9.2b. The ratio between the abundance of two isotopes of a given element, e.g. $^{18}O/^{16}O$ in water, is characteristic of the source of the water and various physical chemical variables such as temperature, precipitation, evaporation, and chemical reactions. Isotope ratios are commonly measured in

TABLE 9.2a. Average Terrestrial Abundances of the Stable Isotopes of Selected Elements of Interest in Environmental Studies

Element	Isotope	Abundance (%)
Hydrogen	^{1}H	99.985
	^{2}H	0.015
Lithium	^{6}Li	7.42
	^{7}Li	92.58
Boron	^{10}B	19.61
	^{11}B	80.39
Carbon	^{12}C	98.893
	^{13}C	1.107
Nitrogen	^{14}N	99.634
	^{15}N	0.366
Oxygen	^{16}O	99.759
	^{17}O	0.0375
	^{18}O	0.2039
Sulfur	^{32}S	95.0
	^{33}S	0.760
	^{34}S	4.22
	^{36}S	0.014
Copper	^{63}Cu	69.09
	^{65}Cu	30.91
Zinc	^{64}Zn	48.89
	^{66}Zn	27.81
	^{67}Zn	4.11
	^{68}Zn	18.57
	^{70}Zn	0.62

(Source: Friedlander et al. (1964)).

per mil (‰) deviation from a standard using the notation,

$$\delta = \left(\frac{R_x}{R_{Std}} - 1 \right) * 1000; \tag{9.4}$$

where R_x and R_{std} are the isotope ratios of the sample and the standard respectively. Standards for isotopic ratios of hydrogen, oxygen, carbon, and nitrogen are Standard Mean Ocean Water (SMOW), Pee Dee Belemnite (PDB), and atmospheric air (Table 9.3). Thus using SMOW as standard, ^{18}O/^{16}O ratios of 0.1995% and 0.1981% are expressed as −5.08 and −12.07 per mil respectively.

For meteoric surface waters, e.g. waters that take direct part in the hydrological cycle as opposed to ocean waters, there is a direct correlation between δ^{18}O and δD, e.g.

$$\delta^2 H = 8 \times \delta^{18} O + 5 \tag{9.5}$$

TABLE 9.2b. Average Terrestrial Abundances of the
Stable Isotopes of Selected Elements of Interest in
Environmental Studies

Element	Isotope	Abundance (%)	Half-life (yr)
Strontium	^{84}Sr	0.56	
	^{86}Sr	9.86	
	^{87}Sr	7.02	
	^{88}Sr	82.56	
Iodine	^{127}I	100	
	^{129}I		1.6×10^7
Mercury	^{196}Hg	0.146	
	^{198}Hg	10.02	
	^{199}Hg	16.81	
	^{200}Hg	23.13	
	^{201}Hg	13.22	
	^{202}Hg	29.80	
	^{204}Hg	6.85	
Lead	^{204}Pb	1.48	
	^{206}Pb	23.6	
	^{207}Pb	22.6	
	^{208}Pb	52.3	
Uranium	^{234}U	0.0056	2.48×10^5
	^{235}U	0.7205	7.13×10^8
	^{236}U		2.4×10^7
	^{238}U	99.274	

(Source: Friedlander et al. (1964)).

TABLE 9.3. Isotopic Compositions of Primary Standards

Primary standard	Isotope ratio	Accepted value ($\times 10^6$) with 95% confidence interval
Standard Mean Ocean Water (SMOW)	$^2H/^1H$	155.76 ± 0.10
	$^{18}O/^{16}O$	2005.20 ± 0.43
	$^{17}O/^{16}O$	373 ± 15
Pee Dee Belemnite (PDB)	$^{13}C/^{12}C$	11237.2 ± 9.0
	$^{18}O/^{16}O$	2067.1 ± 2.1
	$^{17}O/^{16}O$	379 ± 15
Air	$^{15}N/^{14}N$	3676.5 ± 8.1

(Source: Ehleringer and Rundel (1989)).

indicating that the lighter isotope of oxygen occurs along with the lighter isotope of hydrogen, as in the vapors of water evaporated from a lake. At higher altitudes and latitudes, the light isotopes prevail in freshwaters because of repeated cycles of condensation and evaporation, whereas the heavier isotopes are predominant in tropical waters and show small depletions relative to ocean water. This equation was refined by Rozanski et al. (1993) using Vienna Standard Mean Ocean Water (VSMOW) as reference,

$$\delta^2 H = 8.17 \times \delta^{18}O + 10.4 \tag{9.6}$$

During cold periods of the earth's climatic history, the ice caps are large and have low $\delta^{18}O$ values, whereas the oceans have correspondingly higher δO readings. A decrease in δO in the oceans after the meltwater spike from the last glacial period has been found in microfossils of dated ocean sediment cores (Berger et al., 1977; Schrag et al., 1996).

Nitrogen isotope fractionation has been used to study biomagnification in food webs. Rolff et al. (1993) found $\delta^{15}N$ increases of about 1 to 8 and 1 to 11 in a littoral and pelagic food chain respectively, from the Baltic. Thus, metabolic fractionation of stable isotopes of nitrogen can be used to numerically estimate the trophic position of organisms. Since concentrations of many organic contaminants increase up through the food chain, i.e. the organics undergo biomagnification, one could expect a correlation between $\delta^{15}N$ values and concentrations of organic contaminants. Such correlations were found for a pelagic food web in a subarctic lake by Kidd et al. (1995). They studied hexachlorocyclohexanes, HCH, DDT, and chlorinated bornanes (toxaphenes, CHB). The highest correlations occurred for compounds that biomagnify significantly, e.g. DDT and CHB, and that the correlation was less significant for HCH, which undergoes a lower degree of biomagnification. Concentrations of chlorinated hydrocarbons may, therefore, be predicted from $\delta^{15}N$.

A study by Muir et al. (1995) considering a community of walruses on the east coast of Hudson Bay (Inukjuak) with elevated organochlorine concentrations, found that these walruses were feeding at a higher trophic level than others with low concentration levels, and that they were probably using ringed seals for a portion of their diet rather than only bivalves. $\delta^{15}N$ measurements were used to characterize trophic status in the food web. The heavy isotope of nitrogen is enriched in the predator relative to its food, while carbon isotope ratios usually show much less trophic enrichment. Thus $\delta^{15}N$ indicates trophic position while $\delta^{13}C$ reflects the food source (Muir et al., 1995).

Similar conclusions were reached by Kiriluk et al. (1995). They used stable isotopes of nitrogen ($\delta^{15}N$) and carbon ($\delta^{13}C$) to describe the trophic status and interactions of a Lake Ontario pelagic food web. The enrichment of ^{15}N through the food web was correlated with the biomagnification of the persistant lipophilic contaminants p,p'-DDE, mirex, and PCB. $\delta^{13}C$ signatures, on the other hand, were characteristic of the diet of four fish species; lake trout, alewife, rainbow smelt, and shing sculpin.

A recent investigation of the Gulf of the Farallones food web (euphasiids, fish, marine birds, sea lions) considering DDT, HCH (hexachlorocyclohexane), chlordane, hexachlorobenzene, Pb, Hg, and Se demonstrated likewise that $\delta^{15}N$ correlates with trophic level and the concentrations of organochlorine compounds, as well as of Hg (Jarman et al., 1996). The values of $\delta^{13}C$ that ranged from -20.1 per mil in euhasiids to -15 per mil in the northern sea lion were indicative of benthic vs. pelagic feeding.

Stable carbon and oxygen isotope analyses may be used to monitor and quantify biodegradation of diesel fuel or chlorinated hydrocarbons. One of the main advantages of using oxygen isotope analyses for monitoring biodegradation processes is that the $\delta^{18}O$ of oxygen in molecular oxygen, as well as in other common electron acceptors (nitrate and sulfate), changes primarily as a result of microbial respiration rather than inorganic redox reactions (Aggarwal et al., 1997).

Stable lead isotopes have been used to characterize the origin of lead in the environment. While ^{204}Pb is not known to have been created by any radioactive decay process, ^{206}Pb is the endpoint of the ^{238}U series, ^{207}Pb of the ^{235}U series, and ^{208}Pb of the ^{232}Th series (Friedlander et al., 1964). Until 1968, the majority of lead additives in gasoline had a $^{206}Pb/^{204}Pb$ ratio of 17.9 to 18.8 and a $^{206}Pb/^{207}Pb$ ratio of 1.15 to 1.19. Most of the lead was imported at the time from Mexico, Peru, and Canada (Chow et al., 1975). After that time, and until lead in gasoline was gradually phased out in the late 1970s, the proportion of radiogenic lead was significantly higher, i.e. the $^{206}Pb/^{204}Pb$ ratio was higher than 18.8 and the $^{206}Pb/^{207}Pb$ ratio was higher than 1.19. Isotope ratios of aerosol lead in San Diego, California, followed this trend.

The shift towards a higher proportion of radiogenic lead was caused by a reduction in lead imports and an increase in domestic lead production from the Missouri district, where these ratios are high, i.e. 21.78 for $^{206}Pb/^{204}Pb$ and 1.385 for the $^{206}Pb/^{207}Pb$ ratio. Other sources of lead in the environment are coal burning (Chow and Earl 1970) and smelter emissions (Rabinowitz and Wetherill, 1972). The lead isotope ratios in North American coal samples are similar to the lead isotope ratios of lead additives that were used in the 1970s, making a source differentiation difficult for this period. Further discussion of this subject is provided in Chapter 2. Whether today's lead in the environment can be classified according to lead isotope ratios is a matter which is under active investigation.

9.5 APPLICATIONS OF TRACER TECHNOLOGY

9.5.1 Stable Isotope Tracers

Examples of stable isotope tracers used in ecological studies are listed in Table 9.4. In order to evaluate multiple sources of water in aquifers, the use of environmental isotopes and geochemical analysis can be very useful. Kloppmann et al. (2008a) examined the use of ions and stable isotopes as tracers of freshwaters that were produced by four reverse osmosis (RO) desalination plants, including two in Ashkelon and Eilat, Israel. The Ashkelon plant uses seawater as feed (SWRO), and the Eilat plant uses brackish groundwater (BWRO). Ions of boron, lithium, and strontium were measured along with the stable isotope ratios $^2H/^1H$, $^7Li/^6Li$, $^{11}B/^{10}B$, $^{18}O/^{16}O$, and $^{87}Sr/^{86}Sr$.

O, H, and Sr isotopes are not fractionated during the RO process. However, fractionation of Sr can occur during post treatment mainly in the form of carbonate dissolution. ^7Li is preferentially rejected at low pH SWRO. Fractionation of B depends on pH. At low pH, which is typically maintained to prevent scale formation, there is

TABLE 9.4. Examples of Stable Isotope Tracers Used in Studies of Aquatic Systems

Tracers stable isotopes of elements	Application	Reference
B, Li, O, H, Sr	Reverse osmosis desalination	(Kloppmann et al., 2008a)
B, Li, O, H	Artificial recharge of groundwater	(Kloppmann et al., 2008b)
C, N	Hg biomagnification in fish	(Al-Reasi et al., 2007)
Hg	Contribution of anthropogenic Hg in fish from a contaminated lake	(Perrot et al., 2010)
O	Determination of phosphorus sources	(Elsbury et al., 2009) (Young et al., 2009)
Zn and Cu	Origin of contamination in sediments from urban lake	(Thapalia et al., 2010)
Pb	Contribution of natural vs. anthropogenic lead in sediments	Chapter 2 of this book
U	Tracking U of Hanford 300 area in the Columbia River	(Christensen et al., 2010)

little fractionation, whereas at high pH, the ^{11}B of the permeate is increased by 20‰ (SWRO) as a result of selective rejection of the borate ion and preferential permeation of boric acid ^{11}B(OH)$_3$. The δ^{11}B of seawater near Ashkelon was in the range 39.3–39.9‰ vs. NBS951 and the δ^7Li was 31.4‰. Note that the high boron content of RO-generated freshwater can limit its uses for drinking water and irrigation. Effluents from SWRO can be characterized by δ^7Li, δ^{11}B signatures and BWRO may have distinct δ^{11}B values. Assuming that the groundwater without RO recharge has significantly different δ-values, these stable isotopes can serve as geochemical tracers for an RO component of the recharge water in aquifers.

In a follow up study, Kloppmann et al. (2008b) conducted a detailed stable isotope investigation of the artificial recharge (AR) of a coastal dune aquifer in Flanders, Belgium. They considered stable isotopes of B, Li, O, and H in tertiary treated, RO-desalinated domestic wastewater. There was no significant isotope fractionation through the treatment process, which included low pH RO desalination. The treated wastewater was collected after infiltration through ponds and before recovery through pumping wells. The Cl/B ratios were low (3.3 to 5.2) compared to the general study area (130 to 120), and the δ^{11}B ($\sim$ 0‰) and δ^7Li values (10.3 ± 1.7‰) were significantly different from those of the groundwater in the study area. The δ^2H and δ^{18}O values were enriched relative to ambient groundwater as a result of evaporation in the infiltration ponds. This indicates that these stable isotope tracers can complement and constrain AR recoveries obtained by, for example, hydrodynamic modeling.

In order to track mercury biomagnification in the Gulf of Oman, Al-Reasi et al. (2007) measured δ^{13}C and δ^{15}N and total mercury T-Hg and methyl mercury MeHg in 13 fish species. Concentrations of T-Hg in zooplankton measured 21 ± 8.0 ng/g with about 10‰ MeHg, whereas T-Hg concentrations in the fish varied between 3.0 ng/g and 760 ng/g with a MeHg fraction of about 72‰. δ^{13}C is used for tracking carbon flow as in bioaccumulation, whereas $\delta^{15}N$ indicates trophic position or bio-magnification. For $\delta^{13}C$ the researchers determined a higher than expected trophic difference $\Delta\delta^{13}C$ of 3.4‰ between zooplankton and planktivorous fish, and for $\delta^{15}N$ the enrichment at each trophic level reached 3.1‰. The high $\Delta\delta^{13}C$ may suggest that zooplankton may not be the main diet for these fish species. The relatively weak correlation between both T-Hg and MeHg and $\delta^{15}N$ shows that biomagnification in this area is lower than in arctic or temperate areas or African lakes. The correlation may be a result of highly variable T-Hg concentrations in diverse diets consisting of fish with similar $\delta^{15}N$ values. It was determined that the limit of 230 ng/g daily intake of MeHg set by the World Health Organization was met for all fish species except the milk shark (*Rhizoprionodon acutus*), the longtail tuna (*Thunnus tonggol*), the grouper (*Epinephelus epistictus*), and Kawakawa (*Euthynnus affinis*).

The isotopic composition of Hg can also be used to trace sources and bioaccumulation of mercury. Perrot et al. (2010) studied Hg stable isotopes and $\delta^{13}C$ and $\delta^{15}N$ in perch, roach, plankton, and surface sediments from two freshwater systems: Lake Baikal, which is relatively pristine, and a site near the Bratsk Water Reservoir and Angara River, Russia, which is contaminated by mercury pollution from a chlor-alkali plant. The Hg analysis was performed to obtain $\delta^{204}Hg$, $\delta^{202}Hg$, $\delta^{201}Hg$, $\delta^{200}Hg$, $\delta^{199}Hg$ with ^{198}Hg (NIST 3133) as a standard. Both mass-dependent and mass-independent mercury fractionation was considered.

Both regions show correlation of $\delta^{13}C$ and $\delta^{15}N$ with concentrations of methyl mercury in organisms, indicating bioaccumulation and biomagnification. Fish from Lake Baikal show a further correlation of $\delta^{202}Hg$ with the trophic level of organisms. Also, there was a correlation between mass-independent fractionation and total mercury for the Bratsk reservoir. The sediment Hg composition for Bratsk sediment fitted this correlation, pointing to the sediments as a source of the mercury.

Tracing phosphate sources in rivers and lakes is important in order to be able to control the growth of nuisance algae (Zou and Christensen 2012). While this may be undertaken by considering the load from each potential source, there are cases where the loads cannot easily be distinguished, and alternative methods must be explored. Stable isotope techniques can be useful for this purpose. However, because there is only one stable phosphorus isotope ^{31}P, the possibility of using $\delta^{18}O$ of the oxygen in phosphate to serve as a tracer has been considered (Young et al., 2009; Elsbury et al., 2009).

The phosphorus–oxygen bond is resistant to hydrolysis at most ambient temperatures and pressures so that phosphate only exchanges oxygen with water through biological processes. The equilibrium $\delta^{18}O_p$ was calculated by the following empirical equation used in Young et al. (2009),

$$T(^{\circ}C) = 111.4 - 4.3(\delta^{18}O_p - \delta^{18}O_w) \tag{9.7}$$

where $\delta^{18}O_p$ is the oxygen isotope ratio of dissolved phosphate in equilibrium with the water of oxygen isotope ratio $\delta^{18}O_w$, and T is the water temperature. It has been shown that the oxygen isotopic composition of dissolved phosphate is not typically in equilibrium with ambient water, and the measured oxygen ratio $\delta^{18}O_p$ of phosphate may therefore serve as a tracer of phosphate sources.

Young et al. (2009) showed that the mean $\delta^{18}O_p$ of several sources, including fertilizers, aerosols, wastewater treatment effluents, and detergents, varied between +8.4 and +24.9‰, and that there were significant differences between several source categories. They based this on their own analyses and on literature values, and concluded that $\delta^{18}O_p$ values of water bodies have a good potential to serve as tracers of phosphate sources.

A similar conclusion was reached by Elsbury et al. (2009). They studied oxygen isotope ratios of water samples from Lake Erie. $\delta^{18}O_p$ values in the lake measured between +10‰ and +17‰, whereas the equilibrium value was expected to measure approximately +14‰. The riverine weighted average $\delta^{18}O_p$ measured +11‰. The researchers concluded that there must be other sources of phosphorus with higher isotope ratios. One such source could be phosphorus that was released from bottom sediments to the hypolimnion during anoxic conditions. Measurements in the stratified lake in the Central Basin in June 2006 revealed, in fact, high $\delta^{18}O_p$ values at 20 m depth, thus supporting the idea that phosphorus from reducing bottom sediments could be a source. More work on the use of stable oxygen as a phosphate tracer, including characterizing oxygen ratios of potential phosphate sources, is warranted.

Stable isotope techniques have also been used in conjunction with zinc and copper concentrations in dated sediments to identify pollution sources of these metals in an urban lake (Thapalia et al., 2010). The authors studied Lake Ballinger, Washington, United States, which is located approximately 53 km north of a lead and copper smelter which operated between 1900 and 1985 near Seattle. Zinc and Cu isotopic data were collected for the exchangeable, organic, and acid-labile fractions. Isotope ratios were determined for ^{66}Zn, ^{67}Zn, and ^{68}Zn with ^{64}Zn as a standard. For copper, there is only one isotopic ratio, $^{65}Cu/^{63}Cu$. Lighter Zn isotopes are emitted to the atmosphere, whereas tailings and slags are enriched in heavier isotopes. The isotopic composition of copper depends on redox and other chemical and complexation reactions. An important additional source of zinc in the urban environment is automobile tire wear (Christensen and Guinn 1979), while copper in urban runoff can originate, for example, from automobile brake linings. Five tire samples were analyzed for zinc concentrations and stable zinc isotopes, and it was found that concentrations were relatively high after 1945, reflecting increasing urbanization. The average $\delta^{66}Zn$ ratio was $+0.04 \pm 0.2$‰2σ.

The dated sediment core made it possible to consider four time periods: Pre-smelter before 1900, smelter between 1900 and 1945, smelter + urbanization 1945–1985, and post-smelter after 1985. Isotopic ratios for the pre-smelter period were $\delta^{66}Zn = +0.39 \pm 0.09$‰ and $\delta^{65}Cu = + 0.77 \pm 0.06$‰. During smelter operations the ratios were $\delta^{66}Zn = +0.14 \pm 0.06$‰ and $\delta^{65}Cu = +0.94 \pm 0.10$‰, and in the post-smelter period after 1985 they were $\delta^{66}Zn = +0.00 \pm 0.10$‰ and $\delta^{65}Cu = +0.82 \pm 0.12$‰. The lower isotopic ratios for zinc during smelter operation compared to the pre-smelter time are consistent with atmospheric deposition of smelter exhaust. In addition, the lower $\delta^{66}Zn$ values that remained after 1985 reflect increasing urbanization and accompanying automobile tire wear in the post-smelter period. For copper, the reasons for the temporal variation in δ^{65}Cu ratios are less clear. However, it appears that δ^{65}Cu emissions from the smelter reflect the δ^{65}Cu of the feedstock. Zinc isotopes were used as a probe of anthropogenic contamination in the Seine River basin, France (Chen et al. 2008). Although care must be taken to consider background emissions, it appears that stable isotopic ratios for both Zn and Cu are useful indicators of the sources of these metals.

Stable isotopes of lead have successfully been used to characterize sources of anthropogenic lead, which was used as an antiknock agent in gasoline until the 1970s in the United States. Further discussion of this is provided in Chapter 2.

Isotopic fractionation was used to track uranium from the Hanford 300 area in the Columbia River by Christensen et al. (2010). Uranium from the Hanford site has a wide range of ^{235}U/^{238}U and ^{236}U/^{238}U ratios as a result of different degrees of enrichment in nuclear reactors and industrial processes. This contrasts with natural uranium, which has a nearly constant ^{235}/U/^{238}U ratio of 0.0725, close to zero ^{236}U/^{238}U ratio, but variable ^{234}U/^{238}U ratio as a result of alpha recoil effects and chemical reactions. This significant difference in isotopic fractionation makes it possible to determine the fraction of U derived from the Hanford 300 area in the Columbia River. Even for equal atomic ratios of ^{234}U and ^{236}U relative to ^{238}U, the Hanford derived U containing ^{236}U has a ~ 100 times lower activity due to the 96 times longer half-life of ^{236}U compared to that of ^{234}U. Near the Hanford site it was found that the U from the Hanford 300 area measured between 4 and 0.6% respectively, during fall and spring, of the natural dissolved load of Columbia River. It was therefore concluded that the ecological significance of the Hanford-derived load was negligible compared to that of the natural Uranium.

9.5.2 N and O Stable Isotopic Compositions of Nitrate Sources

Nitrate is the dominant form of nitrogen in most waters. It is also toxic to humans in high concentrations (>10 mg/L) as noted in chapter 5, and it is therefore of interest to be able to determine sources of nitrate for the purpose of remediation. Stable isotopic methods for this have been developed and described by, for example Liu et al. (2006), Li et al. (2010), and Huang (2006). Because of relatively high nitrate concentration, karst groundwater can be especially susceptible to nitrate pollution. In their work on the application of δ^{15}N and δ^{18}O to identify nitrate sources in karst groundwater near Guiyang, southwest China, Liu et al. (2006) demonstrated that there is a large variation of nitrate sources between winter and summer, indicating a rapid response to rain or surface water in the karstic groundwater area. The association of δ^{15}N $- $ NO_3^- with nitrate in summer indicates mixing of old groundwater and meteoric water with soil organic nitrogen and fertilizers. Similarly, the NO_3^-/Cl^- ratio vs. chloride indicates mixing of agricultural inputs with nitrate in sewage effluent that is influenced by denitrification. The average values of δ^{15}N and δ^{18}O of nitrate were generally higher in winter than in summer, which may be related to denitrification in winter. For suburban groundwater, nitrification of organic nitrogen in soil and chemical fertilizers were major nitrate sources; whereas nitrate in urban groundwater was controlled by wastewater effluent and denitrification.

Denitrification was shown even more clearly in a study of sources of nitrate in the Changjiang River, China (Li et al., 2010). There was a clear positive correlation between $\delta^{18}O - NO_3^-$ and $\delta^{15}N - NO_3^-$ with a slope of 1.5, which is within the range of previously reported values 1.3–2.1. The ranges of $\delta^{15}N$ and $\delta^{18}O$ measured from 7.3 to 12.9‰ and 2.4–11.2‰ respectively, which is narrower than those identified in karstic groundwater. Furthermore, there was a distinct drop in $\delta^{18}O - NO_3^-$ around the Three Gorges Dam. The authors hypothesized that this may be caused by organic material produced by growth of algae and hydrophytes in the Three Gorges reservoir area. Major sources of nitrate to the Changjiang River were identified as nitrification of organic matter and nitrate in sewage effluents that were influenced by denitrification.

Results of an isotopic nitrate source investigation in the Illinois River, United States, are shown in Figure 9.1 (Huang 2006). Samples were obtained during May 2005 and August 2005, and October 2004 and October 2005. Forty-nine mid-channel sampling points were selected over a 222 km segment from Lockport to Peoria. In October 2004 the nitrate flux decreased from 29 t/d at Lockport to 13 t/d at Peoria. The isotopic compositions changed from $\delta^{15}N = +7.8‰$ and $\delta^{18}O = +0.10‰$ at Lockport to $\delta^{15}N = +13.5‰$ and $\delta^{18}O = +5.4‰$ at Peoria. This is consistent with the major nitrate source being upstream of Lockport, and that denitrification of sewage effluent, manure, and septic waste reducing the nitrate content at Peoria. The slope of a linear correlation between $\delta^{18}O$ and $\delta^{15}N$ is 1.1 in this case (Figure 9.1).

By comparison, during May 2005 the nitrate loading increased from 29 t/d at Lockport to around 131 t/d at Peoria, with isotopic compositions of $\delta^{15}N = +9.7‰$ and $\delta^{18}O = +2.3‰$ at Lockport to $\delta^{15}N = +8.9‰$ and $\delta^{18}O = +6.5‰$ at Peoria. This is indicative of an agricultural nitrate source that significantly increases the nitrate content of the river as it flows downstream from Lockport to Peoria. At Lockport, the isotopic composition is close to that of wastewater effluent from the Stickney Water Reclamation Plant in southwest metropolitan Chicago (SWRP).

9.5.3   Other Physical and Chemical Tracers

In addition to stable isotope tracers, several physical and chemical tracers have been successfully applied, including dyes and fluorescent dyes, pharmaceutically reactive tracers, polylactic acid (PLA) microspheres clay labeled with short strands of synthetic DNA and paramagnetic iron oxide nanoparticles, dissolved gaseous SF_6, and clay labeled with the rare earth metal Ho (Table 9.5).

Effective delivery of remediation amendments, e.g. nutrients, electron donors, and bacteria, for in situ remediation of contaminated clay till requires knowledge of the dispersion patterns of these amendments vertically and radially from injection wells (Christiansen et al., 2012). Remediation via excavation may not be practical or economically justifiable. There is a concern that natural high permeability fractures allow rapid vertical migration of contaminants, and diffusion of contaminant deposits in clay till into fractures that transport groundwater into underlying aquifers. Methods for injection of amendments include pneumatic fracturing, hydraulic fracturing, and direct-push delivery.

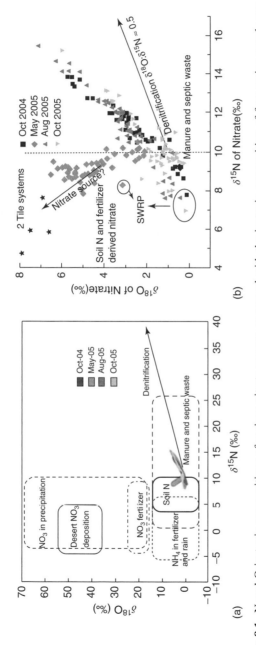

Figure 9.1. N and O isotopic compositions of various nitrate sources compared with the isotopic composition of four major sample sets. Part (a) compares fields of various sources of nitrate. Part (b) shows details of the data from this study. Arrow represents typical denitrification trend. Data from a tile study by (Panno et al., 2006) are also plotted. SWRP = Stickney Water Reclamation Plant (SWRP) in southwest metropolitan Chicago. (Source: Huang (2006)) (*See insert for color representation of the figure.*)

311

TABLE 9.5. Examples of Other Physical and Chemical Tracers Used in Studies of Aquatic Systems

Tracer	Application	Reference
Brilliant blue, fluorescein, and rhodamine WT dyes	Effective delivery of remediation amendments for in situ remediation of contaminated clay till sites	(Christiansen et al., 2012)
Uranine (Fluorescine sodium salt) and rhodamine WT dyes, conservative tracers; and pharmaceutically active substances, reactive tracers	Fate and transport, including attenuation mechanisms, of pharmaceuticals in rivers.	(Kunkel & Radke, 2011)
Blue dye; and polylactic acid (PLA) microspheres with DNA strands and iron oxide nanoparticles	Multiple and potentially interacting hydrological flowpaths.	(Sharma et al., 2012)
Sulfur hexafluoride, SF_6	Pathways and dilution of a point source ocean discharge in the farfield ($\approx 10-66$ km)	(Wanninkhof et al., 2005)
Holmium labeled montmorillonite clay	Fine sediment dynamics in storm water detention ponds.	(Spencer et al., 2011)

In an effort to evaluate the effectiveness of delivery of direct-push amendments, Christiansen et al. (2012) investigated the use of dye tracers for tracking the solute distribution in clay till. A mixture of tracers included brilliant blue and the fluorescent dyes Rhodamine WT and fluorescein, which was injected in depth intervals of 10–25 cm over total depths of 2.5–9.5 m below the surface. Injections occurred in both vadose and saturated zones. Natural fractures are generally more closely spaced above the redox boundary, which was found to be at 3–4 m below surface in the field site in Vadsby near Tåstrup, Denmark. A tracer distribution of at least 1 m was found at all depths. Rhodamine WT was found to be subject to absorption at radii between 0.75 and 1 m. The authors concluded that the direct-push method was effective for distribution of solutes until at least 10 m below the surface with vertical delivery point spacing of 10–25 cm.

Pharmaceutically active substances discharged to rivers can be diluted, degraded, or absorbed. In order to study the fate of such chemicals, a series of tracers were released as short impulses in Säva Brook, a small stream in central Sweden (Kunkel and Radke 2011). Two conservative dye tracers, Uranine and Rhodamine WT, were used along with six pharmaceuticals with expected different fate in rivers: Ibuprufen, declofenac, naproxen, bezafibrate, clofibric acid, and metoprolol. The plumes were followed for 60 h after injection over a 16.2 km stretch of the river. Both Uranine and Rhodamine WT were found to be conservative in these surface waters, although Rhodamine WT may be subject to sorption in porous media as shown, for example, by Christiansen et al. (2012). Of the pharmaceuticals, only ibuprofen and clofibric acid were attenuated along the river stretch with half-lives of 10 h and 2.5 d respectively. It was hypothesized that they were transformed by in-stream biofilms. Based on the shape of the breakthrough curves and the relatively low conductivity of the river bed it was assumed that exchange with the hyporheic zone was minor. Thus reactive tracer tests may be used to evaluate the fate of emerging micropollutants in rivers.

An interesting new development in the study of hydrological tracers is the use of polylactic acid (PLA) microspheres, in which strands of synthetic DNA and paramagnetic iron oxide nanoparticles are incorporated (Sharma et al., 2012). The DNA strands serve as tags or labels and the iron oxide particles facilitate magnetic up-concentration of labeled microspheres for accurate measurement. Attractive features of the microspheres are that the PLA degrades over several weeks or months, and that the rate of degradation can be regulated by changing the composition of the polymers by using suitable additives.

Proof of concept experiments were designed to compare tracer plumes with model results in column and plot experiments, and also to compare tracer and dye in field experiments. The field experiments were conducted in a small (<61 m stretch) stream near Ithaca, New York. This tracer technology appears to work well for the tracking of multiple hydrological flowpaths.

For the purpose of tracing effluent plumes in the farfield from point sources, Wanninkhof et al. (2005) injected gaseous sulfur hexafluoride SF_6 into a 3.5 km long waste discharge pipe carrying wastewater effluent into the Atlantic Ocean near Fort Lauderdale in southeast Florida. One concern is the discharge of nutrients and resultant algal growth in nearshore waters. The injection was carried out as a continuous operation for 6 d in June, 2004. The outfall terminus, which is a single release point without dispersers, is located about 1 m from the bottom at a depth of 27 m. Tracer measurements in the farfield, i.e. beyond 400 m from the outfall, are rare because they require low detection limit of a stable tracer, which naturally should be present in very low concentrations. Sulfur hexafluoride is one such tracer, which, however, has the disadvantage of low solubility in water and tendency to escape to the atmosphere. In addition, it cannot easily be measured in situ, meaning that samples must be brought onboard or onshore for measurement. The measurements were carried out by using a nitrogen stream to remove the SF_6 from water with subsequent gas-chromatographic analysis using an electron capture detector.

The results showed that the plume extended along the Florida coastline northward following the Florida Current, with the plume widening to about 3 km at a distance of 66 km from the discharge point. The discharge was mixed vertically beyond a point about 13 km from the outfall. The resulting dilution measured approximately 200 per kilometer along the plume. Sulfur hexafluoride is effective in tracing the plume in the farfield. However, better quantification of the SF_6 at the injection point and in field samples will allow a more quantitative evaluation of the temporal and spatial extent of contaminant plumes associated with the outfall.

Rare earth elements are potentially useful as tracers as a result of their low detection limits and scarcity in the environment. In addition to yttrium (Y), which can be used as a tracer in connection with X-ray fluorescence, holmium (Ho) has been shown to be an effective tracer of cohesive sediment deposition in stormwater ponds

(Spencer et al., 2011). Stormwater ponds are used as a management tool to control flows and reduce contaminants, e.g. metals, by settling with the sediment. The authors considered simulated storm events consisting of two flushes of a tanker truck containing 1100 L of pond water, each of which was released at a rate of 6 L/sec during 31 min. This was carried out at a stormwater detention pond of a waste management site in Halton, Ontario, Canada.

After release of a mixture of natural pond sediment and Ho-labeled montmorillonite clay at the inlet flume of the pond, the holmium distribution in suspended and bottom sediment was measured using inductively coupled plasma optical emission spectrometry (ICP-OES). Each tanker truck flush took 31 min, and the sediment slurry containing Ho-labeled clay was injected into the inlet flume over a 10 min period, with most tracer being released during the first 1–6 min of discharge.

Measured Ho-distributions ($\mu g/g$) in surface sediments 7 d after the tracer release compared well with 42 h output of net sediment tracer accumulation (kg/m^2) from a three-dimensional hydrodynamic model, indicating that Ho-labeled clay may be an effective tracer of cohesive sediment deposition and transport.

9.5.4 Molecular-Based Biological Tracers

Microbial source tracking is a significant human health consideration, for example for determining the source of bacteria in the environment. Bacteria can originate from the feces of cattle (bovine), swine, birds, and humans, or they can be of non-fecal origin. The occurrence of pathogenic viruses in food and the water environment is another concern. Source identification is important for determining pathogens and designing effective remediation. Traditional methods of identification involve culturing with specific agars followed by plate counts. However, these are time consuming and are frequently insufficiently specific to identify relevant pathogens. By contrast, molecular-based microbial source tracking is more specific, and although initially cumbersome and costly, it is now developing into a cost-effective alternative microbial source tracking tool.

Caldwell et al. (2007) examined multiplex real-time polymerase chain reaction (PCR) amplification of mitochondrial DNA (mtDNA) as a source tracking method in surface waters that were impacted by fecal-contaminated effluents. They designed appropriate primers and probes to distinguish between human, cow, and pig products. The PCR response was specific to human mtDNA in all 24 samples investigated, except that there was a carry-over in the PCR signal from consumed beef, but not pork, in the feces from a group of human volunteers. They had consumed meat in the 24 h preceding stool collection. Thus, mitochondrial real-time PCR effectively identifies eukaryotic bacterial sources. Further work on this is ongoing in order to determine baseline levels and to compare with other source tracking methods. One feature of this method is that primers and probes are specific for each pathogen. Therefore, the method can only identify known specific pathogens for which the assay is designed.

A step in the direction of a more general microbial source tracking method that can identify known pathogenic bacteria as well as provide suggestions for unknown, possibly pathogenic, bacteria is the work of Unno et al. (2012). They used a pyrosequencing-based microbial tracking technique that targets total bacteria or bacteroidetes 16S rDNA. The method assigns contamination sources based on shared operational taxonomic units (OTU) between fecal and environmental bacterial groups. It distinguished effectively between bacteria in a non-fecal contaminated sample, human fecal (HF) contaminated sample, and mixed fecal contaminated sample, except that there was 1‰ OTU specificity carry-over from dairy cattle to humans. Only a small number of barcoded pyrosequences for each environmental sample are needed. The method provides an economical and practical approach to determining sources of fecal contamination in waterways.

Viruses in rivers can be effectively traced by bacteriophage 22 (Shen et al., 2008). Bacteriophages are viruses that invade bacterial cells but have no detrimental effect on animals and humans. Shen et al. (2008) conducted a series of tracer experiments in the Grand River, Michigan using dual tracers: Bacteriophage 22 and Rhodamine WT. The fate of the tracers from the injection point at Ann Street Bridge to 68th Street Bridge, near the mouth of the River flowing into Lake Michigan, was modeled using advection–diffusion reaction equations reflecting transport along the river and interaction with storage zones. The results indicated that bacteriophage 22 was a suitable tracer in the river system as indicated by travel times and storage zone characteristics. Assuming that the inactivation rate during transport is typical for viruses, the bacteriophage 22 parameters should be useful for predictive modeling of other viruses in the river.

9.6   CHEMICAL MASS BALANCE MODELING

Chemical mass balance (CMB) modeling uses trace metals (Rahn et al. 1992) or organic compounds, such as PCB congeners (Rachdawong and Christensen 1997) of PAH compounds (Christensen et al., 1997a, b), as tracers for pollution in an aquatic system. CMB modeling was initially conceived for the modeling of sources of atmospheric aerosols (Miller et al., 1972; Friedlander 1973), but has since been expanded to consider sources of atmospheric pollution in general (Gordon 1988; Hopke 1991; Kenski et al., 1995) and also sources of aquatic pollution as indicated by the above references.

The chemical mass balance receptor model with reaction for organic compounds in sediments may be written,

$$F_j = \sum_{i=1}^{n} \phi_{ji} f_{ji} a_i + \varepsilon_j; \qquad (9.8)$$

where

F_j = measured concentration of compound (congener) j, $1 \leq j \leq$ m, in a given sample,

ϕ_{ji} = known concentration of compound (congener) j in source i,

f_{ji} = fractionation constant for source i and compound (congener) j,

a_i = mass transfer coefficient for source i, and

ε_j = error of concentration j.

While there are a number of possible mathematical forms of the fractionation constant, the following form corresponding to a constant attenuation plus a first order reaction will be considered here (Su et al., 2000),

$$f_{ji} = e^{-(\alpha_j + k_j t)};$$ (9.9)

where $\alpha_j (> 0)$ is an attenuation constant and,

$$k_j = \frac{\ln 2}{T_{1/2}}$$ (9.10)

is the half-life of compound j and t = time of exposure to dissolved oxygen in the top layer of sediment (aerobic biodegradation) or time since entering the section just below this layer (anaerobic degradation). The constants α and k can be derived for aerobic degradation by fitting data from sediment cores with widely different sedimentation rates to a linearized expression for $1/$(apparent half-life) vs. $1/t$.

When the number of marker compounds m exceeds the number of sources n, the mass transfer coefficients a_i can be determined from a least-squares procedure (Su 1997). The fraction of compound j originating from source i is calculated as,

$$p_i = \frac{\phi_{ji} f_{ji} a_i}{\phi_{j1} f_{j1} a_1 + \phi_{j2} f_{j2} a_2 + \cdots + \phi_{jn} f_{jn} a_n}$$ (9.11)

9.6.1 CMB Model for PAHs in Kinnickinnic River, Wisconsin

As an example of the application of this methodology, consider the source apportionment of PAHs in a sediment core VC-6 from the Kinnickinnic River in Milwaukee, Wisconsin (Christensen et al., 1997b). This source is situated near a former coal gasification plant that was operated by Milwaukee Solvay Coke Co. Source profiles of coke oven emissions (CB), coal and wood gasification (CWG), and highway emissions (HWY) are shown in Table 9.6 and Figure 9.2. The average molecular weight (M.W.)$_i$ for each of these sources is defined by

$$(\text{M.W.})_i = \frac{\sum_{j=1}^m \varphi_{ji}}{\sum_{j=1}^m \dfrac{\varphi_{ji}}{(\text{m.w.})_{ji}}};$$ (9.12)

where $(\text{m.w.})_{ji}$ is the molecular weight of compound j in source i.

The fifteen layers of this core have average molecular weights between 165 (second layer from bottom) and 207 (third layer from top). These are within the ranges spanned by the sources from 148 (CWG) to 223 (CB). The results of the

TABLE 9.6. PAH Concentrations in Three Sources (ppm)

PAH	CB[a]	CWG[b]	WB[c]	HWY[d]
NaP	17.2	236.0	12.51	0.26
AcNP		53.6	1.76	
AcN			1.10	0.65
Fl	17.2	28.9	0.40	0.83
PhA	540.6	94.0	2.54	4.75
AN	391.2	32.7	0.44	1.00
FlA	704.4		1.12	5.58
Py	826.4	31.3	0.91	3.95
BaA	3838.1	6.3	0.14	0.84
Chy		3.8		1.42
BaP	2147.9	3.8	0.02	0.82
BghiP		0.7		0.33
C.V.[e]	0.40	0.20	0.61	0.29
M.W.[f]	222.7	147.9	138.9	189.0

Sources:
[a]Air-filter of coke-oven emissions ((National Research Council, 1983), p.2–23).
[b]Coal and wood gasification, Ill#6 coal & wood pellets ((National Research Council, 1983), p.2–13).
[c]Wood burning (Su, 1997).
[d]Highway dust (Singh et al., 1993).
[e]Coefficient of variation.
[f]Average molecular weight (NaP, Fl, PhA, AN, Py, BaA, and BaP).

apportionment are shown in Figure 9.3. This figure shows that CWG was a major source from 1895 to 1943. There is no increase in the number of wood carbon particles during this period, indicating that CWG represents coal gasification and possibly coal tar because the latter has a signature that is very similar to that of CWG. High CWG contributions in the lower half of this core are confirmed by the relatively high NaP values, which are characteristic of CWG (Figure 9.2).

The highway profile (Figure 9.3) reflects the historical development of automobile traffic, except in the bottom two layers where there may have been other sources of oil or petroleum combustion. Analytical results for carbon particles released by the combustion of coal, wood, and oil support the apportionment results (Christensen et al., 1997b).

Other indicators of PAH sources have also been considered. Benner et al. (1995) used 1,7-dimethyl phenanthrene as a tracer for burning of soft wood in distinguishing between PAHs from residential wood combustion and mobile source emission, and Zeng and Vista (1997) used naphthalene alkyl distributions and PAH parent compound distributions to identify and assess the sources of PAH input into the coastal marine environment off the coast of San Diego. PAHs detected in the microlayer and the sediments of San Diego Bay were mainly derived from combustion sources rather than oil spills, despite the many shipping activities in the area. This result is similar to that of Singh et al. (1993). They used a CMB model and concluded that

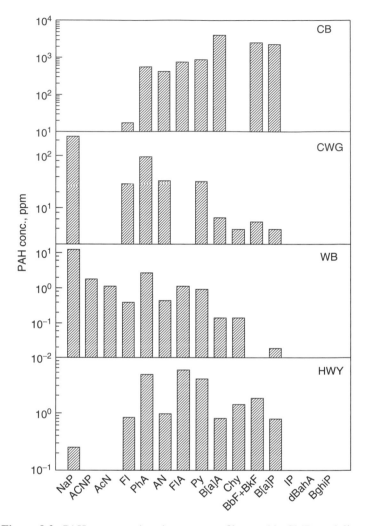

Figure 9.2. PAH concentrations in source profiles used in CMB modeling.

fuel oil was an unlikely PAH source to the Milwaukee Harbor Estuary. The finger-print for fuel oil is heavily weighted by the low molecular weight compounds NaP, Fl, and PhA and is thus not consistent with typical sediment sample profiles from the estuary.

Rather than compound profiles, it is also possible to consider time records of various sources such as PAH inputs to aquatic sediments from the combustion of coal, petroleum, and wood (Christensen and Zhang 1993). US energy consumption from these sources is shown in Figure 9.4 (Rachdawong 1997). In this case, the sample profile is a time record of total PAHs, which is then deemed a linear combination of these source profiles. If the resulting fit is acceptable in terms of a χ^2 criterion, these are then the likely sources. The source contributions in different time intervals

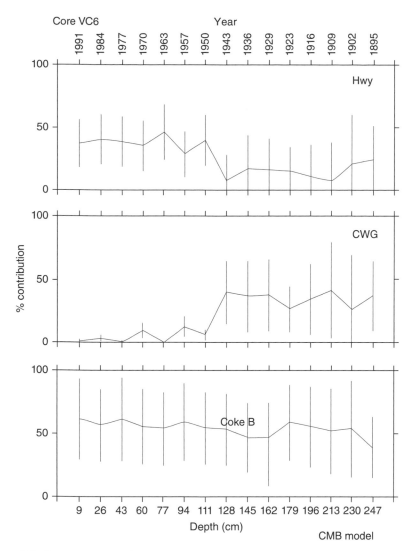

Figure 9.3. Percent contribution by weight to PAHs of HWY, CWG, and Coke-B sources vs. depth and year of core VC-6. (Source: Christensen et al. (1997b) with permission from Elsevier.)

are determined by an equation similar to Equation 9.11. For the input of PAHs to Lake Michigan, Christensen and Zhang (1993) found that the relative contribution from wood burning was greatest (>40%) at remote sites, which is consistent with the smaller diameter (18–22 μm) of carbon particles resulting from wood burning than similar particles produced by the combustion of coal and petroleum (Karls and Christensen 1998). The smaller particles have a lower settling velocity in the atmosphere and a longer residence time; therefore they are transported further before they settle through the water column onto the bottom sediments.

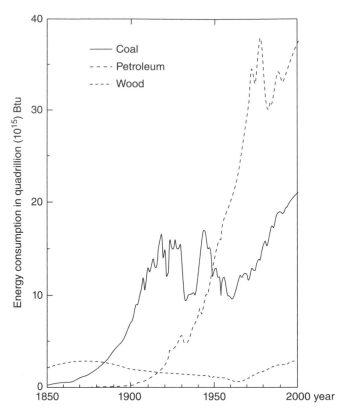

Figure 9.4. US energy consumption for coal, petroleum, and wood from 1850–2000. (Source: Reproduced from Rachdawang (1997), with permission from University of Wisconsin-Milwaukee.)

9.7 FACTOR ANALYSIS

In factor analysis, the fundamental equation is similar to Equation 9.13 (Hopke 1985; Malinowski 1991),

$$D = CR \tag{9.13}$$

$m \times l$ $m \times f$ $f \times l$ ($f \leq m, l$)where m = No. of compounds, l = number of samples, and f = No. of significant factors. In this equation, **D** is a data matrix consisting of m rows with compound concentrations and l columns representing samples. The **C** matrix is a loading matrix containing f source profiles, and the **R** matrix is a score matrix containing f sample contribution profiles. Consider, for example, d_{23},

$$d_{23} = c_{21} r_{13} + c_{22} r_{23} + \cdots + c_{2f} r_{f3} \tag{9.14}$$

Here the total concentration of compound 2 in sample 3 is expressed as a sum of contributions from f sources, where each term is the concentration of compound

2 in a source profile multiplied by the contribution of this source to the third sample. In conventional factor analysis, the elements of matrices $\mathbf{D}$, $\mathbf{C}$, and $\mathbf{R}$ are normalized, i.e. they have zero average and a standard deviation of 1. This formulation presents a difficulty in interpretation in that actual source profiles and source contributions are positive, whereas they are negative in the standard formulation of Equation 9.13.

In order to overcome this problem, factor analysis with non-negative constraints was introduced by Malinowski (1991), and was further developed by Ozeki et al. (1995) and Rachdawong and Christensen (1997). In this version, the concentrations of the data matrix $\mathbf{D}$ are normalized with the average concentrations over all samples, and the loading and score matrices are determined such that they contain only positive elements. In factor analysis with non-negative constraints, the matrices $\mathbf{C}$ and $\mathbf{R}$ may be determined from diagonalization of the covariance matrices $\mathbf{Z} = \mathbf{D}^t\mathbf{D}$ (minor product) and $\mathbf{Z}' = \mathbf{D}^t\mathbf{D}$ (major product). These two matrices have the same eigenvalues, and the eigenvectors for one of these matrices can be calculated from the eigenvectors of the other (Joreskog et al., 1976). In R-mode analysis, the steps are as follows

$$\mathbf{Z} = \mathbf{D}^t\mathbf{D} = \mathbf{C}\ \mathbf{R}^t\mathbf{R}^t\mathbf{C} = \mathbf{C}^t\mathbf{C}, \tag{9.15}$$

assuming that $\mathbf{R}^t\mathbf{R}$ is a unit matrix, $\mathbf{R}^t\mathbf{R} = \mathbf{I}$. This is based on the assumption that source contributions of various factors (sources) are independent and orthogonal. The next step is diagonalization, e.g. by the Jacobi method (Rachdawong 1997), of Z

$$\mathbf{Z}\ \mathbf{U} = \mathbf{U}\Lambda, \tag{9.16}$$

where $\mathbf{U}$ is a matrix of orthonormal eigenvectors and Λ is a diagonal matrix containing the eigenvalues. From Equation 9.16:

$$\mathbf{Z} = \mathbf{U}\Lambda^t\mathbf{U} = (\mathbf{U}\Lambda^{0.5})^t(\mathbf{U}\Lambda^{0.5}) \tag{9.17}$$

Thus, from Equations 9.16 and 9.17,

$$\mathbf{C} = \mathbf{U}\Lambda^{0.5} \tag{9.18}$$

Based on this equation and $\mathbf{Z}' = \mathbf{V}\Lambda^t\mathbf{V}$ where $\mathbf{V}$ is a matrix of orthonormal eigenvectors for $\mathbf{Z}'$, we obtain,

$$\mathbf{R} = {}^t\mathbf{V} \tag{9.19}$$

The score matrix $\mathbf{R}$ can also be obtained from Equation 9.13 as

$$\mathbf{R} = ({}^t\mathbf{C}\ \mathbf{C})^{-1t}\mathbf{C}\mathbf{D} \tag{9.20}$$

Note that Equations 9.18 and 9.19 may be combined to yield,

$$\mathbf{D} = \mathbf{C}\,\mathbf{R} = \mathbf{U}\boldsymbol{\Lambda}^{0.5t}\mathbf{V}, \tag{9.21}$$

which is essentially the singular value decomposition theorem. Q-mode analysis proceeds as above except that $\mathbf{R}$ is determined by diagonalization of the major product $\mathbf{Z}'$. We obtain,

$$\left.\begin{array}{l}\mathbf{C} = \mathbf{U} \\[4pt] \mathbf{R} = \boldsymbol{\Lambda}^{0.5t}\mathbf{V}\end{array}\right\} \tag{9.22}$$

where we have assumed that $^t\mathbf{C}\,\mathbf{C}$ rather than $\mathbf{R}\,{}^t\mathbf{R}$ is a unit matrix. The product $\mathbf{C}\,\mathbf{R}$ is still equal to $\mathbf{D}$ as in Equation 9.13. There are other possible solutions, e.g. a mixed mode where $\mathbf{C} = \mathbf{U}\boldsymbol{\Lambda}^{0.25}$ and $\mathbf{R} = \boldsymbol{\Lambda}^{0.25t}\mathbf{V}$.

9.7.1 Non-negative Constraints Matrix Factorization

Once estimates of $\mathbf{C}$ and $\mathbf{R}$ are determined, negative elements in the matrices, which have no physical meaning, are eliminated through a series of iterative steps in which the basic factor analysis Equation (1-11) is maintained (Ozeki et al., 1995; Rachdawong and Christensen 1997).

Calculated source profiles $\mathbf{C}$ will be compared to candidate profiles, e.g. of PAH sources (Table 1.5) or PCB congeners (Aroclors), through the log Q^2 function,

$$\log Q^2 = \sum_{j=1}^{m} \log^2\left(\frac{\hat{C}_{ji}}{C_{ji}}\right), \tag{9.23}$$

where $\hat{C}_{ji}$ is a scaled factor loading from a predicted source profile (I) and C_{ji} is a weight percent of a congener from a candidate profile. Identification of a given loading vector is established through a minimum for Q^2. The validity of the above FA method with non-negative constraints has been tested successfully by applying it to artificial data sets with known source profiles, created by Monte Carlo simulation (Rachdawong 1997).

Another transformation that is used on occasion is the target transformation factor analysis (TTFA) method, which is similar to the above non-negative constraints FA in that it is based on the diagonalization of a covariance rather than a correlation matrix, and that non-negative elements are sought in loading vectors through several iterations (Hopke 1985).

A version with partially non-negative constraints of the above method was used by Ozeki et al. (1995) to assess the sources of acidity in rainwater from Japan. All ionic concentrations were constrained to positive values, except that of $[\mathrm{H}^+]$, for which negative values were taken as positive values of the hydroxide ion concentration,

$$[\mathrm{OH}^-] = -[\mathrm{H}^+] \tag{9.24}$$

A total of 624 rainwater samples were collected, of which 391 with satisfactory ionic charge balance were used for the analysis. Four major sources of ions were identified: (1) a source of sea salt origin containing Na^+, Cl^-, Mg^{2+}, and SO_4^{2-}, (2) an acidifying source containing H^+, NH_4^+, SO_4^{2-}, and NO_3^-, (3) a source basifying rainwater containing OH^-, NH_4^+, Ca^{2+}, and NO_3^-, and (4) a pollutant source containing K^+ as a major cation.

The sea salt source is small in summer and large in other seasons. During the summer, the amount of rain is small and mostly of local origin, whereas rainclouds during other seasons move over the seas from west to east or from north to south carrying sea salt. The acidifying source is strong during the summer. The reason for this is that petrochemical reactions promoted by strong sunlight during summer transform Nox and SO_2 from automobiles and factories into nitric and sulfuric acid. The basifying source may come from soil dust, Ca^{2+}, and agricultural activity NH_4^+, and the potassium source may come from fertilizers or biomass burning.

PCB Sources in Milwaukee Harbor Estuary.   Rachdawong and Christensen (1997) used principal component analysis (PCA) with non-negative constraints to determine PCB sources in the sediments (cores VC6, VC9, and VC12) of the Milwaukee Harbor Estuary (Figure 9.5). The results of the eigenvalue analysis (Table 9.7) demonstrate that Aroclors 1242, 1254, and 1260 are found in all three cores. In addition, Aroclor 1016 was present in VC9 and VC12, i.e. the Harbor Entrance Channel and the Outer Harbor, but not in VC6 from the Kinnickinnic River. Aroclor 1016 is a distillation product of Aroclor 1242. It was used in capacitors in the early 1970s until the US ban on PCBs in 1977 as a substitute for Aroclor 1242 because of its lower toxicity.

A comparison of literature and predicted PCB profiles, based on congener patterns, is shown in Figure 9.6. As may be seen from this figure, the PCA model is quite effective in predicting the Aroclor profiles. A CMB model, also operating on congeners, was used to support the results of the PCA model. This comparison was successful, especially when only three Aroclors were present (VC6). For example, high contributions of Aroclor1260 were found in the deep (older) layers of VC12 (1942–63) by both the PCA and the CMB model. This is consistent with the fact that Aroclor 1260 was relatively more frequently used in the early years (Christensen and Lo 1986). In no case was Aroclor 1248 identified as a source.

The unique advantage of the PCA model is that it can be applied with no or minimum knowledge of actual sources. For example, the discovery of Aroclor 1016 as a source solely came from use of the PCA model. However, to obtain accurate predictions using the model, the coefficients of variation of each source must be constant. This is not a requirement for the CMB model, for which, however, there must be a set of possible source profiles available.

In order to achieve the most effective source identification and apportionment, one should use a combination of these techniques, preferably in conjunction with an alternative source resolution methodology. Examples of the latter category are alkylated PAHs and $^{14}C/^{13}C$ ratios to distinguish between PAHs from residential wood

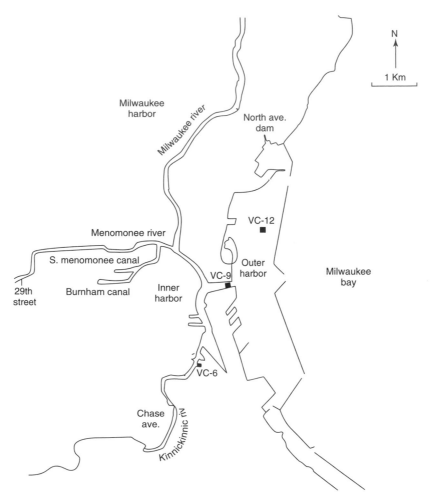

Figure 9.5. Map of the Milwaukee Harbor Estuary with sampling sites. (Source: Reprinted with permission from Rachdawong & Christensen, (1997b), © American Chemical Society.)

burning and mobile sources (Benner et al., 1995), or the use of elemental carbon particles from the burning of coal, oil, and wood to distinguish between PAH sources in lake sediments (Su et al., 1998).

Comparison with CMB Model, Black River and Ashtabula River, Ohio. The results obtained by the non-negative constraints factor analysis (FA) model were compared with CMB model results in a study of PAHs in sediments of the Black River and the Ashtabula River, Ohio (Christensen and Bzdusek 2005). The Black River sampling sites are near the USS/KOBE steel plant, which has contributed to PAH contamination of Black River sediments. Factor analysis confirmed the findings of a CMB

TABLE 9.7. Eigenvalues and log Q² of Factor Loadings for each Candidate Aroclor Profile for VC 6, 9, and 12

Core	Significant eigenvalues	Percent cumulative eigenvalues	Identified aroclor	Log Q² of candidate aroclor series for each identified aroclor				
				1016	1242	1248	1254	1260
VC 6	86.80	86.97	1242[a]	7.19	**1.07**[c]	7.01	1.65	1.91
	1.67	98.84	1254	45.98	36.29	25.97	**9.22**	48.25
	0.86	99.80	1260[b]	9.25	9.68	13.95	6.28	**0.99**
VC 9	100.19	97.06	1242[a]	10.58	**0.43**	6.10	0.62	2.05
	2.10	99.09	1254	11.16	4.53	10.77	**1.91**	2.02
	0.57	99.64	1016[a]	**2.56**	3.02	3.56	7.72	2.98
	0.33	99.96	1260	9.27	18.21	15.41	15.61	**2.16**
VC12	95.26	97.41	1242[a]	10.44	**0.38**	5.82	0.62	2.02
	2.23	99.69	1260	10.38	17.21	13.13	11.75	**0.68**
	0.18	99.88	1254	12.68	8.53	12.44	**4.40**	9.88
	0.10	99.98	1016[a]	**3.90**	9.05	7.12	12.41	6.82

[a]delete outliers (#153 and #180)

[b]delete outlier (#118)

[c]bold-faced = min log Q²

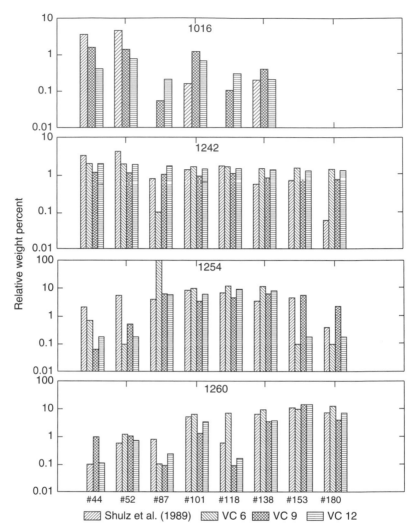

Figure 9.6. Comparison of literature and predicted PCB profiles for VC 6, 9, and 12. (Source: Reprinted with permission from Rachdawong & Christensen, (1997b), © American Chemical Society.)

model identifying traffic (58%, 51%), coke oven (26%, 44%), and wood burning/coal tar (16%, 5%) as the major PAH sources to the Black River and the Ashtabula River respectively. Coal tar was a more likely source in the Black River as a result of relatively high phenanthrene and naphthalene concentrations. As shown in Figure 9.7, three-factor loadings based on seven PAHs compare well with CMB model profiles from the literature when wood burning is represented by the particulate fraction.

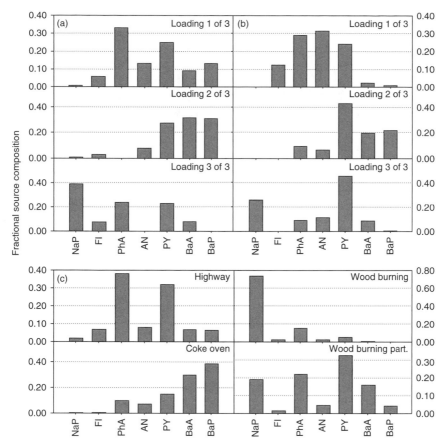

Figure 9.7. Source profiles determined by factor analysis with non-negative rotations (a) for the Black River, and (b) for the Ashtabula River, Ohio. (c) Source profiles, based on literature values used in chemical mass balance (CMB) model. (Source: Christensen and Bzdusek, (2005) with permission from Elsevier.)

From dated cores, the highest PAH concentrations (~125 ppm) were found in 1991 (core BR-6-5) in the Black River. These were likely caused by dredging activities. Other maxima were found in core BR-4 approximately 400 m upstream of the dredged area in 1950 (core BR-4-13) and 1969 (core BR-4-8), coincident with regional atmospheric inputs dominated by coal and traffic respectively (Christensen et al., 1999). Contributions to PAHs of traffic, coke oven, and wood burning/coal tar for the Black River are shown in Figure 9.8. The FA results for traffic and coke oven are in good agreement with the CMB results, whereas FA contributions of wood burning/coal tar exceed those obtained with the CMB model. Note, however, that the wood burning/coal tar source profile used here for the CMB model relates to total rather than particulate PAHs. Ashtabula River sediment with significantly

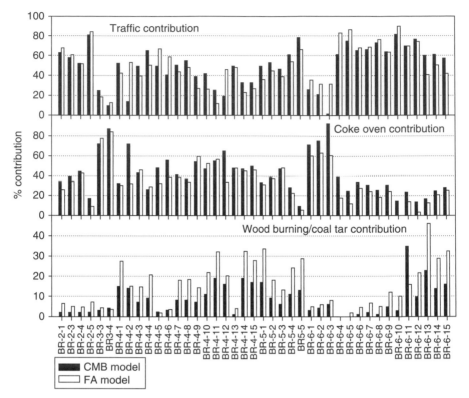

Figure 9.8. PAH contributions from traffic, coke oven, and woodburning/coal tar for the Black River, Ohio. The CMB source profiles are here based on total rather than particulate (part.) PAHs (Figure 9.7). (Source: Christensen and Bzdusek, (2005) with permission from Elsevier.)

lower PAH concentrations (max ~4 ppm) reflected PAH maxima around 1973 (core AR-1-12) and 1972 (AR-3-14), which may be traffic-related.

PAH Sources in Lake Calumet, Chicago. Bzdusek et al. (2004) used a non-negative constraints factor analysis method to determine sources and contributions of PAHs to sediments of Lake Calumet and surrounding wetlands in southeast Chicago. The method follows the steps outlined in Bzdusek (2005). The data matrix $\mathbf{X}$ is approximated as a product of the loading matrix $\mathbf{G}$ and the score matrix $\mathbf{F}$,

$$\mathbf{X} = \mathbf{G} \qquad \mathbf{F} \qquad (9.25)$$

$$(m \times n)\ (m \times p)(p \times n)$$

where m is the number of compounds, n is the number of samples, and p is the number of factors. It is assumed here that $m \leq n$. First, the data matrix is scaled with the average value of each compound in order to obtain comparable values of the compounds. Next, eigenvalues and eigenvectors are extracted from the covariance matrix

$\mathbf{Z}(m \times m) = \mathbf{XX}^T$ where $\mathbf{X}^T$ is the transposed X matrix. By using the singular value decomposition (SVD) we can write,

$$\mathbf{X} = \mathbf{UDV}^T \qquad (9.26)$$

where $\mathbf{U}$ and $\mathbf{V}$ are orthonormal such that $\mathbf{UU}^T = \mathbf{I}$ where I is an identity matrix. The columns of $\mathbf{U}$ are eigenvectors of $\mathbf{XX}^T$ and the columns of $\mathbf{V}$ are the eigenvectors of $\mathbf{X}^T\mathbf{X} \cdot \mathbf{D}(m \times n)$ is a diagonal matrix containing singular values, i.e. the square roots of eigenvalues. By including only the first p eigenvectors in descending order of eigenvalues, we have $\mathbf{G} = \mathbf{UD}$ and $\mathbf{F} = \mathbf{V}^T \cdot \mathbf{F}$ can also be calculated from Equation 9.25 with known $\mathbf{X}$ and $\mathbf{G}$ matrices.

After $\mathbf{G}$ and $\mathbf{F}$ are derived using SVD, rotations are performed to remove negative elements from $\mathbf{G} \cdot \mathbf{G}$ is then backscaled by multiplying the compounds from each source by the average concentration for the respective compounds, and columns of $\mathbf{G}$ are normalized to 100%. $\mathbf{F}$ is then recalculated to maintain the product $\mathbf{X} = \mathbf{GF}$. Rotations maintain the least squares fit but do not require the factor axes to be orthogonal.

Uncertainties in loading and score matrices are estimated by a Monte Carlo method, providing a set of these matrices based on 10 random versions of the data matrix. Diagnostics include coefficient of determination for source profiles and the Exner function (Bzdusek 2005). The validity of solutions was also tested by using known loading and score matrices to produce a data matrix, which was then subjected to factor analysis. Successful reproduction of original loading and score matrices was achieved.

Particulate phase literature PAH profiles of several sources considered for the Lake Calumet study are shown in Figure 9.9. These include power plant, residential coal combustion, coke oven, wood burning, gasoline engine, diesel engine, and traffic tunnel. By comparison, a six-loading factor solution showing PAH profiles obtained by the FA model is shown in Figure 9.10. The loading or source profiles are interpreted as follows: Loading 1 is coke oven, loading 2 is gasoline engine, loading 3 is traffic tunnel, loading 4 is an alternative coke oven profile with lower phenanthrene and pyrene concentrations, and loading 5 is from wood burning or residential coal burning. The contributions are coke oven 47%, traffic 45%, and wood burning/coal residential 2.3%, and an ideno(1,2,3-cd)pyrene source 6%. The results support those previously obtained using a CMB model. However, the FA model may be deemed more fundamental because no assumptions are made about possible sources, and wood burning and the secondary coke oven sources were not recognized in the CMB model.

The wood burning source is present in all cores investigated (D, E, I, and K). It shows a minimum during the 1960s, 1970s, and the early 1980s, but also measurable levels in the most recent layers up to 1997, indicating that wood burning is a major part of the source because residential coal burning was not a significant source after the 1950s. Maximum PAH levels measure ~ 25 ppm in 1988 in core D near interstate I-94. There is a general increase in traffic-related PAH contributions in the 1950s and the early 1960s coincident with the opening of I-94 in 1962. All cores except K have

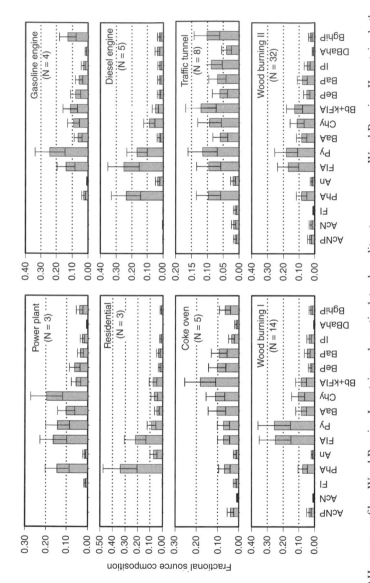

Figure 9.9. PAH source profiles. Wood Burning I contains only particulate phase literature sources, Wood Burning II contains both particulate and gaseous-phase sources. Error bars represent a constant relative error of 40%. (Source: Reprinted with permission from Bzdusek et al. (2004), © American Chemical Society.)

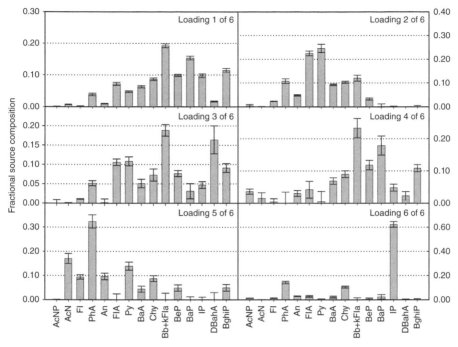

Figure 9.10. Factor loading plot for Lake Calumet six-source factor analysis solution with error bars representing the standard deviation of the mean for nine FA model runs using data sets created by Monte Carlo simulation. (Source: Reprinted with permission from Bzdusek et al. (2004), © American Chemical Society.)

significant coke oven, but slightly higher traffic, contributions. Core K, located about 3.2 km south of a coke oven plant, is dominated by coke oven input.

9.7.2 Positive Matrix Factorization

An alternative and more recent method used to find loading and score matrices based on a known data matrix is positive matrix factorization, which was first developed by Paatero and Tapper (1994) and Paatero (1997). The method differs from the above FA methods in that it does not start with determination of eigenvectors and eigenvalues, and therefore does not require initial orthogonal factor axes. It also allows consideration of uncertainties of measured data. Starting with Equation 9.25, the basic problem is to minimize the following objective function,

$$Q(\mathbf{E}) = \sum_{i=1}^{m} \sum_{j=1}^{n} \left(\frac{E_{ij}}{\sigma_{ij}} \right)^2 \tag{9.27}$$

where,

$$E_{ij} = X_{ij} - X'_{ij} = X_{ij} - \sum_{h=1}^{p} G_{ih} F_{hj} \qquad (9.28)$$

$$i = 1, \dots, m \qquad j = 1, \dots, n$$

In these equations, E_{ij} is the difference between measured and calculated values of the data matrix element i, j for which the experimental uncertainty is σ_{ij}. m is the number of compounds, n is the number of samples, and p is the number of factors. The Q function can now be written as,

$$Q(\mathbf{E}) = \sum_{i=1}^{m} \sum_{j=1}^{n} \left(\frac{X_{ij} - \sum_{h=1}^{p} G_{ih} F_{hj}}{\sigma_{ij}} \right)^2 \qquad (9.29)$$

Initial solutions for **G** and **F**, which can be obtained from normal equations or by setting the derivative of Q with respect to elements of the unknown matrices equal to zero, typically contain several negative elements that have no physical meaning. Thus, rotations are included to eliminate these negative elements either by penalty functions, or as described by Bzdusek (2005), non-negative least squares, which is appropriate if there are several zero elements in the matrices.

Once acceptable **G** and **F** matrices are obtained, it should be realized, however, that PMF suffers from rotational ambiguity if the loading and score matrices do not contain several zeros. This may be seen from a slightly different formulation of Equation 9.25,

$$\mathbf{X} = \mathbf{G}\,\mathbf{R}\,\mathbf{R}^{-1}\mathbf{F} \qquad (9.30)$$

where **R** $(p \times p)$ is rotation matrix. Because the $\mathbf{R}\,\mathbf{R}^{-1} = \mathbf{I}$, the product of $\mathbf{G}' = \mathbf{GR}$ and $\mathbf{F}' = \mathbf{R}^{-1}\mathbf{F}$ still equals **X**; however, the loading and score matrices **G**' and **R**' are different from **G** and **F**. Non-trivial rotations are not allowed if the resulting matrices have negative elements, which is the reason why several zeros in the loading and score matrices produce solutions that are not subject to rotational ambiguity.

A computer program to determine **G** and **F** for a given data matrix **X** that follows Paatero's (1997) method along with diagnostics is available on the Internet (U.S. Environmental Protection Agency 2009a.). The EPA PMF program also includes a parameter F_{peak} that can be adjusted to evenly distribute pairs of factor contributions. Although in some cases this feature may eliminate uncertainty resulting from rotations, there is no assurance that this will always be so. This could be the case, for example, if the pairs of factor contributions are not evenly distributed based on the physical and chemical configuration of sources and receptors in the environment. In order to gain further insight into the method along with full access to loading and score matrices, Bzdusek (2005) independently produced a MATLAB computer program to perform PMF. His program includes both penalty functions and non-negative least squares method for the **G** and **F** solutions as well as several diagnostics metrics, including coefficients of determination for model-produced source profiles. Results obtained by Bzdusek's program are similar to those obtained by EPA PMF (Henry and Christensen 2010).

Since its development in the late 1990s, PMF has been mainly applied to air pollution studies. However, recently it has been found to be useful for the aquatic environment as well. For example, PMF was used to determine PCB congeners and dechlorination in sediments of Lake Hartwell, South Carolina (Bzdusek et al., 2006a), and Sheboygan River, Wisconsin (Bzdusek et al., 2006b). Lake Hartwell sediments were contaminated with PCBs from Sangamo Western facility, which used PCBs as dielectric fluid for power capacitors. The PCBs used were primarily Aroclors 1242, 1254, and 1016. Data sets were collated from cores collected in 1987 and 1998. The total PCB concentration ranged from ∼0 to 58 µg/g. Two factors were determined by PMF. One factor resembled a PCB mixture of 80% Arochlor 1016 and 20% Aroclor 1254, and the other was determined to represent a dechlorinated version of this source following M + Q dechlorination processes, using Bedard notation. Typical processes are 18(25-2) → 4(2-2) and 53(25-26) → 19 (26-2), where congener numbers are indicated with substitution patterns in parenthesis. The underlined chlorine is lost in the process. The profiles of the dechlorinated profiles were similar for the 1987 and 1998 data; however, the contributions were 73% and 87% respectively, indicating that dechlorination had reached a plateau even though the dechlorinated profiles contained several meta and para chlorines. Thus, effective remediation would have to be performed using a process such as capping that would isolate the PCBs from the food chain.

The investigation of PCBs in Sheboygan River focused on sediment cores from the Inner Harbor, where some PCB concentrations were as high as 161 µg/g. Tecumseh Products Company was the likely major source of PCBs, which were used as Aroclors 1248 and 1254 in hydraulic fluids from 1959 to 1971. PMF analysis of the data set produced two major factors: an original source factor resembling a 50% Aroclor 1248 and 50% Aroclor 1254 mixture, and a dechlorinated version of the original source. The contributions constituted 23% of the original source and 77% of the dechlorination factor. In this case, the major dechlorination process was found to be H' with a minor contribution from process M. Typical dechlorination processes here are 66(24-34)→25(24-3) and 18(25-2)→4(2-2), along with many others. Remediation dredging of the Upper Sheboygan River was undertaken in 1989 and 1990, and final removal of highly contaminated sediments from the Inner Harbor was completed in 2012.

A follow-up PMF analysis of the PCB data set from the Inner Harbor was performed by Henry and Christensen (2010). The results were compared with results using the eigenvalue/eigenvector based computer program Unmix. Five factors were found: factor 1 representing the original 50/50% Aroclor 1248/1254 mixture (7.8%) and factors 4 (29.6%) and 5 (27.2%) representing two versions of a dechlorinated profile (Figure 9.11). B-PMF, NNLS indicates Bzdusek's PMF program using non-negative least squares analysis. Furthermore, factors 2 and 3 were recognized among several core samples and appeared to represent various stages of dechlorination. PMF-produced results were closer to the known

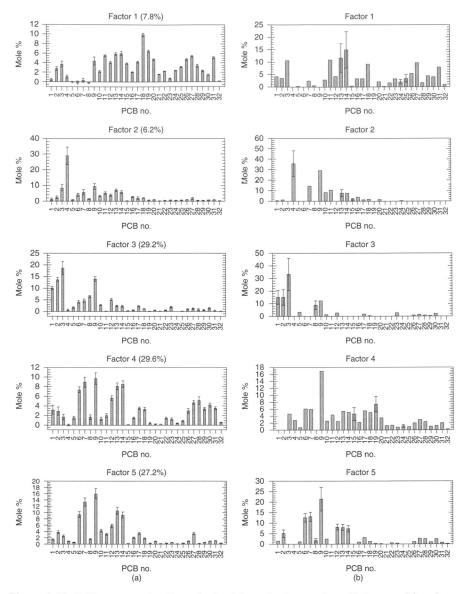

Figure 9.11. PCB congener loadings obtained from the data set from Sheboygan River Inner Harbor, Wisconsin using (a) B-PMF, NNLS and (b) Unmix. (Source: Reprinted with permission from Henry & Christensen, (2010), © American Chemical Society.)

Aroclor source mixture and the measured average dechlorinated profiles than results obtained by Unmix. Because of the many zeros or near zeros in the loading profiles there was no observed rotational ambiguity in the final loading and score matrices.

An application of PMF to the water phase is a study by Rodenburg et al. (2010). They examined a large data base on PCB congener concentrations in effluents, and influents to wastewater treatment plants (WWTPs), for the Delaware River Basin. They identified seven factors. Factor 1 contained congeners 8 and 11 that are present in commercial paint pigments, and it was thought that this factor was characteristic of wastewater effluent or stormwater runoff. Factors 2 and 4 represented advanced and partial dechlorination respectively, with significant concentrations of congeners 4 and 19 (factor 2) and coeluting congeners 44 + 47 + 65 and 45 + 51 (factor 4). Factors 3, 5, and 7 were Aroclors 1248, 1254, and 1260 respectively, while factor 6 indicated Aroclors 206/208/209 produced inadvertently during production of titanium chloride. The dechlorination signals were present in WWTP influents, but not in stormwater runoff, indicating that dechlorination takes place in sewers. The results suggested that PCB dechlorination by anaerobic bacteria may occur in sewers, landfills, and contaminated groundwater.

9.7.3   Unmix

A principal component method for the determination of loading and score matrices that is not dependent on non-negative rotations is the eigenvalue and eigenvector model used in the computer program Unmix (Henry 2003). The program is available on the internet (U.S. Environmental Protection Agency 2009b).

The model plots the coefficients of orthonomal principal components of sample points (columns) of the data matrix in eigenspace, and sources are then identified at the vertices of intersecting hyperplanes that each omit one source. Intersecting hyperplanes defined as "edges", or straight lines in an eigenspace plot normalized to the first principal component, are determined by requiring all sample points to be located within a minimum volume polyhedron. The coordinates of the points of the intersections (vertices) are then converted back into estimated sources by singular value decomposition of the data matrix. This is performed by expanding sources into N eigenvectors, arranged in descending order of eigenvalues. The number of factors (sources) may be determined by the NUMFACT algorithm (Henry et al., 1999). The method is a further development of self-modeling curve resolution described by Lawton and Sylvestre (1971), where the term self-modeling refers to the fact that no assumptions are made regarding the composition of the unknown sources.

We shall outline the method considering three sources, following the approach taken by Zou et al. (2013). The first step is to write the data matrix $\mathbf{X}$ in terms of the theorem of singular value decompositon,

$$\mathbf{X} = \mathbf{U}\mathbf{D}\mathbf{V}^{\mathrm{T}} \tag{9.31}$$

where $\mathbf{D}$ is an $m \times n$ diagonal matrix with the diagonal element filled with the singular values of matrix $\mathbf{X} \cdot \mathbf{U}$ and $\mathbf{V}$ are $m \times m$ and $n \times n$ orthonormal matrices respectively. The columns of $\mathbf{U}$ are eigenvectors of the data matrix $\mathbf{X}$ corresponding to singular values in $\mathbf{D}$ in descending order. Here m = number of chemicals, n = number of samples, We assume again that $m < n$.

Equation 9.31 is now approximated as,

$$\mathbf{X} = \mathbf{U}_q \mathbf{D}_q \mathbf{V}_q^{\mathrm{T}} + \varepsilon = \mathbf{U}_q \mathbf{C}_q + \varepsilon \tag{9.32}$$

with dimensions $\mathbf{U}_q(m \times q)$, $\mathbf{D}_q(q \times q)$, $\mathbf{V}_q(n \times q)$, $\mathbf{C}_q(q \times n)$, and $\varepsilon(m \times n)$, and $\varepsilon(m \times n)$. ε is an error term, and q is the number of eigenvectors selected in descending order of eigenvalues such that $q \leq m$. The coordinates of the measurements in principal component eigenspace are represented by $\mathbf{C}_q(q \times n)$,

$$\mathbf{C}_q = \{c_{k1}, c_{k2}, \dots, c_{kj}, \dots, c_{kn}\} = \mathbf{U}_q^T \mathbf{X} \tag{9.33}$$

Thus eigenspace coordinates can be viewed as dot products of eigenvectors and sample vectors (Henry 1987).

Two methods have been suggested to determine source fingerprints. One is to choose the measurements that are located on the vertices in eigenspace and use the fingerprints of these measurements as source fingerprints (Klingenberg et al., 2009). Another method is to use the estimated coordinates of the factor points in eigenspace represented be a matrix $\mathbf{P}$ where $\mathbf{P} = \{p_{i1}, p_{i2}, \dots p_{iN}\}(i = 1, 2, \dots, q)$ to determine the fingerprints of these factors (G) from (Henry, 2003)

$$\mathbf{G} = \mathbf{U}_q \mathbf{P} \tag{9.34}$$

with dimensions $\mathbf{G}(m \times N)$, $\mathbf{U}_q(m \times q)$, $\mathbf{P}(q \times N)$. N = number of sources or patterns.

Equations 9.32 and 9.34 show that sample and source vectors can be expanded in terms of the first q eigenvectors (Henry 1987).

Graphical Example of Determination of Sources and Source Profiles. We consider here 200 sample points of an artificial air pollution data set (Zou 2011: Appendix E2). The points are located in the eigenspace that is defined by the second and third principal component normalized to the first principal component. The eigenvectors ($\mathbf{E}$) and eigenvalues ($\mathbf{Q}$) of the data matrix are given below:

$$\mathbf{E} = \begin{bmatrix} -0.4360 & 0.1585 & 0.7203 & -0.0709 & -0.2200 & -0.2899 & -0.0651 & 0.3525 \\ 0.5591 & -0.2974 & 0.2499 & -0.5198 & 0.0852 & -0.0210 & 0.3960 & 0.3190 \\ -0.4930 & -0.6914 & -0.1694 & 0.0117 & -0.1175 & 0.3692 & 0.1030 & 0.2989 \\ -0.2782 & 0.4848 & -0.1867 & 0.0070 & 0.4964 & 0.1754 & 0.4916 & 0.3657 \\ 0.1808 & 0.1600 & -0.3806 & 0.2622 & -0.6602 & -0.2520 & 0.2488 & 0.4085 \\ 0.3032 & 0.2319 & 0.2026 & 0.1662 & -0.1168 & 0.7190 & -0.3738 & 0.3395 \\ 0.2097 & -0.2900 & 0.0415 & 0.5667 & 0.5767 & -0.3567 & -0.2242 & 0.3810 \\ -0.0917 & 0.0989 & -0.4094 & -0.5541 & 0.0912 & -0.2017 & -0.5785 & 0.3514 \end{bmatrix}$$

$$Q = 10^5 \times \begin{bmatrix} 0.0000 & 0 & 0 & 0 & 0 & 0 & 0 & 0 \\ 0 & 0.0000 & 0 & 0 & 0 & 0 & 0 & 0 \\ 0 & 0 & 0.0001 & 0 & 0 & 0 & 0 & 0 \\ 0 & 0 & 0 & 0.0001 & 0 & 0 & 0 & 0 \\ 0 & 0 & 0 & 0 & 0.0001 & 0 & 0 & 0 \\ 0 & 0 & 0 & 0 & 0 & 0.0048 & 0 & 0 \\ 0 & 0 & 0 & 0 & 0 & 0 & 0.0122 & 0 \\ 0 & 0 & 0 & 0 & 0 & 0 & 0 & 4.9882 \end{bmatrix}$$

The requirement that all compound or species concentrations must be positive or zero places a constraint on the feasible area of eigenspace. Retaining three eigenvectors $q = 3$, we obtain, according to Equation 9.32, considering the three last columns of $\mathbf{E}$, the following equations for species 3, 5, and 8 for zero values of these species

$$0 = 0.2989 + 0.1030x + 0.3692y \quad \text{(species 3)}$$

$$0 = 0.4085 + 0.2488x - 0.2520y \quad \text{(species 5)}$$

$$0 = 0.3514 - 0.5785x - 0.2017y \quad \text{(species 8)}$$

where x and y are second and third principal components in eigenspace normalized to the first principal component. The feasible region is defined by the space inside the triangle that is defined by the three lines (Figure 9.12). Similar requirements for the remaining five species further narrows the region as may be seen from Figure 9.12.

Equations for the species composition of sample point $\mathbf{P}_i$ and the normalized sample point $\mathbf{p}_i$ may be written,

$$\mathbf{P}_i = a_i \, \mathbf{e}_1 + b_i \cdot \mathbf{e}_2 + c_i \cdot \mathbf{e}_3 \cdot \qquad i = 1, \ldots, n \qquad (9.35)$$

$$\mathbf{p}_i = \mathbf{e}_1 + b_i/a_i \cdot \mathbf{e}_2 + c_i/a_i \cdot \mathbf{e}_3 \cdot \qquad i = 1, \ldots, n \qquad (9.36)$$

where $\mathbf{e}_1, \mathbf{e}_2$, and $\mathbf{e}_3$ are the three principal eigenvectors in the order of the magnitude of the eigenvalues.

In order to determine coordinates of sample point i in eigenspace, normalized to the first principal component, the procedure is as follows. First, sample point i is expanded in terms of the first three eigenvectors according to Equation 9.32

$$\begin{bmatrix} 18.0326 \\ 17.6222 \\ 15.0418 \\ 19.9701 \\ 22.0317 \\ 14.7934 \\ 19.1323 \\ 15.9143 \end{bmatrix} = a_i \begin{bmatrix} 0.3525 \\ 0.3190 \\ 0.2989 \\ 0.3657 \\ 0.4085 \\ 0.3395 \\ 0.3810 \\ 0.3514 \end{bmatrix} + b_i \begin{bmatrix} -0.0651 \\ 0.3960 \\ 0.1030 \\ 0.4916 \\ 0.2488 \\ -0.3738 \\ -0.2242 \\ -0.5785 \end{bmatrix} + c_i \begin{bmatrix} -0.2899 \\ -0.0210 \\ 0.3692 \\ 0.1754 \\ -0.2520 \\ 0.7190 \\ -0.3567 \\ -0.2017 \end{bmatrix} \qquad (9.37)$$

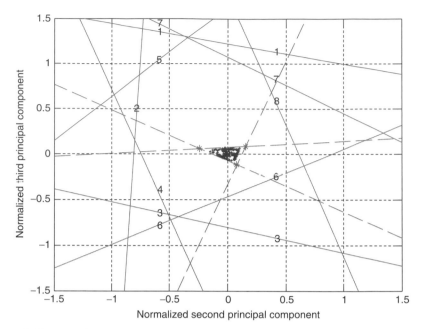

Figure 9.12. Plot of 200 points of artificial data set and 8 species lines in two dimensional plot based on normalized second and third principal components. (Source: Zou (2011) with permission from University of Wisconsin-Milwaukee.) (*See color plate for color version*)

Next, the point vector on the left is dotted with e_1, e_2, and e_3 to obtain a_i, b_i, and c_i respectively. This is equivalent to solving for C_q from Equation 9.33. The normalized second and third principal components are then calculated as $x = b_i/a_i = 0.071563$ and $y = c_i/a_i = -0.02943$ and the sample point i with these coordinates is plotted in eigenspace (Figure 9.13). This procedure is carried out for all 200 sample points.

Estimated coordinates of sources or characteristic patterns located at the vertices of the triangle defining the sampling points in Figure 9.13 are listed in Table 9.8. It should be noted that identification of edges or straight lines defining the triangle in this figure is only possible if there are several sample points that contain low contributions of one or two sources, i.e. points near the three sides or the vertices.

Determination of a source profile proceeds as follows. Consider for example the source profile based on the ERC estimated coordinates for Vertex 3 (factor 3)

$$
\begin{bmatrix} 0.3525 \\ 0.3190 \\ 0.2989 \\ 0.3657 \\ 0.4085 \\ 0.3395 \\ 0.3810 \\ 0.3514 \end{bmatrix} + 0.058 \begin{bmatrix} -0.0651 \\ 0.3960 \\ 0.1030 \\ 0.4916 \\ 0.2488 \\ -0.3738 \\ -0.2242 \\ -0.5785 \end{bmatrix} - 0.106 \begin{bmatrix} -0.2899 \\ -0.0210 \\ 0.3692 \\ 0.1754 \\ -0.2520 \\ 0.7190 \\ -0.3567 \\ -0.2017 \end{bmatrix} = \begin{bmatrix} 0.3795 \\ 0.3442 \\ 0.2657 \\ 0.3756 \\ 0.4496 \\ 0.2416 \\ 0.4058 \\ 0.3392 \end{bmatrix} \tag{9.38}
$$

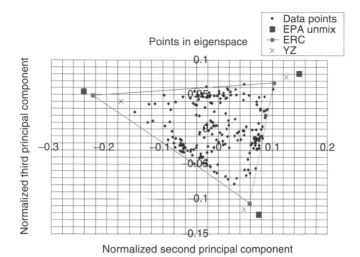

Figure 9.13. Edges of data points and source vertices based on artificial data set. (Source: Zou (2011) with permission from University of Wisconsin-Milwaukee.) (*See color plate for color version*)

TABLE 9.8. Estimated Coordinates for Three Vertices of the Triangle in Figure 9.13 Containing 200 Sample Points

	Vertex 1		Vertex 2		Vertex 3	
Estimated by	x	y	x	y	x	y
YZ	0.125	0.075	−0.175	0.04	0.048	−0.115
ERC	0.103	0.067	−0.225	0.049	0.058	−0.106
Unmix solution	0.149	0.080	−0.241	0.055	0.075	−0.122

Multiplication with a normalization factor of 35.7 provides the profile in the upper right panel of Figure 9.14. This is equivalent to solving for G from Equation 9.34. The results compare well with the EPA Unmix solution. This is also the case for the first two factors, and for the source profiles estimated by YZ.

The basis for Unmix is determination of equations for intersecting hyperplanes so that coordinates of the vertices (sources) can be calculated. Thus, the method benefits from a large number of samples, especially near the vertices, producing results that are as accurate as possible. By contrast, polytopic vector analysis (PVA) is dependent on a single sample point at or near a vertex. However, most environmental data sets do not include samples representing each source. A PVA technique was used by Johnson et al. (2000) to characterize PCBs in San Francisco Bay. They found five factors: Aroclor 1260, two variations of Aroclor 1254, Aroclor 1248, and an unknown factor containing high values of congeners 138 and 153.

Henry and Christensen (2010) compared results of Unmix and PMF applied to two data sets, the Sheboygan River Inner Harbor data set referred to above, and also

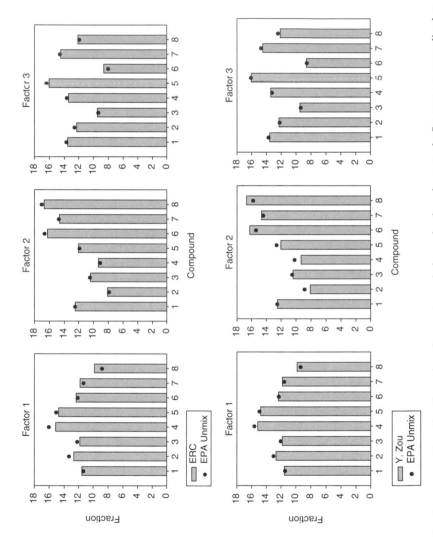

Figure 9.14. Results of source profile estimation using eigenvector and eigenvalue based approach. Sources are normalized to 100 and shown as sources represented by Vertices 1, 2, 3 from left to right. (Source: Zou (2011) with permission from University of Wisconsin-Milwaukee.)

an artificial air pollution data set, which included 8 compounds or species and 200 samples. PMF provided a better solution for the Sheboygan data set because there was no rotational ambiguity, and one of the factors in the Unmix solution was poorly determined because it was dependent on only three sample points. By contrast, the Unmix solution was preferable for the artificial data set because of its well-defined edges and because the PMF solution could be rotated. The basis for a rotation (F_{peak} value) was fairly well established in this case because sample points were randomly distributed in eigenspace, as shown for a similar data set in Figure 9.13. However, this may not be generally true for environmental data sets.

REFERENCES

Aggarwal, PK, Fuller, ME, Gurgas, MM, Maning, JF, Dillon, MA. Use of stable oxygen and carbon isotope analyses for monitoring the pathways and rates of intrinsic and enhanced *in-situ* biodegradation. *Environ. Sci. Technol.* 1997;31:590–596.

Al-Reasi, HA, Ababneh, FA, Lean, DR. Evaluating mercury biomagnification in fish from a tropical marine environment using stable isotopes (δ^{13}C and δ^{15}N). *Environ. Toxicol. Chem.* 2007;26(8):1572–1581.

Benner Jr, BA, Wise, SA, Currie, LA, Klouda, GA, Klinedinst, DB, Zweidinger, B, Stevens, RK, Lewis, CW. Distinguishing the contributions of residential wood combustion and mobile source emissions using relative concentrations of dimethyl phenanthrene isomers. *Environ. Sci. Technol.* 1995;29:382–389.

Berger, WH, Johnson, RF, Killingley, JS Unmixing of the deep-sea record and the deglacial meltwater spike. *Nature* 1977;269:661–663.

Bzdusek, PA, Christensen, ER, Li, A, Zou, Q. Source apportionment of sediment PAHs in Lake Calumet, Chicago: Application of factor analysis with nonnegative constraints. *Environ. Sci. Technol.* 2004;38(1):97–103.

Bzdusek, PA. *PCB or PAH sources and degradation in aquatic sediments determined by positive matrix factorization*. PhD dissertation. University of Wisconsin-Milwaukee; 2005.

Bzdusek, PA, Christensen, ER, Lee, CM, Pakdeesusuk, U and Freedman, DL. PCB congeners and dechlorination in sediments of Lake Hartwell, South Carolina, determined from cores collected in 1987 and 1998. *Environ. Sci. Technol.* 2006a;40(1):109–119.

Bzdusek, PA, Christensen, ER, Lu, J, Christensen, ER. PCB congeners and dechlorination in sediments of sediments of Sheboygan River, Wisconsin, determined by matrix factorization. *Environ. Sci. Technol.* 2006b;40(1):120–129.

Caldwell, JM, Raley, ME, Levine, JF. Mitochondrial multiplex real-time PCR as a source tracking method in fecal-contaminated effluents. *Environ. Sci. Technol.* 2007;41(9): 3277–3283.

Chen, J, Gaillardet, J, Louvat, P. Zinc isotopes in the Seine River waters, France: A probe of anthropogenic contamination. *Environ. Sci. Technol.* 2008;42(17):6494–6501.

Chow, TJ, Snyder, CB, Earl, JL. Isotope ratios of lead as pollutant source indicators. International Atomic Energy Agency, IAEA-SM-191/4; 1975.

Chow, TJ, Earl, JL. Lead isotopes in North American coals. *Science* 1970;176:510–511.

Christensen, ER, Guinn, VP, Zinc from Automobile Tires in Urban Runoff. *ASCE J. Environ. Eng.* 1979;105:165–168.

Christensen, ER, Lo, C-K. Polychlorinated biphenyls in dated sediments of Milwaukee Harbour, Wisconsin, USA. *Environ. Pollut.* (Ser. B), 1986;29:217–232.

Christensen, ER, Zhang, X. Sources of polycyclic aromatic hydrocarbons to Lake Michigan determined from sedimentary records. *Environ. Sci. Technol.* 1993;27:139–146.

Christensen, ER, Phoomiphakdeephan, W, Ab Razak, IA. Water quality in Milwaukee, Wisconsin vs. intake crib location. *ASCE J. Environ. Eng.* 1997a;123(5):492–498.

Christensen, ER, Li, A, Ab Razak, IA, Rachdawong, P, Karls, JF. Sources of polycyclic aromatic hydrocarbons in sediments of the Kinnickinnic River, Wisconsin. *J. Great Lakes Res.* 1997b;23(1):61–73.

Christensen, ER, Rachdawong, P, Karls, JF, Van Camp, RP. PAHs in sediments: unmixing and CMB modeling of sources. *J. Environ. Eng. ASCE* 1999;125(11):1022–1032.

Christensen, ER, Bzdusek, PA. "PAHs in sediments of the Black River and the Ashtabula River, Ohio: Source apportionment by factor analysis." *Wat. Res.* 2005;39:511–524.

Christensen, JN, Dresel, PE, Conrad, MF, Patton, GW, Depaolo, DJ. Isotopic tracking of Hanford 300 area derived uranium in the Columbia River. *Environ. Sci. Technol.* 2010;44(23):3108–3114.

Christiansen, CM, Damgaard, I, Broholm, M, Kessler, T, Bjerg, PL. Direct-push delivery of dye tracers for direct documentation of solute distribution in clay till. *J. Environ. Eng. ASCE,* 2012;138(1):27–37.

Edgington, DN, Robbins, JA. Records of lead deposition in Lake Michigan sediments since 1800. *Environ. Sci. Technol.* 1976;10(3):266–274.

Ehleringer, JR, Rundel, PW. Stable isotopes: history, units, and instrumentation. In: Rundel, PW, Ehleringer, JR, Nagy, KA (Editors). *Stable Isotopes in Ecological Research.* New York: Springer Publishing Co.; 1989.

Elsbury, KE, Paytan, A, Ostrom, NE, Kendall, C, Young, MB, McLaughlin, K, Rollog, ME, Watson, S. Using oxygen isotopes of phosphate to trace phosphorus sources and cycling in Lake Erie. *Environ. Sci. Technol.* 2009;43(9):3108–3114.

Fox, KR, Lytle, DA. *Cryptosporidium and the Milwaukee incident: Critical Issues in Water and Wastewater Treatment.* Proceedings, 1994 ASCE National Conference on Environmental Engineering, Ryan, JN, Edwards, M, Editors. ASCE, New York; 1994, pp. 51–57.

Friedlander, G, Kennedy, JW, Miller, JM. *Nuclear and Radiochemistry,* 2nd Ed. New York: John Wiley & Sons, Inc; 1964.

Friedlander, SK. Chemical element mass balances and identification of air pollution sources. *Environ. Sci. Technol.* 1973;7(3):235–240.

Goldberg, ED. Geochronology with Pb-210. In: *Radioactive Dating, Proceedings of a Symposium on Radioactive Dating.* Vienna, Austria: International Atomic Energy Agency; 1963, pp. 121–131.

Gordon, GE . Receptor Models. *Environ. Sci. Technol.* 1988;22(10):1132–1142.

Jester, WA, Raupach, DC, Haaser, FG. *A comparison of the bromide ion neutron activatable tracer with tritiated water as groundwater tracers.* Third International Conference on Nuclear Methods in Environmental and Energy Research, October 10–13, 1977, Columbia, Missouri.

Henry, RC. Current factor analysis models are ill-posed. *Atmos. Environ.* 1987;21:1815–1820.

Henry, RC. Multivariate receptor modeling by N-dimensional edge detection. *Chemom. Intell. Lab. Syst.* 2003;65:179–189.

Henry, RC, Park, ES, Spiegelman, CH. Comparing a new algorithm with the classic methods for estimating the number of factors. *Chemom. Intell. Lab. Syst.* 1999;48:91–97.

Henry, RC, Christensen, ER. Selecting an appropriate multivariate source apportionment model result. *Environ. Sci. Technol.* 2010;44(7):2474–2481.

Hermanson, M, Christensen, ER. Recent sedimentation in Lake Michigan. *J. Great Lakes Res.* 1991;17(1):33–50.

Hoefs, J. *Stable Isotope Geochemistry*, 3rd Ed. New York: Springer Publishing Co.; 1987.

Hopke, PH. *Receptor Modeling in Environmental Chemistry*. New York: Wiley-Interscience; 1985.

Hopke, PH (Editor). *Receptor Modeling for Air Quality Management*. New York: Elsevier; 1991.

Huang, S. *Seasonal variations in the flux and isotopic composition of nitrate in the Upper Illinois River*. MS thesis. University of Illinois-Chicago, 2006.

Jarman, WM, Hobson, KA, Sydeman, WJ, Bacon, CE, McLaren, EB. Influence of trophic position and feeding location on contaminant levels in the Gulf of the Farallones food web revealed by stable isotope analysis. *Environ. Sci. Technol.* 1996;30:654–660.

Johnson, GW, Jarman, WM, Bacon, CE, Davis, JA, Ehrlich, R, Risebrough, RW. Resolving polychlorinated biphenyl source fingerprints in suspended particulate matter of San Francisco Bay. *Environ. Sci. Technol.* 2000;34(4):552–559.

Joreskog, KG, Klovan, JE, Reyment, RA. *Methods in Geomathematics: Geological Factor Analysis*. Amsterdam, The Netherlands: Elsevier; 1976.

Karls, JF, Christensen, ER. Carbon particles in dated sediments from Lake Michigan, Green Bay, and Tributaries. *Environ. Sci. Technol.* 1998;32(2):225–231.

Kenski, DM, Wadden, RA, Scheff, PA, Lonneman, WA. Receptor modeling approach to VOC emission inventory validation. *ASCE J. Environ. Eng.* 1995;121(7):483–491.

Kidd, KA, Schindler, DW, Hesslein, RH, Muir, DCG. Correlation between stable nitrogen isotope ratios and concentrations of organochlorines in biota from a freshwater food web. *Sci. Total Environ.* 1995;160/161:381–390.

Kiriluk, RM, Servos, MR, Whittle, DM, Cabana, G, Rasmussen, JB. Using ratios of stable nitrogen and carbon isotopes to characterize the biomagnification of DDE, mirex, and PCB in a Lake Ontario pelagic food web. *Can J. Fish. Aquat. Sci.* 1995;52:2660–2674.

Kloppmann, W, Vengosh, A, Guerrot, C, Millot, R, Pankratov, I. Isotope and ion selectivity in reverse osmosis desalinaion: Geochemical tracers for man-made freshwater. *Environ. Sci. Technol.* 2008a;42(13): 4723–4731.

Kloppmann, W, Van Houtte, E, Picot, G, Vandenbohede, A, Lebbe, L, Guerrot, C, Millot, R, Gaus, I, Wintgens, T. Monitoring reverse osmosis treated wastewater recharge into a coastal aquifer by environmental isotopes (B, Li, O, H). *Environ. Sci. Technol.* 2008b;42(23):8759–8765.

Koide, M, Soutar, A, Goldberg, ED. Marine geochronology with 210Pb. *Earth Planet Sci. Lett.* 1972;14:442–446.

Koide, M, Bruland, KW, Goldberg, ED. Th-228/Th-232 and Pb-210 geochronology in marine and lake sediments. *Geochim Cosmochim. Acta* 1973;37:1171–1187.

Kunkel, U, Radke, M. Reactive tracer test to evaluate the fate of pharmaceuticals in rivers. *Environ. Sci. Technol.* 2011;45:6296–6302.

Klingenberg, B, Curry, J, Dougherty, A. Non-negative matrix factorization: ill-posedness and a geometric algorithm. *Pattern Recognition* 2009;42:918–928.

Krishnaswami, S, Lal, D, Martin, J, Meybeck, M. Geochronology of lake sediments. *Earth Planet. Sci. Lett.* 1971;11:407–414.

Krishnaswami, S, Lal, D. Radionuclide limnochronology. In: Lerman, A., Ed. Lakes: Chemistry, Geology, Physics. New York: Springer Publishing Co.; 1978.

Lathrop Jr, RG, Van de Castle, JR, Lillesand, TM. Monitoring river plume transport and mesoscale circulation in Green Bay, Lake Michigan through satellite remote sensing. *J. Great Lakes Res.* 1990;16:471–484.

Lawton, WH, Sylvestre, EA. Self modeling curve resolution. *Technometrics* 1971;13(3): 617–633.

Li, A, Ab Razak, IA, Christensen, ER. *Toxic Organic Contaminants in the Sediments of the Milwaukee Harbor Estuary. Phase III—KinnickinnicRiver sediments. Final Report to U.S. Army Corps of Engineers*, Detroit. University of Wisconsin-Milwaukee; 1995.

Li, S-L, Liu, C-Q, Li, J, Liu, X, Chetelat, B, Wang, B, Wang, F. Assessment of the sources of nitrate in the Changjiang River, China using a nitrogen and oxygen isotopic approach. *Environ. Sci. Technol.* 2010;44(5):1573–1578.

Liu, C-Q, Li, S-L, Lang, Y-C, Xiao, H-Y. Using $\delta^{15}N$ and $\delta^{18}O$-values to identify nitrate sources in karst ground water, Guiyang, Southwest China. *Environ. Sci. Technol.* 2006;40(22):6928–6933.

Malinowski, ER. *Factor Analysis in Chemistry*, 2nd Ed. New York: John Wiley & Sons.; 1991.

Miller, MS, Friedlander, SK, Hidy, GM. A chemical element balance for the Pasadena aerosol. *J. Colloid Interface Sci.* 1972;39(1):165–176.

Mortimer, CH. Discoveries and testable hypotheses from Coastal Zone Color scanner imagery of southern Lake Michigan. *Limn. Oceanog.* 1988;33:203–226.

Muir, DCG, Segstro, MD, Hobson, KA, Ford, CA, Stewart, REA, Olpinski, S. Can seal eating explain elevated levels of PCBs and organochlorine pesticides in walrus blubber from eastern Hudson Bay (Canada)? . *Environ. Pollut.* 1995;90(3):335–348.

Naden, PS, Old, GH, Eliot-Laize, C, Granger, SJ, Hawkins, JMB, Bol, R, Haygarth, P. Assessment of natural fluorescence as a tracer of diffuse agricultural pollution from slurry spreading on intensely-farmed grasslands. *Wat. Res.* 2010;44:1701–1712.

National Research Council. *Polycyclic Aromatic Hydrocarbons: Evaluation of Sources and Effects*. National Academy Press, Washington, DC; 1983.

Osburn, CL, Handsel, LT, Mikan, MP, Paerl, HW, Montgomery, MT. Fluorescence tracking of dissolved and particulate organic matter quality in a river-dominated estuary. *Environ. Sci. Technol.* 2012;46:8628–8636.

Ozeki, T, Koide, K, Kimoto, T. Evaluation of sources of acidity in rainwater using a constrained oblique rotational factor analysis. *Environ. Sci. Technol.* 1995;29:1638–1645.

Panno, SV, Hackley, KC, Kelly, WR, Hwang, HH. Isotopic evidence of nitrate sources and denitrification in the Mississippi river, *Illinois: Journal of Environmental Quality.* 2006;35(2):495–504.

Paatero, P. Least squares formulation of robust nonnegative factor analysis. *Chemom. Intell. Lab. Syst.* 1997;37:23–25.

Paatero, P, Tapper, U. Positive matrix factorization: a nonnegative factor model with optimal utilization of error estimates of data values. *Environmetrics* 1994;5:111–126.

Pennington, W, Tutin, TG, Cambray, RS, Fisher, EM. Observations on lake sediments using fall-out ^{137}Cs as a tracer. *Nature* 1973;242:324–326.

Perrot, V, Epov, VN, Pastukhov, MV, Grebenshchikova, VI, Zouiten, C, Sonke, JE, Husted, S, Donard, OFX, Amouroux, D. Tracing sources and bioaccumulation of mercury in fish of Lake Baikal—Angara River using Hg isotopic composition. *Environ. Sci. Technol.* 2010;44(21):8030–8037.

Rachdawong, P. *Receptor Models for Source Attribution of PAHs and PCBs in Lake Michigan Sediments*. Ph.D. Dissertation. Dept. of Civil Engineering and Mechanics, University of Wisconsin-Milwaukee; 1997.

Rachdawong, P, Christensen, ER. Determination of PCB sources by a principal component method with nonnegative constraints. *Environ. Sci. Technol.* 1997;31(9):2686–2691.

Rabinowitz, MB, Wetherill, GW. Identifying sources of lead contamination by stable isotope techniques. *Environ. Sci. Technol.* 1972;6:705–709.

Rahn, KA, Hemingway, SB, Peacock, MW. A trace-element technique for determining non-point sources of contamination in freshwaters. *Environ. Sci. Technol.* 1992;26(4):788–797.

Rodenburg, LA, Du, S, Fennell, DE, Cavallo, GJ. Evidence for widespread dechlorination of polychlorinated biphenyls in groundwater, landfills, and wastewater collection systems. *Environ. Sci. Technol.* 2010;44(19):7534–7540.

Rolff, C, Broman, D, Naf, C, Zebuhr, Y. Potential biomagnification of PCDD/Fs—new possibilities for quantitative assessment using stable isotope trophic position. *Chemosphere* 1993;27(1–3):461–468.

Rozanski, K, Araguas-Araguas L, Gonfiantini, R. Isotopic patterns in modern global precipitation. In: *Climate Change in Continental Isotopic Records*, Swart, PK, et al. (editors). Washington: American Geophysical Society; 1993, pp. 1–36.

Rundel, PW, Ehleringer, JR, Nagy, KA (Editors). *Stable Isotopes in Ecological Research*. New York: Springer Publishing Co.; 1989.

Schrag, DP, Hampt, G, Murray, DW. Pore fluid constraints on the temperature and oxygen composition of the glacial ocean. *Science* 1996;272:1930–1932.

Schulz, DE, Petrick, G, Duinker, JC. Complete characterization of polychlorinated biphenyl congeners in commercial Aroclor and Clophen mixtures by multidimensional gas chromatrography—electron capture detection. *Environ. Sci. Technol.* 1989;23(7):852–859.

Sharma, AN, Luo, D, Walter, MT. Hydrological tracers using nanobiotechnology: Proof of concept. *Environ. Sci. Technol.* 2012;44(5):1544–1550.

Shen, C, Phanikumar, MS, Fong, TT, Aslam, I, McElmurry, SP, Molloy, SL, Rose, JB. Evaluating bacteriophage P22 as a tracer in a complex surface water system: The Grand River, Michigan. *Environ. Sci. Technol.* 2008;42(7):2426–2431.

Singh, AK, Gin, ME, Christensen, ER. A source-receptor method for determining non-point sources of PAHs to the Milwaukee Harbor Estuary. *Water Sci. Technol.* 1993;28(8–9):91–102.

Spencer, KL, Droppo, IG, He, C, Grapentine, L, Exall, K. A novel tracer technique for the assessment of fine sediment dynamics in urban water management systems. *Wat. Res.* 2011;45:2595–2606.

Su, M-C. *Apportionment of PAH and PCDD/F Sources in Aquatic Sediments by a Chemical Mass Balance Model*. Ph.D. Dissertation, Department Civil Engineering and Mechanics, University of Wisconsin-Milwaukee; 1997.

Su, M-C, Christensen, ER. Apportionment of sources of polychlorinated dibenzo-*p*-dioxins and dibenzofurans by a chemical mass balance model. *Water Research* 1997;31(12):2935–2948.

Su, M-C, Christensen, ER, Karls, JF, Kosuru, S, Imamoglu, I. Apportionment of PAH sources in Lower Fox River sediments by a chemical mass balance model. *Environ. Toxicol. Chem.* 2000;19(6):1481–1490.

Su, M-C, Christensen, ER, Karls, JF. Determination of PAH sources in dated sediments from Green Bay, Wisconsin by a chemical mass balance model. *Environ. Pollut.* 1998;99:411–419.

Thapalia, A, Borrok, DM, Van Metre, PC, Musgrove, M, Landra, ER. Zn and Cu isotopes as tracers of anthropogenic contamination in a sediment core from an urban lake. *Environ. Sci. Technol.* 2010;44(5):1544–1550.

Torgerson, T, Top, Z, Clarke, WB, Jenkins, WJ, Broecker, W. A new method for physical limnology–tritium–helium-3 ages–results for Lakes Erie, Huron, and Ontario. *Limn. Oceanogr.* 1977;22(2):181–193.

Unno, T, Di, DYW, Jang, J, Suh, YS, Sadowsky, MJ, Hur, H-G. Integrated online system for a pyrosequencing-based microbial source tracking method that targets bacteroidetes 16S rDNA. *Environ. Sci. Technol.* 2012;46(5):93–98.

U.S. Environmental Protection Agency. 2009a. EPA Unmix 6.0 Model, http://www. epa.gov/heasd.products/unmix/unmix.htm.

U.S. Environmental Protection Agency. 2009b. EPA Positive Matrix Factorization 3.0, http://www.epa.gov/heasd/products/pmf/pmf_registration.htm.

Wanninkhof, R, Sullivan, KF, Dammann, WP, Proni, JR, Bloetscher, F, Soloviev, AV, Carsey, TP. Farfield tracing of a point source discharge plume in coastal ocean using sulfur hexafluoride. *Environ. Sci. Technol.* 2005;39(22):8883–8890.

Young, MB, McLaughlin, K, Kendall, C, Stringfellow, W, Rollog, M, Elsbury, K, Donald, E, Paytan, A. Characterizing the oxygen isotopic composition of phosphate sources to aquatic ecosystems. *Environ. Sci. Technol.* 2009;40(22):6928–6933.

Zeng, EY, Vista, CL. Organic pollutants in the coastal environment off San Diego, California. 1. Source identification and assessment by compositional indices of polycyclic aromatic hydrocarbons. *Environ. Toxicol. Chem.* 1997;16(2):179–188.

Zou, Y. *PCB and PBDE sources and degradation in aquatic sediments determined by positive matrix factorization*. PhD dissertation. University of Wisconsin-Milwaukee; 2011.

Zou, Y, Christensen, ER. Phosphorus loading to Milwaukee harbor from rivers, storm water, and wastewater treatment. *J. Environ. Eng. ASCE*, 2012;138(2):143–151.

Zou, Y, Christensen, ER, Li, A. Characteristic pattern analysis of polybromodiphenyl ethers in Great Lakes sediments: a combination of eigenspace projection and positive matrix factorization analysis. *Environmetrics* 2013;24:41–50.

10

ECOTOXICOLOGY

10.1 INTRODUCTION

In a broad sense, ecotoxicology covers the harmful effects of various toxic substances, primarily those introduced by humans, on ecosystems. The aquatic environment is our central consideration in this chapter, although land surfaces, groundwater, and the atmosphere are closely linked to the ecology of lakes, rivers, and oceans in many cases. Bioassays with, fish and algae, for example, are used to screen chemicals for their toxic effects. Pathogens can be assessed using classical methods, such as plate counts, but a battery of modern molecular tools, notably polymerase chain reaction (PCR), fluorescence in situ hybridization (FISH) (Wilderer et al., 2002), and gene expression (Marie et al., 2006; Weil et al., 2009), are becoming increasingly popular. The latter techniques are often highly specific to the targeted organisms. In order to simplify environmental toxicology, there is an increased emphasis on biomarkers that may be deemed surrogate responses to several environmental pollutants.

Human health aspects are clearly of major importance. Although traditionally treated under the environmental health label, we will include human health aspects to the extent that environmental pollutants, such as metals and organic compounds, have a significant impact. The influence of pathogens was considered in Chapter 9. Mercury and polychlorinated biphenyls (PCBs) are the contaminants of most concern in Wisconsin's fish (Wisconsin Department of Natural Resources 2009). PCBs and, to some extent, methylated mercury are known to biomagnify, meaning that larger fish eating smaller fish will have a higher contaminant concentration. For the Great

Physical and Chemical Processes in the Aquatic Environment, First Edition. Erik R. Christensen and An Li.
© 2014 John Wiley & Sons, Inc. Published 2014 by John Wiley & Sons, Inc.

Lakes, walleye and bass contain the highest PCB levels. PCBs, now mainly resulting from legacy pollutants in old electrical equipment or sediments, are still a concern for the Great Lakes, whereas mercury, mostly from coal-fired power plants, poses a threat to inland waters in Wisconsin.

Among emerging contaminants, endocrine-disrupting chemicals, such as estradiol and bisphenol A, occupy a special place because their effects are not related to lethality, but rather to changes in reproductive health and male vs. female characteristics in daphnia, amphipods, crustaceans, and fish (Gomes et al., 2004; Wollenberger 2005; Pollino et al., 2007). Toxicity of disinfection by-products from chlorination and ozonation in water and wastewater treatment is another potential threat.

Toxic effects of chemicals can alter as a result of metabolism, detoxification by the organism under consideration, complexation with organics, or degradation (Vaché et al., 2006; McLeod et al., 2007). Degradation of toxic organics, such as PCBs, in the environment can be a major mechanism for reducing toxicity (Field and Sierra-Alvarez 2008). PCBs in concentrations above a few ppm in sediments undergo anaerobic dechlorination. The lower chlorinated congeners can be subjected to aerobic degradation. The increased use of antibiotics for humans and animals is an important environmental problem. It appears that this use has generated antibiotic-resistant genes (ARGs) that can render many antibiotics useless or less effective for infection treatment.

Nanomaterials have recently received significant attention. These materials are usually defined as those with at least two dimensions between 1 and 100 nm (Klaine et al., 2008). Nanoparticles have been known for some time in air pollution as ultrafine particles, and in water and soil as colloids with slightly different size range, they became a concern after 1985 when manufacture of engineered nanoparticles began, following the discovery of carbon nanoparticles or fullerenes (Lovern and Klaper 2006). Nanoparticles can be very reactive and typically have properties that differ from those of their macro counterparts. They have been applied in diverse areas such as electronics, energy, and personal care products, for example sun screens. The environmental effects of nanoparticles are only beginning to be studied (Klaine et al., 2008).

Dose–response models for single and multiple toxicants are an important area because of the statistical description of the response of organisms or populations. In some cases, they can provide insight into mechanisms of toxicity with the subsequent possibility of predicting the response from many toxicants acting simultaneously (Christensen and Chen 1985; Broderius et al., 2005). Common to many recent toxicity models is their reliance on a chemical complexation model to drive the toxic response, as expressed, for example, in Pagenkopf's gill surface interaction model (1983), the Weibull model for algal growth (Christensen and Nyholm 1984), and the biotic ligand model for acute toxicity developed by Di Toro et al. (2001).

Other aspects of ecotoxicology include pulsed toxicity tests (Ashauer et al., 2006) and chronic or sublethal toxicity tests (Robidoux et al., 2004; Valenti et al., 2005). Whereas water quality criteria are typically based on a battery of acute and chronic tests with sensitive organisms (see Chapter 11 for an overview of standards and criteria), sediment quality criteria for metals are linked to water quality criteria as modified by the complexation of metals with acid volatile sulfide (AVS) and organic carbon (Ankley et al., 1996). Sediment quality criteria for organics, such as PCBs or PAHs, are also dependent on complexation with sediment organic carbon (Long et al., 2006; Di Toro and McGrath 2000). Total maximum daily load of pollutants is a relatively new concept (Novotny 2004; Benham et al., 2005; Pedersen et al., 2006), appropriate levels of which are required by the Clean Water Act. It attempts to link loads to water quality criteria in the receiving waters. These topics will be further developed in this chapter.

10.2 BIOASSAYS

Acute and chronic toxicity tests are used to screen or assess the toxicity of various toxic substances in the water environment. Acute testing is usually performed within 48 h and chronic testing is typically undertaken over longer periods because reproductive outcomes or other sublethal effects are being tested. As with most fish bioessays, the endpoint can be mortality. For algal assays, yield or growth rate reduction is more commonly used (Nyholm 1985). In the case of daphnia testing, the endpoint is usually taken to be immobility of the organism as a result of the presence of the toxic compound.

An important and developing concept in toxicity testing is quantitative structure activity modelling, which seeks to relate the toxic response to characteristic functional groups of organic chemicals (Bradbury et al., 2003; Russom et al., 1997). The vast majority of industrial organic chemicals can be characterized by a baseline relationship of the log of lethal concentration causing 50% mortality, LC50, to the log of the octanol–water partition coefficient, Log K_{ow}, a measure of hydrophobicity (Veith and Broderius 1990). A further improvement in predictive ability can be achieved by considering groups of chemicals with similar toxic action. For fathead minnows (*Pimephales promelas*), the main modes of action of organic chemicals were narcosis, oxidative phosphorylation uncoupling, respiratory inhibition, electrophile/proelectrophile reactivity, acetylcholinesterase (AChE) inhibition, and central nervous system seizure responses (Russom et al., 1997). Narcosis is a very common mode of action, characterized by a reversible state of arrested activity, which apparently involves lipid membrane perturbations (Ren and Schultz 2002). Within each group of chemicals with a certain group or mode of action, log K_{ow} will be an effective predictor of log LC50. More specific descriptors for each type of toxicity, such as hydrophobicity (Log (K_{ow})) or electrophilicity, including energy of the highest occupied molecular orbital (EHOMO) and energy of the lowest unoccupied molecular orbital (ELUMO), can be helpful in distinguishing between narcotic and reactive compounds. (Ren and Schultz 2002).

More subtle changes in organisms as a result of exposure to toxic substances can be acquired through biomarkers that can reflect toxicological, immunological, and physiological responses at the organism level. For example, Petala et al. (2009) examined chronic daphnia tests along with a battery of biomarker tests carried out in vitro on liver and kidney of rainbow trout (*Oncorynchus mykiss*) that were exposed to effluents from wastewater treatment. The response of biomarkers was dependent on treatment method and was tissue specific. Biomarkers included glutathione (GSH), glutathione S-transferases (GST), glutathione peroxidase (GPX), lipid peroxidation (LPO), and heme peroxidase. Ozonation increased hepatic peroxidation in terms of heme peroxidase, LPO, and GST, whereas GSH was depleted, reducing the organism's protection against reactive oxidant species attack.

10.2.1 Fish

Fish bioassays are often aimed at comparative studies of the toxicity of many chemicals with different modes of action. For such tests, mortality is an endpoint that is easy to monitor and treat statistically, and the test is completed in a relatively short time period (≤ 48 h). However, more recently there has been an interest in fish full lifecycle tests (FFLT) or tests with other endpoints, such as hatch rate and larval development. For example, Seki et al. (2005) investigated the effect of the endocrine-disrupting chemical 17β estradiol on Medaka (*Oryzias latipes*). They observed concentration-dependent abnormal sex differentiation, induction of hepatic vitellogenin (VTG), and reproductive impairment. The lowest observed effect concentration (LOEC) and no observed effect concentration (NOEC) on parent and progeny generations of medaka were found to be 8.66 and 2.86 ng/L respectively. These low levels suggest that ecotoxic effects occur at much lower concentrations than LC50.

A short-term study of the effects of methyl-*tert*-butyl ether (MTBE) and the metabolite *tert*-butyl alcohol (TBA) on the African catfish (*Clarias gariepinus*) was conducted by Moreels et al. (2006). The time between spawning, egg fertilization, and hatching was about 48 h. Subsequent to hatching, viable larvae were observed 96 h post hatching. MTBE had no significant effect on egg viability, whereas TBA induced a decline in hatch rate. MTBE has toxicological effects on catfish larvae above 50 mg/L, which is much higher than concentrations found in surface water (0.088 mg/L). Whether chronic effects are present at lower than 50 mg/L concentrations remains to seen.

10.2.2 Algae

A typical medium for toxicity testing with freshwater algae is shown in Table 10.1 (Lin et al., 2005). The medium contains macro nutrients, such as nitrate and phosphate, but also several essential micronutrients, i.e. trace metals and iron. Ethylene diamine tetraacetic acid, EDTA, is a synthetic ligand that acts as a metal buffer and thereby minimizes the effects of inadvertently introduced trace metals, and decreases nutrient or metal precipitates, particularly during autoclaving. However, recent studies and protocols tend to eliminate EDTA from the test medium during toxicity testing in order to better approximate the actual environment (Lin et al., 2005; Karner et al., 2006). With trace metal clean methods, use of EDTA in trace metal bioassays can be avoided (Karner et al., 2006).

Carbonate is often present in culture media in relatively high concentrations to restrain pH from increasing beyond approximately 9.0 during the testing period.

TABLE 10.1. Composition of Culture Medium for Algae[a]

Compounds	Concentration (mg/L)	Compounds	Concentration (µg/L)
$NaNO_3$	12.75^b (25.5^c)	H_3BO_3	186
$NaHCO_3$	15.0	$MnCl_2.4H_2O$	415.6
K_2HPO_4	0.52^b (1.04^c)	$ZnCl_2$	3.27
$MgSO_4.7H_2O$	14.7	$CoCl_2.6H_2O$	1.428
$MgCl_2.6H_2O$	12.16	$CuCl_2.2H_2O$	0.012
$CaCl_2.2H_2O$	4.41	$Na_2MoO_4.2H_2O$	7.26
		$FeCl_3.6H_2O$	160.0
		$Na_2EDTA.2H_2O$	30^b (0^c)

(Source: [a]Lin et al. (2005)).
[b]Concentration applied to chemostat incubation
[c]Concentration applied to toxicity testing

An enriched bicarbonate buffer can result in decrease in toxic response of metals and organics (Lin et al., 2005). More moderate bicarbonate concentrations, as shown in Table 10.1 (15 mg/L), are feasible in closed-system algal assays with shorter test duration (30–48 h) and CO_2 enriched headspace (Christensen et al., 2009). Closed systems are also important for the toxicity estimation of volatile compounds.

Cell density or growth rate can be used as the response parameter of algal assays. Whereas cell density usually produces EC50 values that are a factor of 2 lower than growth rate based values, growth rate is deemed a more fundamental parameter with greater reproducibility and is, therefore, often preferred (Nyholm 1985; Lin et al., 2005). Cell numbers, cell volume, dissolved oxygen (Lin et al., 2005), chlorophyll A, or the stable isotope ^{14}C are surrogate measures of biomass. Effective chlorophyll A estimates can be made using fluorescence measurement (Mayer et al., 1997). For special uptake studies, one can use other stable isotopes, such as ^{65}Cu for Copper uptake (Karner et al., 2006), ^{32}P for nutrient phosphorus uptake, and ^{15}N for nitrogen assimilation. Typical test algae include *Pseudokirchneriella subcapitata* (Lin et al., 2005; Christensen et al., 2009) in freshwater and *Thalasiosira weissflogii* (Karner et al., 2006) in seawater. Metal concentrations in media can be measured or calculated from computer programs such as MINEQL + (Karner et al., 2006).

Appropriate statistical procedures are important to characterize growth rates and effective concentrations resulting from algal assays (Nyholm et al., 1992). Estimates of lowest observed effect concentration, no observed effect concentrations, and no effect concentrations can also be made. A recent study has demonstrated that these low effect concentrations can be estimated effectively from the EC50 and slope at 50% response of the dose–response function (Christensen et al., 2009).

10.2.3 Daphnia

Bioassays with *Daphnia magna* are frequently used to evaluate the toxicity of metals and organics (ISO 1996). Typically, five daphnids are placed in each of four or five replicate tubes at each concentration of the toxicant for a period of 48 h. The

TABLE 10.2. Composition of Dilution Water for
Daphnia Magna **Straus Acute Toxicity Test**

Compound	Concentration (mg/L)
$CaCl_2.2H_2O$	294
$MgSO_4 7H_2O$	123.25
$NaHCO_3$	64.75
KCl	5.75

(Source: International Standard Organization ISO 6341:(1996)).
The water must have a pH of 7.8 ± 0.2 and dissolved oxygen concentration ≥ 7 mg/L. Hardness = 250 mg/L ± 25 mg/L as $CaCO_3$

dilution water contains bicarbonate buffer, calcium chloride, magnesium sulfate, and potassium chloride (Table 10.2). The endpoint is deemed to be immobilization of an organism. Reproduction takes place by parthenogenesis. The sensitivity of a Daphnia magna test to several organics is comparable to sensitivities of other test methods. Lin et al. (2005) compared the sensitivity of several bioassay tests to the organics phenol, 2-chlorophenol, 2,4-dichlorophenol, toluene, chlorobenzene, and benzene and found the average ranking in order of relative sensitivity to be as follows: rainbow trout > algae (cell density) > *Daphnia magna* > algae (growth rate) > algae (DO production) > microtox > fathead minnow.

Other cladocerans, *Ceriodaphnia dubia*, *Daphnia ambigua*, and *Daphnia pulex* are either more sensitive, i.e. have lower LC50 values (Cd), or have a higher slope of the dose–response function (Zn) (Shaw et al., 2006). In single metal tests these authors determined 48 h LC50 values ranging from 0.09 to 0.9 µmol/L and 4 to 12.5 µmol/L for cadmium and zinc respectively. However, *Daphnia magna* is widely used and test procedures have been standardized (ISO 1996); therefore its use is generally recommended.

Other daphnia testing has been carried out with arsenic and mercury. For arsenic, it was determined that monomethylarsonous acid (MMA^{III}) was the most toxic substance (120 µg/L), followed by inorganic arsenic (2500–3900 µg/L) and pentavalent methylated arsenicals and phenylarsonic compounds (13800–15700 µg/L), where LC50 values are indicated in parenthesis (Shaw et al., 2007). Tests with first, second, and third generation members of *Daphnia magna* organisms exposed to Hg indicated that subsequent generations of daphnia can become acclimated to Hg, as shown by increased LC50 values. Hg exposure was accompanied by reduced Hg ingestion rate and enhanced induction of metallothionein-like proteins. However, once low level exposure to Hg ceases, the vulnerability can be enhanced as indicated by lower LC50 values (Tsui and Wang 2005). Dissolved organic matter (DOM) tends to complex metals and thereby reduce their toxicity (Karel et al. 2004). This effect was quantified through UV-absorbance as a measure of DOM variability in a biotic ligand model for Cu toxicity to *Daphnia magna* (De Schamphelaere et al., 2004). Absorbance coefficients (ε L/mg/cm) were calculated from

$$\varepsilon_{350} = A_{350}/(d\ [\text{DOC}])\tag{10.1}$$

where A_{350} = absorbance at 350 nm, d = path length and [DOC] = DOC concentration, mgC/L.

10.3 MOLECULAR BIOLOGY TOOLS

Modern molecular tools have made it possible to identify bacteria or bacterial communities using their deoxyribonucleic acid, DNA, and ribosomal ribonucleic acid, rRNA, rather than time consuming or non-specific plate counts. DNA consists of deoxyribonucleotides, which consist of a base, the sugar deoxyribose, and a phosphate group. Two strands are intertwined to form a double helix. Adenine (A), cytosine (C), thymine (T), and guanine (G) are the DNA bases. The sugar and phosphate groups form the core of the two strands, with the bases protruding from the chain. A and T form hydrogen bonds as do C and G, holding the double helix together. DNA contains the necessary information for RNA and protein synthesis in regions known as genes. Gene probes can be used to relate performance of wastewater treatment processes to the basic microbial organisms carrying out the breakdown of, and cell synthesis from, wastewater.

An important category of DNA-specific techniques is polymerase chain reaction (PCR), a DNA amplification method using primers, thermal cycling, and reporters or probes. Denaturing gradient gel electrophoresis (DGGE) is a non-quantitative variant of PCR that has been used to study microbial community complexity, the effect of changing a variable in wastewater treatment, or to follow the expression of genes in the environment. Fluorescent in situ hybridization (FISH) involves gene probes that are designed to hybridize, i.e. form double-stranded structures with rRNA of certain bacteria. The method is quantitative and targets active cells. It is not based on DNA amplification and allows microscopic inspection of intact cells.

Confocal laser scanning microscopy (CLSM) enables the removal of out-of-focus information so that depth-resolved scanning can be performed. The basis for this technique is excitation of a fluorophore, which is a fluorescent dye binding to specific structures in the cell using a focused laser beam. An overview of these and other methods was presented by Wilderer et al. (2002). A further description of PCR, FISH, and gene expression is given in Section below.

10.3.1 Polymerase Chain Reaction (PCR)

PCR involves generation of a DNA fragment from a template strand. This strand is then amplified using the enzyme DNA polymerase from a thermophilic bacterium. Two primers are used to make copies of a given region of DNA located between sequences complementary to the primers. A PCR cycle consists of three steps, which occur at different temperatures and durations: denaturing of the target DNA leading to the formation of single strands, annealing of the primers, i.e. binding to the template DNA, and DNA synthesis. This cycle is repeated 20 to 30 times. During the process the target DNA is amplified exponentially. An example of cycling conditions for three genes that are responsive to endocrine-disrupting compounds were: 94/45 (denaturing), 55/30 (annealing), and 72/90 (synthesis), where the numbers indicate temperature (°C)/duration (s) (Park et al., 2009). PCR is followed by electrophoresis or fluorescence measurement, which can make it quantitative (qPCR) and near real-time. PCR and the non-quantitative related method DGGE are now widely used to characterize bacterial communities or single species. By varying the primer sets in the PCR amplifications, selectivity to specific organisms can be obtained.

10.3.2   Fluorescent in Situ Hybridization (FISH)

Gene probes of single-stranded oligonucleotides consisting of 18 ± 3 nucleotides labeled with a fluorescent dye can be designed to hybridize with ribosomal RNA of certain bacteria, i.e. form double-stranded structures. Hybridization can occur ex situ to isolated nucleic acids or in situ to whole cells. By microscopic detection it is possible to count cells of particular interest, for example because of their presence and role in wastewater treatment. Both methods, dot-blot hybridization and FISH, are quantitative and target rRNA, which is mainly found in active cells.

A study of metal toxicity in activated sludge wastewater treatment showed that biochemical identification of nitrifying bacteria as most vulnerable was confirmed by FISH analysis of the microbial community (Principi et al., 2006). Metal effects were mostly observed for the β-proteobacteria subgroup, which includes ammonia-oxidizing bacteria. Studying the effects of endocrine-disrupting chemicals on Japanese medaka (*Oryzias latipes*), Park et al. (2009) found that FISH combined with histology was able to uncover molecular effects of these chemicals through changes in gene expression.

Application of FISH has been important for the understanding of processes and microorganisms that are responsible for anaerobic oxidation of ammonium, ANAM-MOX, biological phosphorus removal, and nitrification processes (Wilderer et al., 2002).

10.3.3   Gene Expression

Molecular mechanisms of toxic action can be explored by quantifying the expression of messenger RNAs, mRNAs, that encode for proteins that are involved in specific processes of a pathway. FISH is such a technique that can identify relevant mRNA sequences in tissues and cells (Park et al., 2009).

Field and experimental studies have shown the role of metallothioneins (MTs) in sequestration and detoxification of toxic metals, e.g. cadmium Cd and copper Cu, and their role in the metabolism of essential elements, e.g. zinc Zn and copper Cu (Marie et al., 2006). Triploidy was developed in oysters to enhance the production and shelf-life of bivalves. Marie et al. (2006) used quantitative real-time polymerase chain reaction (PCR) to examine differential expression of MT isoform genes as well as biosynthesis of total MT in the gills of triploid and diploid juvenile Pacific oysters (*Crassostrea gigas*) in response to cadmium and zinc exposure. They found comparable response in gill metal bioaccumulation and genetic protein MT response in diploid and triploid oysters. Thus under the exposure conditions considered (15 ug Cd/L and 1 mg Zn/L, 14 d exposure), triploid juvenile oysters have the same capacity to respond to Cd and Zn alone or in combination as diploid oysters.

Perfluorinated compounds, such as perfluorooctanoic acid (PFOA), have been used in textiles to resist staining. Wei et al., (2007) examined the estrogen-like properties of waterborne PFOA to rare minnows (*Gobiocypris rarus*) on the expression of the hepatic estrogen-responsive genes vitellogenin (VTG) and estrogen receptor β (ER β). The results indicated that PFOA can induce testes-ova in mature rare minnows and block female reproduction.

A new European Union regulation, REACH, requires registration, evaluation, authorization, and restriction of chemicals and is therefore expected to lead to an increase in the number of animal tests (Weil et al., 2009). In order to limit the number of animal tests, REACH recommends use of suitable alternative methods. For fish toxicity, the fish embryo test can be performed. However, no replacement is currently available for chronic fish toxicity tests. Analysis of toxic effects using differential gene expression in the zebra fish (*Danio rerio*) embryo test (Gene-*Dar*T) may be such an alternative, however, according to Weil et al. (2009). The results of a comparison between lowest observed effect concentration (LOEC) based on the fish early life stage test and the Gene-*Dar*T Gene expression method indicates that there is a factor of less than 10 between LOECs obtained by the two methods for several common chemicals, such as 1,4-dichlorobenzene, atrazine, and ethylparathion. For other chemicals, for example pentachloroaniline and lindane, the difference is larger.

Onnis-Hayden et al. (2009) presented a real-time gene expression profile method for investigating mechanisms of toxicity and toxicity assessment in environmental samples. They illustrated the method using *Escherichia coli* exposed to mercury Hg^{2+} and mitomycin (MMC), a genotoxin. A cell-array library of 93 *E. coli* K12 strains with transcriptional green fluorescent protein (GFP) fusions covering the best-known stress response genes was used. Genes that are involved in metal toxicity resistance via active cation efflux and metal sequestering were expressed after exposure to mercury, indicating the activation of these defense mechanisms. The field of linking ecotoxicology to genomics is sometimes referred to as toxicogenomics. Although it is a promising field, one should realize that not all forms of toxicity may be expressed through genes and that chemical and physical factors can be of equal or greater importance for the toxic response.

10.3.4 Biomarkers

Biomarkers can be used to assess oil spill episodes and other environmental pollution events (Morales-Caselles et al., 2008; Braune and Scheuhammer, 2008; Wang et al., 2011). Oil spills can increase concentrations of PAHs, PCBs, and toxic metals such as Cd, Cu, Ni, and Co in various compartments of the marine environment. Biomarkers of exposure include several organic compounds that are developed by biological organisms in response to the presence of significant concentrations of pollutants. In their assessment of coastal areas in Spain that were affected by marine pollution, Morales-Caselles et al. (2008) used a suite of biomarkers. The phase one detoxification was assessed by ethoxyresorufin-*O*-dethylase (EROD) activity. Phase two detoxification was monitored by glutathione-*S*-transferase (GST), which is also characteristic of oxidative stress. Other biomarkers included glutathione peroxidase (GPX), glutathione reductase (GR), and ferric reducing ability of plasma (FRAP), which were analyzed to determine the antioxidant status of tissues. Twenty-eight day bioassays were conducted on two invertebrates with different feeding habits: The clam *Ruditapes philippinarum* and the crab *Carcinus maenas*.

The data analysis included principal component analyses of pollutant concentrations and biomarker response in the affected areas; acutely for the Galician coast and chronically for the Bay of Algeciras, and one control area located in the Bay of Cádiz. Biomarker responses were normalized to the total protein content. Results showed that biomarker response is a useful tool in the pollution assessment, providing a response that depends on organic vs. metal pollutant concentrations and is also sensitive to the biological species present; clams or crabs.

An example of application of metallothionein (MT), a biomarker that is sensitive to metal pollution, was provided by Braune and Scheuhammer (2008) who looked at metal pollution in seabirds from the Canadian Arctic. Elevated concentrations of Cd, Hg, Cu, and Zn in liver and kidney tissues had been shown previously to induce synthesis of MT, a low molecular weight cysteine-rich protein. Thick-billed murres (*Uria Lomvia*) from Ivujivik contained the highest mean concentrations of Cd, Zn, and Cu, and, therefore, the highest concentration of MT. However, only Cd and Zn had a positive correlation with MT. The current non-toxic levels of Cu do not induce MT. The highest correlation ($r = 0.719$) was found between kidney Cd and MT.

Biomarkers may be deemed surrogate measures of pollution impact in the environment. Wang et al. (2011) developed an integrated biomarker response (IBR) to pollution with PAHs, PCBs, organochlorine compounds (OCP), and the metals Zn, Cu, Ni, and Pb in six areas of Taihu Lake, China. Goldfish (*Carassius auratus*) brought up in clean water were transferred to the stations under study. In addition to MT, EROD, and GST they used the antioxidant defense enzymes catalase (CAT), reduced glutathione content (GSH), and lipoperoxidation (TBARS) as biomarkers.

High measurements of OCPs were found at stations S3, Dagongshan and S5, Mashan in Gong Bay and Meiliang Bay, and PAHs at station S4, Tuoshan in Meiliang Bay. Metal concentrations were high at station S3. The high values of these pollutants were generally reflected in high IBR values, although the composition by individual biomarker varied between stations. For example, on a relative basis, MT was highest at station S3 as expected, while S4 and S5 showed high values of GST and GSH. The high value of EROD and GSH at station S1, which was not clearly mirrored in the IBR, may reflect PCBs. Station S7 in mid-lake Taihu showed high EROD and GST values and a moderate IBR despite low pollutant concentrations. Thus, whereas the IBR approach has some merit in overall pollution assessment it is a method that must be used with caution.

Another pollution monitoring study has been described by Budka et al. (2010). They used machine learning coding and biomarkers to develop predictive models for water pollution. Biomarkers considered included MT, GST, and CAT along with nine others. Three marine stations of various degrees of pollution and one control station from southwest Norway were considered. Based on this study, the authors claim that biomarkers can successfully distinguish between different pollution levels. However, the evidence was not compelling and there was only a qualitative description of pollution levels rather than a listing of pollutants and concentrations.

Metabolic footprinting is a version of biomarker application (Henriques et al., 2007). Liquid extracellular dissolved substances exuded from activated sludge cultures in response to toxicants are analyzed by liquid chromatography/mass spectrometry to create mass to charge (m/z) signatures for toxicants, which may be helpful in identifying and mitigating their impact.

In their evaluation of the effect of pollution along the Karnataka coast, India, Verlecar et al. (2006) used biological species and species richness, rather than biomarkers, as biological indicators of pollution. Differences in populations of phytoplankton, zooplankton, and benthos between a polluted site, Kulai, in the south, and an undisturbed location in the north, Padubidri, which showed higher phytoplankton species diversity in January 1998, suggested not unexpectedly that industrial and dissolved petroleum hydrocarbons at Kulai had a significant effect on the biota in the area.

10.4   HUMAN HEALTH

The potential effects of environmental contaminants in water are a significant concern. Mercury and PCBs are the contaminants of greatest concern. Some fish may take up contaminants from the water in which they swim and the food they eat. Once caught and eaten, the contaminants in fish can pose a risk to human health. Mercury and PCBs have different origins, with mercury coming from coal-fired power plants, chemical manufacturing plants and incinerators, and from inappropriate disposal of mercury-containing products such as thermometers. PCBs were banned in the United States in 1977 and were previously used as hydraulic fluids, in carbonless copy paper, and in capacitors and transformers.

Other contaminants of concern in fish include dioxins, perfluorooctane sulfonate (PFOS), and contaminants associated with cyanobacteria or blue-green algae. Cyanotoxins recognized to date include hepatoxic microsysteins (MC), nodularins (NOD), the cytotoxin cylindrospermopsin (CYN) and neurotoxic agents such as the saxitoxins (STX) and anatoxin-a (ATXa) (Hunter et al., 2009). These authors showed that an airborne spectrographic imager could be used to monitor season change in the concentration of chlorophyll a and the cyanobacterial biomarker pigment C.phycocyanin in a series of shallow lakes in the UK. The World Health Organization has developed provisional guideline levels of cyanobacteria in recreational waters.

10.4.1   Fisheries Advisories

PCBs are mainly stored in the fat of fish; therefore one can reduce PCB levels in fish by removing fatty areas, such as along the back, dark fatty tissue along the entire length of the fillet or belly fat, and by properly cooking the fish. By contrast, mercury is stored throughout the fish including muscle or fillet, and cooking preparation does not reduce the mercury levels (http://dnr.wi.gov/fish/consumption/). People who frequently eat fish with high concentrations of contaminants, such as mercury and PCBs, can accumulate levels that may be harmful. Small children and children of women who frequently eat fish with high contaminant levels may have lower birth weight and delayed development. For adults, the contaminants can affect the reproductive function, cardiovascular health, immune and nervous systems. and increase the risk of cancer (http://dnr.wi.gov/fish/consumption/). On the other hand, fish is low in saturated fat and high in protein; therefore as long as the contaminant intake is below a certain threshold, the benefits outweigh the potential health risks and one should continue to eat fish. Fish provide omega-3 fatty acids, which reduce cholesterol levels (Burger and Gochfeld 2004). The number of meals and species one can safely eat is often given in fisheries advisories issued by the U.S. Food and Drug Administration (FDA) and individual US states, for example in Wisconsin by the Wisconsin Department of Natural Resources (WDNR).

10.4.2   Mercury

Methylmercury interferes with the architecture of the developing brain, disrupting microtubule assembly and interfering with neuronal migration and connections (Burger and Gochfeld 2004). Mercury can be eliminated slowly over time. By comparison, PCBs are stored in body fat and are more resistant to change. In Wisconsin, mercury is found in darkly stained lakes of northern Wisconsin (http:// dnr.wi.gov/fish/consumption/FishAdv%20Mercury%202010.pdf), for example in Bayfield, Iron, and Marinette counties. Large fish, such as walleye, contain higher mercury concentrations than small fish, such as perch. The guidelines of the Wisconsin fisheries advisories make a distinction between women of childbearing age and children under 15 with most restrictions, and men and older women with fewer restrictions, i.e. the latter group are allowed to eat more fish from the listed lakes.

A study by Burger and Cochfeld of mercury in canned tuna showed that "white" tuna from the relatively large albacore tuna contains more mercury (mean 0.407 ppm) than "light" tuna from the smaller skipjack tuna, which had a mean mercury concentration of 0.118 ppm. Canned mackerel had much lower levels of mercury than tuna.

10.4.3   Polychlorinated Biphenyls (PCBs)

Fisheries advisories for PCBs in Wisconsin include resident species, such as brown trout, chinook salmon, chubs, coho salmon, lake trout, carp, walleye, and smallmouth bass, in Lake Michigan and tributaries, e.g. Fox River, Milwaukee River, and Sheboygan River, see http://dnr.wi.gov/fish/consumption/FishAdvPCBs2010lo.pdf. Advisories limit the number of meals per year, which declines with increased fish length. Cooking preparation methods, including fat removal and heat, can reduce PCB levels. Mean Great Lakes fish consumption (meals/yr), meal size (g/meal), and estimated g/d for Cornwall, Ontario, measured 34.2, 186.1, and 19.9, in the same range as U.S. Environmental Protection Agency estimates used in the development of water quality criteria (Kearney and Cole 2003). However, different species preferences, awareness of advisories, and other fish consumption in other communities have implications for contaminant intake, suggesting a need for site-specific consumption surveys.

Jackson (1997) investigated the potential for stocking changes to reduce PCB concentrations in Lake Ontario's chinook salmon, based on an age-structured chinook salmon-alewife model. Increased stocking rates of the predator salmon can decrease prey survival, skew prey towards smaller fish containing lower PCB levels, and therefore lower predator PCB concentrations. Similarly, decreased stocking rates can increase the salmon PCB levels.

10.5 ENDOCRINE-DISRUPTING CHEMICALS

Environmental endocrine disruption in animals by natural and artificial toxicants has raised considerable concern since the early 1990s (Gomes et al., 2004; Jacobson and Sundelin 2006). Endocrine-disrupting chemicals (EDCs), such as estrone (E1), 17β-estradiol (E2), estiriol, and 17α-ethionylestradiol (EE2), in the aqueous environment can result from improper disposal of pharmaceuticals and personal care products and are often seen in sewage treatment effluents (Gomes et al., 2004). EDCs can affect parameters such as larval development ratio, sex ratio, and egg production in fish and other aquatic organisms such as crustaceans.

EDCs can enter aquatic organisms through bioconcentration from the surrounding water or via food uptake. Gomes et al. (2004) carried out experiments with estrone, the zooplankton *Daphnia magna* and the alga *Chlorella vulgaris*, which served as food for the daphnia. The bioconcentration factor BCF and the partitioning factor are defined by,

$$\text{BCF} = \frac{E_1 \text{ concentration in } D.\,magna}{E_1 \text{ concentration in pond water}}$$

$$\textit{Partitioning factor} = \frac{E_1 \text{ concentration in } D.\,magna}{E_1 \text{ concentration in } C.\,vulgaris} \tag{10.2}$$

The authors found a bioconcentration factor of 228 compared with a partitioning factor of 24, which led them to conclude that bioconcentration is more important than biomagnification via the food source. One must realize, of course, that bioconcentration of EDCs for *C. Vulgaris* (BCF = 37) under similar conditions should also be considered, which would appear to give an even higher concentration factor (37×24). However, other factors play a role too, such as kinetics of uptake and depuration, resulting in E1 distribution in the tissue, and the fact that larger quantities of the chemical can be taken up via bioconcentration through, for example, gills than is possible through food intake.

Jacobson and Sundelin (2006) investigated the effects of food shortage and the fungicide EDC fenarimol on the deposit-feeding amphipod *Monoporeia affinis*. They found that the fertilization frequency of females and the male mating ability was negatively influenced by fenarimol but not significantly by food shortage. However, food shortage decreased weight and female gonad development, whereas exposure to the EDC had little effect. There was no measurable effect on ecdysteroid levels. Toxicity tests with marine crustaceans *Acartia tonsa* and *Nitocra spinipes* for detecting sublethal effects of EDCs were carried out by Wollenberger (2005). Larval development ratio was a sensitive endpoint for both organisms. An OECD draft guideline document was developed for C.Valanoid Copepod Development and Reproduction Test with *Acartia tonsa*.

Yeo and Kang (2006) investigated the relationship between the TiO_2 photocatalytic decomposition of bisphenol A (BPA) and biological toxicity to zebrafish (*Danio rero*). Survival rates of Zebrafish embryos decreased when reared in water that was exposed to the leaching of BPA from epoxy resins. However, after photocatalysis, no toxic effects on hatching rates or morphogenesis of zebrafish were observed. BPA shows estrogenic activity and is deemed an environmental endocrine disruptor. BPA is used as a monomer for the production of polycarbonate and epoxy resins. Leaching of BPA has been observed from baby bottles and reusable containers.

The Australian rainbowfish (*Melanotaenia fluviatilis*) was shown to be an acceptable model for testing the potential effects of 17β-estradiol (E2) (Pollino et al., 2007). Biomarkers of both low (e.g. plasmaestradiol and testosterone) and high (e.g. egg counts and larval length) ecological relevance were examined. In response to E2 exposure, fewer changes were observed in markers of higher ecological relevance. Using fathead minnows, Thorpe et al. (2009) demonstrated that exposure to estrogenic wastewater treatment effluent can result in a reduced reproductive output for fish.

10.6 TYPES OF TOXICITY

Toxic metals in water can poison enzymes by reacting with amino NH_2, imino NH or sulfhydryl SH groups. Other modes of metal toxicity are antimetabolite behavior, e.g. arsenate K_3AsO_4 can occupy sites for phosphates and chlorate $KClO_3$ can occupy sites for nitrate. Metals can also form stable precipitates or chelates with essential metabolites. For example Al, Be, Se, Ti, Y, and Zr react with phosphate, Ba reacts with sulfate, and iron reacts with adenosine triphosphate (ATP).

Cell membrane permeability for Na, K, and Cl can be affected by Au, Cd, and Cu and substitution reactions of structurally or electrochemically important elements can take place as when Li replaces Na, Cs replaces K, and Br replaces Cl. Organic complexation can both enhance and reduce metal toxicity. Methyl mercury CH_3Hg^+ is generally more toxic than Hg^{2+}, Hg^+, and elemental mercury. On the other hand, organically complexed copper is usually less toxic than ionic copper Cu^{2+} or Cu^+. Hydroxyl and bicarbonate ions can also form complexes with metals. Chelating agents such as nitrolotriacetic acid (NTA) can solubilize metals and, therefore, enhance their toxicity. However, if significant hardness is present, Ca or

Mg NTA complexes are easily degraded, thus reducing metals in solution and their toxicity.

Organic compounds can exert toxic effects by causing genotoxicity (mutagenicity) or cancer (certain PAHs). Other toxicity mechanisms include effects on the nervous system (PBDEs), and renal (kidney) and hepatic damage, caused for example by phenol. Other effects include teratological (fetus) effects and compounds affecting the endocrine system (EDCs). Egg shell thinning of the peregrine falcon or bald eagle, which has been previously observed, is a classic example of the environmental effects of DDT. Thanks to the ban on DDT and other chlorinated pesticides, these birds have now been removed from the endangered species list.

10.6.1 Disinfection Byproducts

Water from water filtration plants and wastewater effluent are disinfected to kill or inactivate pathogenic bacteria and viruses. Disinfection can be achieved with chlorine or ultraviolet (UV) light. In the case of water for human consumption, ozone O_3 is frequently used because it is effective not only against bacteria and viruses but also *cryptosporidium* oocysts, which can cause the gastrointestinal illness cryptosporidiosis. An outbreak of cryptosporidiosis occurred in Milwaukee and Wisconsin in 1993. In order to prevent this in the future, ozonation was introduced as disinfectant in conjunction with low doses of chlorine and ammonia, increasing the length of time that combined chlorine would be effective in the water distribution system.

As a result of disinfection with chlorine and ozone, unwanted disinfection byproducts (DBPs) can be formed. Chlorine can react with organic substances, for example methane and humic and fulvic acid, to form trihalomethanes with mutagenic, genotoxic, or carcinogenic activity (Crebelli et al., 2005). An alternative disinfectant, peracetic acid (PAA), has proven to only cause formation of carboxylic acid without mutagenic properties in surface water that is used for human consumption. For wastewater effluent, Crebelli et al. (2005) showed that PAA, as well as sodium hypochlorite NaClO, in concentrations of 2–4 mg/L and contact times of 26–37 min, do not lead to the formation of significant amounts of genotoxic byproducts. This was demonstrated with in vitro bacterial reversion assays and with plant genotoxicity tests.

There is a possibility of formation of the DBP bromate from bromide ion, depending on pH, bromide ion concentration, ozone concentration and reaction time, when ozone is used for disinfection (http://www.health.state.ny.us/environmental/water/drinking/docs/bromate_in_drinking_water.pdf). Bromate can cause health effects in kidneys and may also be linked to human reproductive health.

10.6.2 Detoxification and Degradation

Reduction in toxicity to organisms can occur through the presence of alternate extracellular strong adsorbers such as activated carbon (McLeod et al., 2007). However, there is a concern that activated carbon application to impacted areas, such as aquatic sediments, may have a deleterious effect on aquatic life. Based on laboratory experiments with five sediments collected from different locations in the Netherlands,

Jonker et al. (2009) concluded that powdered activated carbon (AC) can be toxic to aquatic invertebrates, other species such as *Asellus aquaticus* and *Corophium volutator* may avoid AC-enriched sediments, and worms' lipid content may be reduced by AC exposure, which is probably related to a drastic decreased egestion rate of worms. By washing ACs prior to addition, pH effects were minor. There were no discernable effects on the microbiological community in response to the AC exposure.

A different view of AC exposure to benthic communities was offered by Kupryianchyk et al. (2011). These authors conducted experiments with two benthic species, *Gammarus pulex* and *Asellus aquaticus*, in clean and PAH-contaminated sediments. In clean sediments, 2–4% AC amendment only had a minor effect on the survival of *G. pulex* but no effect on growth. In contrast, no survivors were found in PAH-contaminated sediments without AC. However, addition of 1% AC resulted in lowering of water exposure concentrations and increased survival of *G. pulex* and *A. Aquaticus*. The authors concluded that the improvement in habitat quality by AC addition outweighs ecological side effects.

There will undoubtedly be many future studies of the effects of AC addition to aquatic sediments as a remediation technology. AC addition has positive aspects, but it provides only a separation of toxics, meaning also that small highly contaminated AC particles could have a negative effect on the biota. Another question is the long-term behavior of contaminant-laden AC in sediment. If it were possible to physically separate this AC from the rest of the sediment in a cost effective manner, then the methodology would be more acceptable.

Detoxification mechanisms that are intrinsic to organisms include sequestration of the toxic compounds, as in the formation of metallothionin in the presence of toxic metals, or the formation of a transmembrane efflux pump permeability glycoprotein (P-gp) that transports lipophilic compounds, such as PAHs, out of the cells (Vaché et al., 2006). The efflux pump was demonstrated for the fruit fly *Drosophila melanogaster* by the expression of P-gp in cells, embryos, and adult flies in response to increased dose of PAHs. Similar detoxification systems exist in aquatic organisms, for example freshwater ciliates (*Tetrahymena pyriformis*). Toxicity reduction of organic compounds can also occur by microbial degradation; thereby reducing, for example, co-planar dioxin-like toxicity of PCBs (Field and Sierra-Alvarez 2008).

10.6.3   Antibiotics

The increasing spread of antibiotic-resistant pathogens around the world is an important environmental problem (Pruden et al., 2006; Storteboom et al., 2010a, b). This appears to be linked to the widespread use of antibiotic pharmaceuticals in humans and animals. Antibiotics for animals are used as a growth promoter in livestock and poultry and for infection treatment in livestock. Antibiotic resistance genes (ARGs) are deemed emerging environmental contaminants. They may be induced by antibiotics, for example by the poultry growth promoter avoparcin. This appears to be associated with the presence of vancomycin-resistant *enterococci* (Pruden et al., 2006). However, pollution-impacted environments, including animal feed operations and wastewater treatment plants, also play a role in facilitating generation and transport of

ARGs (Storteboom et al., 2010a). These authors also concluded that antibiotics may select for ARGs in native river bacteria. When present in rivers, ARGs can spread to other bacteria, including pathogens, through horizontal gene transfer. However, it appears that transport of ARGs from specific sources is the main mechanism for ARG proliferation in two Colorado rivers compared with selection of ARGs for native river bacteria by antibiotics and possibly other pollutants (Storteboom et al., 2010b). ARGs were analyzed by quantitative PCR (qPCR).

In their study of ARG abundances along the Almendares River, Cuba, Graham et al. (2011) reached a similar conclusion that local pollution sources along the river, such as solid waste landfill, urban agriculture, pharmaceutical and other factories, as well as domestic waste discharges were mainly responsible for the ARGs found. Principal component analysis was used to distinguish between many ARGs, Cu and ampicillin (component 1) and primarily metals (component 2). There were no explicit links identified. Rather the weight of evidence suggested the link between pollution sources and ARGs.

Xu et al. (2007) examined eight selected antibiotics that are mainly used for humans, including chloramphenicol, fluoroquinolones (2), sulfonamides (3) and macrolides (2), in sewage treatment plants along the Pearl River Delta, South China. Removal efficiencies at four plants varied between 76 and 16% with most centered around 52%, emphasizing that sewage treatment plants are generally not designed for antibiotics removal. Thus, continued vigilance and research is necessary in order to control ARGs in the environment.

10.6.4 Nanomaterials

Nanoparticles are defined as materials with at least two dimensions between 1 and 100 nm (Klaine et al., 2008). Nanoparticles have always existed in the environment from both natural and anthropogenic sources. In air pollution, they are known as ultrafine particles and in water and soil pollution as colloids. Natural soil colloids can be vectors for transport of metals through soil. Because of the unique mechanical, electrical, and optical properties of nanoparticles, they have increasingly been used in the manufacture of materials for a variety of applications.

The discovery in 1985 of the spherical C_{60}, buckminsterfullerene, shaped like a soccer ball, was an important step towards the realization and production of carbon nanotubes and related materials. The fullerene was produced by evaporating graphite. Fullerenes and carbon nanotubes are used in plastics, as catalysts, in wastewater treatment, and in sensors, among other applications. Another class of nanomaterials containing metals or metal oxides, such as zinc oxide or TiO_2, are used in sunscreens, cosmetics, and bottle coatings because of their ultraviolet-blocking abilities and transparency in the visible range. Other nanoparticles include quantum dots, which are semiconductor nanocrystals with reactive cores that control their optical properties. Examples are cadmium selenide (CdSe), cadmium telluride (CdTe), and zinc selenide (ZnSe). Quantum dots are used in medical imaging and drug delivery, as well as in solar cells and photonics. Yet another category of nanomaterials are zerovalent metals. Nanoparticulate zero-valent iron has been used as a reducing agent in denitrification and remediation of soil and sediment and in the detoxification of organochlorine pesticides.

Although nanoparticles have many desirable properties, they also have the potential to cause damage to the environment and human health because of their many unique properties, including high reactivity and ability to generate reactive oxygen species (Ma et al., 2009). Lovern and Klaper (2006) studied the mortality of *Daphnia magna* vs. concentrations of titanium dioxide (TiO_2) and fullerene (C_{60}) nanoparticles by filtration in tetrahydrofuran or by sonication. Transmission electron microscopy (TEM) was used to measure diameter, possible aggregation, and concentrations of nanoparticles. The average diameter of TiO_2 particles is 10 to 20 nm compared with only 0.72 nm for C_{60}. Filtered nanoparticles, especially the low diameter fullerenes, have the highest toxicity, with LC50 values between 500 and 600 ppb. Sonicated, but unfiltered, fullerenes have much lower toxicity and show occasional increase in mortality at low concentrations, probably due to less aggregation at lower concentrations.

Buckminsterfullerene aggregates suspended in water inhibited the embryo survival, hatching rate, and heartbeat of zebrafish (*Danio rerio*) at 1.5 mg/L, whereas fullerol, a hydroxylated C_{60} derivative, was not toxic to zebrafish embryos at 50 mg/L (Zhu et al., 2007). The developmental toxicity of C_{60} was attenuated by adding an antioxidant, glutathione. Glutathione contains sulfide functional groups that can capture unpaired electrons and, therefore, remove or reduce free radicals. Also working with zebrafish embryos, King-Heiden et al. (2009) examined the toxicity of quantum dots with CdSe core and ZnS shell. Zebrafish larvae show clear signs of Cd^{2+} toxicity, but nanoparticles have even larger impact and toxicity endpoints that are different from those of Cd^{2+}.

Other aspects of the environmental risk of nanoparticles were discussed by Lowry et al. (2012), Matsuda et al. (2011), and Adams et al. (2006). Lowry et al. (2012) reviewed chemical, physical, and biological transformations of nanoparticles as well as their interactions with macromolecules. Examples of chemical transformations include redox reactions and sulfidation of Ag_2S, for example, which inhibits dissolution. Aggregation of nanoparticles increases the size and decreases the reactivity. A better understanding of these and other transformations is necessary before a full evaluation of the environmental risk of nanoparticles can be carried out. Matsuda et al. (2011) suggest that aqueous suspensions of C_{60} have potential for damge to DNA. Using both Gram-positive (*Bacillus subtilis*) and Gram-negative (*E. coli*) bacteria, Adams et al. (2006) showed that TiO_2, SiO_2, and ZnO water suspensions exhibited antibacterial properties and that *B. subtilis* had the largest growth-inhibiting effect, especially when impacted by sunlight.

10.7   MODELS AND TOXICITY TESTS

10.7.1   Dose–Response Models for Single Toxicants

The response of organisms vs. toxicant concentrations is described by dose–response models. The response can take the form of percent mortality for quantal assays, as with fish. For algae, the response can be taken as cell numbers, biomass or growth rate (Christensen and Nyholm 1984). The growth rate inhibition is then one minus the growth rate relative to a control. Because of the logarithmic nature of cell division during exponential growth, growth rate is especially meaningful. Growth rate may be deemed the number of cell divisions per time unit.

The concentration of a toxicant in aqueous media is not always equal to the nominal concentration obtained by dilution of a given compound. For metals, speciation can play an important role. For example, depending on oxidation state, chromium can be present as chromate CrO_4^{2-}, dichromate $Cr_2O_7^{2-}$, Cr^{3+}, $CrCl_2^+$, and $Cr(OH_2)^+$. Hydroxide or carbonate precipitation can also occur. Metal speciation can be estimated with computer programs such as Minteq. In the case of organic compounds, volatilization is often a consideration, and this is a main reason for the development and application of closed algal assays (Christensen et al., 2009). Distribution of the organic compound onto other phases can be estimated from partition coefficients and application of fugacity concepts.

Typical dose–response models include Weibull, probit, and logit functions (Table 10.3), of which the Weibull model and, to some extent, the logistical or logit model can be interpreted in some cases as having a mechanistic basis (Christensen and Nyholm 1984; Christensen et al., 2009). The slope in the Weibull model η or n can be deemed the number of toxicant molecules A reacting with a receptor of the organism. For quantal assays, the chemical reaction may be written,

$$nA + R \leftrightarrow RA_n \qquad (10.3)$$

Assuming chemical equilibrium with a stability constant K, we obtain,

$$\frac{[RA_n]}{[A]^n \cdot [R]} = K \qquad (10.4)$$

From this expression one obtains the mortality P or survival fraction Q in the form of a Weibull expression (Christensen and Chen 1985) based on the assumption that the probability of death per time unit is proportional to the concentration of blocked receptors $[RA_n]$,

$$\frac{-dN/dt}{N} = f(t) [RA_n] \qquad (10.5)$$

where N is the number of organisms and $f(t)$ is a function of time.

In the case of algal assays, Equation 10.5 becomes,

$$-\frac{d\mu}{\mu} = c \, d[RA_n] \qquad (10.6)$$

where μ is the growth rate and c is assumed to be a constant. Equation 10.5 expresses that the probability of blocking a cell division is proportional to the concentration of blocked receptors, $[RA_n]$. Using Equation 10.4, and integrating, one obtains the probability of non-response (Table 10.3) or relative growth rate Q as,

TABLE 10.3. Single-Variate Dose–Response Functions

Type	Transformation[a]	Probability density	Probability of non-response Q	Transform vs. Q
Weibull	$u = \ln(k) + \eta \ln(z)$ $(A = \ln(k))$	$\exp(t - e^{-t})$	$\exp(-kz^{\eta}) = \exp(-e^{u})$	$u = \ln(-\ln Q)$
Probit	$Y = \alpha + \beta \log(z)$	$\dfrac{1}{\sqrt{2\pi}}\exp\left(-\dfrac{t^2}{2}\right)$	$\displaystyle\int_{Y-5}^{\infty} \dfrac{1}{\sqrt{2\pi}}\exp\left(-\dfrac{t^2}{2}\right)dt = \dfrac{1}{2}\left(1 - \mathrm{erf}\left(\dfrac{Y-5}{\sqrt{2}}\right)\right)$	$Y = 5 + \sqrt{2}\,\mathrm{erf}^{-1}(1 - 2Q)$
Logit	$l = \theta + \phi \ln(z)$	$\dfrac{\frac{1}{4}}{\cosh^2\left(\frac{t}{2}\right)}$	$\dfrac{1}{1 + e^{\theta}z^{\phi}} = \dfrac{1}{1 + e^{l}}$	$l = \ln\left(\dfrac{1-Q}{Q}\right)$

[a] z is a toxicant concentration. α, β, k, η, θ and ϕ are constants.

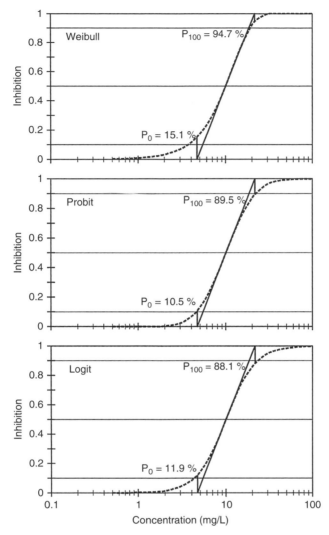

Figure 10.1. Schematic comparison of probit, logit, and Weibull dose–response functions with same EC 50, 10 mg/L, and same slope at EC50. Responses obtained at concentrations where the tangent lines through EC50 intersect the 0 and 100% inhibition lines are also shown. (Source: Christensen et al. (2009) with permission from SETAC.)

$$Q = \frac{\mu}{\mu_0} = \exp\left(-kz^n\right) \tag{10.7}$$

The inhibition P is then $P = 1 - Q$.

The Weibull, probit, and logit models are compared in Figure 10.1. In order to compare the three models, we have assumed the same lethal concentration giving 50%

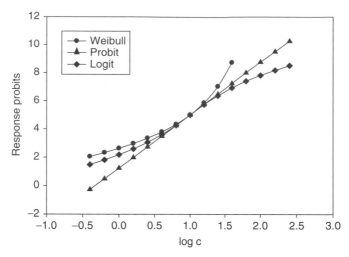

Figure 10.2. Comparison of Probit, logit and Weibull dose–response functions in response probits vs. log concentration plot with same slope at EC50, response probit 5.

response LC50 or effective concentration giving 50% response EC50 (10 mg/L). In addition, without losing generality, we have assumed the same slope at the midpoint (50% response) of the distributions. It can be seen that probit and logit functions are very similar with symmetry around the midpoint, whereas the Weibull model is asymmetric with large response at both low and high concentrations. Differences between the three distributions are seen more clearly in Figure 10.2, in which the y-axis has units of response probits. The probit model is here a straight line, whereas the logit and Weibull models deviate substantially from the probit line at both low and high concentrations.

An example of fitting experimental results for the alga *Pseudokirchneriella subcapitata* to probit and Weibull models is shown in Figure 10.3. As it can also be demonstrated by a more detailed statistical analysis (Christensen et al., 2009), the Weibull model provides the better fit to the test results. The same was found to be true for tetraethlammonium bromide, benzonitrile, and 4-4-(Trifluoromethyl)-phenoxy-phenol, thus suggesting that the Weibull model generally provides the better fit.

The LC50 or EC50 is often the main parameter of interest in toxicity testing, in part because of the fact that it is determined with the least uncertainty. However, other effect concentrations, such as EC10, indicating onset of toxicity, are also important. For algae, EC10 is the effective concentration giving 10% growth rate inhibition (Figures 10.1, 10.3). Ideally, low effect concentrations should be determined from algal assays conducted with appropriate low concentrations of the toxicant, spanning concentrations both above and below the desired response level. A significant difficulty with this approach is the large uncertainty associated with low response levels. In order to avoid this problem, one can estimate low effect concentrations from the EC50 and the slope at the midpoint of the dose–response function as shown by Christensen et al. (2009). The slope (Table 10.4) relates in a simple way to the slope of the

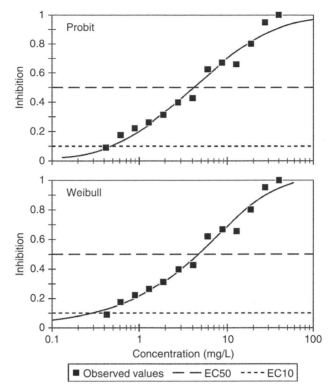

Figure 10.3. Experimental and calculated inhibitions with probit and Weibull models showing growth rate inhibition of *Pseudokirchneriella subcapitata* in response to musculamine. (Source: Christensen et al. (2009) with permission from SETAC.)

linear transformations shown in Table 10.3. Simple expressions for log EC10 in terms of log EC50 and the slope are also displayed in Table 10.4 along with similar expressions for log EC50 based on log EC10 and the slope $Slope_{EC10}$ at the 10% response point of the dose–response function.

10.7.2 Dose–Response Models for Multiple Toxicants

The toxic response of aquatic organisms to two or several toxicants acting simultaneously is an important consideration because toxicants rarely act alone. Based on models proposed by Bliss (1939) and Hewlett and Plackett (1959), Christensen and Chen (1985) developed a multiple toxicity model that can be used for both quantal assays, e.g. fish bioassays, and algae where all organisms come from a single clone.

For two toxicants, the basis is the probability Q of non-response or survival expressed as,

$$Q = P_r \left(\delta_1^{1/\lambda} + \delta_2^{1/\lambda} \le 1 \right) \tag{10.8}$$

TABLE 10.4. Slopes of Weibull, Probit, and Logit Distributions, and EC10 Expressed as a function of EC50 and the Slope; and EC50 Expressed as a Function of EC10 and the Slope at EC10

Type	Slope $= \dfrac{dP}{d\log z}\Big|_{P=0.5}$	EC10 vs. EC50 and slope at EC50	EC50 vs. EC10 and slope at EC10
Weibull	$0.34657 \cdot \ln 10 \cdot \eta = 0.79801 \cdot \eta$	$\log \text{EC10} = \log \text{EC50} - \dfrac{0.65289}{\text{Slope}}$	$\log \text{EC50} = \log \text{EC10} + \dfrac{0.17863}{\text{Slope}_{\text{EC10}}}$
Probit	$\dfrac{1}{\sqrt{2\pi}} \cdot \beta$	$\log \text{EC10} = \log \text{EC50} - \dfrac{0.51126}{\text{Slope}}$	$\log \text{EC50} = \log \text{EC10} + \dfrac{0.22491}{\text{Slope}_{\text{EC10}}}$
Logit	$0.25 \cdot \ln 10 \cdot \phi = 0.57565 \cdot \phi$	$\log \text{EC10} = \log \text{EC50} - \dfrac{0.54931}{\text{Slope}}$	$\log \text{EC50} = \log \text{EC10} + \dfrac{0.19775}{\text{Slope}_{\text{EC10}}}$

where,

P_r = probability; $\delta_i = z_i/\bar{z}_i$

z_i = concentration of toxicant i

$\bar{z}_i$ = concentration tolerance of an individual organism to toxicant i

λ = similarity parameter for the action of two toxicants on two biological systems.

The probability Q is also the survival fraction in quantal assays, whereas it is the number of cell divisions or growth rate for algae. The correlation ρ of toxicant tolerances for different organisms is often, but not always, one. However, if the modes of action of two toxicants are related, then organisms that are very sensitive to one toxicant are also sensitive to the other, and then ρ is close to, but less than one. For algal assays, the correlation is one because all algae originate from the same clone. The parameter λ indicates the degree of overlap of two biological systems affected by the two toxicants. If the two systems are similar, such as certain enzymes, λ is close to one.

Consider now the linear log transforms of concentrations X_i and tolerances u_i,

$$X_i = \alpha_i + \beta_i \log z_i \tag{10-9a}$$

$$u_i = \alpha_i + \beta_i \log \bar{z}_i \tag{10-9b}$$

In the case of single variate toxicity models, the probability of response $P = 1-Q$ can be determined by integrations such as the one shown under the probit model in Table 10.3. Note that Weibull, probit, and logit models for individual toxicants are all of the above type, Equation 10-9b. In the case of two variables, the integration is over a region determined by Equation 10.8 in a u_1, u_2 plane (Christensen and Chen 1989). In such a diagram the actual concentrations X_i are indicated by straight lines perpendicular to the coordinate axes.

A related but different concept is that of isobolograms for two toxicants, which are lines of constant response in a diagram in which the axes represent toxicant concentrations. Typically, the response considered is LC50 or EC50. The points where the isobole intersects the two axes then represent the LC50s or EC50s for the individual toxicants.

In the case of n toxicants and full correlation of toxicant tolerances the isobologram is described by the following equation (Christensen and Chen 1989),

$$\left(\frac{z_1}{Z_1}\right)^{1/\lambda} + \left(\frac{z_2}{Z_2}\right)^{1/\lambda} + \dots + \left(\frac{z_n}{Z_n}\right)^{1/\lambda} = 1 \tag{10.10}$$

A special case of this equation is obtained when $\lambda = 1$ meaning that all n toxicants act on similar biological systems. This case is characterized by the term concentration addition, CA. For two toxicants, the resulting isobole is described by,

$$\left(\frac{z_1}{Z_1}\right) + \left(\frac{z_2}{Z_2}\right) = 1 \tag{10.11}$$

which is a straight line connecting the LC50s or EC50s of the two toxicants. Points below and to the left of this line describe more than additive action, whereas points above and to the right represent less than additive action. Each term in this equation is a toxic unit, and CA is therefore characterized by the sum of toxic units M being equal to one.

Multiple toxicity indices are useful to characterize the nature of the joint action of toxicants at a given response level such as LC50 or EC50. Konemann (1981) realized that although M equals one for CA, it is generally not particularly useful because M for a desired response level depends on the number of chemicals if there is no addition. Also, because of the logarithmic form of the log-linear transformations for individual toxicants, Equation 10-9b, an index that is logarithmic in toxicant concentration would be desirable. As a result, he suggested a multiple toxicity index MTI, a special version of which can be written as,

$$\text{MTI} = 1 - \frac{\log M}{\log n} \tag{10.12}$$

where,

$M = \sum_{i=1}^{n} z_i / Z_i$ = sum of toxic units giving the desired response
n = number of chemicals in the mixture
z_i = concentration of toxicant i in the mixture
Z_i = concentration of toxicant i in the mixture giving the desired response

The multiple toxicity index given by this equation is valid when the concentration of each chemical relative to its effect concentration z_i/Z_i is a constant f for all chemicals. As shown by Christensen and Chen (1989) the MTI defined in this way is equal to the similarity index λ. Equation 10.10 then takes the following form,

$$n\left(\frac{M}{n}\right)^{1/\lambda} = 1$$

which when solved for λ yields,

$$\lambda = 1 - \frac{\log M}{\log n} \tag{10.13}$$

which is the same expression as Equation 10.12 with $\lambda = \text{MTI}$. Concentration addition (CA) is perhaps the most important mode of action that is valid when the correlation ρ and similarity index λ both equal one. As indicated above, the sum of toxic units M then equals one.

Another important case is obtained when these parameters are both zero for quantal assays. In that case the resulting survival fraction Q_{mix} is equal to the product of the individual survival fractions, $Q_1, Q_2, \ldots, Q_n$,

$$Q_{mix} = Q_1 \cdot Q_2 \cdots Q_n \tag{10.14}$$

This case is often referred to as independent joint action.

When studying lethal and sublethal toxicity of organic chemical mixtures to the fathead minnow (*Pimehales promelas*), Broderius et al. (2005) found that bivariate mixtures of chemicals with similar action followed concentration addition, Equation 10.11, while chemicals with different modes of action showed less than additive action with response isoboles between concentration addition and response multiplication, Equation 10.14. The modes of action considered were narcosis I, narcosis II, or uncoupler of oxidative phosphoralation.

In equitoxic multiple chemical lethal and sublethal tests for three groups of chemicals, the sum of toxic units M was, in most cases, within 1 of 95% confidence limits when similar action modes were considered, but was larger than 1 (1.28–1.47) when different modes of action were present. The concentration additive response for multiple chemicals of similar action, and the less than additive response for several different chemicals are in accordance with the results obtained with two chemicals.

Using the alga *Selenastrum capricornutum* (now named *Pseudokirchneriella subcapitata*), Christensen et al. (2001) found that the response to two similar chemicals, nonylamine and decylamine, was near concentration addition, but slightly less than additive, with $\lambda = 0.70 - 0.76$. By contrast, there was strong antagonism between atrazine and decylamine, apparently as a result of atrazine blocking access of the highly toxic decylamine to receptor sites of the algae. The response parameter was growth rate based on biomass, and both EC10 and EC50 were considered as response levels.

They also found that a no-effect concentration EC0 of decylamine could be effectively predicted based on a model that contains EC0 and EC50 as explicit variables,

$$\frac{\mu}{\mu_c} = \frac{1}{1 + \frac{(c-c_0)_+}{c_G}} \tag{10.15}$$

where μ is the growth rate, μ_c is the average growth rate of the controls, c is the toxicant concentration, c_0 is the EC0, and $c_G = \text{EC50} - c_0$. The meaning of $(c - c_0)_+$ is that any negative value of $c - c_0$ should be replaced by zero.

Examples of the application of these models may be found in Chen et al. (2005), Merino-García et al. (2003), and Hsieh et al. (2006). According to Chen et al. (2005) most synergistic (more than additive) actions are observed for reactive toxicants having different mechanisms of toxicity and flat concentration–response curves. In particular, aldehydes and nitriles tend to react synergistically. Aldehydes are electrophilic non-electrolytes and nitriles are cyanogenic toxicants.

These authors worked with 11 aldehydes and two nitriles and the test alga *Raphidocelis subcapitata*, now named *Pseudokirchneriella subcapitata*. Response endpoints considered were growth rate and dissolved oxygen production. Four synergistic joint actions related to malonotrile, and several aldehydes were found. On the other hand, several combined effects between acetonitrile and aldehydes were antagonistic (less than additive). Greater than additive effects were found with flat dose–response curves for individual toxicants, whereas less than additive effects were associated with steep-slope chemicals. Model analyses showed that this mixture toxicity was characterized by response addition ($\rho = -1$, $\lambda = 0$) or response multiplication ($\rho = 0$, $\lambda = 0$).

Merino-García et al. (2003) examined the response of *Daphnia magna* to the amines nonylamine and decylamine, which were nearly concentration additive ($\lambda = 0.55-0.80$), and to decylamine and ethylparathion, which show less than additive or near independent joint action ($\lambda = 0.29-0.40$). Since the daphnia originate from the same clone, it is assumed that the correlation between tolerances of the toxicants is 1. It was found that the joint action, expressed by the λ value, is nearly independent of the response level, IC0-48h, IC10-48h, and IC50-48h, where for example IC50-48h indicates immobilization concentration affecting 50% of the organisms during the test period of 48 h.

When toxicants have similar action, such as non-polar narcotics, the joint response is expected to be concentration additive (CA). Hsieh et al. (2006) found that this was in fact the case for several non-polar narcotic chemicals; aromatics such as benzene and toluene, and aliphatics, for example methanol and ethanol. The test organism was the green alga *Pseudokirchneriella subcapitata* and the toxicity endpoints algal growth rate and dissolved oxygen (DO) production.

10.7.3 Pulsed Toxicity Tests

Concentrations of toxicants in the environment are almost never constant as assumed in standard LC50 or EC50 testing. In pulsed toxicant exposures, the concentration, duration, and pulse intervals can vary. In order to simulate these conditions in laboratory experiments, protocols and models have been developed to quantify the toxic response (Zhao and Newman 2006; Ashauer et al., 2006). Zhao and Newman (2006) exposed amphipods (*Hyalella azteca*) to two different toxicants, copper sulfate ($CuSO_4$) and sodium pentachlorophenol (NaPCP). In exposure duration experiments, the post exposure (latent) mortality was recorded as a function of duration. No significant effect of exposure duration on the latent mortality was determined within the experimental range of exposures (20–60 h).

In the recovery experiments, the amphipods were given four pulse intervals (0, 24, 48, and 72 h) after an initial 12 h exposure, followed by a second exposure of 12 h during which mortality was checked and modeled. Reference animals were not subject to a first exposure. The following equations were used to model time-to-death or survival time and,

$$\ln t_{R_i} = a_R + \varepsilon_{R_i} \tag{10.16}$$

$$\ln t_{T_i} = a_T + b_T \cdot RT_i + \varepsilon_{T_i} \tag{10.17}$$

where t_{R_i} and t_{T_i} are survival times for reference and treatment animals respectively, a_R, a_T, and b_T are constants, and ε_{R_i}, ε_{T_i} are error terms that equal σL where L varies with the proportion dead for which a prediction is made. The t_{R_i} and t_{T_i} can be fitted to Weibull, exponential, or log-normal distributions. For example, for one set of experiments with CuSO4, Zhao and Newman (2006) found these equations corresponding to Equations 10.16 and 10.17:

ln $T = 3.73 + 0.83$ L for reference animals, and

ln $T = 2.58 + 0.01$ RT $+ 0.35$ L for treatment animals,

where T = time-to-death and RT = complete recovery time.

It was found that the Weibull distribution provided the best fit to the data, and that RT = 81 h. The recovery times for NaPCP were considerably shorter (15 h) in one set of experiments. It was concluded that survival analysis provided a more effective way to analyze the response to pulsed exposures.

Ashauer et al. (2006) reviewed models to predict effects on aquatic organisms resulting from time-varying exposure to pesticides. They found that the threshold hazard model (THM) and damage assessment models (DAM) were particularly useful. The THM assumes no detectable effect on organisms up to a certain internal concentration of toxicant, the "no effect" level. Both toxicokinetics, describing uptake and elimination of the chemical, and toxicokinetics, addressing the dynamics of injury and recovery in the organism, are considered in DAM. A standard version of the one-compartment, single first-order kinetics toxicokinetics model is described by the equation,

$$\frac{dC_{int}}{dt} = k_{in}C - k_{out}C_{int} \tag{10.18}$$

where C_{int} (kg/kg) is the internal concentration, C (kg/m^3) is the concentration in the water, k_{in} (m^3/kg/sec) and k_{out} (sec^{-1}) are the uptake and elimination rate constants respectively. Assuming constant environmental concentration C and $C_{int}(0) = 0$, we obtain the solution for the internal concentration at time t:

$$C_{int}(t) = \frac{k_{in}}{k_{out}} \cdot C \left(1 - e^{-k_{out}t}\right) \tag{10.19}$$

The first step in the damage assessment model includes toxicokinetics, represented by Equations 10.18 and 10.19. In the second step, the damage D is deemed proportional to the internal concentration C_{int} and damage repair is proportional to the damage incurred. Because these models have parameters with ecological significance, the authors recommend that THM and DAM be used in conjunction to predict the sublethal and lethal effects of contaminants in response to exposures.

10.7.4 Chronic Toxicity Tests

Even if acute toxicity is low, chronic toxicity, expressed by life-cycle parameters, such as growth and offspring productivity, of aquatic and soil organisms can be substantial. Robidoux et al. (2004) examined acute and chronic toxicity of a new explosive polycyclic nitramine CL-20 (2,4,6,8,10,12-hexanitro-2,4,6,8,10,12-hexaazaisowurtzitane) to the earth worm (*Eisenia andrei*). Acute toxicity was evaluated by 28-d survival of adults, and chronic toxicity by reduction in growth of adults (28 d) and productivity of cocoons (56 d), hatched cocoons (56 d), and number of juveniles (56 d). The soil type was either a natural sandy forest soil, designated RacFor2002, containing 20% organic carbon, or a Sassafras sandy loam soil (SSL) with only 0.33% organic carbon.

Results showed an LC 50 value of greater than 125 mg/kg for survival of adults in RacFac 2002 soil whereas the LC50 measured only 53.4 mg/kg in the SSL soil containing a much lower level of organic carbon. Furthermore, the number of juveniles had lowest observed effect concentrations (LOEC) of 1.59 mg/kg in RacFac2002 soil in contrast to only 0.01 mg/kg in the SSL soil. Similar results were obtained for growth reduction of adults and for both categories of cocoons produced. The results indicate that CL 20 is a reproductive toxicant to the earthworm with lethal effects at higher concentrations. Also, it appears that the high organic content of the Rac-Fac2002 soil decreased the bioavailability of CL 20 because of sorption to the organic carbon in the soil.

An example of chronic toxicity testing in the aquatic environment is the work of Valenti et al. (2005) on the effects of mercury on the early life stages of the rainbow mussel (*Villerosa iris*). Three bioassays, 72-h acute glochidia, 96-h acute juvenile, and 21-d chronic juvenile toxicity tests, were carried out, exposing *Villosa iris* to mercuric chloride salt ($HgCl_2$). Glochidia with a median 48-h LC50 of 39 µug/L Hg were more sensitive to acute exposure then juvenile mussels for which the corresponding 48-h LC50 value was 135 µg/L Hg. In chronic testing there was a significant growth reduction of 2-month-old juvenile mussels at only 8 µg/L Hg, thus emphasizing the value of chronic testing. Their study supported the use of glochidia as a surrogate life stage for juveniles, as acute bioassays are better established for glochidia than for juveniles and because glochidia are readily available for toxicity tests. Glochidia may be used only in short term tests as a result of a decline in their viability during laboratory studies after short periods.

Younger juveniles may be more susceptible to contaminants, but they are then subject to high mortality in the laboratory from predation and are therefore most suitable for testing after about 2 months. One consideration with older juveniles is that they may close their valves for extended periods, which may lead to underestimation of toxicity. Because of these factors, the authors advocated the use of both glochidia and juveniles in ecotoxicological studies.

10.8   QUALITY CRITERIA

10.8.1   Sediment Quality Criteria

During the past 10–15 year there has been an increased awareness of the often significant role of contaminated sediments in negatively impacting water quality of receiving waters (Long et al., 2006; Araúju et al., 2006). This is relevant as point or non-point sources of contaminants are curtailed because sediments can then continue to release pollutants to the water body.

The review of Long et al. (2006) looks at effect-based sediment quality guidelines (SQG) and the mean sediment quality guideline quotients (mSQGQ). The mean quotient is taken over groups of contaminants, such as metals, polycyclic aromatic hydrocarbons (PAHs), or chlorinated organic compounds (COHs), or it can be over different chemicals each with a given SQG. It is shown that there is a good general correlation between organism response and mSQGQ. The method is related to the toxic unit approach as shown in Equations 10.11 and 10.12, and could therefore be expected to apply for concentration additive substances. However, the sum of quotients for different chemicals may be more relevant than mean quotients. Consider, for example, that another chemical is added to the sediment. The mean quotient may stay the same, whereas the toxicity likely will increase as reflected by the sum of the quotients.

Araújo et al. (2006) studied the Rasgão Reservoir close to the metropolitan area of São Paulo, Brazil. The sediments are heavily contaminated with metals and PAHs and show complete absence of benthic life. Toxicity identification in pore water and elutriate with *Ceriodaphnia dubia* indicated that unionized ammonia was the main cause of toxicity, with levels as high as 5.14 mg/L in pore water and 2.06 mg/L in elutriate. Tests with *Vibrio fischeri* were related to the organic fraction of the pore water and elutriate, in which compounds such as benzothiazole and nonylphenol were detected. Mutagenic activity, probably related to the presence of PAHs, was demonstrated with *Salmonella*/microsome assay.

For metals, Di Toro et al. (1992) found that acid volatile sulfide (AVS) is a reactive pool that can bind metals, and therefore render them non-toxic to biota. Acid volatile sulfide is the sulfide that is extracted simultaneously with metals from sediment using cold 1 N HCl (Ankley et al., 1996). For example, Cd^{2+} can be made unavailable to organisms by reaction with FeS(s):

$$Cd^{2+} + FeS(s) \rightarrow CdS(s) + Fe^{2+} \qquad (10.20)$$

The order of reaction when several metals are present is governed by the solubility products (Table 10.5). For the metals shown in this table, the product is smallest for HgS and then for CuS, etc., meaning that HgS will form first, then CuS and then

TABLE 10.5. Metal Sulfide Solubility Products

Metal sulfide	Log K_{sp}
FeS	−22.39
NiS	−27.98
ZnS	−28.39
CdS	−32.85
PbS	−33.42
CuS	−40.94
HgS	−57.25

(Source: Di Toro et al. (1992)).

other metals as long as sufficient sulfide is present. It can be shown that the following inequality holds,

$$\frac{[M^{2+}]}{[M]_A} < \frac{K_{MS}}{K_{FeS}} \tag{10.21}$$

where $[M^{2+}]$ is the metal ion concentration, $[M]_A$ is the concentration of added metal, and K_{MS} and K_{FeS} are solubility products for metal sulfide and iron sulfide. Thus, it can be seen that the metal ion concentration is negligible if [FeS] is greater than $[M]_A$. If reactive metal is present in excess of the AVS, the ionic concentration $[M^{2+}]$ becomes,

$$[M^{2+}] = \gamma_{M^{2+}}\alpha_{M^{2+}}\left([M]_A - [AVS]\right) \tag{10.22}$$

where $\gamma_{M^{2+}}$ and $\alpha_{M^{2+}}$ are activity coefficient of M^{2+} and fraction of metal in ionic form $M^{2+}[M^{2+}]/[\sum M(aq)]$ respectively. $\sum M(aq)]$ is the total dissolved metal (II) concentration.

Binding phases other than AVS may be relevant for metals in sediments. Ankley et al. (1996) considered copper, cadmium, nickel, and lead. They developed sediment quality criteria based on interstitial water criteria, AVS and organic carbon criteria, minimum binding criteria, and AVS criteria. The interstitial water criterion is based on concentration addition, e.g. an extension of Equation 10.11 to n compounds, here metals,

$$\sum_i \frac{[M_{i,d}]}{[FCV_{i,d}]} \le 1 \tag{10.23}$$

where $[FCV_{i,d}]$ = water quality criterion, e.g. EC50, effective concentration giving 50% response of a given endpoint, or the final chronic value for metal i in dissolved form, and $[M_{i,d}]$ is the interstitial water concentration of metal i in dissolved form. While this criterion is reasonable for metals exerting toxicity through similar modes of action, it is difficult to apply because interstitial metal concentrations are often not known and not easy to measure at in situ conditions.

The AVS and organic carbon criterion is a modification of Equation 10.22 to indicate the remaining simultaneous extracted metal i $\Delta[SEM_i]$ after reaction with FeS. Following the order of solubility products (Table 10.5), Cu will react with FeS first, then Pb, Cd, and Ni in that order until the available AVS is exhausted. This criterion can be written as,

$$\sum_i \frac{\Delta[SEM_i]}{K_{d,oc,i}f_{oc}[FCV_{i,d}]} \le 1 \tag{10.24}$$

The minimum partitioning criterion is designed to exclude certain metals from consideration, the sediment concentrations ($\mu g/g_{sed}$) of which would not produce interstitial water concentration above the criteria values when partitioned through materials for which the partition coefficients are quite low, such as chromatographic sand. If these sediment concentrations are partitioned through materials with higher partition coefficients, the interstitial metal concentration would be even lower than the criteria concentration, and the sediment concentrations would therefore be acceptable. The criterion for this case is,

$$\sum_i \frac{[SEM_i]}{K_{d,\min,i}[FCV_{i,d}]} \leq 1 \qquad (10.25)$$

It is here assumed that AVS is small, such that $[SEM_i]$ represents the simultaneous extracted metals of type i. $K_{d,\min,i}$ is the partition coefficient for metal i in the minimal partitioning material with the interstitial water.

The AVS criterion is directly related to the observation that any simultaneously extracted metals are likely to react with AVS as FeS in sediment. Thus if the sum of metal concentrations (μmole/g_{sed}) is less than the total concentration (μmole/g_{sed}) of AVS then all metals are bound to AVS in sediment, and metals are not available to exert toxic action on biota in the aqueous phase. This criterion is, therefore,

$$\sum_i [SEM_i] \leq [AVS] \qquad (10.26)$$

This relationship is illustrated with experimental data in Figure 10.4 (Berry et al., 1996) showing % mortality of the amphipod A. Abdita (R. husoni in the cadmium experiment) vs. SEM/AVS for individual or mixed metals. This was based on 10-d toxicity tests spiked with cadmium, copper, lead, nickel, zinc and a mixture of cadmium, copper, nickel, and zinc. It is seen that the mortality is low (< 20%) for SEM/AVS less than or equal to 1 but that it rises steeply values around 5–8 of this ratio, indicating that concentrations of metal ions are then large enough to affect biota in the aqueous phase.

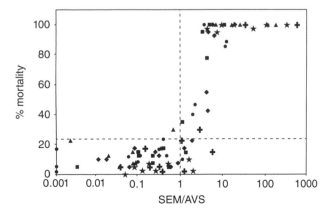

Figure 10.4. Mortality of the amphipod A. abdita (R. hudsoni in the cadmium experiment) vs. the molar ratio of SEM to AVS of sediment. Data below the SEM detection limit are plotted at SEM/AVS = 0.001. ● = Cd; Δ = Cu; * = Ni; □ = Zn; ◇ = Pb; and + = mixed metals. (Source: Berry et al. (1996) with permission from SETAC.)

If any of the criteria represented by Equations 10.23 to 10.26 are violated, toxicity may occur. The above equations can be rewritten in terms of the conceptual framework of the biotic ligand model (Di Toro et al., 2005). The basic assumption in this model is that toxicity is present if a metal–ligand concentration exceeds a critical value. The metal–ligand complex is assumed to be in equilibrium with the free metal ion and the sediment organic carbon concentration of the metal. Competing reactions with cations and inorganic ligands are considered. The model will be explored further in Section 10.8.2.

The sediment concentration at LC50, C_s^*, can be written as a sum of sulfide bound metals, AVS and the partitioned sediment concentration corresponding to the LC50 water concentration, C_w^*:

$$C_s^* = AVS + K_p C_w^* \tag{10.27}$$

Substituting SEM^* for C_s^* we have,

$$SEM^* = AVS + K_p C_w^* \tag{10.28}$$

where K_p (L/kg) is the partition coefficient. Note that in the absence of AVS the critical sediment concentration C_s^* becomes,

$$C_s^* = K_p FCV \tag{10.29}$$

where FCV is the final chronic value, or water quality criterion such as LC50. Considering partitioning to the organic carbon in sediment, K_p can be written as,

$$K_p = f_{oc} K_{oc} \tag{10.30}$$

where f_{oc} ($g_{carbon}/g_{sediment, dry weight}$) is the organic carbon fraction of sediment. By subtracting AVS from SEM^* we obtain the metals available for partitioning from the organic carbon fraction into the water phase,

$$SEM^* - AVS = f_{oc} K_{oc} C_w^* \tag{10.31}$$

where $SEM^* - AVS$ is the excess SEM, SEM_x^*. The organic carbon normalized SEM_x^* is then

$$SEM_{x,oc}^* = \frac{SEM^* - AVS}{f_{oc}} = K_{oc} C_w^* \tag{10.32}$$

If $SEM_{x,oc}$ is greater than $SEM_{x,oc}^*$ a toxic effect can be exerted. Under equilibrium conditions the metal in the organic fraction of sediment is in equilibrium with the free metal ion and the metal bound to the biotic ligand. The concepts of metal partitioning for sediments can also be extended to organics such as narcotic chemicals and PAHs (Di Toro and McGrath 2000).

10.8.2 Water Quality Criteria

The model for toxicity to fish through fish gills developed by Pagenkopf (1983) was generalized to apply to any relevant biotic ligand, replacing the fish gill as the site of action, by Di Toro et al. (2001). For metals, this model may be formulated through a law of mass action,

$$[ML_b^+] = K \ [M^{2+}][L_b^-] \tag{10.33}$$

where L_b^- is a negatively charged biotic ligand, M^{2+} is the metal ionic concentration, ML_b^+ is metal-ligand complex, and K is the equilibrium constant. The idea is that toxicity is triggered when the concentration on biotic ligand exceeds a critical concentration, C_M^*:

$$C_M^* = [ML_b^+] \tag{10.34}$$

The biotic ligand model has been used to predict LC_x values for metals.

Figure 10.5 hows predicted vs. measured LC50 for Fathead Minnows, Rainbow Trout, and Daphnia. There is a good correlation between the LC50s, indicating that the BLM may be used as a predictive tool. To get a better evaluation of the predictive ability of the BLM-predicted LC50s one should also use an alternative, less sophisticated, measure of toxicity such as total metal concentration for comparison. A certain amount of calibration with known LC50s are necessary in both cases, and the question also arises whether the BLM can predict other thresholds such as LC10 or LC90.

In attempting to answer this question, one should note that a biotic ligand-toxicant also forms the basis for the Weibull model developed for dose–response regressions; see Equations 10.3–10.7. The biotic ligand-toxicant is here more general in that it allows other than a one-to-one relationship between toxicant and ligand molecules. This is expressed by the concentration $[RA_n]$ where R is receptor (ligand), A is the toxicant molecule, and n is a number typically between 0.5 and 3.0 indicating a coordination number for the receptor. The value of n or η can be deemed a slope in the Weibull dose–response model (Tables 10.3, 10.4). A main assumption for the Weibull model is that there are many receptors, meaning that the maximum binding capacity has not been reached. In some cases this may not be a valid assumption because, for example, the fish gill copper concentration can show a saturation effect as in Langmuir function for high free copper concentration (Santore et al., 2001).

Because the Weibull model has proven to provide a good fit to many experimental dose–response regressions, see for example Figure 10.3, one may therefore hypothesize that a biotic ligand model allowing n toxicant molecules to react with a receptor has a more general validity and could serve as basis for predicting both toxicity vs. the chemical composition of the water and detailed dose–response regressions.

10.8.3 Total Maximum Daily Loads

The total maximum daily load (kg/d) is an important measure that is cited in the Clean Water Act (Section 303 (d)) and the U.S. EPA's water quality regulations. It requires the states to develop and implement pollution prevention abatement plans

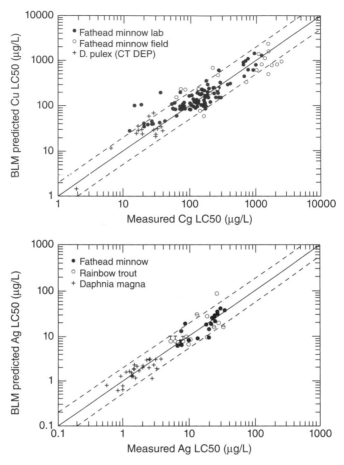

Figure 10.5. Comparison of predicted LC50s using the biotic ligand model (BLM) and measured LC50s for copper and silver. (Source: Di Toro et al. (2001) with permission from SETAC.)

for all waters where the current standards are not met by technology-based standards for point sources (Novotny 2004). TMDLs makes it easier to quantify how much reduction in pollution load is needed to achieve a certain goal in terms of maximum concentration of pollutants in receiving waters. The resulting concentration is not simply a load divided by a water volume, because of hydrodynamics, reactions, sedimentation, and export of pollutants from the water body need to be taken into account.

The TMDL can be written as a sum of waste load allocations of all point loadings and non-point loadings plus a margin of safety (Benham et al., 2005). The margin of safety must be large enough to accommodate uncertainties of estimation as well as acceptable short-term violations. In order to develop TMDLs for watershed, various models have been proposed such as the Generalized Watershed Loading Function

(GWLF, Benham et al., 2005) or use of standard hydrological computer programs, especially the Hydrological Simulation Program - Fortran (HSPF).

Pedersen et al. (2006) investigated organophosphorus insecticides in agricultural and residential runoff and developed guidelines for TMDL determination of chlorpyrifos, diazinon, and malathion. A possible first-flush effect was evaluated by considering load (FF_{25}) delivered by the first 25% of the runoff volume. The FF_{25} were not substantially larger than 0.25, indicating no significant first-flush effect. In this case, the load can be expressed by the event mean concentration multiplied by a summation of the rainfall intensity multiplied by the associated time interval. The lack of a first flush effect may reflect that the compounds are mainly in the soluble phase and that their concentration depends more on solubilization of these compounds from insecticide application rather than accumulation in the watershed. By contrast, a load that shows first-flush effect depends on the antecedent dry period and an area loading function (kg/ha/d) in addition to rainfall intensity and duration (Soonthornnonda and Christensen 2008). Organophosphorus insecticides act by inhibiting the enzyme acetylcholinesterase. The relative potency of the three compounds was multiplied by their concentration to generate a total exposure equivalency reflecting concentration addition.

REFERENCES

Adams, LK, Lyon, DY, Alvarez, PJJ. Comparative eco-toxicity of nanoscale TiO_2, SiO_2, and ZnO water suspensions. *Wat. Res.* 2006;40:3527–3532.

Araújo, RPA, Botta-Paschoal, CMR, Silvério, PF, Almeida, FV, Rodrigues, PF, Umbuzeiro, GA, Jardim, WF, and Mozetto, AA. Application of toxicity identification evaluation to sediment in a highly contaminated water reservoir in southeastern Brazil. *Environ. Toxicol. Chem.* 2006;25(2):581–588.

Ankley, GT, Di Toro, DM, Hansen, DJ, Berry, WJ. Technical basis and proposal for deriving sediment quality criteria for metals. *Environ. Toxicol. Chem.* 1996;15(12):2056–2066.

Ashauer, R, Boxall, A, Brown, C. Predicting effects on aquatic organisms from fluctuating or pulsed exposure to pesticides. *Environ. Toxicol. Chem.* 2006;25(7):1890–1912.

Benham, BL, Brannan, KM, Yagow, G, Zeckoski, RW, Dillaha, TA, Mostaghimi, S, Wynn, JW. Development of bacteria and benthic total maximum daily loads: A case study, Linville Creek, Virginia. *J. Environ. Qual.* 2005;34:1860–1872.

Berry, WJ, Hansen, DJ, Mahony, JD, Robson, DL, Di Toro, DM, Shipley, BP, Rogers, B, Corbin, JM, Boothman, WS. Predicting the toxicity of metal-spiked laboratory sediments using acid-volatile sulfide and interstitial water normalizations. *Environ. Toxicol. Chem.* 1996;15(12):2067–2079.

Bliss, CI. The toxicity of poisons applied jointly. *Annals of Applied Biology.* 1939;26:585–615.

Bradbury, SP, Russom, CL, Ankley, GT, Schultz, TW, Walker, JD. Overview of data and conceptual approaches for derivation of quantitative structure-activity relationships for ecotoxicological effects of chemicals. *Environ. Toxicol. Chem.* 2003;22(8):1789–1798.

Braune, BM, Scheuhammer, AM. Trace element and metallothionein concentrations in seabirds from the Canadian Arctic. *Environ. Toxicol. Chem.* 2008;27(3):645–651.

Broderius, SJ, Kahl, MD, Elonen, GE, Hammermeister, DE, Hoglund, MD. *Environ. Toxicol. Chem.* 2005;24(12):3117–3127.

Budka, M, Gabrys, B, Ravagnan, E. Robust predictive modelling of water pollution using biomarker data. *Wat. Res.* 2010;44:3294–3308.

Burger, J, Gochfeld, M. Mercury in canned tuna: White versus light and temporal variation. *Environ. Res.* 2004;96:239–249.

Chen, C-Y., Chen, S-L, Christensen, ER. Individual and combined toxicity of nitriles and aldehydes to *Raphidocelis subcapitata. Environ. Toxicol. Chem.* 2005;24(5)1067–1073.

Christensen, ER, Kusk, KO, Nyholm, N. Dose-response regressions for algal growth and similar continuous endpoints: Calculation of effective concentrations. *Environ. Toxicol. Chem.* 2009;28(4):826–835.

Christensen, ER, Nyholm, N. Ecotoxicological assays with algae: Weibull dose-response curves. *Environ. Sci. Technol.* 1984;18(9)713–718.

Christensen, ER, Chen, C-Y. A general noninteractive multiple toxicity model including probit, logit and Weibull distributions. *Biometrics* 1985;41:711–725.

Christensen, ER, Chen, C-Y. Modeling of combined toxic effects of chemicals. *Hazard Assessment of Chemicals*, Vol. 6. New York, NY: Hemisphere Publishing Corporation; 1989, Chapter 5, pp. 125–186.

Christensen, ER, Chen, D-X, Nyholm, N, Kusk, KO. Joint action of chemicals in algal toxicity tests: influence of response level and dose-response regression model. *Environ. Toxicol. Chem.* 2001;20(10):2361–2369.

Crebelli, R, Conti, L, Monarca, S, Feretti, D, Zerbini, I, Zani, C, Veschetti, E, Cutilli, D, Ottaviani, M. Genotoxicity of the disinfection by-products resulting from peracetic acid- or hypochlorite-disinfected sewage wastewater. *Wat. Res.* 2005;3:1105–1113.

De Schamphelaere KA, Vasconcelos, FM, Tack, FM, Allen, HE, Janssen, CR. Effect of dissolved organic matter source on acute copper toxicity to Daphnia magna. *Environ. Toxicol. Chem.* 2004;23(5):1248–55.

DM Di Toro, McGrath, JA. Technical basis for narcotic chemicals and polycyclic aromatic hydrocarbon criteria. II. Mixtures and sediments. *Environ. Toxicol. Chem.* 2000;19:1971–1982.

Di Toro, DM, McGrath, JA, Hansen, DJ, Berry, WJ, Paquin, PR, Mathew, R, Wu, KB, Santore, RC. Predicting sediment metal toxicity using a sediment biotic legand model: methodology and initial application. *Environ. Toxicol. Chem.* 2005;24(10):2410–2427.

Di Toro, DM, Allen, HE, Bergman, HL, Meyer, JS, Paquin, PR, Santore, RC. Biotic ligand model of the acute toxicity of metals. 1. Technical basis. *Environ. Toxicol. Chem.* 2001;20(10):2383–2396.

Di Toro, DM, Mahony, JD, Hansen, DJ, Scott, KJ, Carlson, AR, Ankley, GT. Acid volatile sulfide predicts the acute toxicity of cadmium and nickel in sediments. *Environ. Sci. Technol.* 1992;26(1):96–101.

Field, JA, Sierra-Alvarez, R. Microbial transformation and degradation of polychlorinated biphenyls. *Environ. Pollut.* 2008;155:1–12.

Gomes, RL, Deacon, HE, Lai, KM, Birkett, JW, Scrimshaw, MD, Lester, JN. An assessment of the bioaccumulation of estrone in Daphnia magna. *Environ. Toxicol. Chem.* 2004;23(1):105–108.

Graham, DW, Olivares-Rieumont, S, Knapp, CW, Lima, L, Werner, D, Bowen, E. Antibiotic resistance gene abundances associated with waste discharges to the Almendares River near Havana, Cuba. *Environ. Sci. Technol.* 2011;45(2):418–424.

Henriques, IDS, Aga, DS, Mendes, P, O'Connor, SK, Love, NG. Metabolic footprinting: A new approach to identify physiological changes in complex microbial communities upon exposure to toxic chemicals. *Environ. Sci. Technol.* 2007;41(11):3945–3951.

Hewlett, PS, Plackett, RL. A unified theory for quantal responses to mixtures of drugs: Non-interactive action. *Biometrics* 1959;15:566–575.

Hsieh, S-H, Tsai, K-P, Chen, C-Y. The combined toxic effects of nonpolar narcotic chemicals to *Pseudokirchneriella subcapitata. Wat. Res.* 2006;40:1957–1964.

Hunter, PD, Tyler, AN, Gilvear, RJ, Willby, NJ. Using remote sending to aid the assessment of human health risks from blooms of potentially toxic cyanobacteria. *Environ. Sci. Technol.* 2009;43:2627–2633.

International Organization for Standardization. Water Quality—Determination of the inhibition of the mobility of *Dapnia magna* Straus (*Cladocera, Crustacea*)—Acute toxicity test, ISO 6341, third ed; 1996.

Jacobson T, Sundelin, B. Reproductive effects of the endocrine disruptor fenarimol on a Baltic amphipod *Monoporeia affinis. Environ. Toxicol. Chem.* 2006;25(4):1126–1131.

Jackson, LJ. Piscivores, predation, and PCBs in Lake Ontario's pelagic food web. *Ecol. Applications* 1997;7(3):991–1001.

Jonker, MTO, Suijkerbuik, MPW, Schmitt, H, Sinnige, TL. Ecotoxicological effects of activated carbon addition to sediments. *Environ. Sci. Technol.* 2009;43(15):5959–5966.

Karner, DA, Shafer, MM, Overdier, JT, Hemming, JDC, Sonzogni, WC. *Environ. Toxicol. Chem.* 2006;25(4):1106–1113.

Kearney, JP, Cole, DC. Great Lakes and inland sport fish consumption by licensed anglers in two Ontario communities. *J. Great Lakes Res.* 2003;29(3):460–478.

King-Heiden, TC, Wiecinski, PN, Mangham, AN, Metz, KM, Nesbit, D, Pedersen, JA, Hamers, RJ, Heideman, W, Peterson, RE. Quantum dot nanotoxicity assessment using the zebrafish embryo. *Environ. Sci. Technol.* 2009;43:1605–1611.

Klaine, SJ, Alvarez, PJJ, Batley, GE, Fernandes, TF, Handy, RD, Lyon, DY, Mahendra, S, McLaughlin, MJ, Lead, JR. Nanomaterials in the environment: Behavior, fate, bioavailability, and effects. *Environ. Toxicol. Chem.* 2008;27(9):1825–1851.

Konemann, H. Fish toxicity tests with mixtures of more than two chemicals: A proposal for a quantitative approach and experimental results. *Toxicology* 1981;19:229–238.

Kupryianchyk, D, Reichman, EP, Rakowska, MI, Peeters, ETHM, Grotenhuis, JTC, Koelmans, AA. Ecotoxicological effects of activated carbon amendments on macroinvertebrates in nonpolluted and polluted sediments. *Environ. Sci. Technol.* 2011;45:8567–8574.

Lin, JH, Kao, W-C, Tsai, K-P, Chen, C-Y. A novel algal toxicity testing technique for assessing the toxicity of both metallic and organic toxicants. *Wat. Res.* 2005;39:1869–1877.

Long, ER, Ingersoll, CG, Macdonald, DD. Calculation of uses of mean sediment quality guideline quotients: A critical review. *Environ. Sci. Technol.* 2006;40(6):1726–1736.

Lovern, SB, Klaper, R. Daphnia magna mortality when exposed to titanium dioxide and fullerene (C_{60}) particles. *Environ. Toxicol. Chem.* 2006;25(4):1132–1137.

Lowry, GV, Gregory, KB, Apte, SC, Lead, JR. Transformations of nanomaterials in the environment. *Environ. Sci. Technol.* 2012;46:6893–6899.

Ma, H, Bertsch, PM, Glenn, TC, Kabengi, NJ, Williams, PL. Toxicity of manufactured zinc oxide nanoparticles in the nematode *Caenorhabditis elegans. Environ. Toxicol. Chem.* 2009;28(6):1324–1330.

Marie, V, Gonzalez, P, Baudrimont, M, Boutet, I, Moraga, D, Bourdineaud, J.-P, and Boudou, A. Metallothionein gene expression and protein levels in triploid and diploid oysters *Crassostrea gigas* after exposure to cadmium and zinc. *Environ. Toxicol. Chem.* 25(2):412–418.

Matsuda, S, Matsui, S, Shimizu, Y., Matsuda, T. Genotoxicity of colloidal fullerene C_{60}. *Environ. Sci. Technol.* 2011;45:4133–4138.

Mayer P, Cuhel, R, Nyholm, N. A simple in vitro fluorescence method for biomass measurements in algal growth inhibition tests. *Wat. Res.* 1997;31:2525–2531.

McLeod, PB, Van den Heuvel-Greve, MJ, Luoma, SN, Luthy, RG. Biological uptake of polychlorinated biphenyls by *Macoma Balthica* from sediment amended with activated carbon. *Environ. Toxicol. Chem.* 2007;26(5):980–987.

Marie, V, Gonzalez, P, Baudrimont, M, Boutet, I, Moraga, D, Bourdineaud, JP, Boudou, A. Metallothionein gene expression and protein levels in triploid and diploid oysters *Crassostrea gigas* after exposure to cadmium and zinc. *Environ. Toxicol. Chem.* 2006;25(2):412–418.

Merino-García, D, Kusk, KO, Christensen, ER. Joint toxicity of similarly and dissimilarly acting chemicals to *Daphnia magna* at different response levels. *Arch. Environ. Contam. Toxicol.* 2003;45:289–296.

Morales-Caselles, C, Martín-Díaz, M, Riba, I, Sarasquete, C, and Del Valls, TA. The role of biomarkers to assess oil-contaminated sediment quality using toxicity tests with clams and crabs. *Environ. Toxicol. Chem.* 2008;27(6):1309–1316.

Moreels, D, Lodewijks, P, Zegers, H, Rurangwa, E, Vromant, N, Bastians, L, Diels, L, Springael, D, Mercks, R, Ollevier, F *Environ. Toxicol. Chem.* 2006;25(2):514–519.

Novotny, V. Simplified databased total maximum daily loads, or the world is log-normal. *J. Environ. Eng. ASCE* 2004;130(6):674–683.

Nyholm N, Sorensen, PS, Kusk, KO, Christensen, ER. Statistical treatment of data from microbial toxicity tests. *Environ. Toxicol. Chem.* 1992;11:157–167.

Nyholm, N. Response variable in algal growth inhibition toxicity tests - biomass or growth rate? *Wat. Res.* 1985;19:273–279.

Onnis-Hayden, A, Weng, H, He, M, Hansen, S, Ilyin, V, Lewis, K, Gu, A. Prokaryotic real-time gene expression profiling for toxicity assessment. *Environ. Sci. Technol.* 2009;43(12):4574–4581.

Pagenkopf, G. Gill surface interaction model for trace-metal toxicity of fishes: Role of complexation, pH, and water hardness. *Environ. Sci. Technol.* 1983;17:342–347.

Park, J-W, Tompsett, AR, Zhang, X, Newsted, JL, Jones, PD, Au, DWT, Kong, R, Wu, RSS, Giesy, JP, Hecker, M. Advanced fluorescence in situ hybridization to localize and quantify gene expression in Japanese Medaka (*Oryzias latipes*) exposed to endocrine-disrupting compounds. *Environ. Toxicol. Chem.* 2009;28(9):1951–1962.

Pedersen, JA, Yeager, MA, Suffet, IH (Mel). Organophosphorus insecticides in agricultural and residential runoff: Field observations and implications for total maximum daily load development. *Environ. Sci. Technol.* 2006;40:2120–2127.

Petala, M, Kokokiris, L, Samaras, P, Papadopoulos, A, Zouboulis, A. Toxicological and exotoxic impact of secondary and tertiary treated sewage effluents. *Wat. Res.* 2009;43:5063–5074.

Pollino, CA, Georgiades, E, Holdway, DA. Use of the Australian crimson-spotted rainbowfish (*Melanotaenia fluviatilis*) as a model test species for investigating the effects of endocrine disruptors. *Environ. Toxicol. Chem.* 2007;26(10):2171–2178.

Principi, P, Villa, F, Bernasconi, M, Zanardini, E. Metal toxicity in municipal wastewater activated sludge investigated by multivariate analysis and in situ hybridization. *Wat. Res.* 2006;40:99–106.

Pruden, A, Pei, R, Storteboom, H, Carlson, KH. Antibiotic resistance genes as emerging contaminants: studies in northern Colorado. *Environ. Sci. Technol.* 2006;40(23):7445–7450.

Ren, S, Schultz, TW. Identifying the mechanism of aquatic toxicity of selected compounds by hydrophobicity and electrophilicity descritors. *Toxicology letters* 2002;129:151–160.

Robidoux, PY, Sunahara, GI, Savard, K, Berthelot, Y, Dodard, S, Martel, M, Gong, P, Hawari, J. Acute and chronic toxicity of the new explosive CL-20 to the earthwork (*Eisenia andrei*) exposed to amended natural soils. *Environ. Toxicol. Chem.* 2004;23(4):1026–1034.

Russom, CL, Bradbury, SP, Broderius, SJ, Hammermeister, DE, Drummond, RA. Predicting modes of toxic action from chemical structure: acute toxicity in the fathead minnow (Pimephales promelas). *Environ. Toxicol. Chem.* 1997;16(5):948–967.

Santore, RC, Di Toro, DM, Paquin, PR, Allen, HE, Meyer, JS. Biotic ligand model of the acute toxicity of metals. 2. Application to acute copper toxicity in freshwater and fish and daphnia. *Environ. Toxicol. Chem.* 2001;20(10):2397–2402.

Seki, M, Yokota, H, Maeda, M, Kobayashi, K. Fish full life-cycle testing for 17β-estradiol on medaka (*Oryzias Latipes*). *Environ. Toxicol. Chem.* 2005;24(5):1259–1266.

Shaw, JR, Dempsey, TD, Chen, CY, Hamilton, JW, Folt, CL. Comparative toxicity of cadmium, zinc, and mixtures of cadmium and zinc to daphnids. *Environ. Toxicol. Chem.* 2006;25(1):182–189.

Shaw, JR, Glaholt, SP, Greenberg, NS, Sierra-Alvarez, R, Folt, CL. Acute toxicity of arsenic to *Daphnia pulex*: Influence of organic functional groups and oxidation state. *Environ. Toxicol. Chem.* 2007;26(7):1532–1537.

Soonthornnonda, P, Christensen, ER. A load model based on antecedent dry periods for pollutants in stormwater. *Water Environment Research* 2008;80(2):162–171.

Storteboom, H, Arabi, M, Davis, JG, Crimi, B, Pruden, A. Identification of antiobiotic resistance-gene molecular signatures suitable as tracers of pristine river, urban, and agricultural sources. *Environ. Sci. Technol.* 2010a;44(6):1947–1953.

Storteboom, H, Arabi, M, Davis, JG, Crimi, B, Pruden, A. Tracking antibiotic resistance genes in the South Platte River basin using molecular signatures of urban, agricultural, and pristine sources. *Environ. Sci. Technol.* 2010b;44(19):7397–7404.

Thorpe, KL, Maack, G, Benstead, R, Tyler, CR. Estrogenic wastewater treatment works effluents reduce egg production in fish. *Environ. Sci. Technol.* 2009;43(8):2976–2982.

Tsui, MTK, Wang, W-X. Multigenerational acclimation of *Daphnia magna* to mercury: Relationships between biokinetics and toxicity. *Environ. Toxicol. Chem.* 2005;24(11):2927–2933.

Vaché, C, Camares, O, De Graeve, F, Dastugue, B, Meiniel, A, Vaury, C, Pellier, S, Leoz-Garziandia, E, Bamdad, M. *Drosophila melanogaster* p–glycoprotein: A membrane detoxification system toward polycyclic aromatic hydrocarbon pollutants. *Environ. Toxicol. Chem.* 2006;25(2):572–580.

Valenti, TW, Cherry, DS, Neves, RJ, Schmerfeld, J. Acute and chronic toxicity of mercury to early life stages of the rainbow mussel, *Villosa iris* (Bivalvia: Uniondae). *Environ. Toxicol. Chem.* 2005;24(5):1242–1246.

Veith, GD, Broderius, SJ. Rules for distinguishing toxicants that cause type I and type II narcosis syndromes. *Environ. Health Perspect.* 1990;87:207–211.

Verlecar, XN, Desai, SR, Sarkar, A, Dalal, SG. Biological indicators in relation to coastal pollution along Karnataka coast, India. *Wat. Res.* 2006;40:3304–3312.

Wang, C, Lu, G, Wang, P, Wu, H, Qi, P, Liang, Y. Assessment of environmental pollution of Taihu Lake by combining active biomonitoring and integrated biomarker response. *Environ. Sci. Technol.* 2011;45:3746–3752.

Wei, Y, Dai, J, Liu, M, Wang, J, Xu, M, Zha, J, Wang, Z. Estrogen-like properties of perfluorooctanoic acid as revealed by expressing hepatic estrogen-responsive genes in rare minnows (*Gobiocypris rarus*). *Environ. Toxicol. Chem.* 2007;26(11):2440–2447.

Weil, M, Scholz, S, Zimmer, M, Sacher, F, Duis, K. Gene expression analysis in zebrafish embryos: A potential approach to prodict effect concentrations in the fish early life stage test. *Environ. Toxicol. Chem.* 2009;28(9):1970–1978.

Wilderer, PA, Bungartz, H-J, Lemmer, H, Wagner, M, Keller, J, Wuertz, SM. Modern scientific methods and their potential in wastewater science and technology. *Wat. Res.* 2002;36:370–393.

Wisconsin Department of Natural Resources. 2009. Fish consumption advisories. http://dnr. wi.gov/fish/consumption/.

Wollenberger, L. *Toxicity tests with Crustaceans for detecting sublethal effects of potential endocrine disrupting chemicals*. PhD dissertation. Environment and Resources, Technical University of Denmark; 2005.

Xu, W, Zhang, G, Li, X, Zou, S, Li, P, Hu, Z, Li, J. Occurrence and elimination of antibiotics at four sewage treatment plants in the Pearl River Delta (PRD), South China. *Wat. Res.* 2007;41:4526–4534.

Yeo, M-K, Kang, M. Photodecomposition of bisphenol A on nanometer-sized TiO$_2$ thin film and the associated biological toxicity to zebrafish (*Danio rerio*) during and after photocatalysis. *Wat. Res.* 2006;40:1905–1914.

Zhao, Y, Newman, MC. Effects of exposure duration and recovery time during pulsed exposures. *Environ. Toxicol. Chem.* 2006;25(5):1298–1304.

Zhu, X, Zhu, L, Li, Y, Duan, Z, Chen, W, Alvarez, PJJ. Developmental toxicity in zebrafish (*Danio rerio*) embryos after exposure to manufactured nanomaterials: Buckminsterfullerene aggregates (nC$_{60}$) and fullerol. *Environ. Toxicol. Chem.* 2007;26(5):976–979.

11

AMBIENT WATER QUALITY CRITERIA

11.1 INTRODUCTION

The quality of natural waters—how "clean" they are—became a serious concern in the eighteenth century with the advent of rapid urbanization and industrialization. A literature search found hundreds of scientific papers published before 1900 that focused on the water quality of natural water bodies. Water quality remains a top priority issue for society as a result of population growth, industrial and economic development, the change in global climate, and other factors.

With regard to water quality, the concept "the purer, the better" is incorrect. Pure water, H_2O, is neither naturally available nor suitable to the needs of maintaining the health of the ecosystem and humans. The quality of natural waters depends largely on natural conditions and processes. However, urbanization and industrialization have caused the deterioration of water quality with significant adverse impact on the ecosystem and human health. The challenges with regard to water quality are well summarized by the United Nations Environment Program (2010).

The discussion in this chapter is limited to the criteria and standards for natural ambient waters, such as rivers and lakes, in accordance with the focus of this book. The regulations and criteria for public water supplies, wastewater discharges and other effluents, and special use waters are not parts of the discussions, although all of these interact with the ambient water quality in a complex manner.

Physical and Chemical Processes in the Aquatic Environment, First Edition. Erik R. Christensen and An Li.
© 2014 John Wiley & Sons, Inc. Published 2014 by John Wiley & Sons, Inc.

11.2   A PRIMER ON AMBIENT WATER QUALITY REGULATIONS

At the global level, the United Nations (UN) has made great efforts to address issues regarding ambient water quality. UN-Water, in which many UN agencies and programs participate, designs global policy on water quality based on analyses of the challenges that arise from natural processes and human activities, and the impact of declining water quality on the integrity of the ecosystem and human health. UN-Water presents four fundamental strategies: pollution prevention, treatment of polluted water, safe use of wastewater, and the restoration and protection of the ecosystem. A series of specific recommendations on policy interventions are proposed, calling for better monitoring data collection and analysis, more effective communication, education and advocacy, improvements in financial and economical approaches, legal and institutional arrangements, and improved technology and infrastructure (UN-Water, 2011). The UN CEO Water Mandate addresses the issues of water access and sustainability (UN, 2007). The Global Environmental Monitoring System (GEMS) water component, GEMS/Water, which is managed by the UNEP in collaboration with the World Health Organization (WHO) and others, collects water quality data from numerous rivers, lakes, wetlands, and groundwater in 106 countries and other participations, with some data dated back to 1965 (UNEP, 2010). Guidelines and manuals are provided to the participants for the system's implementation and operation (Bartram and Balance, 1996; Allard, 1992; Chapman, 1996). Given the complexity in water uses and water quality issues, there are no worldwide, well-defined, enforceable ambient water quality criteria (AWQC).

In the United States, legislative action on water quality took place at national level in 1948, with the comprehensive Water Pollution Control Act. A non-exhaustive list of legislations since that time includes

1948: Water Pollution Control Act

1956: Federal Water Pollution Control Act (FWPCA)

1965: Water Quality Act

1968: The Wild and Scenic Rivers Act

1969: National Environmental Policy Act

1972: Federal Water Pollution Control Act Amendments, Clean Water Act (CWA)

1972: Federal Insecticide, Fungicide and Rodenticide Act

1973: Endangered Species Act

1976: Resource Conservation and Recovery Act

1980: Comprehensive Environmental Response, Compensation, and Liability Act

1984: Superfund Authorization and Renewal Act

1987: Water Quality Act

2000: Beach Environmental Assessment and Coastal Health (BEACH) Act

2000: Wet Weather Water Quality Act

2007: Water Quality Financing Act

2009: Water Quality Investment Act

For surface waters, the CWA is probably the most important piece of legislation. It establishes quality standards and regulates the discharges of pollutants. Under the CWA, the United States Environmental Protection Agency (U.S. EPA) has implemented pollution control programs, including the establishment of the National Pollutant Discharge Elimination System (NPDES) permit program to control industrial and municipal wastewater discharges.

The complexity of the ambient water quality issue was recognized by the regulations and their amendments. Therefore, the principle of state–federal cooperation has been followed. It is the responsibility of the federal government (U.S. EPA) to provide criteria based solely on sound scientific data and judgment, without considering the technical feasibility and economic impacts. These national recommended AWQC are often called "304(a) criteria" because they were established by the U.S. EPA pursuant to Section 304(a)(1) of the CWA. The 304(a) criteria include both numerical and narrative values (U.S. EPA, 2013a). They are guidelines rather than enforceable standards. Much information can be found in the *Water Quality Standards Handbook, second edition* (U.S. EPA, 1994). Updated versions of this book (U.S. EPA, 2012a) provide guidance issued in support of the Water Quality Standards (WQS) Regulation (40 CFR 131, as amended).

The state government and authorized Native American peoples and territories are responsible for establishing enforceable WQS for their waters. The national AWQC can be adopted or modified to reflect specific conditions or feasible and scientifically defensible techniques. The water quality standards set by the state and peoples must be reviewed and approved by the U.S. EPA, before they become law. A repository of water quality standards issued by states, peoples, and territories can be found on the EPA website (U.S. EPA, 2013b).

For each water body or system, the WQS must contain four components. First, the uses of the water must be defined. For example, if the water is to be used solely for agriculture, the water quality standards can be very different from those pertaining to domestic water supply for human consumption. Other designated uses include fishery, primary (e.g. swimming and other activities with full body contact) and secondary (e.g. boating, wading and other activities with limited body contact) recreations, and protection of aquatic life. A body of water is often used for multiple purposes. In addition to the designated uses and the water quality criteria that protect those uses, the established water quality standards must also include an antidegradation policy, to maintain and protect existing uses and high quality waters, and general policies addressing implementation issues.

11.3   CURRENT US WATER QUALITY CRITERIA

The current recommended national AWQC criteria list defines approximately 150 pollutants, including more than 100 priority pollutants (P) and more than 40 non-priority pollutants (NP). The criteria were separately designed to protect aquatic life (Table 11.1), protect human health (Table 11.2), and prevent organoleptic (e.g. taste

TABLE 11.1. Ambient Water Quality Criteria for the Protection of Aquatic Life

Pollutant	CAS Number	P/NP	Freshwater CMC 1 (acute) (μg/L)	Note[a]	Freshwater CCC 1 (chronic) (μg/L)	Note[a]	Saltwater CMC 1 (acute) (μg/L)	Note[a]	Saltwater CCC 1 (chronic) (μg/L)	Note[a]	Publication year
Acrolein	107028	P	3		3						2009
Aesthetic qualities		NP	Narrative statement—see document								1986
Aldrin	309002	P	3	G			1.3	G			1980
Alkalinity		NP	20,000	C							1986
alpha-Endosulfan	959988	P	0.22	G,Y	0.056	G,Y	0.034	G,Y	0.0087	G,Y	1980
Aluminum pH 6.5–9.0	7429905	NP	750	I	87	I,S					1988
Ammonia	7664417	NP	Freshwater criteria are pH, temperature and life-stage dependent / Saltwater criteria are pH and temperature dependent								2013 / 1989
Arsenic	7440382	P	340	A,D	150	A,D	69	A,D	36	A,D	1995
Bacteria		NP	For primary recreation and shellfish uses—see document								1986
beta-Endosulfan	33213659	P	0.22	G,Y	0.056	G,Y	0.034	G,Y	0.0087	G,Y	1980
Boron		NP	Narrative statement—see document								1986
Carbaryl	63252	NP	2.1		2.1		1.6				2012
Cadmium	7440439	P	2	D,E	0.25		40	D	8.8	D	2001
Chlordane	57749	P	2.4	G	0.0043	G	0.09	G	0.004	G	1980
Chloride	16887006	NP	860,000		230,000						1986
Chlorine	7782505	NP	19		11		13		7.5		1986
Chloropyrifos	2921882	NP	0.083		0.041		0.011		0.0056		1986
Chromium (III)	16065831	P	570	D,E	74	D,E					1995
Chromium (VI)	18540299	P	16	D	11	D	1,100	D	50	D	1995
Color		NP	Narrative statement—see document								1986

Pollutant	CAS	P/NP	Freshwater CMC	Freshwater CCC	Saltwater CMC	Saltwater CCC	Year
Copper	7440508	P	Freshwater criteria calculated using the BLM mm—see document		4.8 D,cc	3.1 D,cc	2007
Cyanide	57125	P	22 Q	5.2 Q	1 Q	1 Q	1985
Demeton	8065483	NP		0.1 C		0.1 C	1985
Diazinon	333415	NP	0.17	0.17	0.82	0.82	2005
Dieldrin	60571	P	0.24	0.056 O	0.71	0.0019 G	1995
Endrin	72208	P	0.086	0.036 O	0.037	0.0023 G	1995
gamma-BHC (Lindane)	58899	P	0.95		0.16		1995
Gases, total dissolved		NP	Narrative statement—see document C				1986
Guthion	86500	NP		0.01 C		0.01 C	1986
Hardness		NP	Narrative statement—see document				1986
Heptachlor	76448	P	0.52 G	0.0038 G	0.053 G	0.0036 G	1980
Heptachlor epoxide	1024573	P	0.52 G,V	0.0038 G,V	0.053G,V	0.0036G,V	1981
Iron	7439896	NP		1,000 C			1986
Lead	7439921	P	65 D,E	2.5 D,E	210 D	8.1 D	1980
Malathion	121755	NP		0.1 C		0.1 C	1986
Mercury	7439976	P	1.4 D,hh	0.77 D,hh	1.8 D,ee,hh	0.94 D,ee,hh	1995
Methylmercury	22967926	P					
Methoxychlor	72435	NP		0.03 C		0.03 C	1986
Mirex	2385855	NP		0.001 C		0.001 C	1986
Nickel	7440020	P	470 D,E	52 D,E	74 D	8.2 D	1995
Nonylphenol	84852153	NP	28 b	6.6	7	1.7	2005
Nutrients		NP					
Oil and grease		NP	Narrative statement—see document C				1986
Oxygen, dissolved freshwater	7782447	NP	Warmwater and coldwater matrix—see document				1986
Oxygen, dissolved saltwater	7782447	NP	Saltwater—see document				1986

(continued)

TABLE 11.1. (*Continued*)

Pollutant	CAS Number	P/NP	Freshwater CMC 1 (acute) (µg/L)	Note[a]	CCC 1 (chronic) (µg/L)	Note[a]	Saltwater CMC 1 (acute) (µg/L)	Note[a]	CCC 1 (chronic) (µg/L)	Note[a]	Publication year
Parathion	56382	NP	0.065	I	0.013	I					1995
Pentachlorophenol	87865	P	19	F	15	F	13		7.9	C,P	1995
pH		NP			6.5–9	C			6.5–8.5		1986
Phosphorus elemental	7723140	NP							0.03	N	1986
PCBs		P			0.014	N			0.03	N	
Selenium	7782492	P		L	5.0		290	D,dd	71	D,dd	1995
Silver	7440224	P	3.2	D,E			1.9	D			1980
Solids suspended and turbidity		NP	Narrative statement—see document			C					1986
Sulfide-hydrogen sulfide	7783064	NP	Narrative statement—see document		2.0	C			2.0	C	1986
Tainting Substances		NP	Narrative statement—see document								1986
Temperature		NP	Species-dependent criteria—see document			M					
Toxaphene	8001352	P	0.73		0.0002		0.21		0.0002		1986
TBT		NP	0.46		0.072	D,E	0.42		0.0074		2004
Zinc	7440666	P	120	D,E	120	D,E	90	D	81	D	1995
4,4′-DDT	50293	P	1.1	G,ii	0.001	G,ii	0.13	G,ii	0.001	G,ii	1980

PCBs, polychlorinated biphenyls; TBT, tributyltin; DDT, dichlorodiphenyltrichloroethane.

P/NP Indicates either a priority pollutant (P) or a non-priority pollutant (NP).

[a] See Appendix 11.A for code descriptions.

[b] See EPA's Ecoregional criteria for total phosphorus, total nitrogen, chlorophyll a, and water clarity (Secchi depth for lakes; turbidity for streams and rivers) (& Level III Ecoregional criteria).

TABLE 11.2. Ambient Water Quality Criteria for the Protection of Human Health

Pollutant	CAS Number	P/NP	Human health for the consumption of — Water + Organism µg/L	Note[a]	Organism only µg/L	Note[a]	Publication Year
Acenaphthene	83329	P	670	B,U	990	B,U	2002
Acrolein	107028	P	6	ll	9	ll	2009
Acrylonitrile	107131	P	0.051	B,C	0.25	B,C	2002
Aldrin	309002	P	0.000049	B,C	0.00005	B,C	2002
alpha-BHC	319846	P	0.0026	B,C	0.0049	B,C	2002
alpha-Endosulfan	959988	P	62	B	89	B	2002
Anthracene	120127	P	8,300	B	40,000	B	2002
Antimony	7440360	P	5.6	B	640	B	2002
Arsenic	7440382	P	0.018	C,M,S	0.14	C,M,S	1992
Asbestos	1332214	P	7 million fibers/L	I			1991
Barium	7440393	NP	1,000	A			1986
Benzene	71432	P	2.2	B,C	51	B,C	2002
Benzidine	92875	P	0.000086	B,C	0.0002	B,C	2002
Benzo(a) anthracene	56553	P	0.0038	B,C	0.018	B,C	2002
Benzo(a) pyrene	50328	P	0.0038	B,C	0.018	B,C	2002
Benzo(b) fluoranthene	205992	P	0.0038	B,C	0.018	B,C	2002
Benzo(k) fluoranthene	207089	P	0.0038	B,C	0.018	B,C	2002
Beryllium	7440417	P		Z			
beta-BHC	319857	P	0.0091	B,C	0.017	B,C	2002
beta-Endosulfan	33213659	P	62	B	89	B	2002
Bis(2-chloroethyl) ether	111444	P	0.03	B,C	0.53	B,C	2002
Bis(2-chloroisopropyl) ether	108601	P	1,400	B	65,000	B	2002

(*continued*)

395

TABLE 11.2. (*Continued*)

| Pollutant | CAS Number | P/NP | Human health for the consumption of | | | | Publication Year |
| | | | Water + Organism | | Organism only | | |
			μg/L	Note[a]	μg/L	Note[a]	
Bis(2-ethylhexyl) phthalateX	117817	P	1.2	B,C	2.2	B,C	2002
Bromoform	75252	P	4.3	B,C	140	B,C	2002
Butylbenzyl phthalateW	85687	P	1,500	B	1,900	B	2002
Cadmium	7440439	P		Z			
Carbon tetrachloride	56235	P	0.23	B,C	1.6	B,C	2002
Chlordane	57749	P	0.0008	B,C	0.00081	B,C	2002
Chlorobenzene	108907	P	130	Z,U	1,600	U	2003
Chlorodibromomethane	124481	P	0.4	B,C	13	B,C	2002
Chloroform	67663	P	5.7	C,P	470	C,P	2002
Chlorophenoxy herbicide (2,4-D)	94757	NP	100	Z			1986
Chromium (III)	16065831	P		Z Total			
Chromium (VI)	18540299	P		Z Total			
Chrysene	218019	P	0.0038	B,C	0.013	B,C	2002
Copper	7440508	P	1,300	U			1992
Cyanide	57125	P	140	jj	140	jj	2003
Dibenzo(a,h)anthracene	53703	P	0.0038	B,C	0.018	B,C	2002
Dichlorobromomethane	75274	P	0.55	B,C	17	B,C	2002
Dieldrin	60571	P	0.000052	B,C	0.00C054	B,C	2002
Diethyl phthalateW	84662	P	17,000	B	44,000	B	2002
Dimethyl phthalateW	131113	P	270,000		1,100,000	B	2002

Name	CAS		First		Second		Year
Di-n-butyl phthalateW	84742	P	2,000	B	4,500	B	2002
Dinitrophenols	25550587	NP	69	B	5,300	B	2002
Endosulfan sulfate	1031078	P	62	B	89	B	2002
Endrin	72208	P	0.059	B	0.06		2003
Endrin Aldehyde	7421934	P	0.29	B	0.3	B,H	2002
Ether, bis(chloromethyl)	542881	NP	0.0001	C	0.00029	C	2002
Ethylbenzene	100414	P	530	B	2,100		2003
Fluoranthene	206440	P	130	B	140	B	2002
Fluorene	86737	P	1,100	B	5,300	B	2002
gamma-BHC (lindane)	58899	P	0.98		1.8		2003
Heptachlor	76448	P	0.000079	B,C	0.000079	B,C	2002
Heptachlor epoxide	1024573	P	0.000039	B,C	0.000039	B,C	2002
Hexachlorobenzene	118741	P	0.00028	B,C	0.00029	B,C	2002
Hexachlorobutadiene	87683	P	0.44	B,C	18	B,C	2002
Hexachlorocyclo-hexane-Technical	608731		0.0123	H	0.0414	H	
Hexachlorocyclopentadiene	77474	P	40	U	1,100	U	2003
Hexachloroethane	67721	P	1.4	B,C	3.3	B,C	2002
Ideno(1,2,3-cd)pyrene	193395	P	0.0038	B,C	0.018	B,C	2002
Isophorone	78591	P	35	B,C	960	B,C	2002
Manganese	7439965	NP	50	O	100	A	2002
Methylmercury	22967926	P		A,Z	0.3 mg/kg	J	2001
Methoxychlor	72435	NP	100	B			1986
Methyl bromide	74839	P	47	B	1,500	B	2002
Methylene chloride	75092	P	4.6	B,C	590	B,C	2002

(continued)

397

TABLE 11.2. (*Continued*)

Pollutant	CAS Number	P/NP	Human health for the consumption of				Publication year
			Water + Organism		Organism only		
			μg/L	Note[a]	μg/L	Note[a]	
Nickel	7440020	P	610	B	4,600	B	1998
Nitrates	14797558	NP	10,000	A			1986
Nitrobenzene	98953	P	17	B	690	B,H,U	2002
Nitrosamines		NP	0.0008		1.24		1980
Nitrosodibutylamine, *N*	924163	NP	0.0063	C	0.22	C	2002
Nitrosodiethylamine, *N*	55185	NP	0.0008	C	1.24	C	2002
Nitrosopyrrolidine, *N*	930552	NP	0.016	C	34	C	2002
N-Nitrosodimethylamine	62759	P	0.00069	B,C	3	B,C	2002
N-Nitrosodi-*n*-propylamine	621647	P	0.005	B,C	0.51	B,C	2002
N-Nitrosodiphenylamine	86306	P	3.3	B,C [b]	6	B,C	2002
Nutrients		NP					
Pathogen and pathogen indicators			See EPA's 2012 recreational water quality criteria				2012
Pentachlorobenzene	608935	NP	1.4	E	1.5	E	2002
Pentachlorophenol	87865	P	0.27	B,C	3	B,C,H	2002
pH		NP	5–9				1986
Phenol	108952	P	10,000	ll,U	860,000	ll,U	2009
PCBs		P	0.000064	B,C,N	0.000064	B,C,N	2002
Pyrene	129000	P	830	B	4,000	B	2002
Selenium	7782492	P	170	Z	4,200		2002
Solids dissolved and salinity	7782492	NP	250,000	A			1986

398

Chemical	CAS		Value	Code	Value	Code	Year
Tetrachlorobenzene,1,2,4,5-	95943	NP	0.97	B	1.1	B	2002
Tetrachloroethylene	127184	P	0.69	C	3.3	C	2002
Thallium	7440280	P	0.24	Z	0.47		2003
Toluene	108883	P	1,300		15,000		2003
Toxaphene	8001352	P	0.00028	B,C	0.00028	B,C	2002
Trichloroethylene	79016	P	2.5	C	30	C	2002
Trichlorophenol,2,4,5-	95954	NP	1,800	B	3,600	B	2002
Vinyl chloride	75014	P	0.025	C,kk	2.4	C,kk	2003
Zinc	7440666	P	7,400	U	26,000	U	2002
1,1,1-Trichloroethane	71556	P		Z			
1,1,2,2-Tetrachloroethane	79345	P	0.17	B,C	4	B,C	2002
1,1,2-Trichloroethane	79005	P	0.59	B,C	16	B,C	2002
1,1-Dichloroethylene	75354	P	330		7,100		2003
1,2,4-Trichlorobenzene	120821	P	35		70		2003
1,2-Dichlorobenzene	95501	P	420		1,300		2003
1,2-Dichloroethane	107062	P	0.38	B,C	37	B,C	2002
1,2-Dichloropropane	78875	P	0.5	B,C	15	B,C	2002
1,2-Diphenylhydrazine	122667	P	0.036	B,C	0.2	B,C	2002
1,2-*trans*-Dichloroethylene	156605	P	140	Z	10,000		2003
1,3-Dichlorobenzene	541731	P	320		960		2002
1,3-Dichloropropene	542756	P	0.34	C	21	C	2003
1,4-Dichlorobenzene	106467	P	63		190		2003
2,3,7,8-TCDD (dioxin)	1746016	P	*5.00E-09*	C	*5.10E-09*	C	2002

(*continued*)

TABLE 11.2. (*Continued*)

Pollutant	CAS Number	P/NP	Human health for the consumption of					Publication year
			Water + Organism		Organism only			
			µg/L	Note[a]	µg/L	Note[a]		
2,4,6-Trichlorophenol	88062	P	1.4	B,C	2.4	B,C,U		2002
2,4-Dichlorophenol	120832	P	77	B,U	290	B,U		2002
2,4-Dimethylphenol	105679	P	380	B	850	B,U		2002
2,4-Dinitrophenol	51285	P	69	B	5,300	B		2002
2,4-Dinitrotoluene	121142	P	0.11	C	3.4	C		2002
2-Chloronaphthalene	91587	P	1,000	B	1,600	B		2002
2-Chlorophenol	95578	P	81	B,U	150	B,U		2002
2-Methyl-4,6-dinitrophenol	534521	P	13		280			2002
3,3'-Dichlorobenzidine	91941	P	0.021	B,C	0.028	B,C		2002
3-Methyl-4-chlorophenol	59507	P		U		U		
4,4'-DDD	72548	P	0.00031	B,C	0.00031	B,C		2002
4,4'-DDE	72559	P	0.00022	B,C	0.00022	B,C		2002
4,4'-DDT	50293	P	0.00022	B,C	0.00022	B,C		2002

PCBs, polychlorinated biphenyls; TBT, tributyltin; DDD, dichlorodiphenyldichloroethane; DDE, dichlorodiphenyldichloroethylene; DDT, dichlorodiphenyl-trichloroethane.

P/NP Indicates either a priority pollutant (P) or a non-priority pollutant (NP).

[a]See Appendix 11-B for code descriptions.

[b]See EPA's Ecoregional criteria for total phosphorus, total nitrogen, chlorophyll a and water clarity (Secchi depth for lakes; turbidity for streams and rivers) (and Level III ecoregional criteria).

TABLE 11.3. Ambient Water Quality Criteria for Organoleptic Effects

Pollutant	CAS number	Organoleptic effect criteria (µg/L)	FR cite/ source[*]
Acenaphthene	83329	20	Gold Book
Color		NP	Gold Book
Iron	7439896	NP	Gold Book
Monochlorobenzene	108907	20	Gold Book
Tainting Substance		NP	Gold Book
3-Chlorophenol		0.1	Gold Book
4-Chlorophenol	106489	0.1	Gold Book
2,3-Dichlorophenol		0.04	Gold Book
2,5-Dichlorophenol		0.5	Gold Book
2,6-Dichlorophenol		0.2	Gold Book
3,4-Dichlorophenol		0.3	Gold Book
2,4,5-Trichlorophenol	95954	1	Gold Book
2,4,6-Trichlorophenol	88062	2	Gold Book
2,3,4,6-Tetrachlorophenol		1	Gold Book
2-Methyl-4-chlorophenol		1800	Gold Book
3-Methyl-4-chlorophenol	59507	3000	Gold Book
3-Methyl-6-chlorophenol		20	Gold Book
2-Chlorophenol	95578	0.1	Gold Book
Copper	7440508	1000	Gold Book
2,4-Dichlorophenol	120832	0.3	Gold Book
2,4-Dimethylphenol	105679	400	Gold Book
Hexachlorocyclopentadiene	77474	1	Gold Book
Manganese	7439965		
Nitrobenzene	98953	30	Gold Book
Pentachlorophenol	87865	30	Gold Book
Phenol	108952	300	Gold Book
Zinc	7440666	5000	45 FR79341

Note: These criteria are based on organoleptic (taste and odor) effects. Because of variations in chemical nomenclature systems, this listing of pollutants does not duplicate the listing in Appendix A of 40 CFR Part 423. Also listed are the Chemical Abstracts Service (CAS) registry numbers, which provide a unique identification for each chemical.
[*]Gold Book (Source: U.S. EPA, (1986)).

and odor) effects (Table 11.3). The lengthy footnotes for Tables 11.1 and 11.2 are presented as Appendix 11.A and Appendix 11.B respectively. Tables 11.1–11.3, and their footnotes were directly downloaded on April 29, 2013, from the U.S. EPA (2013b). In addition, the "Additional Notes" section (Appendix 11.C), further explaining the terms used in the criteria tables, has been downloaded.

The priority pollutants include metals such as antimony, arsenic, beryllium, cadmium, chromium (III), chromium (VI), copper, lead, mercury (and methylmercury), nickel, selenium, silver, thallium, and zinc. Organic pollutants include phenols, organic pesticides, selected polycyclic aromatic hydrocarbons, and polychlorinated biphenyls (PCBs). PCBs are deemed a single pollutant, despite the fact that different congeners have different ecological and human toxicities. The insecticide toxaphene, which is in fact a mixture of 177 individual organic compounds with similar molecular structures containing camphene, is similarly classified (see Chapter 7). Asbestos is a carcinogenic material and is included as a single priority pollutant. Cyanide is the only inorganic anion priority pollutant. Some "pollutants", such as color and hardness, are actually physicochemical parameters of natural waters. For some pollutants, no numerical criteria are calculated but they are addressed using narrative criteria. In addition, some non-priority pollutants are regulated under the Safe Drinking Water Act (SDWA) with the maximum contaminant level (MCL) values (U.S. EPA, 2009), and are thus not given numerical AWQC.

11.3.1 Aquatic Life Criteria

The criteria for aquatic life are intended to protect the vast majority of the aquatic communities in the country. Approximately 40 out of the 60 pollutants in the current AWQC list have numerical aquatic life criteria, as summarized in Table 11.1. Efforts have been made in recent years toward regulating contaminants of emerging concern, particularly endocrine disruptor chemicals (EDCs) such as pharmaceuticals and personal care products or PPCPs (see Chapter 7). Although the concentrations of PPCPs in natural waters are generally very low, their potentially high impact on the evolution of ecological systems is a serious concern.

Aquatic life criteria were developed based on scientific knowledge of accumulation and the toxicity (acute and chronic) of chemicals in aquatic organisms. The toxicity consideration is designed to protect the organisms, and the data on bioaccumulation are used to determine whether the contaminated organisms pose risks to their consumers, including humans. The guidelines for establishing such criteria were developed by the U.S. EPA in 1980 and 1985, and they are still being used today although revisions are expected (U.S. EPA, 1998).

The numerical aquatic life criteria are expressed by the criteria maximum concentration (CMC), the criterion continuous concentration (CCC), acute averaging period, chronic averaging period, acute frequency of allowed exceedance, and chronic frequency of allowed exceedance. The CMC is an estimate of the highest concentration of a material in surface water to which an aquatic community can be briefly exposed without resulting in an unacceptable effect, and the CCC is an estimate of the highest concentration of a material in surface water to which an aquatic community can be exposed indefinitely without resulting in an unacceptable effect. Understandably, the CMC values are higher than or equal to the CCC values. In addition, the CMC and CCC values for freshwater are different from those for saltwater.

The acute and chronic "averaging periods" are defined as the periods of time over which the receiving water concentration is averaged for comparison with criteria concentrations. These specifications limit the duration of concentrations above the criteria. The acute and chronic "frequencies of allowed exceedance" define how often criteria can be exceeded without unacceptably affecting the community (U.S. EPA, 1996).

11.3.2 Human Health Criteria

There are more than100 priority, plus about 16 non-priority, pollutants in the list for human health protection; double the numbers for aquatic life. Two sets of AWQC were established for human health, depending on the uses of water (see Table 11.2). Generally, the criteria for waters that are to be used as the sources of both drinking water and dietary aquatic organisms are stricter than those for the waters that only provide dietary organisms (such as fish and shell fish). Both sets of criteria were established with either cancer or non-cancer effects as health end point.

The methodology for deriving AWQC for the protection of human health is described in detail by the U.S. EPA (2000). For known, probable, or possible human carcinogens (U.S. EPA, 1999), the AWQC generally correspond to lifetime excess cancer risk levels of 10^{-7} to 10^{-5}, which means the risk of having one additional case of cancer in a population of 10 million to that in a population of 100,000.

AWQC values (in mg/L) for human health with cancer as the end point were calculated by extrapolation from animal tests and human epidemiological studies. The two equations below use the nonlinear (Equation 11.1) and linear (Equation 11.2) extrapolations respectively.

Nonlinear

$$AWQC = \frac{POD}{UF} \cdot RSC \cdot \left(\frac{BW}{DI + \sum_{i=2}^{4} \left(FI_i \cdot BAF_i \right)} \right) \quad (11.1)$$

Linear

$$AWQC = RSD \cdot \left(\frac{BW}{DI + \sum_{i=2}^{4} \left(FI_i \cdot BAF_i \right)} \right) \quad (11.2)$$

where

AWQC = ambient water quality criterion (in mg/L)

POD = point of departure based on a nonlinear low-dose extrapolation (in mg/(kg d)).

UF = Uncertainty Factor based on a nonlinear low-dose extrapolation (unitless)

RSC = Relative source contribution factor to account for non-water sources of exposure.

RSD = risk-specific dose based on a linear low-dose extrapolation (in mg/(kg d))

BW = human body weight (default = 70 kg for adults)

DI = drinking water intake (default = 2 L/d for adults)

FI_i = fish intake at trophic level (TL) I (I = 2, 3, and 4) (defaults for total intake = 0.0175 kg/d for general adult population and sport anglers, and 0.1424 kg/d for subsistence fishers). Trophic level breakouts for the general adult population and sport anglers are:

TL2 = 0.0038 kg/d; TL3 = 0.0080 kg/d; and TL4 = 0.0057 kg/d.

BAF_i = bioaccumulation factor at trophic level I ($I = 2$, 3, and 4), lipid normalized (L/kg) (see Chapter 7). To establish AWQC, a single BAF value is used for each trophic level of aquatic organisms.

For a non-carcinogen contaminant, a criterion was derived from a threshold concentration for non-cancer adverse effects, using the equation

$$AWQC = RfD \cdot RSC \cdot \left(\frac{BW}{DI + \sum_{i=2}^{4} \left(FI_i \cdot BAF_i \right)} \right)$$

where

RfD = reference dose for non-cancer effects (in mg/(kg d))

Other terms are the same as for cancer effects.

Apparently, the methodology for deriving the AWQC values is rather complex. It involves decision-making on science policies and management, and needs scientific knowledge on bioaccumulation, exposure assessment, and cancer or non-cancer risk assessment. Exposure assessment is a quantitative description of the mass of a chemical pollutant that is uptaken by humans per unit of time via various routes, such as inhalation (breathing), ingestion (eating, drinking, etc.), and skin contact. Risk assessment on human health takes into account both exposure level and the toxicological effects of the chemicals.

11.3.3 Organoleptic Effects

In addition to cancer and non-cancer effects, the organoleptic effects (taste and odor) are concerns because they can impair the intended use of water. The criteria for the organoleptic impacts of these pollutants were intended to control undesirable taste and/or odor without considering the toxicological effect on humans. These pollutants are given in Table 11.3. Most of them are chlorinated phenols.

In some cases, the criteria based on organoleptic effect can be much more stringent than those based on toxicologic end points. For example, the criterion for phenol is 300 μg/L, which is much lower than the 10^5 μg/L criterion for human health. The same can be found for a number of other pollutants, including copper, zinc, chlorophenols, chlorobenzene, acenaphthene, acenaphthylene, and hexachlorocyclopentadiene.

11.4 WATER QUALITY DATABASES

Worldwide, the UN GEMS Water Program provides environmental water quality data and information. These are publically available and are used around the world by governments, environmental groups, and individuals. From the data, the status and the trend of global water quality can be depicted and used for decision-making in managing the sustainability of world water resources. See http://www.gemstat.org/ for details.

In the United States, water quality information is readily available from the USDA National Agricultural Library, EPA STORET (short for storage and retrieval) Database (U.S. EPA, 2012b), the Water Quality Assessment and Total Maximum Daily Loads Information network (U.S. EPA, 2013c), USGS Water Data for the Nation (USGS, 2013), National Aquatic Resource Surveys (U.S. EPA, 2013d), and many other sources. MyWaters Mapper (U.S. EPA, 2013e) is an EPA web site for the general public to learn about the basic quality of their local waters. EnvironMapper (U.S. EPA, 2013f) allows public access of information regarding the quality of air, water, and land, as well as the contaminations by wastes, toxic chemicals, and radiation at specified US locations.

APPENDIX 11.A   FOOTNOTE FOR TABLE 11.1

A This recommended water quality criterion was derived from data for arsenic (III), but is applied here to total arsenic, which might imply that arsenic (III) and arsenic (V) are equally toxic to aquatic life and that their toxicities are additive. No data are known to be available concerning whether the toxicities of the forms of arsenic to aquatic organisms are additive. Please consult the criteria document for details.

C The derivation of this value is presented in the Red Book (EPA 440/9-76-023, July, 1976). The CCC of 20 mg/L is a minimum value except where alkalinity is naturally lower, in which case the criterion cannot be lower than 25% of the natural level.

D Freshwater and saltwater criteria for metals are expressed in terms of the dissolved metal in the water column. See "Office of Water Policy and Technical Guidance on Interpretation and Implementation of Aquatic Life Metals Criteria (PDF)," (49 pp, 3 MB) October 1, 1993, by Martha G. Prothro, Acting Assistant Administrator for Water, available on NSCEP's web site and 40CFR §131.36(b)(1). Conversion Factors applied in the table can be found in Appendix A to the Preamble–Conversion Factors for Dissolved Metals.

E The freshwater criterion for this metal is expressed as a function of hardness (in mg/L) in the water column. The value given here corresponds to a hardness of 100 mg/L. Criteria values for other hardness may be calculated per the equation presented in the criteria document.

F Freshwater aquatic life values for pentachlorophenol are expressed as a function of pH. Values displayed in table correspond to a pH of 7.8.

G This Criterion is based on 304(a) aquatic life criterion issued in 1980 and was issued in one of the following documents: Aldrin/Dieldrin (PDF) (153 pp, 7.3 MB) (EPA 440/5-80-019), Chlordane (PDF) (68 pp, 3.1 MB)

(EPA 440/5-80-027), DDT (PDF) (175 pp, 8.3 MB) (EPA 440/5-80-038), Endosulfan (PDF) (155 pp, 7.3 MB) (EPA 440/5-80-046), Endrin (PDF) (103 pp, 4.6 MB) (EPA 440/5-80-047), Heptachlor (PDF) (114 pp, 5.4 MB) (EPA 440/5-80-052), Hexachlorocyclohexane (PDF) (109 pp, 4.8 MB) (EPA 440/5-80-054), Silver (EPA 440/5-80-071). The Minimum Data Requirements and derivation procedures were different in the 1980 Guidelines than in the 1985 Guidelines (PDF) (104 pp, 3.3 MB). If evaluation is to be done using an averaging period, the acute criteria values given should be divided by 2 to obtain a value that is more comparable to a CMC derived using the 1985 Guidelines.

I This value for aluminum is expressed in terms of total recoverable metal in the water column.

J This value was derived using the GLI (Great Lakes Initiative, meaning provided by authors) Guidelines (60 FR 15393–15399, March 23, 1995; 40CFR132 Appendix A); the differences between the 1985 Guidelines and the GLI Guidelines are explained on page iv of the 1995 Updates. No decision concerning this criterion was affected by any considerations that are specific to the Great Lakes (Footnote J not cited in EPA web posting per 5/23/2014).

L The CMC $= 1/[(f1/CMC1) + (f2/CMC2)]$ where f1 and f2 are the fractions of total selenium that are treated as selenite and selenate respectively, and CMC1 and CMC2 are 185.9 and 12.82 µg/L respectively. However, based on findings from a February 2009 SETAC Pellston Workshop on Ecological Assessment of Selenium in the Aquatic Environment, diet is the primary pathway of selenium exposure to aquatic life, and traditional methods for predicting toxicity on the basis of exposure to dissolved concentrations are not appropriate for selenium. (To view a summary of the SETAC Pellston workshop including key findings, visit http://www.setac. org/resource/resmgr/publications_and_resources/selsummary.pdf).

M U.S. EPA. *Water Quality Criteria 1972*. EPA-R3-73-033. Springfield, VA: National Technical Information Service, 1973; U.S. EPA. *Temperature Criteria for Freshwater Fish: Protocol and Procedures*. EPA 600/3-77-061. Springfield, VA: National Technical Information Service, 1977.

N This criterion applies to total PCBs, (e.g. the sum of all congener or all isomer or homolog or Aroclor analyses).

O The derivation of the CCC for this pollutant (Endrin) did not consider exposure through the diet, which is probably important for aquatic life occupying upper trophic levels.

P According to page 181 of the Red Book: For open ocean waters where the depth is substantially greater than the euphotic zone, the pH should not be changed more than 0.2 units from the naturally occurring variation or any case outside the range of 6.5–8.5. For shallow, highly productive coastal and estuarine areas where naturally occurring pH variations approach the lethal limits of some species, changes in pH should be avoided but in any case should not exceed the limits established for fresh water, i.e. 6.5–9.0.

Q This recommended water quality criterion is expressed as microgram of free cyanide (as CN) per liter.

R EPA is in the process of updating this criterion to reflect the latest scientific information. As a result, this criterion might change substantially in the near future.

S There are three major reasons why the use of Water–Effect Ratios might be appropriate.

The value of 87 µg/L is based on a toxicity test with the striped bass in water with pH = 6.5–6.6 and hardness <10 mg/L. Data in "Aluminum Water–Effect Ratio for the 3M Plant Effluent Discharge, Middleway, West Virginia" (May 1994) indicate that aluminum is substantially less toxic at higher pH and hardness, but the effects of pH and hardness are not well quantified at this time.

In tests with the brook trout at low pH and hardness, effects increased with increasing concentrations of total aluminum even though the concentration of dissolved aluminum was constant, indicating that total recoverable is a more appropriate measurement than dissolved, at least when particulate aluminum primarily consists of aluminum hydroxide particles. In surface waters, however, the total recoverable procedure might measure aluminum associated with clay particles, which might be less toxic than aluminum associated with aluminum hydroxide.

EPA is aware of field data indicating that many high quality waters in the United States contain more than 87 g aluminum/L, when either total recoverable or dissolved is measured.

V This value was derived from data for heptachlor and the criteria document provides insufficient data to estimate the relative toxicities of heptachlor and heptachlor epoxide.

Y This value was derived from data for endosulfan and is most appropriately applied to the sum of alpha-endosulfan and beta-endosulfan.

cc When the concentration of dissolved organic carbon is elevated, copper is substantially less toxic and use of water–effect ratios might be appropriate.

dd The selenium criteria document (EPA 440/5-87-006, September 1987) provides that if selenium is as toxic to saltwater fishes in the field as it is to freshwater fishes in the field, the status of the fish community should be monitored whenever the concentration of selenium exceeds 5.0 µg/L in salt water, because the saltwater CCC does not take into account uptake via the food chain.

ee This recommended water quality criterion was derived on page 43 of the mercury criteria document (PDF) (144 pp, 6.4 MB) (EPA 440/5-84-026, January

1985). The saltwater CCC of 0.025 µg/L, given on page 23 of the criteria document, is based on the Final Residue Value procedure in the 1985 Guidelines. Since the publication of the Great Lakes Aquatic Life Criteria Guidelines in 1995 (60 FR 15393–15399, March 23, 1995), the Agency no longer uses the Final Residue Value procedure for deriving CCCs for new or revised 304(a) aquatic life criteria.

hh This recommended water quality criterion was derived from data for inorganic mercury (II), but is applied here to total mercury. If a substantial portion of the mercury in the water column is methylmercury this criterion will probably be under-protective. In addition, even though inorganic mercury is converted to methylmercury and methylmercury bioaccumulates to a great extent, this criterion does not account for uptake via the food chain because sufficient data were not available when the criterion was derived.

ii This criterion applies to DDT (dichlorodiphenyltrichloroethane) and its metabolites (i.e. the total concentration of DDT and its metabolites should not exceed this value).

mm The available toxicity data, when evaluated using the procedures described in the "Guidelines for Deriving Numerical National Water Quality Criteria for the Protection of Aquatic Organisms and Their Uses," indicate that freshwater aquatic life should be protected if the 24-h average and 4-d average concentrations do not respectively exceed the acute and chronic criteria concentrations calculated by the Biotic Ligand Model.

APPENDIX 11.B   FOOTNOTE FOR TABLE 11.2

A This human health criterion is the same as originally published in the Red Book, which predates the 1980 methodology and did not utilize the fish ingestion BCF approach. This same criterion value is now published in the Gold Book.

B This criterion has been revised to reflect The Environmental Protection Agency's q1* or RfD, as contained in the Integrated Risk Information System (IRIS) as of May 17, 2002. The fish tissue bioconcentration factor (BCF) from the 1980 AWQC document used to derive the original criterion was retained in each case.

C This criterion is based on carcinogenicity of 10-6 risk. Alternate risk levels may be obtained by moving the decimal point (e.g. for a risk level of 10-5 move the decimal point in the recommended criterion one place to the right).

D According to the procedures described in the Guidelines for Deriving Numerical National Water Quality Criteria for the Protection of Aquatic Organisms and Their Uses, except possibly where a very sensitive species is important at a site, freshwater aquatic life should be protected if both conditions specified in Appendix C to the Preamble—Calculation of Freshwater Ammonia Criterion—are satisfied.

F The derivation of this value is presented in the Red Book (EPA 440/9-76-023, July, 1976).

H No criterion for protection of human health from consumption of aquatic organisms excluding water was presented in the 1980 criteria document or in the 1986 Quality Criteria for Water. Nevertheless, sufficient information was presented in the 1980 document to allow the calculation of a criterion, even though the results of such a calculation were not shown in the document.

I This criterion for asbestos is the MCL developed under the SDWA.

J This fish tissue residue criterion for methylmercury is based on a total fish consumption rate of 0.0175 kg/d.

M EPA is currently reassessing the criteria for arsenic.

N This criterion applies to total PCBs (e.g. the sum of all congener or all isomer or homolog or Aroclor analyses).

O This criterion for manganese is not based on toxic effects, but rather is intended to minimize objectionable qualities such as laundry stains and objectionable tastes in beverages.

P Although a new RfD is available in IRIS, the surface water criteria will not be revised until the National Primary Drinking Water Regulations: Stage 2 Disinfectants and Disinfection Byproducts Rule (Stage 2 DBPR) is completed, because public comment on the relative source contribution (RSC) for chloroform is anticipated.

R U.S. EPA. *Water Quality Criteria 1972*. EPA-R3-73-033. Springfield, VA: National Technical Information Service, 1973; U.S. EPA. *Temperature Criteria for Freshwater Fish: Protocol and Procedures*. EPA 600/3-77-061. Springfield, VA: National Technical Information Service, 1977.

S This recommended water quality criterion for arsenic refers to the inorganic form only.

T U.S. EPA. *Ambient Water Quality Criteria for Dissolved Oxygen*. EPA 440/5-86-003. Springfield, VA: National Technical Information Service; 1986.

U The organoleptic effect criterion is more stringent than the value for priority toxic pollutants.

Z A more stringent MCL has been issued by the EPA under the SDWA. Refer to drinking water regulations 40CFR141 or Safe Drinking Water Hotline (1-800-426-4791) for values.

jj This recommended water quality criterion is expressed as total cyanide, even though the IRIS RFD we used to derive the criterion is based on free cyanide. The multiple forms of cyanide that are present in ambient water have significant differences in toxicity due to their differing abilities to liberate the CN-moiety. Some complex cyanides require even more extreme conditions than refluxing with sulfuric acid to liberate the CN-moiety. Thus, these complex cyanides are expected to have little or no "bioavailability" to humans. If a substantial fraction of the cyanide present in a water body is present in a complexed form (e.g. $Fe_4[Fe(CN)_6]_3$), this criterion may be over conservative.

kk This recommended water quality criterion was derived using the cancer slope factor of 1.4 (LMS exposure from birth).

ll This criterion has been revised to reflect the Environmental Protection Agency's cancer slope factor (CSF) or reference dose (RfD), as contained in the IRIS as of (date of publication of Final FR Notice). The fish tissue BCF from the 1980 Ambient Water Quality Criteria document was retained in each case.

APPENDIX 11.C ADDITIONAL NOTES

1. *Criteria Maximum Concentration and Criterion Continuous Concentration.* The CMC is an estimate of the highest concentration of a material in surface water to which an aquatic community can be exposed briefly without resulting in an unacceptable effect. The CCC is an estimate of the highest concentration of a material in surface water to which an aquatic community can be exposed indefinitely without resulting in an unacceptable effect. The CMC and CCC are just two of the six parts of an aquatic life criterion; the other four parts are the acute averaging period, chronic averaging period, acute frequency of allowed exceedance, and chronic frequency of allowed exceedance. Because 304(a) aquatic life criteria are national guidance, they are intended to be protective of the vast majority of the aquatic communities in the United States.

2. *Criteria Recommendations for Priority Pollutants, Non-Priority Pollutants and Organoleptic Effects.* This compilation lists all priority toxic pollutants and some non-priority toxic pollutants, and both human health effect and organoleptic effect criteria issued pursuant to CWA §304(a). Blank spaces indicate that the EPA has no CWA §304(a) criteria recommendations. For a number of non-priority toxic pollutants not listed, CWA §304(a) "water + organism" human health criteria are not available, but the EPA has published MCLs under the SDWA that may be used in establishing water quality standards to protect water supply designated uses. Because of variations in chemical nomenclature systems, this listing of toxic pollutants does not duplicate the listing in Appendix A of 40 CFR Part 423. Also listed are the Chemical Abstracts Service CAS registry numbers, which provide a unique identification for each chemical.

3. *Human Health Risk.* The human health criteria for the priority and non-priority pollutants are based on carcinogenicity of 10-6 risk. Alternate risk levels may be obtained by moving the decimal point (e.g. for a risk level of 10-5 move the decimal point in the recommended criterion one place to the right).

4. *Water Quality Criteria published pursuant to Section 304(a) or Section 303(c) of the CWA.* Many of the values in the compilation were published in the

California Toxics Rule. Although such values were published pursuant to Section 303(c) of the CWA, they represent the Agency's most recent calculation of water quality criteria and are thus the Agency's 304(a) criteria.

5. *Calculation of Dissolved Metals Criteria.* The 304(a) criteria for metals, shown as dissolved metals, are calculated in one of two ways. For freshwater metals' criteria that are hardness-dependent, the dissolved metal criteria were calculated using a hardness of 100 mg/L as $CaCO_3$ for illustrative purposes only. Saltwater and freshwater metals' criteria that are not hardness-dependent are calculated by multiplying the total recoverable criteria before rounding by the appropriate conversion factors. The final dissolved metals' criteria in the table are rounded to two significant figures. Information regarding the calculation of hardness-dependent conversion factors is included in the footnotes.

6. *Maximum Contaminant Levels.* The compilation includes footnotes for pollutants with MCLs more stringent than the recommended water quality criteria in the compilation. MCLs for these pollutants are not included in the compilation, but can be found in the appropriate drinking water regulations (40 CFR 141.11-16 and 141.60-63), or can be accessed through the Safe Drinking Water Hotline (800-426-4791) or online.

7. *Organoleptic Effects.* The compilation contains 304(a) criteria for pollutants with toxicity-based criteria as well as nontoxicity-based criteria. The basis for the nontoxicity-based criteria are organoleptic effects (e.g. taste and odor) which would make water and edible aquatic life unpalatable but not toxic to humans. The table includes criteria for organoleptic effects for 23 pollutants. Pollutants with organoleptic effect criteria more stringent than the criteria based on toxicity (e.g. included in both the priority and non-priority pollutant tables) are footnoted as such.

8. *Gold Book.* The "Gold Book" is Quality Criteria for Water: 1986. EPA 440/5-86-001.

9. *Correction of Chemical Abstract Services Number.* The Chemical Abstract Services number (CAS) for bis(2-chloroisoprpyl) ether, has been revised in IRIS and in the table. The correct CAS number for this chemical is 108-60-1. The previous CAS number for this pollutant was 39638-32-9.

10. *Contaminants with Blanks.* EPA has not calculated criteria for contaminants with blanks. However, permit authorities should address these contaminants in NPDES permit actions using the States' existing narrative criteria for toxics.

11. *Specific Chemical Calculations.* Selenium—Aquatic Life: This compilation contains aquatic life criteria for selenium that are the same as those published in the proposed CTR (California Toxics Rule, meaning provided by authors). In the CTR, EPA proposed an acute criterion for selenium based on the criterion proposed for selenium in the Water Quality Guidance for the Great Lakes System (61 FR 58444). The GLI and CTR proposals take into account data showing that selenium's two prevalent oxidation states in water, selenite and selenate, present differing potentials for aquatic toxicity, as well as new data

indicating that various forms of selenium are additive. The new approach produces a different selenium acute criterion concentration, or CMC, depending upon the relative proportions of selenite, selenate, and other forms of selenium that are present. EPA is currently undertaking a reassessment of selenium, and expects the 304(a) criteria for selenium will be revised based on the final reassessment (63 FR 26186). However, until such time as revised water quality criteria for selenium are published by the Agency, the recommended water quality criteria in this compilation are EPA's current 304(a) criteria.

REFERENCES

Allard, M, editor. *GEMS/Water Operational Guide*, 3rd ed. UMEP/WHO/UNESCO/WMO Programme on Global Water Quality Monitoring and Assessment. GEMS/W. 92.1; 1992.

Bartram J, Balance R, editors. *Water Quality Monitoring—A Practical Guide to the Design and Implementation of Freshwater Quality Studies and Monitoring Programmes*. 1996. Available at http://www.who.int/water_sanitation_health/resources/wqmonitor/en/. Accessed April 29, 2014.

Chapman, D, editor. *Water Quality Assessments—A Guide to Use of Biota, Sediments and Water in Environmental Monitoring*, 2nd ed. Cambridge: University Press; 1996.

UN. CEO water mandate. 2007. Available at http://ceowatermandate.org/. Accessed April 29, 2014.

UNEP. *Clearing the waters: A focus on water quality solutions*. United Nations Environment Program; 2010. ISBN: 978-92-807-3074-6. Available at http://www.unep.org/PDF/Clearing_the_Waters.pdf. Accessed April 29, 2014.

UN-Water. Policy Brief on Water Quality. March 2011.

U.S. EPA. Quality criteria for water: 1986. "Gold Book". EPA 440/5-86-001. Office of Water, United States Environmental Protection Agency; 1986.

U.S. EPA. Water quality standards handbook, second edition. EPA-823-B94-005a (4305). Office of Water, United States Environmental Protection Agency; 1994.

U.S. EPA. Water quality criteria documents for the protection of aquatic life in ambient water: 1995 updates. EPA Number: 820B96001. U.S. EPA; September 1996.

U.S. EPA. Draft revisions to the methodology for deriving ambient water quality criteria for the protection of human health. Fact Sheet. EPA-822-F-98-004. Office of Water. United States Environmental Protection Agency; 1998.

U.S. EPA. Guidelines for Carcinogen Risk Assessment. Review Draft. Office of Research and Development. Washington, DC. 1999. NCEA-F-0644.

U.S. EPA. Methodology for deriving ambient water quality criteria for the protection of human health. EPA-822-B-00-004; October 2000.

U.S. EPA. National recommended water quality criteria. United States Environmental Protection Agency, Office of Water, Office of Science and Technology 2009 (4304T); 2009. Available at http://www.epa.gov/ost/criteria/wqctable/. Accessed April 29, 2014.

U.S. EPA. Water: Handbook. EPA-823-B-12-002; 2012a. Available at http://water.epa.gov/scitech/swguidance/standards/handbook/index.cfm. Accessed April 29, 2014.

U.S. EPA. STORET and WQX, EPA's repository and framework for sharing water monitoring data. 2012b. Available at http://www.epa.gov/storet/. Accessed April 29, 2014.

U.S. EPA. National recommended water quality criteria. Office of Water, Office of Science and Technology, United State Environmental Protection Agency; 2013a. Available at http://water.epa.gov/scitech/swguidance/standards/criteria/current/index.cfm. Accessed April 29, 2014.

U.S. EPA. Water: State, tribal & territorial standards. 2013b. Available at http://water.epa.gov/scitech/swguidance/standards/wqslibrary/index.cfm. Accessed April 29, 2014.

U.S. EPA. Watershed assessment, tracking & environmental results—Water quality assessment and total maximum daily loads information. 2013c. Available at http://www.epa.gov/waters/ir/. Accessed April 29, 2014.

U.S. EPA. National Aquatic Resource Surveys. 2013d. Available at http://water.epa.gov/type/watersheds/monitoring/aquaticsurvey_index.cfm. Accessed April 29, 2014.

U.S. EPA. MyWATERSMapper. 2013e. Available at http://watersgeo.epa.gov/mwm. Accessed April 29, 2014.

U.S. EPA. EnvironMapper. 2013f. Available at http://www.epa.gov/emefdata/em4ef.home. Accessed April 29, 2014.

USGS. USGS Water Data for the Nation. 2013. Available at http://waterdata.usgs.gov/nwis. Accessed April 29, 2014.

INDEX

Physical and Chemical Processes in the Aquatic Environment, First Edition. Erik R. Christensen and An Li.
© 2014 John Wiley & Sons, Inc. Published 2014 by John Wiley & Sons, Inc.